# Lecture Notes in Computer Science 790

Edited by G. Goos and J. Hartmanis

Advisory Board: W. Brauer   D. Gries   J. Stoer

Jan van Leeuwen (Ed.)

# Graph-Theoretic Concepts in Computer Science

19th International Workshop, WG '93
Utrecht, The Netherlands, June 16-18, 1993
Proceedings

Springer-Verlag

Berlin Heidelberg NewYork
London Paris Tokyo
Hong Kong Barcelona
Budapest

Series Editors

Gerhard Goos
Universität Karlsruhe
Postfach 69 80
Vincenz-Priessnitz-Straße 1
D-76131 Karlsruhe, Germany

Juris Hartmanis
Cornell University
Department of Computer Science
4130 Upson Hall
Ithaca, NY 14853, USA


Volume Editor

Jan van Leeuwen
Department of Computer Science, Utrecht University
Padualaan 14, 3584 CH Utrecht, The Netherlands

CR Subject Classification (1991): G.2.2, F.2, F.1.2-3, F.3, F.4, E.1


ISBN 3-540-57899-4 Springer-Verlag Berlin Heidelberg New York
ISBN 0-387-57899-4 Springer-Verlag New York Berlin Heidelberg


CIP data applied for

© Springer-Verlag Berlin Heidelberg 1994
Printed in Germany

Typesetting: Camera-ready by author
SPIN: 10132100        45/3140-543210 - Printed on acid-free paper

# Preface

The 19th International Workshop on Graph-Theoretic Concepts in Computer Science (WG'93) was held near Utrecht (the Netherlands), in the conference facilities of the 'National Sports Centre Papendal' of the Dutch National Sports Federation. The workshop featured sessions on structural graph theory, dynamic graph algorithms, 'hard' problems on special classes of graphs, structure-oriented graph algorithms, circuit and net theory, interconnection networks, distributed algorithms on graphs, and graph embedding and layout.

The WG workshops are an annual event for all researchers interested in the study and application of graph-theoretic concepts. The successful tradition of the workshops is well documented in the series Lecture Notes in Computer Science, as evidenced by the following overview of the past five WG's:

| WG | Year | Organizer | Organizing Site | Proceedings |
|----|------|-----------|-----------------|-------------|
| 15 | 1989 | M. Nagl | Aachen | LNCS Vol 411 |
| 16 | 1990 | R.H. Möhring | Berlin | LNCS Vol 484 |
| 17 | 1991 | G. Schmidt and R. Berghammer | München | LNCS Vol 570 |
| 18 | 1992 | E.W. Mayr | Frankfurt | LNCS Vol 657 |
| 19 | 1993 | J. van Leeuwen | Utrecht | this volume |

A complete listing of all prior WG workshops and their proceedings was included in the proceedings of WG'89 (page 374). The program committee for WG'93 consisted of:

| | | | |
|---|---|---|---|
| Ernst W. Mayr | (München) | Paul Spirakis | (Patras) |
| Rolf H. Möhring | (Berlin) | Roberto Tamassia | (Providence) |
| Manfred Nagl | (Aachen) | Gottfried Tinhofer | (München) |
| Hartmut Noltemeier | (Würzburg) | Jan van Leeuwen | (Utrecht, chairman) |
| Gunther Schmidt | (München) | Jiri Wiedermann | (Prague) |

In response to the Call for Papers for WG'93 the program committee received 92 submissions, indicating a strong and ever growing interest for the WG workshops. In view of the very good overall quality of the submissions and after a careful refereeing process, the committee decided to accept as many as 35 papers into the scientific program of the workshop. The selection reflects many of the current directions of research in the area of graph-based algorithms, both within graph-theory and in more applied contexts like circuits and networks. The larger number of selected papers could be accommodated by setting varying lengths for the presentations, and allowed for an optimal exchange of information on current research.

The present volume contains all papers presented in the workshop. All papers have been carefully revised based on the comments and suggestions which were received by the authors during the workshop. The material thus represents an up-to-date account of the work in the various directions. We hope that the papers in this volume stimulate further research in the area of applied graph theory and its many ramifications from algorithm design to software engineering.

WG'93 enjoyed a very international and stimulating participation. We are grateful to the 'National Sports Centre Papendal' of the Dutch National Sports Federation for allowing the use of its excellent conference facilities for the workshop, to the Department of Computer Science of Utrecht University for supporting the organization of the WG'93 workshop, and to Margje Punt and Goos Kant for invaluable assistance in all matters related to the workshop.

Utrecht, February 1994                                                    Jan van Leeuwen

# Contents

# Near-Optimal Dominating Sets
# in Dense Random Graphs
# in Polynomial Expected Time[*]

*Sotiris E. Nikoletseas*[(1),(2)]    *Paul G. Spirakis*[(1),(2)]

(1) Dept of Computer Science and Eng, Patras Univ, Greece
(2) Computer Technology Institute, P.O.Box 1122, 26110, Patras, Greece

### Abstract

The existence and efficient finding of small dominating sets in dense random graphs is examined in this work. We show, for the model $G_{n,p}$ with $p = 1/2$, that:

1. The probability of existence of dominating sets of size less than $\log n$ tends to zero as $n$ tends to infinity.

2. Dominating sets of size $\lceil \log n \rceil$ exist almost surely.

3. We provide two algorithms which construct small dominating sets in $G_{n,1/2}$ and run in $O(n \log n)$ time (on the average and also with high probability). Our algorithms almost surely construct a dominating set of size at most $(1 + \epsilon) \log n$, for any fixed $\epsilon > 0$.

Our results extend to the case $G_{n,p}$ with $p$ fixed to any constant $< 1$.

## 1   Introduction

The *dominating set* problem is the following: Given a graph $G(V, E)$ and a positive integer $k < |V|$, is there a subset $V' \subseteq V$, (called a dominating set), with $|V'| \leq k$ such that for all $u$ in $V - V'$ there is a $v$ in $V'$ for which $\{u, v\} \in E$? The problem is NP-Complete (see [GJ,79]), even for planar graphs with maximum degree 3. We examine here the "average case" complexity of the problem, following the line of research of [DF,89]. The input graph is selected randomly from the distribution $G_{n,p}$ of random graphs with $p = 1/2$. Our results can easily be generalized for the case of $p$ having any constant value.

The model $G_{n,1/2}$ has been used extensively (see e.g. [Sp,87], [Ku,91], [DF,89]) in order to study the average complexity of problems such as the *clique number w* or the *chromatic number χ*. For these problems it has been shown that $G_{n,1/2}$ exhibits a certain *concentration property*. i.e. that *almost surely* the values of $w$ and $\chi$ tend to concentrate at a particular function of $n$ (e.g. $w(G) \sim 2 \log n$ and $n/(2 \log n) \leq \chi(G) \leq (3/2 + o(1))n/\log n$ with probability tending to 1 as $n$ tends to infinity). In addition, efficient algorithms for finding cliques or colorings whose sizes are very close to the lower bounds known, have been demonstrated. These algorithms run in (small) polynomial time (on the average and also with high probability) and are surprisingly simple. In most cases, a variation of the greedy method produces a clique (or a coloring) whose size has a very good ratio to the lower bound. Thus, as [Ku,91] remarks, the problem of *near optimal* coloring (or clique finding) for $G_{n,p}$ might

---

[*]This research is partially supported by the ESPRIT Basic Research Action Nr 7141 (ALCOM II) and by the Ministry of Education of Greece.

be easier than problems that are NP-hard (or hard "on average" in the sense of Levin [Le,86] or also Gurevitch [Gu,91]). Of course it is not clear at the present time whether there are expected polynomial time algorithms for getting an *optimal* coloring or clique size.

Similar considerations have been raised for other problems which are hard in the worst case. For example, when $p$ is at least $(c \log n)/n$ ($c$ a constant $> 2$) [AV,79] have presented algorithms which in time $O(n(\log n)^2)$ almost certainly find directed or undirected Hamiltonian circuits in $G_{n,p}$ (and also in time $O(n \log n)$ almost certainly find a perfect matching). Already in 1976 Posa had shown (in [Po,76]) that Hamiltonian lines exist almost surely in random graphs $G_{n,N}$ with number of edges $N \geq cn \log n$ (for a suitable constant $c$), via a non-constructive proof. See [Ka,89] for an excellent overview of research concerning the efficient on the average solution of hard problems (which includes even efficient algorithms for matroid intersections, see [RS,80]).

The dominating set problem has been motivated by practical network considerations. The question of studying the average case complexity of the dominating set problem was posed in [DF,89]. Our work is the first to provide answers for the case of *dense* random graphs. Our results are surprisingly *tight*. More specifically, we show that:

1. The probability of existence of dominating sets of size less than $\log n$ in $G_{n,1/2}$ tends to zero as $n$ tends to infinity.

2. On the contrary, for any fixed $\epsilon > 0$, $G_{n,1/2}$ has many dominating sets of size at most $(1 + \epsilon) \log n$ almost surely!

3. We present algorithms which in time $O(n \log n)$ almost surely construct dominating sets of size at most $(1 + \epsilon) \log n$ for any desirable fixed $\epsilon > 0$, thus achieving an approximation ratio of $1 + \epsilon$.

From the work of Grimmett, McDiarmid ([GM,75]) and McDiarmid ([Mc,84]) (see also Bollobás ([Bo,85])) it is known that if $\alpha(G)$ is the size of the largest independent set in a graph $G$, then for $G$ a random graph drawn from $G_{n,1/2}$ it is true that (almost surely): $2 \log n - 6 \log \log n \leq a(G) \leq 2 \log n$.

In addition, that work provides a greedy algorithm that constructs in small polynomial time a *maximal* independent set of size *at least* $\log n - w(n) \log \log \log n + O(1)$, where $w(n)$ is any monotone increasing unbounded function, with probability tending to 1 as $n \to \infty$. Thus Grimmett and McDiarmid's technique constructs an independent set of size at least $1/2$ of the most likely maximum size with high probability.

Since a maximal independent set is also a dominating set, it is interesting to compare our results with the implications of Grimmett and McDiarmid's results.

1. We show in this work that dominating sets of size less than $\log n$ in $G_{n,1/2}$ *do not exist* almost surely. This *improves* Grimmett and McDiarmid's approximation ratio for independent sets (and cliques) since it follows that the independent set their algorithm constructs *will actually be at least of size* $\log n$ (and not $\log n - w(n) \log \log \log n$).

2. McDiarmid's result provides a lower bound (almost surely) on the size of the maximal independent set constructed but not an upper bound other than $2 \log n$. The independent set produced could actually have any size between $\log n$ (in the light of our results) and $2 \log n$ (due to McDiarmid). Thus, *as far as dominating sets are concerned*, McDiarmid's method provides (indirectly) a dominating set *at most twice* the optimal ($\log n$) size, almost surely. We improve this by showing explicitly how to construct in $O(n \log n)$ time dominating sets of size at most $(1 + \epsilon) \log n$ for any (constant) $\epsilon > 0$. Note also that the probability of success of our techniques is of the form $1 - n^{-\alpha}$, where $\alpha > 0$ can be controlled by the algorithm designer (the first of our methods actually allows any constant $\alpha > 1$).

Our techniques employ the second moment method in a non-trivial way for the existential results. The algorithms need to develop certain methods of taking into account the dependencies caused by "exposed" edges in the course of the evolution of the algorithm. Our results indicate that the problem of near-optimal construction of dominating sets in random graphs might not be NP-complete on the average in the Levin or Gurevitch sense. In the sequel, let $G$ be the random graph on $n$ vertices with edge probability $p = 1/2$.

# 2   Small Dominating Set Size Concentration

## 2.1   A lower bound

**Lemma 1** *For any $D < \log n$ the probability of the existence of a dominating set of size $D$ in $G_{n,1/2}$ tends to zero as $n$ tends to infinity.*

**Proof:** Let $Q(D)$ be the probability of existence of a dominating set of size $D$. Let $T$ be a particular vertex set of size $D$. Then:

$$Prob\{T \text{ is a dominating set}\} = \left(1 - \frac{1}{2^D}\right)^{n-D}$$

(to see this consider each vertex $v$ in $V - T$ separately. $T$ fails to dominate $v$ with probability $(1/2)^D$ and the events are independent).

Now:

$$
\begin{aligned}
Q(D) &= Prob\{\text{There is a dominating set of size } D\} \\
&\leq \sum_{T,|T|=D} Prob\{T \text{ is a dominating set}\} \\
&\leq \binom{n}{D}\left(1 - \frac{1}{2^D}\right)^{n-D}
\end{aligned}
$$

But $\binom{n}{D} \leq n^D$ and $1 - 1/2^D \leq e^{-1/2^D}$. (to see this recall that $\forall x : e^{-x} \geq 1 - x$ and use $x = 1/2^D$).

It follows that $Q(D) \leq n^D e^{-\frac{n-D}{2^D}}$ i.e.

$$Q(D) \leq e^{\frac{D}{2^D}}\left(e^{D\log n - \frac{n}{2^D}}\right) \tag{1}$$

Consider the second exponent $F = D\log n - n/2^D$. Since $D < \log n$, put $D = (\log n)/c$ for some $c > 1$. Then $F = (\log n)^2/c - n^{1-1/c} < (\log n)^2 - n^{1-1/c}$. Let $\epsilon = 1 - 1/c > 0$. For $F$ to tend to $-\infty$ it is enough to have $(\log n)^2 \ll n^\epsilon$ or $\epsilon \gg 2\log\log n/\log n$, which is true for all $\epsilon$ as $n \to \infty$. But when $F \to -\infty$ then $Q(D) \to 0$. $\square$

## 2.2   Abundance of dominating sets of size $(1+\epsilon)\log n$

**Lemma 2** *Let $T$ be any particular vertex set of size $(1 + \epsilon)\log n$, where $\epsilon > 0$ is any fixed constant. Then $Prob\{T \text{ is a dominating set}\} \to 1$ as $n \to +\infty$.*

**Proof:**

$$Prob\{\text{T is dominating}\} = \left(1 - \frac{1}{2^{(1+\epsilon)\log n}}\right)^{n-(1+\epsilon)\log n} = \left(1 - \frac{1}{n^{(1+\epsilon)}}\right)^{n-(1+\epsilon)\log n}$$

By the Bernoulli inequality this is at least:

$$1 - \frac{n - (1+\epsilon)\log n}{n^{1+\epsilon}} = 1 - n^{-\epsilon} + \frac{(1+\epsilon)\log n}{n^{1+\epsilon}} \geq 1 - n^{-\epsilon}$$

*Remark:* Note that the above Lemma implies that almost any vertex set of size $(1 + \epsilon)\log n$ is a dominating set.

4

## 2.3 Concentration at the size log n

In order to show the sharp concentration around the size $\log n$, we shall employ a second moment method.

**Definition 1** *Let $X^{(k)}$ be the number of dominating sets of size $k$.*

**Lemma 3** *The expected value of $X^{(k)}$ is:*

$$E\left(X^{(k)}\right) = \binom{n}{k}\left(1 - \frac{1}{2^k}\right)^{n-k}$$

**Proof:** Let $T \subseteq V$ a particular set of $k$ vertices. Let:

$$X_T = \begin{cases} 1 & \text{if } T \text{ is a dominating set} \\ 0 & \text{otherwise} \end{cases}$$

Clearly $Prob\{X_T = 1\} = E(X_T) = \left(1 - \frac{1}{2^k}\right)^{n-k}$.
But $X^{(k)} = \sum X_T$ over all $T$ such that $|T| = k$, i.e. (by linearity of expectation)
$E\left(X^{(k)}\right) = \sum_{|T|=k} E(X_T) = \binom{n}{k} E(X_T)$ i.e. $E\left(X^{(k)}\right) = \binom{n}{k}\left(1 - \frac{1}{2^k}\right)^{n-k}$ $\square$

**Corollary 1** *When $k = \log n$ then $E\left(X^{(k)}\right) \to +\infty$.*

**Proof:** $E\left(X^{(\log n)}\right) = \binom{n}{\log n}\left(1 - \frac{1}{n}\right)^{n-\log n}$

But $\lim_{n\to+\infty}\left(1 - \frac{1}{n}\right)^{n-\log n} = e^{-1}$ and $\lim_{n\to+\infty}\binom{n}{\log n} = +\infty$. $\square$

However, Corollary 1 is not enough to show existence of $\log n$-sized dominating sets almost surely.

To proceed, we recall that:

**Fact 1 ([Sp,87])** *If the random variable $X$ has mean $m$ and variance $Var(X)$ then*

$$Prob\{X = 0\} \leq \frac{Var(X)}{m^2}$$

It then follows that:

**Fact 2** *If $E(X) \to +\infty$ and $Var(X) = o(E^2(X))$ then $Prob\{X = 0\} \to 0$ (here convergence is with respect to the number of vertices $n$).*

In order to apply Fact 2, it is enough to show that $Var(X^{(k)}) = o(E^2(X^{(k)}))$ when $k = \log n$ (since Corollary 1 shows that $E(X^{(\log n)}) \to +\infty$). In the sequel, by using $X^{(k)} = \sum_{|S|=\log n} X_S$ (as in Lemma 3) we can easily get (as in [Sp,87]) that, by fixing a set $S$ of $|S| = \log n$ :

$$\frac{Var(X^{(k)})}{E^2(X^{(k)})} = \sum_{i=0}^{k} Prob\{|S \cap T| = i\}[f(S,T) \text{ given } |S \cap T| = i] \tag{2}$$

where:

$$f(S,T) = \frac{E(X_T/X_S = 1)}{E(X_T)} - 1$$

and $X_T, X_S$ are indicator variables of the sets $T, S$ with respect to being dominating sets. Clearly:

$$Prob\{|S \cap T| = i\} = \frac{\binom{k}{i}\binom{n-k}{k-i}}{\binom{n}{k}}$$

**Definition 2** *Let*

$$\alpha(i) = \frac{\binom{k}{i}\binom{n-k}{k-i}}{\binom{n}{k}}$$

*when* $k = \log n$

**Definition 3** *Let* $f_i(S,T) = f(S,T)$ *when* $|S \cap T| = i$ *and* $k = \log n$

**Lemma 4** *There is a constant* $\gamma$ *and an* $n_0 > 0$ *:* $\forall n \geq n_0$ *and* $\forall i = 1, \ldots, \log n$ *:* $f_i(S,T) \leq \gamma$.

**Proof:** Since $X_T = 0$ or $1$ we get $E(X_T/X_S = 1) \leq 1$. Thus

$$f_i(S,T) \leq \frac{1}{E(X_T)} - 1$$

where $|T| = \log n$. But, by the proof of Lemma 3

$$\begin{aligned}
E(X_T) &= \left(1 - \frac{1}{2^{\log n}}\right)^{n-\log n} \\
&= \left(1 - \frac{1}{n}\right)^{n-\log n} \\
&\geq \left(1 - \frac{1}{n}\right)^{n}
\end{aligned}$$

But $\forall n \geq 2 : \left(1 - \frac{1}{n}\right)^n \geq 1/4$, thus $E(X_T) \geq 1/4$, hence $f_i(S,T) \leq 3$ and choose $\gamma = 3$. $\square$

**Lemma 5** $\alpha(i)$ *is monotone decreasing on* $i$ *for* $i = 1, \ldots, \log n$.

**Proof:** Let

$$\Lambda = \frac{\alpha(i+1)}{\alpha(i)} = \frac{k-i}{i+1} \frac{k-i}{n-2k+i+1} \leq \frac{k^2}{n-2k+1}$$

(for $i \geq 0$). For $k = \log n \Rightarrow$

$$\Lambda \leq \frac{(\log n)^2}{n - 2\log n + 1} < 1 \text{ for all } n > 64$$

*Remark:* Since we are interested in asymptotic results for large graphs ($n \to \infty$), small finite graphs (of a few nodes) are not of interest and can be (in any case) processed by brute force.

**Lemma 6** *For* $i = 2, 3, \ldots, \lceil \log n \rceil$ *we have* $\alpha(i) \leq \frac{4}{n}$

**Proof:** From Lemma 5 we have $\alpha(i) \leq \alpha(2)$. But

$$\alpha(2) = \frac{\binom{k}{2}\binom{n-k}{k-2}}{\binom{n}{k}}$$

$$\leq \ k^2 \frac{\frac{(n-k)!}{(k-2)!(n-2k+2)!}}{\frac{n!}{k!(n-k)!}}$$

$$\leq \ k^2(k-1)k \ \frac{(n-2k+3)}{(n-k+1)} \ \frac{(n-2k+4)}{(n-k+2)} \ \cdots \ \frac{(n-k)}{(n)}$$

$$\leq \ k^4 \frac{1}{n-k+1} \frac{1}{n-k+2}$$

(since all other ratios are of the form $\frac{n-2k+j}{n-k+j} < 1$).
i.e. (for $k = \log n$)

$$\alpha(2) \leq \frac{(\log n)^4}{(n-\log n+1)^2} \leq \frac{n}{(n/2)^2} = \frac{4}{n}$$

$\square$

**Lemma 7**

$$\alpha(1) \leq \frac{(\log n)^2}{n-\log n+1}$$

**Proof:**

$$\alpha(1) \ = \ \frac{\begin{pmatrix} k \\ 1 \end{pmatrix} \begin{pmatrix} n-k \\ k-1 \end{pmatrix}}{\begin{pmatrix} n \\ k \end{pmatrix}}$$

$$= \ k^2 \frac{(n-2k+2)}{(n-k+1)} \ \frac{(n-2k+3)}{(n-k+2)} \ \cdots \ \frac{(n-k)}{(n)}$$

$$\leq \ \frac{k^2}{n-k+1}$$

(since all other ratios are of the form $\frac{n-2k+j}{n-k+j} < 1$).
We know get:
**Theorem 1**

$$\frac{Var\left(X^{(k)}\right)}{E^2\left(X^{(k)}\right)} \to 0 \ as \ n \to +\infty \ for \ k = \log n$$

**Proof:** By equation (2) we have for $k = \log n$

$$\frac{Var(X^{(k)})}{E^2(X^{(k)})} = \sum_{i=0}^{k} \alpha(i) f_i(S,T)$$

But for $i = 0$, $f_i(S,T) = 0$ (since $S \cap T = \emptyset$). Thus

$$\frac{Var(X^{(k)})}{E^2(X^{(k)})} \ \leq \ \gamma\alpha(1) + \gamma \sum_{i=2}^{\log n} \alpha(i) \ \text{(by Lemma 4)}$$

$$\leq \ \gamma\frac{(\log n)^2}{n-\log n+1} + \gamma\frac{4\log n}{n} \ \text{(by Lemma 6)}$$

$$\leq \ \frac{18(\log n)^2}{n} \to 0 \ as \ n \to +\infty$$

$\square$

**Corollary 2** For $k = \log n$, $Prob\{X^{(k)} = 0\} \leq 18\frac{(\log n)^2}{n} \to 0 \ as \ n \to +\infty$

Thus we get:

**Theorem 2** *There exist dominating sets of size $\lceil \log n \rceil$ in $G_{n,1/2}$ with probability at least*
$1 - 18\frac{(\log n)^2}{n}$ *i.e. almost surely.*

$\square$

# 3 Efficient algorithms for finding small dominating sets

Note that Theorem 2 (and its proof) provides to us an existential result through a non-constructive argument. It is, therefore, of interest to discover efficient (i.e. polynomial time) algorithms to construct a dominating set as close to $\log n$ as possible. Although Lemma 2 indicates that by choosing any vertex set of size $(1 + \epsilon) \log n$ ($\epsilon > 0$ fixed) we may succeed in finding it to be a dominating set with high probability, however the repetition of this experiment provides trials which are *not independent* since the edges between previously and currently tried sets have been exposed already. In the sequel, we provide algorithms whose probability of successfully discovering small dominating sets is controlled by the implementor and whose analysis overcomes the dependency problems.

## 3.1 The algorithm of repeated trials

The main idea of this algorithm is to exploit the abundance of dominating sets of size $(1 + \epsilon) \log n$ while "backward" dominance is guaranteed by construction.

**Definition 4** *A vertex set $A$ dominates a vertex set $B$ $(A \cap B = \emptyset)$ iff each vertex of $B$ is neighbour of at least one vertex of $A$.*

The input to the algorithm below is an instance $G$ of $G_{n,1/2}$.

**ALGORITHM "REPEATED TRIALS"**
(Comment: Let $\epsilon > 0$ be any fixed constant i.e. a parameter of our algorithm).
1. $i \leftarrow 1$
2. Choose a set $T_i$ of $(1 + \epsilon) \log n$ vertices at random.
3. Let $V_i \leftarrow V - \{T_1 \cup T_2 \cup \cdots \cup T_i\}$.
4. Test whether $T_i$ dominates $V_i$. If yes then output $T_i$ as a dominating set and exit. Else proceed to step 5.
5. For each $v \in T_i$ select a *distinct* vertex $f(v) \in V_i$ such that $\{v, f(v)\}$ is an edge, by calling the fuction SELECT.
6. If the construction of step 5 fails, then exit and output "failure".
7. Let $F$ be the set of $f(v)$. Clearly $|F| = (1 + \epsilon) \log n$.
8. $i \leftarrow i + 1$.
9. $T_i \leftarrow F$.
10. Go to 3.
**END OF "REPEATED TRIALS"**

**FUNCTION SELECT** $(T_i, F)$
**Input:** A set of vertices $T_i$.
**Output:** A set of vertices $F$ or "failure".
1. Let $T_i = \{v_1^i, v_2^i, \ldots, v_\lambda^i\}, \lambda = (1 + \epsilon) \log n$.
2. Choose a neighbour $f(v_1^i)$ of $v_1^i$ in $V_i$ so that $f(v_1^i)$ is also a neighbour of $v_1^{i-1}, v_1^{i-2}, \ldots, v_1^1$. If such a neighbour does not exist, return "failure"; $j \leftarrow 1$.
3. Let $F = \{f(v_1^i)\}$.
4. $j \leftarrow j + 1$; if $j > \lambda$ then exit with $F$.
5. Choose a neighbour $f(v_j^i)$ of $v_j^i$ in $V_i$ so that $f(v_j^i) \in V_i - F$ and $f(v_j^i)$ is also a neighbour of $v_j^{i-1}, v_j^{i-2}, \ldots, v_j^1$; if this is impossible then exit with "failure".
6. $F \leftarrow F \cup \{f(v_j^i)\}$.
7. Go to 4.
**END OF "SELECT"**

**Lemma 8** *For every $i$ the probability that $T_i$ dominates $V_i$ is at least $1 - n^{-\epsilon}$, i.e. step 4 of "REPEATED TRIALS" succeeds with probability at least $1 - n^{-\epsilon}$.*

**Proof:** Each time a new $T_i$ is considered, its possible edges to the current $V_i$ are "fresh" (not looked up by the algorithm up to this point). Thus the probability that $T_i$ dominates $V_i$ is:

$$(1 - \frac{1}{2^{|T_i|}})^{|V_i|} = (1 - \frac{1}{2^{(1+\epsilon)\log n}})^{|V_i|} = (1 - \frac{1}{n^{(1+\epsilon)}})^{|V_i|}$$

By the Bernoulli inequality this is at least

$$1 - \frac{|V_i|}{n^{1+\epsilon}} \geq 1 - \frac{n}{n^{1+\epsilon}} = 1 - n^{-\epsilon}$$

$\square$

**Lemma 9** *Given $|V_i| > \frac{n}{2}$ and given that SELECT was called at most a constant $k$ number of times, then the probability that SELECT returns failure is at most $2^{k+1} exp(-\frac{\beta^2}{2^{k+1}}n)$ for $\beta \in (0,1)$.*

**Proof:** Choose a $\beta \in (0,1)$. By Chernoff bounds, the number of neighbours in $V_i$ of any particular $v_j^i \in T_i$ is at least $(1 - \beta)\frac{n}{4}$ with probability at least $1 - exp(-\frac{\beta^2}{2}\frac{n}{4})$. From these neighbours, those who are also neighbours of $v_j^{i-1}, v_j^{i-2}, \ldots, v_j^1$ (i.e. belong to the intersection of neighbours of $v_j^1, v_j^2, \ldots, v_j^i$) are about $\frac{n}{2^k}$ when SELECT is called $k$ times. The probability that the number of neighbours of $v_j^i$ who are also neighbours of $v_j^{i-1}, v_j^{i-2}, \ldots, v_j^1$ is less than $(1 - \beta)\frac{n}{2^k}$ is at most $2^k exp(-\frac{n}{2^k}\frac{\beta^2}{2})$ i.e. extremely small (we accumulate the probability of the bad event each time SELECT is called). Let $\epsilon_j$ be the event "$v_j$ has at least $(1 - \beta)\frac{n}{4}$ neighbours in $V_i$". The probability that any $\epsilon_j$ fails is:

$$Prob\{\cup \overline{\epsilon_j}\} \leq \sum_j Prob\{\overline{\epsilon_j}\} \leq (1 + \epsilon)\log n \; exp(-\frac{\beta^2}{8}n) \leq exp(-\frac{\beta^2}{16}n)$$

But when $\epsilon_j$ holds $\forall v_j \in T_i$ then the construction of $F$ succeeds with certainty, since the selection of previous $f(v_j)$'s cannot exhaust any set of neighbours in $V_i$ of $v_j$'s not examined yet. Hence the total probability of error is $\leq 2^k exp(-\frac{n}{2^k}\frac{\beta^2}{2}) + exp(-\frac{\beta^2}{16}n) \leq 2^{k+1} exp(-\frac{n}{2^{k+1}}\beta^2)$

**Lemma 10** *The expected number of trials of "REPEATED TRIALS" is at most 2. In fact, for any constant $k > 1$ the probability that the number of trials (i.e. repetitions of the loop of statements 3-10 of "REPEATED TRIALS") exceeds $k$ is at most $n^{-\epsilon k}$, conditioned on non-failure of SELECT.*

**Proof:** Since each trial fails (at step 3) with probability at most $n^{-\epsilon}$ and since trials are independent because each time fresh edges are considered we have:
$Prob\{$ number of trials before success $> k\} \leq n^{-\epsilon k}$ $\square$

**Theorem 3** *Let $\epsilon > 0, \alpha > 1$ any constants. The algorithm REPEATED TRIALS constructs a dominating set of size $(1 + \epsilon)\log n$ in expected time $O(n \log n)$ and with probability of success at least $1 - n^{-\alpha}$.*

**Proof:** Choose $k = \frac{\alpha+1}{\epsilon}$. Then, by Lemma 10, $Prob\{$ number of trials $> k\} \leq n^{-(\alpha+1)}$, given no failure in SELECT. But, by Lemma 9, for each of the (fixed) $k$ trials (since $|V_i| \geq n - k(1 + \epsilon)\log n > \frac{n}{2}$) we have that $Prob\{$ SELECT fails $\} \leq 2^{k+1} exp(-\frac{n}{2^{k+1}}\beta^2)$. Thus, "REPEATED TRIALS" succeeds in at most $k$ trials without failure in SELECT with probability at least $1 - max\{n^{-(\alpha+1)}, k \; 2^{k+1} exp(-\frac{n}{2^{k+1}}\beta^2)\}$ i.e. at least $1 - n^{-\alpha}$. $\square$

## 3.2 The greedy approach

The greedy approach each time chooses a vertex to be put in the dominating set and then it deletes all its neighbours (and the corresponding edges) from the graph. Note that each remaining graph is random since its edges are not exposed. The method achieves dominating sets of size $(1 + \epsilon) \log n$ almost surely and runs in expected $O(n \log n)$ time.

**ALGORITHM "GREEDY"** (Input $G(V, E)$ of $G_{n,1/2}$)

$i \leftarrow 0$ ; $V_i \leftarrow V$. Let $\epsilon > 0$ a fixed constant, to be chosen.

$D \leftarrow \emptyset$.

**Until** $|V_i| \leq \epsilon \log n$ **do**

**Begin**

Choose a vertex $u_i \in V_i$.

Let $N_i$ the set of neighbours of $u_i$ in $G(V_i, E_i)$, where $E_i$ are the edges induced by $V_i$ on $G(V, E)$. Let $N_i$ include $u_i$ by convention.

Let $V' \leftarrow V_i - N_i$.

$i \leftarrow i + 1$ ; $D \leftarrow D \cup \{u_i\}$.

$V_i \leftarrow V'$.

**End**

$D \leftarrow D \cup V_i$.

Output $D$.

**END OF ALGORITHM "GREEDY"**

**Lemma 11** *Fix a $\beta \in (0, 1)$. Choose $\epsilon > 0$. Provided that $|V_i| \geq \epsilon \log n$ we have*

$$Prob\{|N_i| \geq (1 - \beta)\frac{|V_i|}{2}\} \geq 1 - n^{-\frac{\beta^2}{4}\epsilon}$$

**Proof:** Consider the sequence of $|V_i|$ Bernoulli trials with probability of success $1/2$ (of each vertex in $V_i$ being a neighbour of $u_i$). Of course $u_i \in N_i$ with probability 1, which is even better). By Chernoff bounds then:

$$Prob\{|N_i| \geq (1 - \beta)\frac{|V_i|}{2}\} \geq 1 - e^{-\frac{\beta^2}{4}|V_i|} \geq 1 - n^{-\frac{\beta^2}{4}\epsilon}$$

$\square$

**Definition 5** *Let the event $\varepsilon_i$ be "at the $i^{th}$ execution of the loop, $|N_i| \geq (1 - \beta)\frac{|V_i|}{2}$"*

**Definition 6** *Let $\varepsilon = \varepsilon_1 \cap \varepsilon_2 \cap \cdots \cap \varepsilon_t$ until $|V_t| < \epsilon \log n$*

**Definition 7** *Let $\gamma = \frac{1 - \beta}{2}$*

**Lemma 12** *Conditioned on $\varepsilon$, the number of iterations $t$ is $\log n / \log(\frac{1}{1-\gamma}) + \Theta(\log \log n)$*

**Proof:** Clearly $|V_{i+1}| \leq (1 - \gamma)|V_i|$ i.e. $|V_t| \leq (1 - \gamma)^t n$. Thus we must have $(1 - \gamma)^t n \leq \epsilon \log n$ implying $t \geq \frac{\log n}{\log(\frac{1}{1-\gamma})} + \Theta(\log \log n)$. Note that $\frac{1}{1-\gamma} = \frac{2}{1+\beta}$. By choosing (for any $\epsilon' > 0$) $\beta = 2^{\frac{\epsilon'}{1+\epsilon'}} - 1$ we get a $t \leq (1 + \epsilon') \log n$.

$\square$

**Lemma 13** $Prob\{\varepsilon\} \geq 1 - n^{-\frac{\beta^2}{8}\epsilon}$

**Proof:** $Prob\{$not $\varepsilon\} \leq \sum_j Prob\{\overline{\varepsilon_j}\} \leq t n^{-\frac{\beta^2}{4}\epsilon}$.

By Lemmas 11, 12, i.e. $Prob\{$ not $\varepsilon$ $\} \leq (1 + \epsilon') \log n \, n^{-\frac{\beta^2}{4}\epsilon} \leq n^{-\frac{\beta^2}{8}\epsilon}$

$\square$

**Theorem 4** *Algorithm GREEDY constructs a dominating set of size $(1 + \epsilon' + \epsilon) \log n$ with probability at least $1 - n^{-\frac{\beta^2}{8}\epsilon}$ and terminates in time $O((1 + \epsilon')n \log n)$ with the same probability.*

**Proof:** By the previous Lemmas.

$\square$

# 4 Future Work and Extensions

Algorithm "REPEATED TRIALS" has a stronger probability of success. Both algorithms achieve an approximation ratio of $1 + \epsilon$ for any $\epsilon > 0$ (fixed). In this sense we actually have provided an approximation scheme.

We pose as an open problem the question of finding in expected polynomial time an *optimal* dominating set in $G_{n,1/2}$.

It is easy to extend the analysis to any constant $p$ in $G_{n,p}$. Analyzing dominating sets for sparse $G_{n,p}$ is still open and we are currently working on it.

Our results can also be extended to answer the problem of efficiently finding near-optimal $k$-dominating sets in random graphs. A $k$-dominating set is a set of vertices such that every other vertex of the graph is at distance at most $k$ from at least one vertex of this set. Note that this extension may lead to new and more efficient techniques for finding shortest paths in random graphs.

**Acknowledgement:** We wish to thank G.Tinhofer for his insightful remarks on our work. His remarks motivated us to significantly improve the approximation ratio of our algorithms.

# References

[AV,79] *"Fast Probabilistic Algorithms for Hamiltonian Circuits and Matchings"*, by D.Angluin and L.Valiant, Journal of Computer and System Sciences, vol.18, p.155-193 (1979)

[Bo,85] *"Random Graphs"*, by B. Bollobás, Academic Press (1985)

[DF,89] *"The Solution of Some Random NP-hard Problems in Polynomial Expected Time"*, by M.Dyer and A.Frieze, Journal of Algorithms, vol.10, p.451-489 (1989)

[GJ,79] *"Computers and Intractability: A Guide to the Theory of NP-Completeness"*, by M.Garey and D.Johnson, W.H.Freeman and Co., NY (1979)

[GM,75] *"On colouring random graphs"*, by G. Grimmett and C. McDiarmid, Math. Proc. Cambridge Phil. Soc. 77, p.313-324 (1975)

[Gu,91] *"Average Case Completeness"*, by Y.Gurevitch, Journal of Computer and System Sciences, vol.42, p.346-348 (1991)

[Ka,89] *"Probabilistic Analysis"*, by R.Karp, Lecture Notes CS 292F, U.C.Berkeley (1989)

[Ku,91] *"Parallel coloring of graphs with small chromatic numbers"*, by L.Kučera, Technical Report, Charles Univ. Prague (1991)

[Le,86] *"Average Case Complete Problems"*, by L.Levin, vol.15, SIAM J. Comp., p.285-286 (1986)

[Mc,84] *"Colouring random graphs"*, by C. McDiarmid, Ann. Oper. Res. 1 (1984)

[Po,76] *"Hamiltonian Circuits in Random Graphs"*, by L.Posa, Discrete Mathematics, vol.14, p.359-364 (1976)

[RS,80] *"Random Matroids"*, by J.Reif and P.Spirakis, Proc. 12th Annual ACM Symp. Theory of Computing, p.385-397 (1980)

[Sp,87] *"Ten Lectures On The Probabilistic Method"*, by J.Spencer, SIAM CBMS-NSF Regional Conference Series (1987)

# Approximating Minimum Weight Perfect Matchings for Complete Graphs Satisfying the Triangle Inequality

N.W. Holloway, S. Ravindran and A.M. Gibbons*

Department of Computer Science, University of Warwick, Coventry CV4 7AL, UK.

**Abstract.** We describe an $O(log^3 n)$ time $NC$ approximation algorithm for the CREW P-RAM, using $n^3 / \log n$ processors with a $2 \log_3 n$ performance ratio, for the problem of finding a minimum-weight perfect matching in complete graphs satisfying the triangle inequality. The algorithm is conceptually very simple and has a work measure within a factor of $log^2 n$ of the best exact sequential algorithm. This is the first $NC$ approximation algorithm for the problem with a sub-linear performance ratio. As was the case in the development of sequential complexity theory, matching problems are on the boundary of what problems might ultimately be described as *tractable* for parallel computation. Future work in this area is likely to decide whether these ought to be regarded as those problems in $NC$ or those problems in $RNC$.

*keywords*: approximation algorithms, parallel algorithms, matching

## 1  Introduction

As [19] has emphasised, matching problems have played an important rôle in the foundations of sequential algorithmic complexity theory. This is because they are important problems that arise in many guises that can be solved in polynomial time, but for which all naive algorithms take exponential time. In fact, it was in Edmonds' celebrated paper [3], on matching algorithms that the connection between *tractable problems* and *polynomial time solvable problems* was first made. It is likely that matching problems will play a similar rôle in the development of parallel algorithmic complexity theory.

An efficient parallel algorithm is generally defined by the complexity class $NC$ which is the class of problems which can be solved in polylogarithmic time using a polynomial number of processors (see [8, 15]). There are practically no extant algorithms placing matching problems in $NC$ with the notable exception of the *maximal matching problem* [12] and certain algorithms for special classes of graphs which we cite later. The class of problems which can be solved in polylogarithmic *expected* time using a polynomial number of processors is called

---

* Partially supported by SERC grant GR/H/76487 and the ESPRIT BRA programme contract No. 7141, ALCOM II.

*RNC* (see, for example, chapter 3 of [9] and section 2.5.5 of [19]). Most matching problems can be solved in parallel using randomness ([16, 20]) and so belong to *RNC*. It has been stated [19] that whether a modern definition of a *tractable* problem in parallel computation is one can that can be solved rapidly with randomisation or one that can be solved rapidly without randomisation may ultimately depend upon whether fast parallel algorithms for matching require randomisation.

Recall that a matching in a graph is a subset of edges, so that no two elements have a common vertex. If every vertex of a graph is an end-point of such an element then the matching is a *perfect matching*. Given a weighted graph, a *minimum-weight perfect matching* is a perfect matching whose sum of edge weights is a minimum.

This paper addresses the problem of finding minimum-weight perfect matchings in complete weighted graphs satisfying the *triangle inequality*. In a graph satisfying the triangle inequality, the weight of any single edge forming a triangle with two other edges is less than or equal to the sum of the weights of these other two edges. Such an inequality is satisfied in many natural problems. The problem that we address is an important sub-task for many problems of combinatorial optimisation and features, for example, in solutions to Chinese Postman Problems and in Approximations to the Travelling Salesman Problem (see, for example, [7]). Even with restrictions on the graph such as completeness and triangle inequality satisfaction, the problem seems very difficult to place in *NC*. We have therefore addressed the problem of finding an *NC approximation algorithm*. This is an algorithm running in polylogarithmic time using a polynomial number of processors and which finds a perfect matching whose weight $M$ does not exceed $R_A M^*$, where $M^*$ is the weight of a minimum-weight perfect matching and $R_A$ is called the *approximation ratio*.

Specifically, we describe an *NC* approximation algorithm for the minimum-weight perfect matching problem for graphs satisfying the triangle inequality for which $R_A = 2 \log_3 n$. This is the first such deterministic algorithm with a sub-linear performance ratio. Previously, it was known (see chapter 3 of [9]) that there is an *NC* approximation algorithm for the maximum- (equivalently, minimum-) weight perfect matching problem such that $R_A = n$.

Karp, Upfal and Wigderson [16] described an *RNC* algorithm for the minimum-weight perfect matching problem, which runs in $O(\log n \log^2(Wn))$ time (after the improvements of [6]) using $O(Wn^{3.5})$ processors, where $W$ is the maximum weight of any edge. A faster *RNC* algorithm was obtained in [20] which runs in $O(\log^2 n)$ time using $O(mWn^{3.5})$ processors, where $m$ is the number of edges. These algorithms are in *RNC* only if $W$ is relatively small (that is, $W = n^{O(1)}$).

Although finding an *NC* algorithm for minimum-weight perfect matching seems to be hard, there are *NC* algorithms for finding a perfect matching in special classes of unweighted graphs. Examples are dense graphs [1], bipartite graphs with a bounded permanent [5], complements of transitive oriental graphs [10] and line graphs [21]. However, there is no known deterministic *NC* algorithm for minimum weight perfect matching for complete graphs. The best known deterministic parallel algorithm for complete graphs runs in $O(n^3/p + n^2 \log n)$ polynomial time using p ($\leq n$) processors [22].

An exact solution for minimum weight perfect matching can be computed in $O(n^3)$ sequential time by the intricate algorithm of Edmonds' [3] and this provides a target for the work measure of parallel algorithms. There are sequential approximation algorithms for special graphs ([11, 23, 24, 26]). The algorithms of [23] find an approximate minimum-weight perfect matching for graphs satisfying the triangle inequality. However, it is not clear that these algorithms can be effectively parallelised. Even if they could be, they would provide a much more intricate solution to the problem solved by this paper. One of the virtues of our algorithm is its simplicity.

## 2 Approximate Minimum Weight Perfect Matching in a Complete Weighted Graph

We start by providing an overview of the algorithm whose input is a complete weighted graph $G = (V, E)$ with edge set $E$ and vertex set $V$, where $|V| = n$ and $n$ is even.

The first part of the algorithm concerns the construction of a graph $T = (V, E')$ where $E' \subset E$, thus $T$ is a spanning subgraph of $G$. The essential properties of $T$ will be that each component is a tree containing an even number of vertices, and the total sum of its edge weights will be less than $(2 \log_3 n)M^*$, where $M^*$ is the sum of edge weights of a minimum weight perfect matching in $G$.

The second part of the algorithm first constructs an approximate matching of weight $M$. This is done by constructing, for each component of $T$, a Hamiltonian circuit with an even number of edges. The sum of the edge weights in all such circuits is less than $(4 \log_3 n)M^*$. A perfect matching in $G$ is then obtained by taking alternate edges on each such circuit and such that (of the two possibilities presented by each circuit) the lightest weight possibility is chosen. The weight, $M$, of the perfect matching constructed in this way satisfies the inequality: $M \leq 2M^* \log_3 n$.

We now consider the two parts of the algorithm in more detail. Afterwards we consider precise details of its parallel execution, justify the bounds on the approximation and consider the complexity parameters.

### 2.1 The Algorithm

The algorithm consists of two stages, the construction of $T$ and the construction of the approximate matching of weight $M$ from $T$.

**Construction of $T$**

The construction is performed over at most $\log_3 n$ phases.

At the beginning of the $i^{th}$ phase, the vertices of $G$ have been partitioned into disjoint subsets whose union is $V$. Each such subset forms a *super-vertex* of the graph $G_i$. If such a super-vertex contains an even number of vertices of $G$, then it is an *even* super-vertex, otherwise it is an *odd* super-vertex. To determine the weight of an edge between two super-vertices $V_i$ and $V_j$ of $G_i$, choose the smallest of all possible edges $(v_i, v_j)$ of $G$, such that $v_i \in V_i$ and $v_j \in V_j$. The actual edge of $G$ whose weight is used for $G_i$ is noted. At the start of the iteration, $G_1$ is made up of $n$ super-vertices, each containing a single vertex from the graph $G$ (and thus the edges in $G_1$ are simply the edges from $G$). The graph $T$ initially contains the vertices of $G$, but has no edges.

The action of each phase is as follows. For every odd super-vertex of $G_i$, we construct the shortest path from it to another (any other) odd super-vertex. Note that these are not just simple edges, since the path may pass through even super-vertices. The result of this construction is a set of directed paths, where each component is a tree with an extra path forming a cycle. For complexity reasons we want to avoid having a cycle containing more than two paths. This is achieved if we place an extra order on the possible paths. It is sufficient that if there are two paths of equal weight, we choose the path leading to a vertex with the lowest index. We do not need to differentiate between two paths of the same weight leading to the same vertex, and in this case, we choose one arbitrarily. This will ensure that there will only be two paths involved in cycle, which means that they can be identified in constant time. We then arbitrarily drop one path in the cycle of each component to leave a forest of trees, in which we are only interested in the edges of the paths, not their direction. These remaining path edges have corresponding edges in the graph $G$ (noted in the construction of edges for graph $G_i$), and these edges are added to the graph $T$. The super-vertices that are to be coalesced are those which are joined by the generated paths. Essentially, these can be determined using a connected components algorithm. The super-vertices that are in the same component are now coalesced to form a new super-vertex, and the edge weights adjusted accordingly. Each component of $T$ corresponds to a super-vertex of $G_i$. The edges that are added to $T$ link super-vertices of $G_i$, they must also link components of $T$. These edges link trees in $T$ together with no cycles, so $T$ remains a forest with each component corresponding to a super-vertex of $G_i$.

## Construction of the Matching $M$ from $T$

The input to this stage of the algorithm is the graph $T$, a forest with components $T_i$, output from the first stage. The components of $T$ represent a partition of the vertices of $G$ in which each part contains an even number of vertices. For each $T_i$ we find a pre-order numbering of the vertices. Now, for each $T_i$, such a numbering defines an even-length circuit in $G$ obtained by visiting the vertices in the order of their pre-order indices. For each such circuit, there are two subsets formed by taking alternate edges. We choose the subset with the smaller weight to form the approximate minimum-weight perfect matching edges, $M_i$, of the component $T_i$. Our approximation to the minimum-weight perfect matching is the union of all the $M_i$. The total weight (Lemma 2) of edges chosen to belong to the approximate minimum-weight perfect matching, $M$, is less than $2M^* \log_3 n$

## 2.2 Validity of the Approximation Ratio Claim

We now prove that the total weight of the edges of the graph $T$ is less than $2M^* \log_3 n$. In the construction of $T$, we only continue iterating when there are odd super-vertices remaining, and so we need only consider these when calculating the number of iterations. The number of odd super-vertices will always decrease, since each odd super-vertex will be coalesced with at least one other. For continued iteration, some super-vertices must have coalesced to form new odd super-vertices, each containing at least three previous odd super-vertices. So, at every iteration, the number of odd super-vertices will be reduced by at least a factor of three, and thus $\log_3 n$ iterations will be are sufficient to remove all odd super-vertices (notice that by an elementary theorem of graph theory, there will always be an even number of odd super-vertices). All we now need for our proof is the following Lemma.

**Lemma 1.** *The sum of the edge weights of the edges added to $T$ in each iterated phase of its construction is less than or equal to $2M^*$.*

*Proof.* In any of the iterated phases used in the construction of $T$, the edges added to $T$ are those belonging, for every odd super-vertex, to a shortest path from such a super-vertex to another. We first show that, in $G_i$, there exists a path from any odd super-vertex to some other odd super-vertex only using edges of a minimum-weight perfect matching. Consider then, any odd super-vertex $V_k$ of $G_i$. Now because there are an odd number of vertices of $G$ in $V_k$, not all of these vertices can be matched by edges of a minimum-weight perfect matching of $G$ which connect pairs of vertices contained in $V_k$. There must therefore be an edge of a minimum-weight perfect matching connecting $V_k$ to some other super-vertex that is a vertex of $G_i$. This is the first edge of the path whose existence we wish to prove. If this edge takes us to an vertex corresponding to another odd super-vertex then we have finished. If it takes us to a even super-vertex, then it will match one of its constituent vertices and there will be an odd number remaining to be matched by a minimum-weight perfect matching. The implication is that there is another edge from this even super-vertex which takes us on to yet another super-vertex. Continuing in this way, we see that there must exist a path in $G_i$ from an odd super-vertex to another odd super-vertex and that such a path only uses edges of a minimum-weight perfect matching. Notice, of course, that any two such paths may have edges in common. Let $S_k^*$ be the sub-set of edges defining the shortest path from the super-vertex $V_k$ (which corresponds to some odd super-vertex) to some other similar vertex which uses edges of a minimum-weight perfect matching only and let $S_k$ be the subset defining the path from vertex $V_k$ to some similar vertex as constructed by the algorithm.

We need, to obtain a worst case bound on the weight of the *union* of the $S_k$ in terms of the weight of the union of the $S_k^*$. This is because we already have a worst case bound on the weight of the union of the $S_k^*$ provided by $M^*$, the weight of a minimum-weight perfect matching of $G$. Notice that for all super-vertices $V_k$, $weight(S_k) \leq weight(S_k^*)$ because the algorithm chooses minimum-weight paths for $S_k$ using all edges of $G$, and $S_k^*$ can only use edges of a minimum-weight perfect matching. However, it does not follow that the weight of the *union* of the $S_k$ will be less than the weight of the *union* of the $S_k^*$ because there may be an entirely different sharing of edges between paths in the two cases. To obtain a worst bound, we need to consider the cases in which the $S_k$ have a minimum union and the $S_k^*$ have a maximum union. In the latter case, notice that no two of the $S_k^*$ may share edges whose combined weight is more than half the weight of the lightest set of the two, otherwise the $S_k^*$ would not be shortest paths of their type. The situation of the union of the $S_k^*$ having a minimum weight corresponds to every path sharing half its weight with every other path. In the case of maximising the weight of the union of edges of the $S_k$ is essentially that of practically no sharing (although the detail is a little more subtle, this observation suffices to achieve the bound we seek). Thus, the weight of the union of the $S_k$ is bound, in the worst case, by twice the weight the union of the $S_k^*$, but the weight of the union of the $S_k^*$ is bound by $M^*$ and so the lemma follows.

We have proved that the sum of the weights of the edges of $T$ is bound by $2M \log_3 n$. The following Lemma provides a similar bound on the approximation ratio of the algorithm.

**Lemma 2.** *The weight, $M$, of the perfect matching found by the algorithm is bounded as follows:*

$$M \leq 2M^* \log_3 n$$

*where, $M^*$ is the weight of a minium-weight perfect matching and $n$ is the number of nodes of $G$. The input graph $G$ is a complete weighted graph satisfying the triangle inequality.*

*Proof.* The total weight of the edges of the graph $T$ is, by Lemma 2, bounded by $2M^* \log_3 n$. Consider each tree component $T_i$ of $T$. Now consider the standard *twice-around-the-spanning-tree* circuit (see, for example, [7]) obtained by visiting the nodes in pre-order and making short-cuts to avoid re-visiting nodes that have already been visited. Such short-cuts are always possible because $G$ is complete and they will be *short*-cuts because we have the triangle-inequality holding. Thus the weight of such a circuit is bounded by twice the weight of the spanning tree edges. However, the weight of the edges chosen for the matching from this circuit constitute at most half the weight of the circuit and so at most the weight of the spanning tree. Thus, over all such circuits, we choose a weight of edges for the matching which is less than or equal to the weight of $T$ and the lemma is proved.

## 2.3 Parallel Execution and Complexity of the Algorithm

Consider first the construction of the graph $T$. There are at most $\log_3 n$ phases and within each, the activities dominating the computation time are the construction of all shortest paths and the coalescing of super-vertices which can essentially be achieved by an algorithm for finding the connected components of a graph. As we cite later, there are well known polylogarithmic time parallel algorithms performing these tasks using a polynomial numbers of processors. Other tasks are trivially solved by less costly parallel algorithms. Thus, we may express the time-complexity for constructing $T$ as $O((SP(n)+CC(n))\log n)$ using $max(p_{SP(n)}+p_{CC(n)})$ processors, where $SP(n)$ is the parallel time for the all shortest paths problem using $p_{SP(n)}$ processors and $CC(n)$ is the parallel time for the connected components problem using $p_{CC(n)}$ processors.

Since we have ensured that the construction of $T$ is such that it is a forest of trees, the tasks in the second stage of the algorithm all have highly efficient parallel solutions. This includes finding a pre-order numbering of tree vertices (for which we can use the Euler tour technique of [27]) and the problem of choosing a set of alternate edges of least weight from a circuit (which has very easy solutions employing ranking and summing). Thus, the parallel time and work required for the construction of the approximate minimum-weight perfect matching from $T$ are small compared with what are required for the construction of $T$.

Thus overall, we see that the problem of finding an approximate minimum-weight perfect matching has a parallel solution taking $O((SP(n)+CC(n))\log n)$ time using $max(p_{SP(n)}, p_{CC(n)})$ processors. We can now see what this means in terms of variants of the P-RAM and using the best extant parallel algorithms for the problems of finding all shortest paths and connected components.

Consider first the best practical (in terms of modest constants hidden by the order notation) extant solutions for the all pairs shortest path problem. The problem can be solved in $O(\log^2 n)$ time using $n^3/\log n$ processors on a CREW P-RAM or on an EREW P-RAM. These algorithms are adaptations (see, for example, [8]) of a *common* CRCW P-RAM algorithm of Kucera ([17]). Clearly, the same algorithm will run within the same complexity bounds on an *arbitrary* CRCW P-RAM. On the model for which it was described, Kucera's algorithm runs in $O(\log n)$ time using $n^4$ processors. Now consider the best extant solutions for the connected components problem. The problem can be solved on an *arbitrary* CRCW P-RAM in $O(\log n)$ time using $(m+n)$ processors [25]. For the EREW P-RAM [14] (and therefore for the CREW P-RAM, although an earlier algorithm [13] already existed for this model) the problem can be solved in $O(\log^{3/2} n)$ time using $O(m+n)$ processors.

From the preceding information, we have the following theorem.

**Theorem 3.** *There is an algorithm to find an approximate minimum-weight perfect matching of a complete weighted graph satisfying the triangle inequality, with $n$ nodes, having a performance ratio $R_A = 2\log_3 n$ which:*

*1. runs in $O(\log^2 n)$ time using $n^4$ processors on an arbitrary CRCW P-RAM.*

2. *runs in $O(\log^3 n)$ time using $n^3/\log n$ processors on either a CREW P-RAM or an EREW P-RAM.*

For all P-RAM models, the problem of finding all shortest paths (both in terms of time complexity and work) dominates the computation time and the work measure (that is, the processor number, computation time product). The best sequential time-complexity for an exact solution on complete graphs, $O(n^3)$, can still be achieved by the primal-dual algorithm of Edmonds [2] as improved by Gabow [4] and Lawler [18]. Thus, our algorithm is within a factor of $\log^2 n$ of the work measure of Edmonds' algorithm. The faster computation afforded by implementation on the CRCW P-RAM comes (as in commonly the case for P-RAM implementations) at a high cost in terms of numbers of processors required.

## 3 Conclusion and Open Problems

We have described a P-RAM computation that places the problem of finding an approximate minimum-weight perfect matching in a complete weighted graph satisfying the triangle inequality in $NC$ with a performance ratio of $2\log_3 n$. The algorithm is conceptually very simple and comes within a $\log^2 n$ factor of the work measure of the best sequential algorithms. It is also the first NC-approximation algorithm for the task with a sub-linear performance ratio.

The question of whether the problem of finding an exact solution to the minimum-weight perfect matching problem can be placed in the complexity class $NC$ remains open. The problem is still open for complete weighted graphs, even those that satisfy the triangle inequality. Indeed, the seemingly simpler task of finding an $NC$ approximation algorithm for this matching problem with a constant performance ratio remains un-solved.

Resolution of the existence or otherwise of appropriate algorithms in this area may ultimately help place more precise boundaries around what ought to be regarded as *tractable* problems for parallel computation.

## References

1. E. Dahlhaus and M. Karpinski, "Parallel Construction of Perfect Matching and Hamiltonian Cycles on Dense Graphs", *Theoretical computer science* 61 (1988) 121-136
2. J. Edmonds, "Matching and Polyhedrons with 0,1 Vertices", *Journal of Research of the National Bureau of Standards B*, 125-130 (1965).
3. J. Edmonds, "Paths, trees and flowers", *Canadian Journal of Mathematics* 17 (1965) 449-67.
4. H.N. Gabow, "Implementations of Algorithm for Maximum Matching on Nonbiparitte Graphs", *Ph.D. Dissertation*, Dept. of Computer Science, Stanford University, 1974.

5. D.Y. Grigoriev and M. Karpinski, "The Matching Problem for Bipartite Graphs with Polynomially Bounded Permanents is in NC", *Proceedings of the Annual IEEE Symposium on Foundations of Computer Science* (1987), 166-172.

6. Z. Galil and V. Pan, "Improved Processor Bounds for Combinatorial Problems in RNC", *Combinatorica* 8 (1988) 189-200.

7. A.M. Gibbons, *Algorithmic Graph Theory*, Cambridge University Press (1985).

8. A.M. Gibbons and W. Rytter, *Efficient Parallel Algorithms*, Cambridge University Press (1988).

9. A.M. Gibbons and P.G. Spirakis (eds), *Lectures on Parallel Computation*, Cambridge University Press (1993).

10. D. Hembold and E. Mayer, "Two-processor Scheduling is in NC", *VLSI Algorithm and Architectures*, editors: Makedon et al., Lecture Notes in Computer Science, Vol. 227 (1986) 12-25.

11. M. Iri, K. Murota and S. Matsui, "Linear-time Approximation Algorithms for Finding the Minimum-Weight Perfect Matching on a Plane", *Information Processing Letters* 12 (1981) 206-209.

12. A. Israeli and Y. Shiloach, "An improved algorithm for maximal matching", *Information Processing Letters"* 33 (1986) 57-60.

13. D.B. Johnson and P. Metaxas, "Connected components in $O(\log^{3/2}|V|)$ parallel time for the CREW PRAM", FOCS, 1991

14. D.R. Karger, N. Nisan and M. Parnas, "Fast Connected Components Algorithms for the EREW PRAM", *4th Annual ACM Symposium on Parallel Algorithms and Architectures*, 373-381 (1992)

15. R. Karp and V. Ramachandran, "Parallel algorithms for Shared Memory Machines", Handbook of Theoretical Computer Science, J. van Leeuwen (editor), vol.1, Elsevier and MIT Press (1991).

16. R. Karp, E. Upfal and A. Wigderson, "Constructing a Perfect Matching is in Random NC", *Proceedings of the Annual ACM Symposium on Theory of Computing* (1985), 22-32.

17. L. Kucera, "Parallel computation and conflicts in memory access", *Information Processing Letters*, Vol. 14, (1982) 93-96

18. E.L. Lawler, *Combinatorial Optimization: Networks and Matroids*, Holt-Rinehart-Winston, New York, 1976.

19. F. Thomson Leighton, *Introduction to Parallel Algorithms and Achitectures: Arrays • Trees • Hypercubes*, Morgan Kaufmann, California (1992).

20. K. Mulmuley, U. Vazirani and V. Vazirani, "Matching is as Easy as Matrix Inversion", *Proceedings of the Annual ACM Symposium on Theory of Computing* (1987), 345-354.

21. J. Naor, "Computing a Perfect Matching in a Line Graph", *Proceedings of the $9^{th}$ Conference on the Foundations of Software Technology and Theoretical Computer Science*, (1989) 139-148.

22. C.N.K. Osiakwan and S.G. Akl, "The Maximum weight perfect matching problem for complete weighted graphs is in PC", *Proceedings of the $2^{nd}$ IEEE Symposium on Parallel and Distributed Processing* (1990) 880-887.

23. D.A. Plaisted, "Heuristic Matching for Graphs Satisfying the Triangle Inequality", *Journal of Algorithms* 5 (1984) 163-179.

24. E.M. Reingold and R.E. Tarjan, "In a greedy Heuristic for Complete Matching", *SIAM Journal of Computing* 10 (1981) 676-681.

25. Y. Shiloach and U. Vishkin, "An $O(\log n)$ parallel connectivity algorithm", *Journal of Algorithms*, Vol. 3, (1982) 57-67.

26. K.J. Supowit, D.A. Plaisted and E.M. Reingold, "Heuristics for Weighted Perfect Matching", *Proceedings of the Annual ACM Symposium on Theory of Computing*, (1980) 398-419.

27. R.E. Tarjan and U. Vishkin, "Finding biconnected components and computing tree functions in logarithmic parallel time", *Proceedings of the 25$^{th}$ Annual IEEE Symposium on the Foundations of Computer Science*, (1984), 12-20, also SIAM *Journal of Computing*, 14, 4(1985) 862-74.

# Hierarchically Specified Unit Disk Graphs[1]

## (Extended Abstract)

*M. V. Marathe*      *V. Radhakrishnan*      *H. B. Hunt III*      *S. S. Ravi*

### Abstract

We characterize the complexity of several basic optimization problems for unit disk graphs specified hierarchically as in [LW87a, Le88, LW92]. Both PSPACE-hardness results and polynomial time approximations are presented for most of the problems considered. These problems include *minimum vertex coloring, maximum independent set, minimum clique cover, minimum dominating set and minimum independent dominating set.*

Our PSPACE-hardness results imply the PSPACE-hardness of the geometric location problems in [MS84, WK88], when sets of points are specified hierarchically as in [BOW83] or [LW92]. Also, each of our PSPACE-hardness results holds, when the hierarchical specifications are 1-level restricted (Definition 2.2) and the graphs are specified hierarchically either as in [BOW83] or as in [LW92].

For $k$-level restricted hierarchical specifications, where $k$ is fixed, we have also developed a polynomial time algorithm to solve the maximum clique problem, and a polynomial time relative approximation algorithm for minimum coloring.

# 1   Introduction, Motivation and Summary of Results

Over the last decade several theoretical models have been put forward to succinctly represent objects hierarchically [BOW83, GW83, Le88, Wa84]. Two well known hierarchical specification languages are those of Lengauer et al. [LW87a, Le88] and of Bentley et al. [BOW83]. These specification languages were motivated by VLSI circuit design. Other areas where hierarchical specifications and consequently hierarchically specified graphs arise are finite element analysis [HLW92], software engineering [GJM91], material requirement planning and manufacturing resource planning in a multistage production system [MTM92] and processing hierarchical Datalog queries [Ul88].

For graphs specified hierarchically using their model, Lengauer et al. [Le88, Le89] have given efficient algorithms to solve several graph theoretic problems including minimum spanning forests and planarity testing. The language of Bentley et al.[BOW83] defines sets of geometric objects and can be interpreted naturally as specifying the intersection graphs of the sets of objects defined. It is in this sense that we view the specification language of [BOW83] as defining graphs. Intersection graphs of unit disks have been very useful in modeling

---

[1]This research was supported by NSF Grants CCR-89-03319 and CCR-90-06396. The authors are with the Department of Computer Science, University at Albany – SUNY, Albany, NY 12222. Email addresses : {madhav, rven, hunt, ravi}@cs.albany.edu.

problems in such diverse areas as image processing [HM85], VLSI circuit design [MC80], broadcast networks [Ha80, Ka84], and geometric location theory [MF90]. Consequently, the complexity of problems for unit disk graphs has been studied extensively in the literature [CCJ90, FPT81, MHR92, MS84, WK88].

In this paper, we combine these two lines of research and study the complexity of problems for hierarchically specified intersection graphs of unit disks/squares.

Our complexity results hold for graphs specified either as in Bentley et al. [BOW83] or as in Lengauer et al. [LW87a, Le88, Le89]. Most of the problems considered are shown **both** to be PSPACE-hard and to have polynomial time approximations. Our results are summarized as follows:

1. The following problems are PSPACE-hard for hierarchically specified unit disk graphs: maximum independent set, 3-coloring, minimum dominating set, minimum independent dominating set and minimum clique cover. Moreover, these PSPACE-hardness results hold even when the hierarchical specifications are 1-level restricted. (See Definition 2.2.)

2. The following location problems are PSPACE-hard, when the sets of points are specified hierarchically using the hierarchical specification language of [BOW83]: Euclidean $p$-$Center$ problem, Euclidean $p$-$Median$ problem, Rectilinear $p$-$Center$ problem, Rectilinear $p$-$Median$ problem, the Euclidean $p$-$Dispersion$ problem and the bottleneck version of Euclidean $p$-$Dispersion$ problem. Again, these results hold even when the hierarchical specifications are 1-level restricted. (Each of these geometric location problems is known to be NP-complete when inputs are not specified hierarchically [MS84, WK88].)

3. We present polynomial time relative approximation algorithms for the problems maximum independent set, minimum dominating set and minimum independent dominating set, when unit disk graphs are specified hierarchically using Lengauer's model. Each of these approximation algorithm has a performance guarantee of 5.

4. For each fixed $k \geq 1$, we have developed a polynomial time relative approximation algorithm with a performance guarantee of 6 for the minimum coloring problem and a polynomial time algorithm for the maximum clique problem when unit disk graphs are specified using a $k-$level restricted hierarchical specification.

What we mean by a polynomial time approximation algorithm for hierarchically specified instances is illustrated by the following example.

**Example:** Consider the minimum vertex cover problem, where the input is a hierarchical specification of a graph $G$, and we wish to compute the size of a minimum vertex cover for $G$. Our polynomial time approximation algorithm for the vertex cover problem computes the size of an approximate vertex cover and runs in time *polynomial in the size of the hierarchical description*, rather than the *size* of $G$. Moreover, it also solves in polynomial time (in the size of the hierarchical specification) the following **query problem**: Given any vertex $v$ of

$G$ and the path from the root to the node in the *hierarchy tree* in which $v$ occurs, determine whether $v$ belongs to the approximate vertex cover so computed. The algorithms can also be modified to output a hierarchical specification of the solution. □

# 2  Definitions and Description of the Model

We begin with a brief review of the hierarchical graph model introduced by Lengauer.

**Definition 2.1** *[Le89, LW92] A hierarchical specification $\Gamma = (G_1, ..., G_p)$ of a graph is a sequence of undirected simple graphs $G_i$ called cells. The graph $G_i$ has $m_i$ edges and $n_i$ vertices. $p_i$ of the vertices are distinguished and are called pins. The other $(n_i - p_i)$ vertices are called inner vertices. $r_i$ of the inner vertices are distinguished and are called nonterminals. The $(n_i - r_i)$ vertices are called terminals.*

Note that there are $n_i - p_i - r_i$ vertices defined explicitly in $G_i$. We call these *explicit vertices*. Each pin of $G_i$ has a unique label, its *name*. The pins are assumed to be numbered between 1 and $p_i$. Each nonterminal in $G_i$ has two labels, a *name* and a *type*. The type is a symbol from $G_1, ..., G_{i-1}$. If a nonterminal vertex $v$ is of the type $G_j$, then the terminal vertices which are the neighbors of $G_j$ are in one-to-one correspondence with the pins of $G_j$. (Note that all the neighbors of a nonterminal vertex must be terminals. Also, a terminal vertex may be a neighbor of several nonterminal vertices.) Without loss of generality we assume that for each $G_i$ there is a nonterminal node associated with it. The size of $\Gamma$ is $n = \sum_{1 \le i \le p} n_i$, and the edge number is $m = \sum_{1 \le i \le p} m_i$.

The expansion $E(\Gamma)$ is the graph associated with the hierarchical definition $\Gamma$. With $\Gamma$ one can associate a natural tree structure depicting the sequence of calls made by the successive levels. We call it the *hierarchy tree* of $\Gamma$.

*We note again that our approximation algorithms answer query problems without explicitly expanding the hierarchical specification.*

Let the length of the encoding of $\Gamma$ be $n$, and let the numbers be represented in binary. Then the total number of nodes in $E(\Gamma)$ can be $2^{\Omega(n)}$.

**Definition 2.2** *A hierarchical graph specification $\Gamma = (G_1, ..., G_n)$ of a graph $G$ is* **1-level-restricted** *if $\forall(u, v) \in E$, one of the following conditions holds :*

1. *$u$ and $v$ are the explicit vertices in the same instance of $G_i$ $(1 \le i \le n)$.*

2. *$u$ is an explicit vertex in an instance of $G_i$ and $v$ is a explicit vertex in an instance of $G_j$ and the instance of $G_i$ calls instance of $G_j$ $(1 \le j < i \le n)$.*

The above definition can be extended to define *k-level restricted*, for any fixed $k \ge 1$. Such descriptions still can lead to exponentially large graphs. Moreover, we can show that many practically occurring hierarchical descriptions (see [LW87a, Le88, HLW92]) are *k-level restricted* for small values of $k$. Hence it is natural to investigate if problems are easier for such specifications.

**Definition 2.3** *An instance $F = (F_1(X^1), \ldots, F_{n-1}(X^{n-1}), F_n)$ of Hierarchical 3 SAT (H3SAT) is of the form*

$$F_i(X^i) = (\bigwedge_{1 \le j \le l_i} F_{i_j}(X_j^i, Z_j^i)) \bigwedge f_i(X^i, Z^i)$$

*for $1 \le i \le n$ where $f_i$ are 3CNF formulae, $X^i, X_j^i, Z^i, Z_j^i$, are vectors of boolean variables such that $X_j^i \subseteq X^i$, $Z_j^i \subseteq Z^i$, $0 \le i_j < i$. (Thus, $F_1$ is just a 3CNF formula.) An instance of H3SAT specifies a 3CNF formula $f$, that is obtained by expanding the $F_j$, $2 \le j \le n$, as macros where the variables $Z$'s introduced in any expansion are considered distinct. The problem H3SAT is to decide whether the formula $f$ specified by $F$ is satisfiable.*

### The Language of Bentley, Ottmann and Widmayer

The hierarchical language used here to describe a set of unit disks is almost identical to that used by Bentley, Ottmann and Widmayer [BOW83] to describe a set of isothetic rectangles. The only difference is that instead of the BOX command we have the DISK command. The syntax of the DISK command is

$$\text{DISK } (x, y, r)$$

where $(x, y)$ is the center of the disk and $r$ is the radius.

With the set of unit disks defined as above, we associate an *intersection graph* which has one vertex per unit disk and two vertices are joined by an edge iff the corresponding disks to intersect. Without loss of generality, tangential disks are assumed to intersect [CCJ90]. Let us denote a hierarchical specification given in the specification language of Bentley Ottmann and Widmayer as a BOW-specification and that given in the specification language of Lengauer as an L-specification. A unit disk graph specified hierarchically using a BOW-specification will be referred to as a hierarchical unit disk (HUD) graphs.

## 3 Complexity Results

In this section we discuss our hardness results for hierarchically specified intersection graphs. We first give a transformation which enables us to prove complexity results, for graphs specified using $L$-specifications, by proving the results, for graphs specified by BOW-specifications. Next, we state a theorem from [HRS93], that H3SAT is PSPACE-complete. We then use this problem to prove the PSPACE-hardness of the maximum independent set problem for hierarchically specified unit disk graphs.

**Theorem 3.1** $\forall k \ge 1$, *there is a polynomial time transformation which converts a k-level-restricted BOW-specification of a unit disk graph $G$ into a k-level-restricted L-specification of the same unit disk graph $G$.* $\square$

**Theorem 3.2** *The problem H3SAT is PSPACE-complete.* $\square$

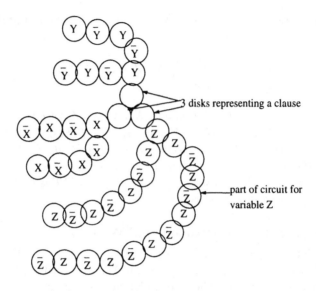

3 disks representing a clause

part of circuit for variable Z

Figure 1: Clause configuration for the clause $(\bar{x} \vee \bar{y} \vee z)$

The above theorem can be further strengthened so as to apply to more restricted forms of hierarchical specifications.

**Lemma 3.1** *H3SAT is PSPACE-complete even when restricted to instances of the following form:*

$$F_1(X,Y) = C_1 \wedge C_2 \wedge C_3 \wedge ... \wedge C_m$$

$$F_2(X,Y) = F_1(X,Z) \bigwedge F_1(Z,Y)$$

$$F_3(X,Y) = F_2(X,Z) \bigwedge F_2(Z,Y)$$

$$\vdots$$

$$F_{n-1}(X,Y) = F_{n-2}(X,Z) \bigwedge F_{n-2}(Z,Y)$$

$$F_n(X,Y) = F_{n-1}(X,Z) \bigwedge F_{n-1}(Z_r,Y) \bigwedge f_1(X) \bigwedge f_2(Y)$$

*where $C_1, C_2, ..., C_m$, are 3 literal clauses, $f_1$ and $f_2$ are 3CNF formulas, $X, Y$ and $Z$ are vectors of boolean variables with $|X| = |Y| = |Z| = n$.* □

Let $RH3SAT$ denote the restricted form of $H3SAT$ in which instances are of the form given in Lemma 3.1.

Next we sketch the PSPACE-hardness proofs of the maximum independent set problem for 1-level-restricted hierarchically specified unit disk (UD) graphs.

**Independent Set:** Before we give the PSPACE-hardness proof for the independent set problem for hierarchically specified unit disk graphs, it is instructive to recall the proof of the fact that the independent set problem for unit disk graphs

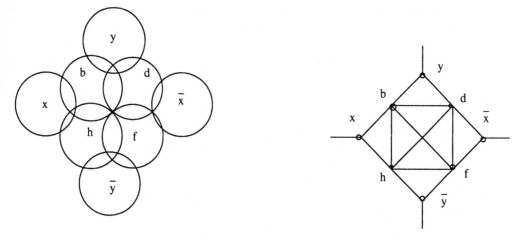

Figure 2: Crossover box

in the non-hierarchical case is NP-complete. The proof appears in [FPT81] as well as in [WK88]. Starting from a 3CNF formula $F$, they construct a unit disk graph $G$ and an integer $k_1$ such that $F$ has an independent set of size $k_1$ iff $F$ is satisfiable.

In this construction, each variable $x$ is represented by a cycle of disks of even length. The even numbered vertices will denote the literal $\bar{x}$ and the odd numbered disks will denote the literal $x$. The clauses are represented by the clause configuration shown in Figure 1. Each clause is represented by a set of three mutually intersecting disks. Each of the disks corresponds to a literal. If the literal appears negated, it is connected to the disk which represents the unnegated variable in the cycle and if the literal is unnegated, it is connected to the disk which represents the negated variable. A schematic diagram of the layout of the clause and the cycle configuration is shown in Figure 3. Now each crossover between two cycles is replaced by a junction which is constructed so as to remove any interference between two cycles. A schematic diagram of the junction is shown in Figure 2. Let $k_1 = 1/2 \sum r_i + J + M$, where $r_i$ denotes the length of the even length cycle for the variable $x_i$, $M$ denotes the number of clauses in the instance of 3SAT, and $J$ denotes the number of junctions in the UD graph.

It can be easily seen from the construction that for each cycle of length $r_i$ corresponding to a variable $x_i$, one can choose only $r_i/2$ circles in any independent set. Further, for each junction, only one node can be included in any independent set. Similarly, for each clause configuration exactly one vertex can be included in any independent set. With these observations in mind, it is easy to verify that $F$ is satisfiable iff the graph $G$ has an independent set of size $k_1$.

**Theorem 3.3** *The maximum independent set problem for 1-level-restricted BOW-specified unit disk graphs is PSPACE-hard.*

**Proof Sketch:** Starting from an instance $F = (F_1, \cdots F_n)$ of $RH3SAT$, we construct in polynomial time a hierarchical specification of a unit disk graph $G = (G_1, \cdots G_n)$ and an integer $k$, such that the $RH3SAT$ instance is satisfiable iff the UD graph $E(G)$ has an independent set of size $k$. The construction is done bottom up, level by level. For each $F_i$, $(1 \leq i \leq n)$ we create a graph $G_i$ whose definition contains two calls of $G_{i-1}$ and some additional DISK commands.

The graph has $n$ levels and the levels are constructed as follows:

**Specification for $G_1$:** Except for a minor modification, $G_1$ is the graph resulting from the construction of [FPT81, WK88] described above applied to $F_1$. The only modification needed is that if a variable $x$ occurs only in $F_1$ then its corresponding even length cycle occurs entirely in $G_1$. Otherwise only a part of the cycle occurs in $G_1$. (The part of the cycle appearing in $G_1$ for such a variable is an even length chain as shown in Figure 3. This chain will be joined with other partial chains corresponding to the variable in $G_i$, $i > 1$.) Thus, given a $3CNF$ formula $F$, we can construct in polynomial time, and hence by a polynomial number of primitive statements, a layout of the unit disks such that $F$ is satisfiable iff the corresponding UD graph has an independent set of size $k_1$. Here, $k_1$ is a easily computable function of number of clauses and number of variables.

All the disks lie within a box of width $W_1$ and height $H_1$, where $W_1$ and $H_1$ = $O(m_1 * n_1)$, where $n_1$ denotes the number of variables and $m_1$ denotes the number of clauses in $F_1$. They can be laid out in $O(m_1 * n_1)$ time. Note that the values of $W_1$ and $H_1$ are bounded by a polynomial in $n_1$ and hence by a polynomial in the size of the specification of $F$.

**Specification for $G_i$, $(1 < i < n)$:** Next we explain how to obtain a specification for $G_i$, $2 \leq i \leq n$. This corresponds to the formula $F_i$ which is defined by

$$F_i(X, Y) = F_{i-1}(X, Z) \bigwedge F_{i-1}(Z, Y)$$

where $|X| = |Y| = |Z| = n$. The construction of $G_i$ is shown in Figure 3. Note that $G_i$ contains two calls to $G_{i-1}$ corresponding to each $F_{i-1}$ and the open ended chains for the corresponding variables in the vector $Z$ in each copy of $G_{i-1}$ are joined to form one chain to denote that they are the same variables. The second call to $G_{i-1}$ is translated so that the set of disks defined in the two different calls to $G_{i-1}$ do not intersect. The open ended chains for $X$ and $Y$ need to be extended upwards so that they can be joined with their corresponding chains at a later stage. Also observe that the when the chains corresponding to individual $z_i's$ are joined, they introduce crossovers and these are removed by the junctions described earlier. It is clear from our construction that all these steps consist of a polynomial number of DRAW commands. Thus the definition of $G_i$ will appear as follows:

DEFINE $G_i$
    DRAW $G_{i-1}$ at (0,0)
    DRAW $G_{i-1}$ at ($2^{i-1} * W_1$ , 0 ) /* $W_1$ is the width of $G_0$ */

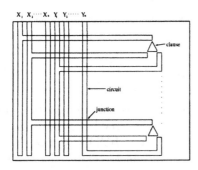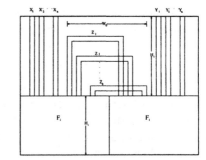

Figure 3: Clauses and cycles in Level 0 and $i$.

The above sequence of DRAW commands is followed by a sequence of DRAW commands that extend the vectors $X$ and $Y$ and join the two partial chains corresponding to vector $Z$.

At this stage, we point out a crucial property of our specification. We note that the bounding box for $G_i$ will have width $2^{i-1} \cdot W_1$ while its height is $O(i*H_1)$. This crucial observation allows us to get a specification which is only polynomial in the size of the specification of $F$. Specifically, in the definition of each $G_i$, we have only polynomially many calls to DRAW commands. This is because the height of the box grows polynomially in $H_1$ and the partial chains for each $z_i$ are at most a distance $W_1$ away.

**Specification for $G_n$:**

The UD graph $G_n$ has the following components. It has two calls to $G_{n-1}$ which correspond to the two calls to $F_{n-1}$ in the instance of $RH3SAT$. Then we have two more UD graphs $Initial$ and $Final$ which correspond to $f_1$ and $f_2$ respectively. Lastly, we need to ensure that the chains for the variables in vector $X$ are joined to chains of the corresponding variables used in $f_1$. We have to make sure that the cycles associated with each variable are of even length. A sketch of the actual definition of $G_n$ is given below.

DEFINE $G_n$

> DRAW $G_{n-1}$ at $(0,0)$
> DRAW $G_{n-1}$ at $(2^{n-1} * W_1 , 0 )$ /* $W_1$ is the width of $G_1$ */
> Sequence of DRAW commands to join the two partial chains corresponding to the vector $Z$.
> Sequence of DRAW commands corresponding to $Initial$.
> Sequence of DRAW commands corresponding to $Final$.

$E(G_n)$ defines the UD graph which would be produced by the construction of [FPT81, WK88] starting from an expanded form of the hierarchical $3SAT$.

Choosing all the even vertices from the chains for the variables corresponds to setting the variable true and choosing all the odd vertices corresponds to setting the variable false. From each of the junctions, we can pick exactly one vertex. The way a clause configuration is joined to variable cycles forces us to choose exactly one node from a clause configuration. Let $k_1$ be the same as given in the flat case. Define $k_i$ as follows.

$$k_i = 2 * k_{i-1} + 1/2(\sum_v l_v) + C_i + Q_i$$

where $k_{i-1}$ denotes the number obtained for $F_{i-1}$ called in the definition of $F_i$, $l_v$ denotes the length of the even length chain for the variable $v$ appearing in $F_i$, $C_i$ denotes the number of clauses in the the instance RH3SAT, and $Q_i$ denotes the number of crossover boxes introduced explicitly in $F_i$. Finally, let $k = k_n$.

Note that $k$ can be computed easily in a bottom-up level-by-level manner in polynomial time. By an argument similar to that used in [FPT81] it can be shown that the UD graph $E(G_n)$ has an independent set of size $k$ iff the corresponding $RH3SAT$ instance is satisfiable. $\Box$

## 4 Approximation Algorithms

In this section we discuss our approximation algorithm for the maximum independent set problem for hierarchically specified intersection graphs. Although we will sketch our algorithm for unit disk graphs it will be clear that a similar algorithm can be used for obtaining an approximate independent set for hierarchically specified intersection graphs of other regular polygons. We assume that the unit disk graphs are specified using the L-specification. When the specifications are k-level-restricted for some fixed $k$, our approximation algorithms for L-specified unit disk graphs can be modified to obtain approximation algorithms for BOW-specified unit disk graphs. We assume that the L-specification is such that there are no edges between pins defined at the same level. We can transform in polynomial time an arbitrary L-specification to a new L-specification such that the above property holds.

Most of our heuristics are based on a forbidden subgraph property of unit disk graphs. The proof of this property relies on a geometric observation concerning packing of unit disks in the plane given in [MHR92].

**Lemma 4.1** *[MHR92] Let $C$ be a circle of radius $r$ and and let $S$ be a set circles of radius $r$ such that every circle in $S$ intersects $C$ and no two circles in $S$ intersect each other. Then, $|S| \leq 5$.* $\Box$

Lemma 4.1 implies that the unit disk graphs are $K_{1,6}$ free. Because of this the following simple approximation algorithm guarantees a performance ratio of 5. The algorithm simply consists of picking an arbitrary vertex in the graph and adding it to the independent set. Next we delete the picked vertex and its neighbors from the graph and repeat the process. This process guarantees a performance of 5 and can be verified using Lemma 4.1. We refer to this heuristic as FIND-SET. Therefore, our main task here is to show how we can run the above heuristic in polynomial time when the input is a L-specification of a unit disk graph $G$. As will be seen, our algorithm computes the number of vertices in the approximate independent set in time polynomial in the size of the hierarchical description. Again, given a vertex, we can tell in polynomial time if the vertex belongs to the approximate independent set computed.

We use a variant of *bottom-up* method for processing hierarchical graphs. The bottom-up method has been used in [LW87a], [Le88], and [Wi90] for designing efficient algorithms for hierarchical graphs. This method aims at finding a small graph $G_i^b$ called the *burnt graph* which can replace each occurrence of $G_i$ in such a way that $G_i$ and $G_i^b$ behave identically with respect to the problem under consideration. The bottom up method should produce such burnt graphs efficiently. Note that the bottom-up method has previously been applied only to obtain solutions for problems solvable in polynomial time for L-specifications. Here and in [MHR93], the bottom-up approach is used to obtain approximate solutions for PSPACE-hard problems for hierarchical specifications.

**Heuristic HIND-SET**

**Input:** *A hierarchical L-specification* $\Gamma = (G_1, ..., G_n)$ *of a unit disk graph G.*
**Output:** *A hierarchical L-specification* $\beta = (H_1, ..., H_n)$ *of a near-optimal independent set for G.*

1. Repeat the following steps for $1 \leq i \leq n$.

   (a) Let $A_i$ denote the set of all the explicit vertices in $G_i$. Starting from the set $A_i$, we create a new set $B_i$ as follows. For each vertex $v \in A_i$, we place it in the set $B_i$ iff $v$ is *not* adjacent to any of the pins marked *removed* in the burnt graphs of $G_j$, where $G_j$, $j < i$ appears in the definition of $G_i$. (**Note:** vertex $v$ is placed in the set $B_i$ iff none of its neighbors in $G_j$, $j < i$ have been placed in $V_j$. Let $G(B_i)$ denote the subgraph induced on the nodes in $B_i$.)

   (b) Use Algorithm FIND-SET on the graph on the set of vertices $B_i$ to obtain the set $X_i$. (Again realize that we need not consider any edges which are from these explicit vertices to the pins.)

   (c) Let $V_i = X_i \cup \bigcup_j V_j$ where $G_j$, $j < i$ appears in the definition of $G_i$.
   
   (Observe that the sets are created implicitly.)

   (d) Now construct the burnt graph $G_i^b$ as follows: The pins in $G_i^b$ are the same as the pins in $G_i$. A pin in $G_i$ is marked *removed* iff the pin is *either* adjacent to one of vertices in the set $X_1$ *or* it is adjacent to one of the pins in $G_j$, $j < i$, which appears in the definition of $G_i$.

We note that in Step 2(c) of the above algorithm, the set $V_i$ is updated after the addition of each vertex. Also, we do not maintain these sets explicitly. We just need the number of vertices in $V_i$. This information can be maintained by keeping track of the partition of explicit vertices in $G_i$.

Next, we show (Theorem 4.1) that the algorithm indeed computes a near optimal independent set for the given hierarchical specification in polynomial time in the size of the specification. The following proposition is used in the proof of the theorem.

**Proposition 4.1** *Consider the burnt graph created by the algorithm at any given stage i. Then the pins of the burnt graph are marked* removed *iff there exists a vertex in the hierarchy tree* $HT(\Gamma_i)$ *which is chosen in the independent set and has an edge to the pin.* □

**Theorem 4.1** *Let* $\Gamma$ *be a hierarchical specification of a unit disk graph G. Then we can compute in time polynomial the size of* $\Gamma$, *an approximate independent set whose size is within a factor of five of the size of an optimal independent set for G.*

*Proof:* Follows from Proposition 4.1 and the discussion preceding the details of Heuristic HIND-SET. □

# References

[BOW83] J.L. Bentley, T. Ottmann, P. Widmayer, "The Complexity of Manipulating Hierarchically Defined set of Intervals," *Advances in Computing Research, ed. F.P. Preparata* Vol. 1, (1983), pp. 127-158.

[CCJ90] B.N. Clark, C.J. Colbourn, D.S. Johnson, "Unit Disk Graphs" *Discrete Mathematics*, 86(1990), pp. 165-177.

[HRS93] H.B. Hunt III, V. Radhakrishnan, R.E. Stearns "On The Complexity of Generalized Satisfiability and Hierarchically Specified Generalized Satisfiability Problems," in preparation.

[FPT81] R.J. Fowler, M.S. Paterson, S.L. Tanimoto, "Optimal Packing and Covering in the Plane are NP-Complete," *Information Processing Letters*, Vol 12, No.3, June 1981, pp. 133-137.

[GW83] H. Galperin and A. Wigderson, "Succinct Representation of Graphs," *Information and Control* , Vol.56, 1983, pp. 183-198.

[GJM91] C. Ghezzi, M. Jazayeri, D. Mandrioli, *Fundamentals of Software Engineering*, Prentice Hall, Englewood Cliffs, NJ.

[Ha80] W.K. Hale, "Frequency Assignment: Theory and Applications," *Proc. IEEE*, Vol. 68, 1980, pp 1497-1514.

[HM85] D. S. Hochbaum, W. Mass, "Approximation Schemes for Covering and Packing Problems in Image Processing and VLSI," *JACM*, Vol. 32,No. 1, 1985, pp 130-136.

[HLW92] F. Höfting, T. Lengauer and E. Wanke, "Processing of Hierarchically Defined Graphs and Graph Families", in *Data Structures and Efficient Algorithms* (Final Report on the DFG Special Joint Initiative), LNCS 594, Springer-Verlag, 1992, pp. 44-69.

[Ka84] K. Kammerlander, "C 900 – An Advanced Mobile Radio Telephone System with Optimum Frequency Utilization," *IEEE Trans. Selected Areas in Communication*, Vol. 2, 1984, pp 589-597.

[LW87a]  T. Lengauer, E. Wanke, "Efficient Solutions for Connectivity Prob-
         lems for Hierarchically Defined Graphs ," *SIAM J. Computing*, Vol.
         17, No. 6, 1988, pp. 1063-1080.

[Le88]   T. Lengauer, "Efficient Algorithms for Finding Minimum Spanning
         Forests of Hierarchically Defined graphs", *Journal of Algorithms*, Vol.
         8, 1987, pp. 260-284.

[Le89]   T. Lengauer, "Hierarchical Planarity Testing," *J.ACM* , Vol.36, No.3,
         July 1989, pp. 474-509.

[LW92]   T. Lengauer, K.W. Wagner, "The correlation between the complexi-
         ties of non-hierarchical and hierarchical versions of graph problems",
         *JCSS*, Vol. 44, 1992, pp. 63-93.

[MF90]   P. Mirchandani and R.L. Francis, *Discrete Location Theory*, John Wi-
         ley and Sons, 1990.

[MHR93]  M.V. Marathe, H.B. Hunt III, and S.S. Ravi, "The Complexity of
         Approximating PSPACE-Complete Problems for Hierarchical Speci-
         fications", in the proceedings of *ICALP'93*, July 1993, pp 76-87.

[MHR92]  M.V. Marathe H.B. Hunt III and S.S. Ravi, "Geometric Heuristics for
         Unit Disk Graphs", in the proceedings of *4th Canadian Conference
         on Computational Geometry*, 1993, pp 244-249.

[MTM92]  J.O. McClain, L.J. Thomas and J.B. Mazzola, *Operations Manage-
         ment,* Prentice Hall, Englewood Cliffs, 1992.

[MC80]   C. Mead and L. Conway, *Introduction to VLSI systems* Addison Wes-
         ley, 1980.

[MS84]   N. Meggido, K Supowit, "On The Complexity Of Some Common Ge-
         ometric Location Problems," *SIAM Journal Of Computing*, Vol 13,
         No.1, February 1984, pp. 182-196.

[Wa84]   K.W. Wagner, "The complexity of problems concerning graphs with
         regularities", *Proc. 11th Symposium on Math. Foundations of Com-
         puter Science*, LNCS 176, Springer-Verlag, 1984, pp. 544-552.

[WK88]   D.W. Wang, Y.S. Kuo, "A Study On Geometric Location Problems,"
         *Information Processing Letters*, Vol.28, No.6, August 1988, pp. 281-
         286.

[Wi90]   M. Williams, "Efficient Processing of Hierarchical Graphs ," *TR 90-
         06*, Dept of Computer Science, Iowa Sate University. (Parts of the re-
         port appeared in WADS'89 and SWAT'90 coauthored with Fernandez-
         Baca.)

[Ul88]   J.D. Ullman, *Principles of Database and Knowledge Base Systems*,
         Vol.1, Computer Science Press, Rockville, MD, 1988.

# Bounded Tree-Width and LOGCFL

Egon Wanke

German National Research Center for Computer Science, GMD
D-53757 Sankt Augustin, Germany

**Abstract.** We show that (1) the recognition of tree-width bounded graphs and (2) the decidability of graph properties—which are defined by finite equivalence relations on $h$-sourced graphs—on tree-width bounded graphs belong to the complexity class LOGCFL. This is the lowest complexity class known for these problems. Our result complements the research in a series of papers [1, 2, 3, 5, 8, 9, 12, 15, 16] by Arnborg, Bodlaender, Chandrasekharan, Courcelle, Hedetniemi, Lagergren, Proskurowski, Reed, Robertson, Seymour, Seese, and many others.

## 1 Introduction and Summary

Many authors have designed algorithms for graph problems on tree-width bounded graphs. Most of these algorithms firstly compute a tree-decomposition of a given graph $G$ and, after that, they process the graph with respect to its tree structure. This tree-oriented processing of the graph is based on the existence of a finite equivalence relation on so-called $h$-*sourced graphs* (or $h$-*graphs* for short). $h$-graphs are graphs with $h$ distinct so-called *source vertices*. Two $h$-graphs $G$ and $J$ can be composed to form a graph $G\|J$ by identifying the $i$th source vertex from $G$ with the $i$th source vertex from $J$. If $\Pi$ is a graph property, then two $h$-graphs $G$ and $J$ are called *replaceable with respect to* $\Pi$, denoted by $G \sim_\Pi J$, if and only if for all $h$-graphs $H$: $\Pi(G\|H) = \Pi(J\|H)$, see also [14]. Replaceability is always an equivalence relation, and if there are finitely many equivalence classes for all $h$-graphs and each $h$, then $\Pi(G)$ is called a *finite state graph property* (or *finite graph property* for short) and decidable in linear time on any tree-width bounded graph $G$, if a decomposition of $G$ is given, see for example [14]. It is known from Courcelle [9] that all graph properties expressible in *monadic second order logic* with quantifications over vertex sets and edge sets (MS properties) define such finite equivalence relations. Thus, all MS properties have linear time decision algorithms on tree-width bounded graphs.

The best time complexity known for finding a tree-decomposition of a graph is due to Bodleander [5], who gives a linear time algorithm. Parallel algorithms for finding a tree-decomposition and for deciding graph properties on tree-width bounded graphs are also known; see for example [12, 4, 8]. However, finding a tree-decomposition of a graph need not be the bottleneck for solving efficiently graph problems on tree-width bounded graphs. Arnborg, Courcelle, Proskurowski, and Seese [2] have shown that each set of graphs of bounded tree-width defined by an MS property can be defined by a certain *graph reduction system*, and each set of graphs defined by such a graph reduction system can be recognized in linear time. That is, there are linear time algorithms for deciding MS properties on tree-width bounded graphs that do not require a tree-decomposition of the input graph.

This paper shows that the recognition problem for tree-width bounded graphs and the decidability of finite graph properties on tree-width bounded graph are contained in the complexity class LOGCFL. This is (1) the class of languages defined by log-space reductions to context-free languages or, equivalently, (2) the class of decision problems solvable by nondeterministic auxiliary pushdown automata in logarithmic space and polynomial time [18] or, equivalently, (3) the class of languages defined by a uniform family of unbounded fan-in circuits of polynomial size and $O(\log(n))$ depth in which no AND-gate has fan-in exceeding 2 [6] or, equivalently, (4) the class of decision problems solvable by alternating Turing machines obeying an $O(\log(n))$ space bound and a polynomial bound on the total size of the computation tree [17]. In this paper we use the fourth characterization of LOGCFL. From a complexity-theoretical point of view, this is the lowest complexity class known for the recognition problem of tree-width bounded graphs and for deciding finite graph properties on tree-width bounded graphs.

## 2  Tree-width

We consider *finite directed loop-free graphs*[1] $G = (V_G, E_G)$, where $V_G$ is a finite set of *vertices* and $E_G = V_G \times V_G$ is a finite set of *directed edges*. For an edge from $u$ to $v$, we call $v$ a *child* of $u$ and $u$ a *parent* of $v$. Vertices without parents are called *root vertices*, vertices without children are called *leaves*. The *size* of $G$, denoted by $size(G)$, is its total number of vertices and edge.

A *directed (undirected) path* from $u$ to $v$ is a sequence of vertices $(u_1, \ldots, u_n)$ such that $u = u_1$, $v = u_n$, and $(u_i, u_{i+1}) \in E_G$ $((u_i, u_{i+1}) \in E_G$ or $(u_{i+1}, u_i) \in E_G$, respectively), for $i = 1, \ldots, n-1$. For a directed path from $u$ to $v$, we call $v$ a *successor* of $u$ and $u$ a *predecessor* of $v$.

A *rooted tree* $T$ is a graph with exactly one root vertex, denoted by $root(T)$, and the property that all non-root vertices have exactly one parent and are successors of $root(T)$. A *rooted forest* is the disjoint union of rooted trees. A *rooted tree* $T'$ is a *rooted sub-tree* of a rooted tree $T$ if $V_{T'} \subseteq V_T$ and $E_{T'} \subseteq E_T$. $T'$ is a *complete rooted sub-tree* of $T$ if, additionally, each successor of $root(T')$ in $T$ is also contained in $T'$.

A graph $G$ is a *subgraph* of a graph $H$ if $V_G \subseteq V_H$ and $E_G \subseteq E_H$. For a graph $G$ and an edge set $F \in E_G$, the graph $G - F$ is obtained by removing all edges of $F$ from $G$. For a vertex $u \in V_G$, the graph $G - u$ is obtained by removing $u$ and all its incident edges from $G$. For a vertex set $U$, the graph $G - U$ is obtained by removing all vertices of $U$ and their incident edges from $G$. $induce(G, U) := G - (V_G - U)$ denotes the subgraph of $G$ *induced* by the vertices in $U$. A graph is called *connected* if between each pair of vertices there is an undirected path. A graph $H$ is called a *connected component* of a graph $G$, if $H$ is a connected induced subgraph of $G$ and there is no edge in $G$ that connects a vertex from $H$ with a vertex from $G - V_H$.

**Definition 1.** [16] Let $G$ be a graph and $T$ a rooted tree whose vertices $t$ are labeled by vertex sets $X(t) \subseteq V_G$ such that $V_G = \bigcup_{t \in V_T} X(t)$. The system $(G, T)$ is a *tree-decomposition* of width $k$ if and only if

1. for each edge $(u, v)$ in $G$ there is a vertex $t$ in $T$ such that $u, v \in X(t)$,

---

[1] All results are easy to extend to undirected graphs with loops and multiple edges.

2. for each vertex $u$ in $G$, the subgraph of $T$ induced by the vertices in $\{t | u \in X(t)\}$ is a rooted tree, and
3. for each vertex $t$ in $T$: $1 \leq |X(t)| \leq k+1$.

We say, $G$ has *tree-width* $k$ if there is a tree-decomposition $(G, T)$ of width $k$.

The *size* of a tree-decomposition $(G, T)$, denoted by $size((G, T))$, is $size(G) + size(T) + \sum_{t \in V_T} |X(t)|$.

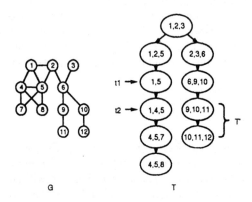

**Fig. 1.** A tree-decomposition $(G, T)$ of width 2. The vertices of $G$ are numbered. The edges of $G$ are drawn as undirected lines, because their directions are unimportant for the tree-decomposition. The vertices $t$ of $T$ are labeled by the vertices in $X(t)$.

Figure 1 shows a tree-decomposition of width 2. The next definition restricts the notion of a tree-decomposition to connected graphs.

**Definition 2.** A tree-decomposition $(G, T)$ of width $k$ is a *rooted tree-decomposition* of width $k$ if

1. for all edges $(t_1, t_2)$ in $T$: $X(t_2) - X(t_1) \neq \emptyset$ and $X(t_1) - X(t_2) \neq \emptyset$ and
2. for each complete rooted sub-tree $T'$ of $T$ the subgraph $G'$ of $G$ induced by the vertices $\cup_{t \in T'} X(t)$ is connected.

We say, $G$ has *rooted tree-width* $k$ if there is a rooted tree-decomposition $(G, T)$ of width $k$.

The tree-decomposition of Figure 1 is not a rooted tree-decomposition, because $T$ has (1) an edge $(t_1, t_2)$ such that $X(t_1) - X(t_2) = \emptyset$ and (2) a complete sub-tree $T'$ such that the vertex set $\cup_{t \in T'} X(t)$ induces a disconnected subgraph $G'$ of $G$.

**Lemma 3.** *A connected graph has tree-width $k$ if and only if it has rooted tree-width $k$.*

Due to space restrictions, we omit the proof. Definition 2 implies also that the size of $T$ (together with the vertex labeling $X$) in each rooted tree-decomposition $(G, T)$ of width $k$ is in $O(|V_G| \cdot k)$.

# 3  $h$-Graphs

An $h$-sourced graph, or $h$-graph for short, is a pair $J = (G, P_G)$, where $G = (V_G, E_G)$ is a graph and $P_G$ is a sequence of $h$ distinct vertices of $G$. The vertices of $P_G$ are called *source vertices*. The graph $G$ underlying the sourced graph $J$ is denoted by $graph(J)$. The set of all sourced graphs with *at most* $h$ source vertices is denoted by $\mathcal{G}(h)$. The set of all sequences of *at most* $h$ distinct vertices from a vertex set $V$ is denoted by $V^h$. If $P$ is a sequence of vertices, then $set(P)$ denotes the set of all vertices from $P$.

Let $G = (V_G, E_G)$ be a graph and $u, v$ be two vertices of $G$. The graph obtained by fusing vertex $u$ to vertex $v$, denoted by $fuse(G, u, v)$, has vertex set $V_G - \{u\}$ and edge set

$$(E_G - \{(p, q) \in E_G | p = u \vee q = u\}) \cup \{(w, v) | (w, u) \in E_G\} \cup \{(v, w) | (u, w) \in E_G\}.$$

A graph $G$ is the *parallel composition* of two $h$-graphs $(H, P_H)$ and $(J, P_J)$, denoted by $(H, P_H) \| (J, P_J)$, if and only if $G$ is obtained by fusing all vertices of $P_H$ to the corresponding vertices of $P_J$ in the disjoint union of $H$ and $J$.

Figure 2 shows an example of a parallel composition.

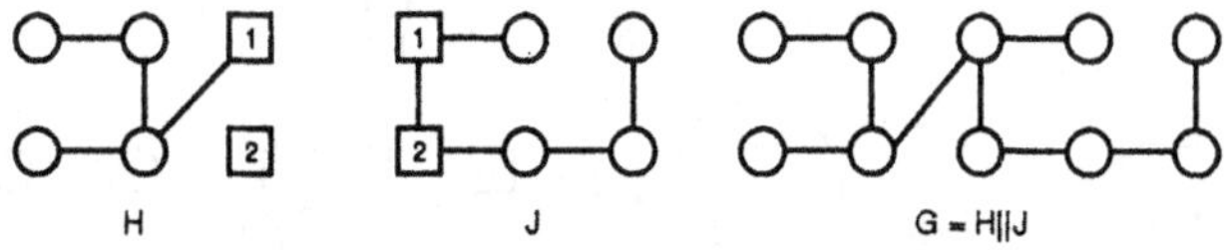

**Fig. 2.** An example of a parallel composition. Source vertices are drawn as squares. The integer in a square is the position of the vertex in the source list.

**Definition 4.** [13, 14] Let $\Pi$ be a graph property.

1. Two $h$-graphs $G$ and $J$ are *replaceable* with respect to $\Pi$ and the parallel composition $\|$, denoted by $G \sim_\Pi J$, if and only if for all $h$-graphs $H$:

$$\Pi(G \| H) = \Pi(J \| H)$$

2. $\Pi$ is called a *finite state graph property*, or *finite graph property* for short, with respect to the parallel composition $\|$ if and only if for each $h \geq 0$ there is a finite set $M$ of $h$-graphs such that for each $h$-graph $G$ there is some $h$-graph $J \in M$ replaceable with $G$ with respect to $\Pi$.

Henceforth, we omit the phrase "with respect to the parallel composition $\|$". Replaceability is an equivalence relation on $h$-graphs compatible with respect to changing the sequence of source vertices, as the next lemma states.

**Lemma 5.** *[14] Let $G$ and $J$ two $h$-graphs replaceable with respect to a graph property $\Pi$. Let $H$ be a $h$-graph, and $Q$ be a sequence of $h'$ distinct vertices from $H$. Then the two $h'$-graphs $(G \| H, Q)$ and $(J \| H, Q)$ are replaceable with respect to $\Pi$.*

If $\Pi$ is a decidable finite graph property and if a finite set $\mathcal{M}$ of $h$-graphs as defined in Definition 4 is explicitly given, then replaceability with respect to $\Pi$ is decidable, see [14]. This is because then $G \sim_\Pi J$ if and only if for all $h$-graphs $H$ from $\mathcal{M}$:

$$\Pi(G\|H) = \Pi(J\|H).$$

If additionally $G$ and $J$ have constant size, then $G \sim_\Pi J$ is decidable in constant time. Note also that $G \sim_\Pi J$ implies $\Pi(graph(G)) = \Pi(graph(J))$.

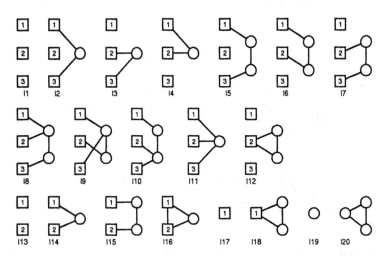

**Fig. 3.** 20 sourced graphs pairwise not replaceable with respect to two-colorability.

Figure 3 shows a possible set $\mathcal{M}$ for the property whether a graph $G$ is two-colorable, i.e., whether there is a mapping $f : V_G \to \{1,2\}$ such that $f(u) \neq f(v)$ for each edge $(u,v) \in E$. The graphs in $\mathcal{M}$ are all pairwise not replaceable with respect to two-colorability. For instance, $I_8$ and $I_9$ are not replaceable because $I_8\|H$ is not two-colorable but $I_9\|H$ is two-colorable for $H = I_2$. $\mathcal{M}$ can also be used for the property non-two-colorability.

We have chosen the notion of a finite state graph property, because the definition of finite equivalence classes for $h$-graphs offers some nice algorithmic properties and does not need the formalism of an algebraic framework. Additionally, the notion of a finite state graph property can easily be extended to other graph compositions mechanisms, see for example [19, 20].

## 4   The Algorithm

Now we design an algorithm which recognizes all graphs $G$ of tree-width $k$ that satisfy a finite graph property $\Pi$. Henceforth, the notion replaceability is always used with respect to the property $\Pi$. If $\Pi(G)$ is "true" for all $G$, the algorithm decides whether $G$ has tree-width $k$. The integer $k$ is assumed to be fixed, i.e, it is constant for the time complexity of the algorithm. However, the time complexity will be exponentially in $k$ even if $\Pi$ is the trivial graph property that is always true.

The algorithm is designed to be executable by an alternating Turing machine (ATM) with logarithmic work-space and polynomial computation tree-size for any fixed $k$. ATMs [7] are a generalization of nondeterministic Turing machines, described informally as follows. The states are partitioned into *existential* and *universal* states. A tree is a *computation tree*[2] of an ATM on an input $G$ if its vertices are labeled with configurations of the machine on $G$, such that the children of any non-leaf vertex labeled by a universal configuration $Y$ include all successor configurations of $Y$ and the single child of any non-leaf vertex labeled by an existential configuration $Y$ includes one successor configuration of $Y$. A computation tree is *accepting* if the root is labeled with the initial configuration, and all the leaves are accepting configurations.

The input graph $G$ for the algorithm is assumed to be given as a sequence of vertices and edges. This assumption corresponds to a representation of $G$ on a read-only Turing machine input-tape. The sequence defines a total order on all vertices of $G$. We denote this order by $\prec$. It is immediate to test in logarithmic work-space whether $u \prec v$ for two given vertices $u$ and $v$.

The algorithm uses a nondeterministic statement named CHOOSE that performs an existential computation step of an ATM. The input graph is accepted by the ATM if and only if at least one of all possible choices leads to an accepting computation. The algorithm starts also new processes with a statement named START that performs an universal computation step of an ATM. Such a statement branches the computation tree. The input graph is accepted by an ATM if and only if all of the started processes and the process that starts all the processes lead to an accepting computation.

The algorithm needs some notation that is described in advance. We explain our notation and parts of the algorithms by examples. All the examples use the graph $G$ from Figure 1, $k = 2$, and the property "non-two-colorability".

$\mathcal{M}$ denotes a set of sourced graphs, where
1. each sourced graph in $\mathcal{M}$ has at most $k + 1$ source vertices and
2. each sourced graph with at most $k + 1$ source vertices is replaceable with exactly one sourced graph from $\mathcal{M}$.

Remark: The size of $\mathcal{M}$ is constant, because $\Pi$ is a finite graph property and not part of the input, and $k$ is fixed. Although the size of $\mathcal{M}$ is independent of the input graph $G$, it is in general exponentially in $k$. We assume that $\mathcal{M}$ is explicitly given. Under this assumptions, it is decidable in constant time whether or not the graph underlying a sourced graph from $\mathcal{M}$ fulfills $\Pi$.

Example: The sourced graphs from Figure 3 form a set $\mathcal{M}$ for non-two-colorability and $k = 2$.

$maxsize_{\mathcal{M}}$ denotes the maximal number of vertices in the sourced graphs of $\mathcal{M}$.

Remark: $maxsize_{\mathcal{M}} \geq k + 1$ is constant, because it is independent of $G$.

Example: In our example from Figure 3 $maxsize_{\mathcal{M}} = 5$.

$g : \mathcal{G}(k + 1) \rightarrow \mathcal{M}$ is a mapping that associates each sourced graph $G$ from $\mathcal{G}(k + 1)$ with a replaceable sourced graph $g(G)$ from $\mathcal{M}$.

We write $g(H, P_H)$ instead of $g((H, P_H))$, if $H$ is a graph with source list $P_H$.

---

[2] A computation tree of an ATM can be considered as an accepting sub-tree of the complete configuration tree.

Remark: The complexity of $g$ is the complexity of $\Pi$, because $\Pi(graph(g(G))) = \Pi(graph(G))$, and $\Pi(graph(g(G)))$ is decidable in constant time. Remember that $maxsize_{\mathcal{M}}$ is constant, since $\mathcal{M}$ itself is fixed with respect to the size of $G$.

Example: $g(J) = I_{15}$ and $g(H) = I_{13}$ for the graphs $J$ and $H$ from Figure 2.

$+_{\|} : \mathcal{M} \times \mathcal{M} \to \mathcal{M}$ is the operation defined by

$$(G, P_G) +_{\|} (J, P_J) := g((G, P_G)\|(J, P_J), P_J).$$

Remark: $+_{\|}$ is associative and computable in constant time, because the sourced graphs in $\mathcal{M}$ have bounded size.

Example: $I_4 +_{\|} I_5 = I_8$.

$\Phi_G(W, v)$, for a vertex set $W \subseteq V_G$ and a vertex $v \in V_G - W$, denotes the vertex set of the connected component of $G - W$ that includes vertex $v$.

If $W$ is a sequence of vertices, then we write $\Phi_G(W, v)$ instead of $\Phi_G(set(W), v)$.

Remark: It is decidable nondeterministically on logarithmic work-space whether or not a given vertex is contained in $\Phi_G(W, v)$. Note that NL is closed under complementation, see [11].

Example: For the graph $G$ from Figure 1, $G - \{u_2\}$ has two connected components $\Phi_G(\{u_2\}, u_{10}) = \{u_3, u_6, u_9, u_{10}, u_{11}, u_{12}\}$ and $\Phi_G(\{u_2\}, u_1) = \{u_1, u_4, u_5, u_7, u_8\}$. .

The algorithm consists of a main part and a process named RECOGNIZE. Its task is to compute $\Pi(G)$ by first computing $g(G, \emptyset)$ and then $\Pi(g(G, \emptyset))$. The main part processes the connected components $G_1, \ldots, G_m$ of the input graph $G$ step by step. That is, $g(G, \emptyset)$ is computed by

$$g(G_1, \emptyset) +_{\|} \cdots +_{\|} g(G_m, \emptyset).$$

Each such $G_j$ for $j = 1 \ldots m$ is identified by its smallest vertex $u_j$ with respect to the order $\prec$. That is,

$$G_j = induce(G, \Phi_G(\emptyset, u_j))$$

and $G_j$ has no vertex $u$ with $u \prec u_j$. If $G_j$ has at most $maxsize_{\mathcal{M}}$ vertices, then $g(G_j, \emptyset)$ is computable in constant time, since the size of $G_j$ is constant. If $G_j$ has more than $maxsize_{\mathcal{M}}$ vertices, then $g(G_j, \emptyset)$ is computed with the help of the process RECOGNIZE. We explain later how RECOGNIZE works. The main algorithm chooses a 0-graph $H_2$ from $\mathcal{M}$ and starts the process RECOGNIZE with the parameter list $(H_2, \emptyset, u_j)$. Then, RECOGNIZE verifies whether $g(G_j, \emptyset) = H_2$. Remember that each $G_j$ is explicitly defined by its smallest vertex $u_j$, the third parameter of RECOGNIZE. Finally, in the main part, all computed 0-graphs $g(G_j, \emptyset)$ for $j = 1 \ldots m$ are combined to a 0-graph $H_1$ by $H_1 := H_1 +_{\|} g(G_j, \emptyset)$ for $j = 1, \ldots, m$, where $H_1$ was initialized by the empty 0-graph. The definition of replaceability, Lemma 5, and the assumption about the process RECOGNIZE imply that the graph underlying the resulting 0-graph $H_1$ fulfills $\Pi$ if and only if the input graph $G$ fulfills $\Pi$.

For the graph $G$ from Figure 1, the main algorithm calls the process RECOGNIZE only once with parameter list $(H_2, \emptyset, u_1)$ for some $H_2 \in \mathcal{M}$.

## MAIN ALGORITHM

1: LET $H_1 := g(\emptyset_0)$;

```
2:  FOR ALL u ∈ V_G SUCH THAT ∀u' ∈ Φ_G(∅, u) − {u} : u ≺ u' DO
3:      IF |Φ_G(∅, u)| ≤ maxsize_M
4:          THEN BEGIN
5:              LET H := induce(G, Φ_G(∅, u));
6:              LET H₁ := H₁ +∥ g(H, ∅); END
7:          ELSE BEGIN
8:              CHOOSE H₂ ∈ M;
9:              START RECOGNIZE(H₂, ∅, u);
10:             LET H₁ := H₁ +∥ H₂; END
11: IF Π(graph(H₁))
12:         THEN HALT AND ACCEPT;
13:         ELSE HALT AND REJECT;
```

We explain the main algorithm line by line to see that it is executable by an ATM with logarithmic work-space.

**Line 1:** The 0-graph $H_1$ is initialized by the 0-graph of $\mathcal{M}$ replaceable with the empty 0-graph, i.e., the graph with an empty vertex set, an empty edge set, and an empty source list.

**Line 2:** The smallest vertex $u$ of each connected component of $G$ is considered. This is possible on logarithmic work-space because membership in $\Phi_G(\emptyset, u)$ is decidable nondeterministically on logarithmic work-space.

**Line 3, 4, 5, and 6:** If $\Phi_G(\emptyset, u)$ has at most $maxsize_\mathcal{M}$ vertices, then variable $H$ is used to denote the subgraph of $G$ induced by $\Phi_G(\emptyset, u)$. Now $g(H, \emptyset)$ is computable in constant time. In Line 6, $g(H, \emptyset)$ is combined with $H_1$.

**Line 7, 8, 9, and 10:** If $\Phi_G(\emptyset, u)$ has more than $maxsize_\mathcal{M}$ vertices, then the algorithm chooses a $h$-graph from $\mathcal{M}$ in variable $H_2$ and starts a process RECOGNIZE with parameter list $(H_2, \emptyset, u)$. This process halts in an accepting state if and only if $g(H, \emptyset) = H_2$, where $H$ is the subgraph of $G$ induced by $\Phi_G(\emptyset, u)$. In Line 10, $H_2$ is combined with $H_1$.

**Line 11, 12, and 13:** If all connected components of $G$ are analyzed, the graph underlying $H_1$ is verified to fulfill $\Pi$.

The basic part of the main algorithm is the process RECOGNIZE. In general, the second parameter of RECOGNIZE is a sequence of distinct vertices from $G$. Let $(H_2, P, v)$ be a possible parameter list for RECOGNIZE. Let $V_J := \Phi_G(P, v) \cup set(P)$ and $J$ be the subgraph of $G$ induced by the vertices in $V_J$ without the edges between any pair of vertices from $P$. Then, RECOGNIZE is designed to halt in an accepting state if and only if $g(J, P) = H_2$.

The process RECOGNIZE works as follows: If $J$ has at most $maxsize_\mathcal{M}$ vertices then $g(J, P)$ is computable in constant time. Assume $J$ has more than $maxsize_\mathcal{M}$ vertices. Then RECOGNIZE computes $g(J, P)$ by guessing recursively a rooted tree-decomposition $(J, T)$ for $J$ such that $X(root(T)) = set(P)$ and $root(T)$ has exactly one child $r$. That is, RECOGNIZE chooses a sequence of at most $k + 1$ vertices $Q$ $(= X(r))$ of $J$ such that $set(P) − set(Q) \neq \emptyset$, $set(Q) − set(P) \neq \emptyset$, and if there is an edge between two vertices $u_1, u_2$ in $G$ such that $u_1 \in \Phi_G(P, v) − set(Q)$ then $u_2 \in \Phi_G(P, v) \cup set(Q)$. Intuitively speaking, both vertices $u_1, u_2$ of the edge are either in $set(Q)$ or in a vertex set defined in a succeeding step. These are the prerequisites for a possible rooted tree-decomposition, see Definition 1 and 2.

Now, RECOGNIZE has to find a rooted sub-tree that represents the subgraph of $G$ induced by the vertices $\Phi_G(P, v) \cup set(Q)$. Let $J'$ be the subgraph of $G$ induced by the vertices $\Phi_G(P, v) \cup set(Q)$ without the edges between vertices from $Q$. RECOGNIZE computes $g(J', Q)$ by

$$g(J_1, Q) +_{\parallel} \cdots +_{\parallel} g(J_m, Q),$$

where $J_1, \ldots, J_m$ are the subgraphs induced by the vertices of the connected components of $J' - set(Q)$ and the vertices of $Q$ but without the edges between vertices from $Q$. That is, the process RECOGNIZE determines the smallest vertex $w$ from each connected component in $J' - set(Q)$ and computes

$$g(induce(\Phi_G(Q, w) \cup set(Q)) - \{(p, q) \in E_G | p, q \in set(Q)\}, Q).$$

This will be done again with a process RECOGNIZE. A sourced graph from $\mathcal{M}$ is chosen in a variable $H_3$ and a new process RECOGNIZE with the parameter list $(H_3, Q, w)$ is started for each $w$. All $|Q|$-graphs $H_3$ are combined to a $|Q|$-graph $H_1$ by $H_1 := H_1 +_{\parallel} H_3$, where $H_1$ was initialized by the empty $|Q|$-graph. This empty $|Q|$-graph has $|Q|$ vertices, no edges, and $|Q|$ source vertices. Then, by the definition of replaceability and Lemma 5 the resulting sourced graph $H_1$ is $g(J', Q)$.

Now, RECOGNIZE can compute $g(J, P)$ in constant time as follows. Let

$$K = induce(G, set(P) \cup set(Q)) - \{(p, q) \in E_G | p, q \in set(P)\}$$

be the subgraph of $G$ induced by the vertices in $P$ and $Q$ without the edges between vertices from $P$, then $g(J, P) = g(H_1 \| (K, Q), P)$ is verified with $H_2$. If both sourced graphs are equal, the process halts in an accepting state. Note that the main algorithm starts the process RECOGNIZE with an empty vertex list. That is, RECOGNIZE has also to take the start condition into consideration. We explain the the process RECOGNIZE line by line.

**PROCESS RECOGNIZE($H_2 \in \mathcal{M}, P \in V_G^{k+1}, v \in V_G$)**

```
1: IF |Φ_G(P,v) ∪ set(P)| ≤ maxsize_M THEN BEGIN
2:     LET H := induce(G, Φ_G(P,v) ∪ set(P)) - {(p,q) ∈ E_G|p,q ∈ set(P)};
3:     IF H_2 = g(H, P)
4:         THEN HALT AND ACCEPT;
5:         ELSE HALT AND REJECT; END
6: CHOOSE A NONEMPTY Q ∈ (Φ_G(P,v) ∪ set(P))^{k+1};
7: IF P ≠ ∅ AND [set(Q) ⊆ set(P) OR set(P) ⊆ set(Q) OR
8:     ∃(u_1, u_2) or (u_2, u_1) ∈ E_G : u_1 ∈ Φ_G(P,v) - set(Q) ∧ u_2 ∉ Φ_G(P,v) ∪ set(Q)]
9:     THEN HALT AND REJECT;
10: LET H_1 := g(∅_{|Q|});
11: FOR ALL w ∈ Φ_G(P,v) - set(Q) SUCH THAT ∀w' ∈ Φ_G(Q,w) - {w} : w ≺ w' DO
12:     BEGIN
13:         CHOOSE H_3 ∈ M;
14:         START RECOGNIZE(H_3, Q, w);
15:         H_1 := H_1 +_∥ H_3; END
16: LET K := induce(G, set(Q) ∪ set(P)) - {(p,q) ∈ E_G|p,q ∈ set(P)};
17: IF H_2 = g(H_1 ∥ (K, Q), P)
18:     THEN HALT AND ACCEPT;
19:     ELSE HALT AND REJECT;
```

**Line 1, 2, 3, 4, and 5:** If $\Phi_G(P, u) \cup set(P)$ has at most $maxsize_{\mathcal{M}}$ vertices, then variable $H$ is used to denote the subgraph of $G$ induced by the vertices in $\Phi_G(P, u) \cup set(P)$ without the edges between the vertices in $P$. Now $g(H, P)$ is computable in constant time and can be compared with $H_2$.

**Line 6:** If $\Phi_G(P, u) \cup set(P)$ has more than $maxsize_{\mathcal{M}}$ vertices, then a nonempty sequence $Q$ of at most $k + 1$ vertices from $\Phi_G(P, u) \cup set(P)$ is chosen.

**Line 7,8, and 9:** The algorithm tests whether $set(P) = \emptyset$ (the start condition) or whether $set(Q) \not\subseteq set(P)$, $set(P) \not\subseteq set(Q)$, and there is no edge $(u_1, u_2)$ or $(u_2, u_2)$ such that $u_1 \in \Phi_G(P, v) - set(Q)$ and $u_2 \notin \Phi_G(P, v) \cup set(Q)$. If this condition does not hold, the process halts in a rejecting state. Note that all this can be verified nondeterministically on logarithmic work-space, because NL is closed under complementation.

**Line 10:** The variable $H_1$ is initialized by the empty $|Q|$-graph $g(\emptyset_{|Q|})$.

**Line 11:** The smallest vertex $w$ of each connected component in the subgraph of $G$ induced by vertex set $\Phi_G(P, u) - set(Q))$ is determined.

**Line 12, 13, 14, and 15:** For each $w$ determined in Line 11, a $|Q|$-graph is chosen from $\mathcal{M}$ into variable $H_3$, a new process RECOGNIZE with parameter list $(H_3, Q, w)$ is started, and $H_3$ is combined with $H_1$.

**Line 16** Variable $K$ is defined to be the subgraph of $G$ induced by the vertices in $P$ and $Q$ but without the edges between vertices from $P$.

**Line 17, 18, and 19:** The final $H_1$ is combined with $(K, Q)$. The result together with the vertex list $P$ is compared with $H_2$.

For the graph $G$ from Figure 1 and the non-two-colorability property the process RECOGNIZE with parameter list $(I_{20}, \emptyset, u_1)$ could work as follows: In Line 1 we have $|\Phi_G(\emptyset, u_1) \cup set(\emptyset)| = 12$ which is greater than $maxsize_{\mathcal{M}}$. Suppose the algorithm chooses in Line 6 $Q = (u_1, u_2, u_3)$. Since $P$ is empty, the next step is Line 10, where we initialize $H_1$ by $I_1$. Then in Line 11 we consider $w = u_4$ and $w = u_6$. Suppose the algorithm chooses for $H_3$ the sourced graphs $I_{12}$ and $I_3$, respectively. When the FOR iteration has finished we get $H_1 = I_{12} (= I_1 +_{\parallel} I_{12} +_{\parallel} I_3)$.

In Line 16 we compute the graph $K = (\{u_1, u_2, u_3\}, \{(u_1.u_2)\})$. In Line 17 we check whether $H_2 = g(H_1 \| (K, Q), P)$. Since $H_2 = I_{20}$, $H_1 = I_{12}$ and thus $H_2 = g(H_1 \| (K, Q), P)$, the process accepts the parameter list $(I_{20}, \emptyset, u_1)$.

However, the ATM running the complete algorithm accepts $G$ only if RECOGNIZE accepts also $(I_{12}, (u_1, u_2, u_3), u_4)$ and $(I_3, (u_1, u_2, u_3), u_6)$ started in Line 14.

For parameter list $(I_{12}, (u_1, u_2, u_3), u_4)$, in Line 1, we have $|\Phi_G((u_1, u_2, u_3), u_4) \cup set((u_1, u_2, u_3))| = 7$ which is greater than $maxsize_{\mathcal{M}}$. Suppose the algorithm chooses in Line 6 $Q = (u_1, u_2, u_5)$. Then $set(Q) - set(P) \neq \emptyset$, $set(P) - set(Q) \neq \emptyset$, and there is no edge between two vertices such that one of them is in $\Phi(P, u_4) - set(Q)$ $(= \{u_4, u_7, u_8\})$ and the other is not in $\Phi_G(P, u_4) \cup set(Q)$ $(=. \{u_1, u_2, u_4, u_5, u_7, u_8\})$. For instance, the choice $Q = (u_2, u_3, u_4)$ is not possible, because there is an edge $(u_1, u_5)$ where $u_5$ is in $\Phi_G(P, u_4) - set(Q)$ $(= \{u_5, u_7, u_8\})$ and $u_1$ is not in $\Phi_G(P, u_4) \cup set(Q)$ $(= \{u_2, u_3, u_4, u_5, u_7, u_8\})$.

Let us continue with the choice $Q = (u_1, u_2, u_5)$. In Line 10, $H_1$ is initialized by $I_1$. In Line 11, the algorithm considers $w = u_4$. Suppose the algorithm chooses $H_3 = I_{12}$. When the FOR iteration has finished we get
$H_1 = I_{12}$, $K = (\{u_1, u_2, u_5\}, \{(u_1, u_5), (u_2, u_5)\})$, $g((K, Q)) = I_8$, and $g(H_1 \| (K, Q), P) = I_{12}$. That is, the process accepts $(I_{12}, (u_1, u_2, u_3), u_4)$.

For parameter list $(I_3, (u_1, u_2, u_3), u_6)$, in Line 1, we have $|\Phi_G(P, u_6) \cup set(P)| = 8$ which is greater than $maxsize_{\mathcal{M}}$. Suppose the algorithm chooses in Line 6 $Q = (u_2, u_3, u_6)$. Then $set(Q) - set(P) \neq \emptyset$, $set(P) - set(Q) \neq \emptyset$, and there is no edge between two vertices such that one of them is in $\Phi_G(P, u_6) - set(Q)$ $(= \{u_9, u_{10}, u_{11}, u_{12}\})$ and the other is not in $\Phi_G(P, v) \cup set(Q)$
$(= \{u_2, u_3, u_6, u_9, u_{10}, u_{11}, u_{12}\})$. In Line 10, $H_1$ is initialized by $I_1$. In Line 11, the algorithm considers $w = u_9$ and $w = u_{10}$. Suppose the algorithm chooses $H_3 = I_1$ for both parameter lists. When the FOR iteration has finished we get $H_1 = I_1$, $K = (\{u_2, u_3, u_6\}, \{(u_2, u_6), (u_3, u_6)\})$, $g(K, Q) = I_8$, and $g(H_1 \| (K, Q), P) = I_3$. That is, the process accepts also $(I_3, (u_1, u_2, u_3), u_6)$.

The correctness of the complete algorithm is based on the definition of replaceability and Lemma 5. It is also immediate that the main algorithm and each process RECOGNIZE uses logarithmic work-space. Each process has a constant number of variables. The variables $H_1$, $H_2$, $H_3$ are used to store sourced graphs with a constant number of vertices. The variables $H$ and $K$ are used to store graphs with a constant number of vertices. The variables $P$ and $Q$ are used for sequences with a constant number of vertices. The variables $u$, $v$, and $w$ are used to store single vertices.

The complete algorithm uses logarithmic work-space, because each vertex can be stored on logarithmic work-space. Note that the space used by the complete algorithm is not the sum of the space used in all started processes. Since alternating Turing machines have universal computation steps, the total size of the work-space is the maximum of the work-space used in all started processes, including the main process.

The size of the computation tree is linear in the size of the input $G$, because the size of each rooted tree-decomposition of fixed width is linear in the size of $G$. Thus we have proved the following theorem:

**Theorem 6.** *The set of all graphs of tree-width $k$, that satisfy a finite graph property can be recognized by an ATM with logarithmic work-space and linear computation tree-size.*

Due to the result obtained by Ruzzo [17] that a language is in LOGCFL if and only if it is recognized by an ATM with logarithmic work-space and polynomial computation tree-size, we get the following corollary:

**Corollary 7.** *The problem of deciding a finite graph property on a tree-width bounded graph is contained in LOGCFL.*

Since MS properties define finite equivalence relations on $h$-sourced graphs, see [9], the corollary above implies the membership to LOGCFL for many graph properties on tree-width bounded graphs. Such graph properties are for example: domatic number for fixed $k$, graph $k$-colorability for fixed $k$, achromatic number for fixed $k$, monochromatic triangle, partition into connected subgraphs with finite property P, induced subgraph with finite property P, cubic subgraph, Hamiltonian circuit, Hamiltonian path, graph isomorphism for fixed $H$, graph contractability for fixed $H$, graph homomorphism for fixed $H$, graphs with forbidden pairs for fixed $k$, disjoint connecting paths for fixed $k$, chordal graph completion for fixed $k$, graph minor for fixed $H$, kernel, and many others, see for example [3] and [12]. The problem definitions can be found in [10].

# References

1. S. Arnborg, D.G. Corneil, and A. Proskurowski. Complexity of finding embeddings in a k-tree. *SIAM Journal of Algebraic and Discrete Methods*, 8(2):227–284, April 1987.
2. S. Arnborg, B. Courcelle, A. Proskurowski, and D. Seese. An algebraic theory of graph reduction. In H. Ehrig, H.J. Kreowski, and G. Rozenberg, editors, *Graph-Grammars and Their Application to Computer Science*, volume 532 of *Lecture Notes in Computer Science*, pages 70–83. Springer-Verlag, Berlin/New York, 1991. To appear in Journal of the ACM.
3. S. Arnborg, J. Lagergren, and D. Seese. Problems easy for tree-decomposable graphs. *Journal of Algorithms*, 12:308–340, 1991.
4. H.L. Bodlaender. NC-algorithms for graphs with bounded tree-width. In J. van Leeuwen, editor, *Proceedings of Graph-Theoretical Concepts in Computer Science*, volume 344 of *Lecture Notes in Computer Science*, pages 1–10. Springer-Verlag, Berlin/New York, 1988.
5. H.L. Bodlaender. A linear time algorithm for finding tree-decompositions of small tree-width. In *Annual ACM Symposium on Theory of Computing*, 1993.
6. A. Borodin, S.A. Cook, P.W. Dymond, W.L. Ruzzo, and M.L. Tompa. Two applications of inductive counting for complementation problems. *SIAM Journal of Computing*, 18:559–578, 1989.
7. A.K. Chandra, D.C. Kozen, and L.J. Stockmeyer. Alternation. *Journal of the ACM*, 28:114–133, 1981.
8. N. Chandrasekharan and A. Hedetniemi. Fast parallel algorithms for tree decomposition and parsing partial k-trees. In *26th Annual Allerton Conference on Communication, Control, and Computing*, 1988.
9. B. Courcelle. The monadic second-order logic of graphs I: Recognizable sets of finite graphs. *Information and Computation*, 85:12–75, 1990.
10. M.R. Garey and D.S. Johnson. *Computers and Intractability, A Guide to the Theory of NP-Completeness*. W.H. Freeman and Company, San Francisco, 1979.
11. N. Immerman. Languages that capture complexity classes. *SIAM Journal of Computing*, 16(4):760–778, August 1987.
12. J. Lagergren. Efficient parallel algorithms for tree-decomposition and related problems. In *Annual ACM Symposium on Foundations of Computer Science*, pages 218–223. IEEE, 1990.
13. T. Lengauer and E. Wanke. Efficient solution of connectivity problems on hierarchically defined graphs. *SIAM Journal of Computing*, 17(6):1063–1080, December 1988.
14. T. Lengauer and E. Wanke. Efficient analysis of graph properties on context-free graph languages. *Journal of the ACM*, 40(2):368–393, 1993.
15. B. Reed. Finding approximate separators and computing tree width quickly. In *Annual ACM Symposium on Theory of Computing*, pages 221–228, 1992.
16. N. Robertson and P.D. Seymour. Graph minors II. Algorithmic aspects of tree width. *Journal of Algorithms*, 7:309–322, 1986.
17. W.L. Ruzzo. Tree-size bounded alternation. *Journal of Computer and System Sciences*, 21:218–235, 1980.
18. I.H. Sudborough. On the tape complexity of deterministic context-free languages. *Journal of the ACM*, 25:405–414, 1978.
19. E. Wanke. Algorithms for graph problems on BNLC structured graphs. *Information and Computation*, 94(1):93–122, September 1991.
20. E. Wanke. k-NLC graphs and polynomial algorithms. *Special Issue in Annals of Discrete Mathematics*, 1993. To appear.

# On reduction algorithms for graphs with small treewidth[*]

Hans L. Bodlaender

Department of Computer Science, Utrecht University
P.O. Box 80.089, 3508 TB Utrecht, the Netherlands

**Abstract.** Some new ideas are presented on graph reduction applied to graphs with bounded treewidth. It is shown that the method can not only be applied to solve decision problems, but also some optimization problems and construction variants of several of these problems on graphs with some constant upper bound on the treewidth.

Also, the exisistence is shown of finite, safe, complete, and terminating sets of reduction rules, such that on any graph $G$ with treewidth at most some constant $k$, $\Omega(n)$ applications of reduction rules can be applied simultaneously. This result is used to obtain a class of randomized parallel algorithms that solve many problems on graphs with bounded treewidth and bounded degree in $O(\log n)$ expected time with $O(n)$ processors on a EREW PRAM. Among others, such an algorithm is obtained to recognize the class of graphs with treewidth at most $k$ and maximum vertex degree at most $d$, for constant $k$ and $d$.

## 1 Introduction

In this paper, new ideas and results are presented on graph reduction, applied to graphs with bounded treewidth.

We consider reduction rules, where a connected subgraph of a graph $G$ is to be replaced by another smaller subgraph (under some additional rules, see Section 2 for the precise definitions.) Arnborg et al [2] showed that for each property $P$, which is 'finite index' and each constant $k$, there exists a finite, complete, safe, and terminating set of reduction rules for graphs with treewidth at most $k$: a graph $G$ is reduced by a series of applications of reduction rules from the set to a graph from some finite set of 'small' graphs, if and only if $P(G)$ holds and the treewidth from $G$ is at most $k$. This result is used to show the existence of linear time algorithms, that decide whether property $P$ holds for a given graph $G$ with bounded treewidth, without the need of using a tree-decomposition of $G$. It should be noted that the algorithm uses more than linear memory. (The set of finite index properties includes many interesting properties, including all properties expressible in monadic second order logic.)

In this paper, we extend these results from [2] in three ways:

- We show that a variant of the method can be used to solve several optimization problems.

---

[*] This work was partially supported by the ESPRIT Basic Research Actions of the EC under contract 7141 (project ALCOM II).

- We discuss a method to solve in many cases also the construction variants of the problem.
- We show the existence of finite, complete, safe, and terminating sets of reduction rules such that on any graph with treewidth at most $k$ in the class to be recognized, $\Omega(n)$ applications of a reduction rule from the set can be applied simultaneously.

The latter result leads to a class of randomized parallel algorithms, that decide on finite index properties, or solve the optimization problems mentioned above, in $O(\log n)$ expected time with $O(n)$ processors on a CRCW PRAM, on graphs with bounded treewidth and bounded degree. These include an algorithm with this time and processor bounds that recognizes the class of graphs with treewidth at most $k$ and degree at most $d$. This latter result should be compared with the following existing results:

- an algorithm that uses $O(\log n)$ time, but $O(n^{3k+1})$ processors [4],
- an algorithm that uses $O(\log^2 n)$ time and $O(n/\log n)$ processors (which can be obtained by a parallel implementation of the algorithm in [9]),
- a sequential algorithm, that uses $O(n)$ time [6].

Each of the three algorithms mentioned above does not require a bound on the degree of the input graph. So, although the result presented in this paper has its clear limitations, it is the first parallel algorithm for the problem that uses $O(\log n)$ expected time and loses only a (poly-)logarithmic factor in its processor-time product, compared with the time of the best sequential algorithm.

See [5, 7] for more backgrounds on graph reduction, and graphs of bounded treewidth.

## 2 Preliminaries

In this paper, the graphs we consider are undirected, do not contain self-loops or multiple edges. (Similar results can be derived for directed graphs. For simplicity, we concentrate on undirected graphs.)

With $d(G)$, we denote the maximum degree over all vertices of $G$.

The notion of treewidth was introduced by Robertson and Seymour [10].

**Definition.** A tree-decomposition of a graph $G = (V, E)$ is a pair($\{X_i \mid i \in I\}$, $T = (I, F)$) with $\{X_i \mid i \in I\}$ a family of subsets of $V$, one for each node of $T$, and $T$ a tree such that

- $\bigcup_{i \in I} X_i = V$.
- for all edges $(v, w) \in E$, there exists an $i \in I$ with $v \in X_i$ and $w \in X_i$.
- for all $i, j, k \in I$: if $j$ is on the path from $i$ to $k$ in $T$, then $X_i \cap X_k \subseteq X_j$.

The treewidth of a tree-decomposition $(\{X_i \mid i \in I\}, T = (I, F))$ is $\max_{i \in I} |X_i| - 1$. The treewidth of a graph $G$, denoted $tw(G)$ is the minimum treewidth over all possible tree-decompositions of $G$.

**Definition.** A *terminal graph* is a triple $(V, E, X)$ with $(V, E)$ an undirected graph, and $X \subseteq V$ is an ordered subset of the vertices, called the set of *terminals*. Vertices in $V - X$ are called *inner vertices*. Terminal graph $(V, E, X)$ is called a $k$-terminal graph, if $|X| = k$. A terminal graph $(V, E, X)$ is said to be *open*, if there are no edges between terminals ($X \times X \cap E = \emptyset$).

The usual undirected graphs (i.e., without terminals) will be simply called *graph*.

**Definition.** The operation $\oplus$ maps two terminal graphs with the same number of terminals to a graph, by taking the disjoint union of the two graphs and then identifying the corresponding terminals, i.e., for $i = 1 \cdots k$, the $i$th terminal of the first terminal graph is identified with the $i$th terminal of the second graph ($k$ the number of terminals).[2]

For an example, see figure 1.

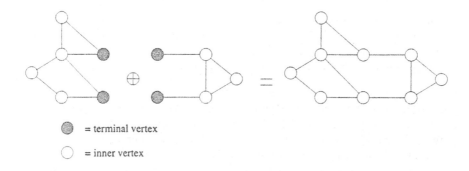

     ● = terminal vertex

     ○ = inner vertex

**Fig. 1.** Example of $\oplus$ operation

Two terminal graphs $(V_0, E_0, < x_1, \cdots, x_k >)$ and $(V_1, E_1, < y_1, \cdots, y_l >)$ are said to be *isomorphic*, if $k = l$ and there exists a function $f : V_0 \to V_1$ with for all $v, w \in V_0 : (v, w) \in E_0 \Leftrightarrow (f(v), f(w)) \in E_1$ and for all $i$, $1 \le i \le k : f(x_i) = y_i$. (The main difference between the usual definition of graph isomorphism is that we require that the corresponding terminals are mapped to each other.)

A *reduction rule* $A$ is an ordered pair of two open terminal graphs with the same number of terminals. It is denoted as $G_0 \overset{A}{\to} G_1$, or if $A$ is clear from the context as $G_0 \to G_1$. An application of rule $G_0 \to G_1$ is the operation, that takes a graph $H$ of the form $H_0 \oplus H_2$, with $H_0$ isomorphic to $G_0$, and replaces it by the graph $H_1 \oplus H_2$, with $H_1$ isomorphic to $G_1$. We write $H \overset{A}{\to} H_1 \oplus H_2$. An example is given in figure 2.

For a set of reduction rules $\mathcal{A}$, we write $H \overset{\mathcal{A}}{\to} H'$, if there exists an $A \in \mathcal{A}$ with $H \overset{A}{\to} H'$.

Let $P$ be a property of graphs. We say that the set of reduction rules $\mathcal{A}$ is *safe* for $P$, if whenever $H \overset{\mathcal{A}}{\to} H'$, then $P(H) \Leftrightarrow P(H')$. The set is said to be *complete*

---

[2] The $\oplus$ operation could create a graph with multiple edges between a pair of vertices. However, by using the $\oplus$ operation only on pairs of terminal graphs of which at least one is open, we will make sure, that this never happens.

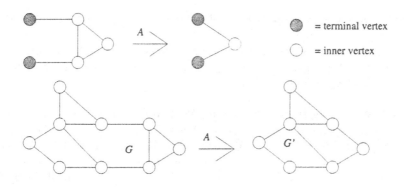

**Fig. 2.** Applying rule A to G yields G'

for $P$, if the set of graphs $\{H \mid P(H) \wedge \neg \exists H' : H \overset{A}{\to} H'\}$ is finite. It is said to be *terminating*, if there does not exist an infinite sequence $H_1 \overset{A}{\to} H_2 \overset{A}{\to} H_3 \overset{A}{\to} \cdots$. It is said to be *decreasing*, if whenever $H \overset{A}{\to} H'$, then $H'$ contains fewer vertices than $H$. Clearly, a decreasing set of rules is terminating.

A set of reduction rules $\mathcal{A}$, that is finite, safe, complete, and terminating for a property $P$ corresponds to an algorithm that decides whether property $P$ holds on a given graph: repeat applying rules from $\mathcal{A}$ starting with the input graph, until no rule from $\mathcal{A}$ can be applied anymore. If the resulting graph belongs to the finite set $\{H \mid P(H) \wedge \neg \exists H' : H \overset{A}{\to} H'\}$, then $P$ holds on the input graph, otherwise it does not. In [2] it is shown how when the set is decreasing, this algorithm can be implemented, such that it takes linear time and polynomial space.

For any property $P$ of graphs, we define the equivalence relation $\equiv_{P,l}$ on $l$-terminal graphs, as follows:

$$H_1 \equiv_{P,l} H_2 \Leftrightarrow (\forall \ l\text{-terminal graphs } K : P(G \oplus K) \Leftrightarrow P(H \oplus K))$$

We say that a property $P$ is of *finite index*, if for all $l$: $\equiv_{P,l}$ has a finitely many equivalence classes. It appears that many important graph properties are finite index. For instance, all properties that can be formulated in monadic second order logic are finite index. These include Hamiltonicity, $k$-colorability (for fixed $k$), and many others. (See e.g. [3].)

For a property $P$, the property $P_{tw,K}$ is defined as $P_{tw,K}(G) = P(G) \wedge tw(G) \leq K$.

**Lemma 1.** *If $P$ is finite index, then $P_{tw,K}$ is finite index.*

(A similar lemma holds, if we pose an additional constant upper bound on the maximum degree of vertices in the graph.)

Finite index corresponds to 'finite state': there exists a linear time algorithm that decides the property on graphs, given with a tree-decomposition of bounded treewidth. Moreover, this algorithm is of a special, well described structure. See e.g. [1].

Safe reduction rules are implied by the equivalence relation $\equiv_{P,l}$: if for $l$-terminal graphs $G_1$, $G_2$ it holds that $G_1 \equiv_{P,l} G_2$, then it directly follows from the definitions that the reduction rule $G_1 \to G_2$ is safe.

*Reduction rules for optimization problems* We now extend the idea of graph reduction to optimization problems. Let $R$ be a function, mapping the set of graphs to $\mathbf{Z}$ $\cup\{$ false$\}$. Typically, $R$ will be an optimization problem, like independent set, vertex cover, etc. The value **false** is used to denote that a certain condition does not hold, especially it is used to deal with graphs that have treewidth more than the fixed upper bound $K$. Denote $\mathcal{Z} = \mathbf{Z} \cup\{$ false$\}$. Define addition on $\mathcal{Z}$ as follows: if $i, j \in \mathbf{Z}$, then we take for $i + j$ the usual sum, and for all $i \in \mathcal{Z}$ : $i+$ **false** $=$ **false**$+i =$ **false**.

We say that $R$ is *finite integer index*, if for each constant $l$, there exists a finite set $S_l$ and a function $\kappa_l$ that maps each $l$-terminal graph to a pair $(s, i) \in S_l \times \mathcal{Z}$, such that for all $l$-terminal graphs $G_1$, $G_2$, and for all $s \in S_l$, $i, j \in \mathcal{Z}$: if $\kappa_l(G_1) = (s, i_1)$ and $\kappa_l(G_2) = (s, i_2)$, then for all $l$-terminal graphs $H$,

$$R(G_1 \oplus H) - i_1 = R(G_2 \oplus H) - i_2$$

As a shorthand notation, we write $G_1 \equiv_{R,k}^{+i} G_2$, if there exist $s \in S_k$, $i_0 \in \mathcal{Z}$ with $\kappa_k(G_1) = (s, i_0 + i)$ and $\kappa_k(G_2) = (s, i_0)$. We call $\kappa_k(G)$ the integer index of $k$-terminal graph $G$.

The idea is now to maintain with the (possibly reduced) input graph an integer variable (futher denoted as: 'the counter'). We now take reduction-counter rules $G_1 \to^{+i} G_2$, that not only carry out the rule $G_1 \to G_2$ as described above, but also add $i$ to the counter. (Formally, a reduction-counter rule is a pair, consisting of an integer, and an ordered pair of open terminal graphs with the same number of terminals.) Notations as for reduction rules are used here in the same way.

We say a set of reduction-counter rules B is *safe*, if for all rules $B \in \mathcal{B}$, for all graphs $H_1$, $H_2$, if $H_1 \overset{B}{\to}^{+i} H_2$, then $R(G_1) = R(G_2) + i$. The set is said to be *complete*, if the set $\{H \mid R(H) \neq$ false $\land \neg \exists H', i \in \mathbf{Z}: H \overset{B}{\to}^{+i} H'\}$ is finite. The definitions of *terminating* and *decreasing* are similar to the finite index case.

If for a finite integer index optimization function $R$, we have for two $k$-terminal graphs, $G_1 \equiv_{R,k}^{+i} G_2$, then the rule $G_1 \to^{+i} G_2$ is safe. The rule corresponds to changing the $k$-terminal subgraph isomorphic to $G_1$ to a subgraph, isomorphic to $G_2$, and simultaneously adding $i$ to the counter.

It is important to note that the sum of $R(G)$ and the counter does not change under safe reduction-counter rules. Thus, when $G$ has been rewritten to a small graph $G'$, one can determine $R(G)$ by calculating $R(G')$ and adding the counter to this number.

For a finite integer index function $R$, the funtion $R_{tw,K}$ is defined as

$$R_{tw,K} = \begin{cases} \text{false,} & \text{if } tw(G) > K \\ R(G) & \text{otherwise} \end{cases}$$

**Lemma 2.** *If $R$ is finite integer index, then $R_{tw,K}$ is finite integer index.*

*Some useful lemmas on graphs with bounded treewidth* Below, we give two lemmas on the existence of subgraphs of a certain size and type in graphs with bounded treewidth. These lemmas will be used later in the paper.

**Lemma 3.** *Let $k, r$ be positive integers. If $G = (V, E)$ is a graph with $n$ vertices and treewidth $\leq k$, $n \geq r+1$, then $G$ can be written as $G_1 \oplus G_2$, with $G_1$ and $G_2$ terminal graphs with at most $k + 1$ terminals, and $G_1$ has at least $r + 1$ and at most $2r + k$ vertices.*

The following lemma basically sais that in a graph with small treewidth, one can find 'many' terminal graphs as subgraph, that have size and number of terminals between certain bounds, and do not share inner vertices.

**Lemma 4.** *Let $k, r$ be positive integers. If $G = (V, E)$ is a graph with $n$ vertices and treewidth $\leq k$, $n \geq r+1$, then there exists a collection $C$ of at least $\frac{|V|}{36(k+1)r}$ pairs of sets of vertices $(W, Z)$, such that*

1. *If $(W_1, Z_1), (W_2, Z_2) \in C$ and $(W_1, Z_1) \neq (W_2, Z_2)$, then $W_1 \cap W_2 = \emptyset$.*
2. *If $(W, Z) \in C$, then $r \leq |W \cup Z| \leq 18r(k + 1)$.*
3. *If $(W, Z) \in C$, then $|Z| \leq 6k + 6$.*
4. *If $(W, Z) \in C$, then for all $(v, w) \in E : v \in W \Rightarrow w \in W \cup Z$.*
5. *If $(W, Z) \in C$, then $G[W \cup Z]$ is a connected subgraph of $G$.*

# 3 Finite, safe, complete, and terminating sets of reduction rules

In this section, we show that for each finite index property $P$ there exists a finite, safe, complete and decreasing set of reduction rules for $P$, and for each finite integer index function $R$, there exists a finite, safe, complete, and decreasing set of reduction-counter rules for $R$. The first of these results has been shown already by Arnborg et al [2]; the proof given below may be somewhat easier to follow.

**Theorem 5.** *Let $k$ be a constant.*
*(i) [2] If $P$ is finite index, then there exists a finite, safe, complete and decreasing set of reduction rules $A$ for $P_{tw,k}$. Moreover, all terminal graphs in a left-hand-side of a rule in $A$ are connected.*
*(ii) If $R$ is finite integer index, then there exists a finite, safe, complete and decreasing set of reduction-counter rules $B$ for $P_{tw,k}$. Moreover, all terminal graphs in a left-hand-side of a rule in $B$ are connected.*

*Proof.* (i) For every $l \leq k+1$, and every equivalence class $Q$ of $\equiv_{P_{tw,k},l}$ that contains at least one open connected $l$-terminal graph, choose a 'representing' open connected $l$-terminal graph $G_Q \in Q$. Let $r$ be the maximum number of vertices of a graph $G_Q$ over all $l \leq k + 1$ and all these equivalence classes $Q$. Now, for all $l \leq k + 1$, for all open connected $l$-terminal graphs $G$ with at least $r + 1$ and at most $2r + k$ vertices, add the reduction rule $G \rightarrow G_Q$ to a set of reduction rules $A$.

Note that, as each right-hand-side of a rule in $\mathcal{A}$ is open, applying a rule in $\mathcal{A}$ can never give multiple edges between a pair of vertices. It is also easy to see that $\mathcal{A}$ is finite: there are finitely many $l$-terminal graphs with at most $2r + k$ vertices. Safeness of the resulting set $\mathcal{A}$ follows directly from the fact that each left- and right-hand-side of a rule in $\mathcal{A}$ belong to the same equivalence class of a relation $\equiv_{P_{tw,k},l}$. As, by lemma 3, each graph with treewidth at most $k$ and at least $r + 1$ vertices, has an applicable rule from the set $\mathcal{A}$, completeness follows directly: the set $\{H \mid P(H) \wedge \neg \exists H' : H \overset{\mathcal{A}}{\to} H'\}$ contains only vertices with at most $r$ vertices. It is obvious that $\mathcal{A}$ is decreasing.

(ii) For every $l \le k + 1$, consider set $S_l$ and function $\kappa_l$ as in the definition of finite integer index, for the problem $R_{tw,k}$. For each $s \in S_l$ such that there exists an open connected $l$-terminal graph $G_s$ with $\kappa_l(G_s) = (s, i)$ with $i \in \mathbf{Z}$, choose such a 'representing' open connected $l$-terminal graph $G_s$. Again, let $r$ be the maximum number of vertices over all representing graphs $G_s$. For all $l \le k + 1$, for all open connected $l$-terminal graphs $G$ with at least $r + 1$ and at most $2r + k$ vertices with $\kappa_l(G) = (s, i)$ for some $s \in S_l$, $i \in \mathbf{Z}$, add the reduction-counter rule $G \to^{+i-i'} G_Q$ to a set of reduction-counter rules $\mathcal{B}$. As above, it follows that the resulting set is finite, safe, and decreasing. It is complete, because when for a graph $H$ it holds that $\neg \exists H', i \in \mathbf{Z} : H \overset{\mathcal{B}}{\to}^{+i} H'$, then either $H$ has at most $r$ vertices, or $H$ can be written as $H_1 \oplus H_2$, with $\kappa_l(H_1) = (s, \mathbf{false})$ for some $l \le k + 1$, $s \in S_l$. In the latter case, $R(H) = \mathbf{false}$. $\qquad\square$

Note that the sets A and B correspond to linear time algorithms that solve $P$ or $R$ on graphs with treewidth at most $k$ [2]. It should be remarked that these algorithms need more than linear memory, and that linear time and linear space algorithms can be obtained by first finding a tree-decomposition of treewidth at most $k$ [6].

## 4   Finite integer index problems

In this section we give some examples of problems that are finite integer index, and of some problems that can be shown not to be finite integer index.

**Theorem 6.** *The Maximum Independent Set problem is finite integer index.*

*Proof.* Recall that the size of a maximum independent set in a graph $G$ is denoted by $\alpha(G)$. The integer index of a $k$-terminal graph $G = (V, E, X)$ is the pair

$$(f : \bar{\mathcal{P}}(X) \to \{0, 1, \ldots, |X|\}, \alpha(G[V - X]))$$

with for all $Y \subseteq X$:

$$f(Y) = \alpha(G[V - X \cup Y]) - \alpha(G[V - X])$$

We first have to prove correctness of this definition. Clearly, for all $Y \subseteq X$, $f(Y) \ge 0$. Next note, that if $Z$ is an independent set in $G[V - X \cup Y]$, then $Z - Y$ is an independent set in $G[V - X]$, hence $\alpha(G[V - X \cup Y]) - \alpha(G[V - X]) \le |Y| \le |X|$.

Consider $k$-terminal graphs $G_1 = (V_1, E_1, X)$ and $G_2 = (V_2, E_2, X)$ with integer index $(f, r_1)$ and $(f, r_2)$ respectively. We must show that for all $k$-terminal graphs

$H = (W, F, X)$, $\alpha(G_1 \oplus H) - r_1 = \alpha(G_2 \oplus H) - r_2$. Suppose $Z$ is a maximum independent set in $G_1 \oplus H$. Note that there exists an independent set $Z'$ of size $r_1 + f(Z \cap X_1)$ with $Z' \cap (X - Z) = \emptyset$. The size of $Z \cap V$ must be precisely $r_1 + f(Z \cap X)$: it cannot be more, by definition of $r_1$ and $f$, and it cannot be less, because then $Z - (V \cap Z) \cup Z'$ would be an independent set in $G \oplus H$ of size, larger than the size of $Z$.

There also exists an independent set $Z''$ of size $r_2 + f(Z \cap X)$ in $G_2$ with $Z'' \cap (X - Z) = \emptyset$. Now $Z - (Z \cap V) \cup Z''$ is an independent set in $G_2 \oplus H$ of size $|Z| + r_2 - r_1$. Hence $\alpha(G_1 \oplus H) - r_1 \leq \alpha(G_2 \oplus H) - r_2$. In the same way one can prove that $\alpha(G_1 \oplus H) - r_1 \geq \alpha(G_2 \oplus H) - r_2$. □

Without proof, we mention that each of the following problems is finite integer index: Partition into Cliques, Vertex Cover, Dominating Set, Covering by Cliques, Hamiltonian Completion Number. The Maximum Cut is finite integer index for graphs with bounded degree. It is also possible to show for some problems that they are not finite integer index. Notable examples are Longest Path, Longest Cycle, Steiner Tree.

# 5 Simultaneous applications of reductions and parallel algorithms

Two applications of reduction rules (or reduction-counter rules) on the same graph are said to be *concurrent*, if the subgraphs to be rewritten do not share any vertex that is non-terminal in at least one of the subgraphs. A collection of applications of reduction(-counter) rules is said to be concurrent, if the applications are pairwise concurrent.

The idea behind concurrent applications of rules is that in a parallel algorithm, all reduction steps from a concurrent set can be carried out simultaneously. This is very useful in order to obtain fast parallel algorithms, based on reduction. The following theorem shown that there exist sets, which always allow a linear number of concurrent reductions.

**Theorem 7.** *Let $K$ be a constant. If $P$ is finite index ($R$ is finite integer index), then there exists a finite, safe, complete, and decreasing set of reduction rules (reduction-counter rules) $\mathcal{A}$ for $P_{tw,K}$ ($R_{tw,K}$), such that*

- *All terminal graphs in a left-hand-side of a rule in $\mathcal{A}$ are connected.*
- *There exist constants $c \in \mathbf{R}^+$, $c' \in \mathbf{N}$ such that for all graphs $G = (V, E)$ with treewidth at most $K$, there exists a concurrent set of $c \cdot |V|$ applications of rules from $\mathcal{A}$, or $|V| \leq c'$.*

*Proof.* Let $P$ be finite index. (The proof for finite integer index is very similar.) For every equivalence class $Q$ of $\equiv_{P_{tw,k},l}$ ($l \leq 6k + 6$), choose a representing $l$-terminal graph $G_Q \in Q$. Let $r$ be the maximum size of a representing $G_Q$ over all these equivalence classes. For every $l \leq 6k + 6$, and for every connected $l$-terminal graph $G$ with at least $r + 1$, and at most $18(r+1)(k+1)$ vertices, add the rule $G \to G_Q$ to a set of reduction rules $\mathcal{A}$, ($G_Q$ the representing terminal graph of the equivalence class to which $G$ belongs.) Similar as in theorem 5, one can prove that the resulting set $\mathcal{A}$ is finite, safe, complete and decreasing. Lemma 4 shows that there exist at least $\frac{|V|}{36(k+1)(r+1)}$ independent applications of a reduction rule in $A$, when $|V| \geq r+1$. □

To transform a set of reduction rules (or reduction-counter rules) as described in theorem 7 into a parallel algorithm, we first must have a method to find in parallel a large enough set of independent applications of applicable rules. At present, the only methods we know of yield slower parallel algorithms than other presently known parallel algorithms to solve problems on graphs with bounded treewidth. However, when we impose a degree bound on the input graph, then the following method works, and compares favorably with existing solutions.

Our algorithm uses $O(\log n)$ expected time, and $O(n)$ processors on an EREW-PRAM. (Note that the algorithm is randomized; it never produces a wrong answer.) We describe the algorithm for finite index properties. In case of finite integer index, some additional counting must be done in steps 5 and 6 of the algorithm.

Let $G = (V, E)$ be the input graph. Let $n$ be the smallest power of 2, that is at least $|V|$. We have $2n - 1$ processors, $n$ of which may represent a vertex, (the *vertex processors*) and $n - 1$ of which are used for counting purposes (the *counting processors*). The processors are numbered $p_1, \ldots, p_{2n-1}$, with $p_1, \ldots, p_{n-1}$ the counting processors. Suppose we have a fixed set $\mathcal{A}$ of reduction rules (as in theorem 7), each rewriting a subgraph with at least $c_1$ vertices and at most $c_2$ vertices to a smaller subgraph. $\mathcal{A}$ is safe for the property $P(G) \wedge tw(G) \leq k \wedge d(G) \leq d$. If a graph with $n'$ vertices belongs to the class, recognized by $\mathcal{A}$, then there exists a concurrent set of reduction rules, such that at least $c_3 \cdot n'$ vertices are an inner vertex of a rule in the set, or $n' \leq c_4$. ($c_1$, $c_2$, $c_4 \in \mathbf{N}$; $c_3 \in \mathbf{R}^+$.) Vertex processors may be alive (when the processor that they represent still is in the reduced graph) or dead (when they do (no longer) represent a vertex.)

The algorithm consists of a number of rounds; each round takes constant time. In each round, the following steps are successively done:

1. Each vertex processor, that is alive, checks whether there exists an applicable reduction rule that has the vertex represented by the processor as inner vertex. If so, the processor is called *applicable* in this round. (This step can be done in constant time, as the degree of the graph is bounded by a constant.)

2. Define the confict graph as follows: each applicable processor forms a vertex in the confict graph. There is an edge between two applicable processors, if at least one the corresponding rule applications contains a vertex of the other processor as inner vertex. Compute the confict graph. (Note that there is a constant upper bound on the degree of vertices in the conflict graph, say $d'$. E.g., one can take $d' = d^{c_2}$.)

3. Each applicable processor picks a random number in $\{1, \ldots, d'\}$. It *wins* in this round, if its number is higher than the number, picked by all other vertices. (It follows from well established results on parallel algorithms, that with high probability, at least a constant fraction of the processors wins. Note that applications of reduction rules, corresponding to winning processors are concurrent.)

4. Carry out all applications of reduction rules corresponding to winning processors. Each new vertex in a right hand side of a rule in this set is given to a processor

that owned a (now obsolete) vertex in the left hand side of this rule, such that no processor receives more than one vertex. Note that with high probability, a constant fraction of the processors lose the vertex they owned and do not get a new vertex: they die, and do not perform any steps anymore.

5. Each vertex processor $p_i$ sets $al_i = 1$, if $p_i$ is alive, and $al_i = 0$, if $p_i$ is dead. It sets $ap_i = 1$, if $p_i$ is applicable, and $ap_i = 0$, if $p_i$ is not applicable.

6. Each counting processor $p_i$ executes:
$$al_i := al_{2i} + al_{2i+1};$$
$$ap_i := ap_{2i} + ap_{2i+1}.$$
If the round number is at least $\log n$, then $al_1$ $(ap_1)$ denotes the number of processors, that were alive (applicable) $\log n$ rounds ago. Processor $p_1$ does the following steps: Check whether $ap_1 \geq c_3 \cdot al_1$. If this does not hold, then we can conclude that the input graph does not belong to the class to be recognized (too little concurrent rule applications were possible). In that case, we stop. Otherwise, check whether $al_1 \leq c_4$. If so, the remaining graph is of constant size: solve the problem in $O(1)$ time.

Note: if the graph we deal with is in the class to be recognized, then at least a constant fraction of the alive processors is applicable. With high probability, at least a constant fraction of these wins, and hence with high probability, at least a constant fraction of the active processors dies in the round. Standard counting arguments show that the average number of rounds is $O(\log n)$.

**Theorem 8.** *Let $k$, $d$, be constants. Let $P$ (R) be finite (integer) index. Then there exists a randomized algorithm, that solves $P$ (R) on graphs $G$ with treewidth at most $k$ and maximum degree at most $d$ in $O(\log n)$ expected time on an EREW-PRAM with $O(n)$ processors.*

**Corollary 9.** *For constant $k$, $d$, the class of graphs with treewidth $\leq k$ and maximum vertex degree $\leq d$ is recognizable in $O(\log n)$ expected time on an EREW-PRAM with $O(n)$ processors.*

Using similar techniques one can also obtain sequential algorithms using graph reduction which need only linear space and time.

**Theorem 10.** *If for property $P$ there exists a finite, safe, complete, and decreasing set of reduction rules $A$ for $P$, such that all terminal graphs in a left-hand-side of a rule in $A$ are connected, and there exist constants $c \in \mathbf{R}^+$, $c' \in \mathbf{N}$ such that for all graphs $G = (V, E) \in P$, there exists a concurrent set of $c \cdot |V|$ applications of rules from $A$, or $|V| \leq c'$, then there exists a recognition algorithm for $P$ that uses linear time and linear space.*

# 6 Final remarks

## 6.1 Constructing solutions

The algorithms discussed so far are decision algorithms: no solutions are constructed. For instance, we have established an algorithm to determine the size of a maximum independent set of a graph with bounded treewidth, but this algorithm does not yield the independent set of maximum size itself. In several cases, it is possible to solve the construction versions of problems, by first performing the reduction algorithm (as described before in this paper) and then undoing the reduction steps one by one, while keeping a solution of the graph that we are working with.

**Lemma 11.** *Suppose $H_2 \to^{+i} H_1$ is a safe reduction-counter rule for the Independent Set problem, $H_1 = (V_1, E_1)$, $H_2 = (V_2, E_2)$ l-terminal graphs. There exists a mapping $f : \mathcal{P}(V_1) \to \mathcal{P}(V_2)$, such that for all l-terminal graphs $G$, if $S$ is a maximum independent set in $H_1 \oplus G$, then $S - (S \cap V_1) \cup f(S \cap V_1)$ is a maximum independent set in $H_2 \oplus G$.*

It follows that when we undo the reduction step $H_2 \to^{+i} H_1$, we can compute the solution for the independent set problem for the larger graph from the solution for the smaller graph in constant time: we need only to look at how the solution looks like in the part of the graph that is rewritten. In total, construction of the solution for the original input graph takes $O(n)$ time. The same approach can be applied to the parallel algorithms of section 5.

It seems that this approach works for many other problems: Hamiltonian circuit, graph coloring, vertex cover, etc. We conjecture that there are general descriptions of classes of problems that can be dealt with in this way. An interesting candidate for such a class would be the class of problems that can be expressed in monadic second order logic (see e.g. [8]). Another interesting open problem is whether it is possible to find tree-decompositions and/or pathdecompositions with small treewidth or pathwidth with this approach.

## 6.2 Graph reduction as an heuristic to solve problems on arbitrary graphs

Graph reduction may possibly also be a useful tool which helps to solve hard problems on arbitrary (sparse) graphs. A possible approach to many such problems could be:

- Fix some safe and decreasing set of reduction or reduction-counter rules.
- Apply reduction rules on the input graph and reduced versions, until no rule application is possible.
- Use another method to solve the problem on the graph after the reductions.
- Possibly, construct a solution for the original problem from the solution for the reduced problem by using the method of section 6.1.

The hope is, of course, that the graph after the reductions is smaller than the original graph, and hence, that the running time of the third step is much smaller than the running time when this algorithm would have been applied directly to the input graph. Provided that the reductions can be carried out quick enough, this approach may be a nice tool to reduce the time needed to solve some graph problems on arbitrary sparse graphs.

## 6.3 Open problems

This paper leaves several directions for further research open. Are there general characterizations of large classes of finite integer index problems? What classes of graphs (without any bound on the treewidth) can be recognized using graph reduction? For instance, the class of graphs with maximum vertex degree $d$ ($d$ fixed) is recognizable with graph reduction. Are there other, interesting classes of graphs with this properties? Is the class of planar graphs recognizable in this way? Classes of graphs with bounded genus? (The latter problem seems particular interesting, as this may be an approach to obtain much faster algorithms for recognition of graphs with bounded genus than known so far.) Can we find efficient parallel algorithms for graphs with bounded treewidth without a degree bound? Is it possible to decrease the number of processors used in the algorithm of section 5 to $O(n/\log n)$?

## Acknowledgement

The author thanks Ton Kloks and Hans Zantema for useful discussions and remarks on the subject of this paper.

# References

1. K. R. Abrahamson and M. R. Fellows. Finite automata, bounded treewidth and well-quasiordering. In *Graph Structure Theory, Contemporary Mathematics vol. 147*, pages 539–564. American Mathematical Society, 1993.
2. S. Arnborg, B. Courcelle, A. Proskurowski, and D. Seese. An algebraic theory of graph reduction. In H. Ehrig, H. Kreowski, and G. Rozenberg, editors, *Proceedings of the Fourth Workshop on Graph Grammars and Their Applications to Computer Science*, pages 70–83. Springer Verlag, Lecture Notes in Computer Science, vol. 532, 1991. To appear in J. ACM.
3. S. Arnborg, J. Lagergren, and D. Seese. Easy problems for tree-decomposable graphs. *J. Algorithms*, 12:308–340, 1991.
4. H. L. Bodlaender. NC-algorithms for graphs with small treewidth. In J. van Leeuwen, editor, *Proc. Workshop on Graph-Theoretic Concepts in Computer Science WG'88*, pages 1–10. Springer Verlag, Lecture Notes in Computer Science, vol. 344, 1988.
5. H. L. Bodlaender. A tourist guide through treewidth. Technical Report RUU-CS-92-12, Department of Computer Science, Utrecht University, Utrecht, 1992. To appear in Acta Cybernetica.
6. H. L. Bodlaender. A linear time algorithm for finding tree-decompositions of small treewidth. In *Proceedings of the 25th Annual Symposium on Theory of Computing*, pages 226–234. ACM Press, 1993.
7. B. Courcelle. Graph-rewriting: an algebraic and logical approach. In J. van Leeuwen, editor, *Handbook of Theoretical Computer Science, volume B*, pages 192–242, Amsterdam, 1990. North Holland Publ. Comp.
8. B. Courcelle. The monadic second-order logic of graphs I: Recognizable sets of finite graphs. *Information and Computation*, 85:12–75, 1990.
9. B. Reed. Finding approximate separators and computing tree-width quickly. In *Proceedings of the 24th Annual Symposium on Theory of Computing*, pages 221–228, 1992.
10. N. Robertson and P. D. Seymour. Graph minors. II. Algorithmic aspects of tree-width. *J. Algorithms*, 7:309–322, 1986.

# Algorithms and Complexity of Sandwich Problems in Graphs (extended abstract)

Martin Charles Golumbic[1], Haim Kaplan[2] and Ron Shamir[2]

[1] IBM Israel Scientific Center, Technion City, Haifa, Israel and Bar-Ilan University
Ramat Gan, Israel.
email: golumbic@israearn.bitnet

[2] Department of Computer Science, Sackler Faculty of Exact Sciences, Tel Aviv
University, Tel-Aviv 69978, Israel.
email: haimk@math.tau.ac.il, shamir@math.tau.ac.il

**Abstract.** Given two graphs $G^1 = (V, E^1)$ and $G = (V, E^2)$ such that $E^1 \subseteq E^2$, is there a graph $G = (V, E)$ such that $E^1 \subseteq E \subseteq E^2$ which belongs to a specified graph family? Such problems generalize recognition problems and arise in various applications. Concentrating mainly on subfamilies of perfect graphs, we give polynomial algorithms for several families and prove the NP-completeness of others.

## 1 Introduction

The graph $G^2 = (V^2, E^2)$ is a *supergraph* of the graph $G^1 = (V, E^1)$ if $V^2 = V$ and $E^1 \subseteq E^2$. Given such two graphs, the graph $G = (V, E)$ is called a *sandwich graph* for the pair $G^1, G^2$ if $E^1 \subseteq E \subseteq E^2$. In other words, $G$ must be "sandwiched" between $G^1$ and $G^2$. We define the *graph sandwich problem for property $\Pi$* (denoted $\Pi$-SP) as follows:

GRAPH SANDWICH PROBLEM FOR PROPERTY $\Pi$ ($\Pi$-SP):
INPUT: Two graphs, $G^1$ and $G^2$, such that $G^2$ is a supergraph of $G^1$.
QUESTION: Does there exist a sandwich graph for the pair $G^1, G^2$ which
satisfies property $\Pi$ ?

Define also $E^0 = E^2 \setminus E^1$, and let $E^3$ be the set of all edges in the complete graph with vertex set $V$, which are not in $E^2$. In this notation, the edge set of a sandwich graph must include all of $E^1$, no edge from $E^3$ and any subset of the edges in $E^0$.

In this paper we study graph sandwich problems for various properties $\Pi$. Such problems arise as natural generalizations of recognition problems. The *recognition problem* is to determine if a given graph $G$ satisfies a desired property (connectedness, chordality, perfectness, etc.), or, equivalently, to decide if $G$ belongs a specific family of graphs. Many graph families were shown to have important applications. Moreover, optimization problems which are NP-hard on general graphs (e.g., clique, coloring or independent set) were shown to be polynomial on many important graph families. For a systematic study of such families and many fascinating applications see [12, 21]. [18] and [3] review many additional results.

In practice, it may happen that the input graph may not belong to the desired family but is "close" to the family in some sense, so one may wish to slightly relax the condition for accepting a given input. One type of relaxation is *completion problems*: Given a graph and an integer $k$, can one add to the original graph at most $k$ edges in order to obtain a graph in the desired family? Such problems have been studied for interval graphs, edge graphs, path graphs (cf. [9, pp. 198-199]) and chordal graphs [27].

Sandwich problems may be viewed as a different kind of relaxation of the recognition problems: Certain edges must definitely be included in the graph, and certain edges are disallowed, but there is freedom in deciding to include any subset of the (possibly many) other edges. Sandwich problems have been studied explicitly in [16], and implicitly in [17], [2] and [25]. Below we give several examples of important sandwich problems arising in practice.

**Physical Mapping of DNA** [5]: In molecular biology, information on intersection or non-intersection of pairs of segments originating from a certain DNA chain is known experimentally. The problem is how to arrange the segments as intervals along a line (the DNA chain), so that their pairwise intersections match the experimental data. In the graph presentation vertices correspond to segments, two vertices are connected by an $E^1$-edge (resp., $E^3$-edge) if their segments are known to intersect (resp., not to intersect). Typically information on intersections is known only in part. That ambiguity introduces $E^0$-edges into the graph. In that case, the decision problem is equivalent to the interval sandwich problem which was shown to be NP-complete [16].

**Temporal reasoning**: Given is a set of events, and for each pair of events a specification whether (i) they are disjoint, (ii) they share some common timepoint, or (iii) both cases are possible. Is this information consistent, i.e., can one assign a time interval to each event so that all pairwise relations hold? This problem is equivalent to the interval sandwich problem [16].

**Phylogenetic Trees**: Buneman [4] showed that the perfect phylogeny (PP) problem in evolution reduces to the graph theoretical problem of triangulating colored graphs (TCG). Kannan and Warnow [19] showed that TCG reduces to PP. Bodlaender et al. [2] and Steel [25] recently proved that TCG is NP-complete. It is easy to see that TCG is a restriction of the chordal sandwich problem, which is therefore also NP-complete.

**Sparse Systems of Linear Equations**: Consider the system of equations $Ax = b$ where $A$ is sparse, symmetric and positive definite. When performing Gaussian elimination on $A$, an arbitrary choice of pivots may result in the *fill-in* of some zero positions with nonzeroes, thereby reducing sparsity. Given $A$, define a graph $G(A) = (V, E)$ where $|V| = n$ and $[v_i, v_j] \in E$ iff $a_{ij} \neq 0$ and $i \neq j$. Rose [22] proved that finding a sequence of pivots which induces minimum fill-in is equivalent to finding a minimum set of edges whose addition to $G(A)$ make the graph chordal. This problem was proved to be NP-complete by Yannakakis [27]. Asking for a sequence of pivots such that the fill-in they induce may occur only in specific positions of $A$ is equivalent to solving a chordal sandwich problem in which $G^1 = G$ and $[v_i, v_j] \in E^0$ iff $i \neq j$, $a_{ij} = 0$, and position $(i, j)$ in the matrix is allowed to become nonzero during factorization. The problem arises in practice when one wants to maintain (and exploit) a special structure of the zeroes in the matrix throughout the elimination.

The paper is organized as follows: Section 2 contains definitions and basic results on relative complexity of sandwich problems. Section 3 contains polynomial algorithms for the sandwich problem for split graphs, threshold graphs and cographs. Section 4 contains NP-completeness proofs of the sandwich problem for comparability, permutation, circle, interval, circular arc, path graphs and related problems. Many proofs are sketched or omitted. The interested reader may find complete proofs in [14] and [13], where we also show additional results, including polynomial and NP-complete variants of the Eulerian sandwich problem.

## 2  Definitions and Basic Results

For standard graph theoretical definitions see [12, 21]. The input to a sandwich problem can be specified by the vertex set $V$ and the partition $(E^1, E^0, E^3)$, or by $V$ together with any non-complementary pair out of the four sets $E^0, E^1, E^2, E^3$. Hence, we may denote a problem instance by $(V, E^1, E^2)$, or $(V, E^1, E^0)$, or $(V, E^1, E^3)$. The superscripts of the two edge sets will indicate in which form the instance is given. Define also $G^i = (V, E^i), i = 0, 1, 2, 3$.

For a graph $G = (V, E)$ and a vertex subset $X \subseteq V$, denote by $E_X$ the set of edges of $G$ with both endpoints in $X$. $G_X = (X, E_X)$ is the *induced subgraph* of $G$ on the vertex set $X$. If $(V, E^i, E^j)$ is a sandwich instance, the *induced sandwich instance* on $X$ is $(X, E_X^i, E_X^j)$ denoted $(V, E^i, E^j)_X$.

For two disjoint graphs $G = (V, E)$ and $H = (W, F)$, their *union* is $G \cup H = (V \cup W, E \cup F)$, and their *join* $G + H$ is the graph obtained from $G \cup H$ by adding all the edges between vertices from $V$ and vertices from $W$. If $H$ is a single vertex $p$ we will also denote the union of $G$ and $\{p\}$ by $G \cup p$ and the join by $G + p$.

A graph property $\Pi$ is *hereditary on subgraphs* if, when a graph $G$ satisfies $\Pi$, every subgraph of $G$ satisfies $\Pi$. Every graph property which can be characterized in terms of forbidden subgraphs is hereditary on subgraphs. A graph property $\Pi$ is *ancestral* if when a graph $G$ satisfies $\Pi$, every supergraph of $G$ satisfies $\Pi$.

**Proposition 1.** *(a) Let $\Pi$ be a hereditary property on subgraphs. There exists a sandwich graph for $(V, E^1, E^2)$ with the property $\Pi$ iff $G^1 = (V, E^1)$ has the property $\Pi$.*
*(b) Let $\Pi$ be an ancestral graph property. There exists a sandwich graph for $(V, E^1, E^2)$ with property $\Pi$ iff $G^2 = (V, E^2)$ has property $\Pi$.* □

Hence, for each property which is hereditary on subgraphs (resp. ancestral) the sandwich problem reduces to the recognition problem of this property on the single graph $G^1$ (resp. $G^2$). For example, sandwich problems for the properties planarity, bipartiteness and acyclicity can be checked on $G^1$, and the SP for properties 'k connected' and 'containing an k-clique' can be checked on $G^2$.

For a property $\Pi$ of graphs, we define the *complementary property* $\overline{\Pi}$ as follows: For every graph $G$, $G$ satisfies $\overline{\Pi}$ iff $\overline{G}$ satisfies $\Pi$. Some well known examples are co-chordality and co-comparability.

**Proposition 2.** *There is a sandwich graph with property $\Pi$ for the instance $(V, E^1, E^0)$ iff there is a sandwich graph with property $\overline{\Pi}$ for the instance $(\tilde{V}, \tilde{E}^1, \tilde{E}^0)$, where $\tilde{V} = V$, $\tilde{E}^1 = E^3$ and $\tilde{E}^0 = E^0$.* □

**Theorem 3.** *For every property $\Pi$, the complexities of $\Pi$-SP and $\overline{\Pi}$-SP are linearly equivalent.* $\qquad\qquad\qquad\square$

$\Pi$-SP is clearly at least as hard as the problem of recognizing graphs with property $\Pi$, since recognizing if a graph $G = (V, E)$ satisfies property $\Pi$ is equivalent to solving $\Pi$-SP for $E^1 = E^2 = E$. Note that the complexity of the sandwich problem for a certain graph property is *not* implied by the complexity of the problem for weaker or stronger properties.

Note that every problem which asks for the existence of certain subgraphs in a given graph (e.g., spanning tree, perfect matching and Hamiltonian path) can be viewed as a sandwich problem: Take $E^1 = \emptyset$, $E^0 = E$, and define the property $\Pi$ appropriately.

In the rest this paper we shall concentrate on the complexity of the sandwich problem for various properties $\Pi$. In view of the discussion above, we shall concentrate on properties $\Pi$ for which (A) the recognition problem is polynomial, and (B) are not hereditary on subgraphs or ancestral.

## 3 Polynomial Sandwich Problems

**3.1 Split Graphs:** A graph $G = (V, E)$ is a *split graph* if there is a partition of the vertex set $V = K + I$ where $K$ induces a clique in $G$ and $I$ induces an independent set. We now show that the split sandwich problem is polynomial. Given an input $(V, E^1, E^3)$ for the split sandwich problem we need to determine for each vertex if it belongs to $K$ or $I$. Clearly, if $[x, y] \in E^1$, then the situation $[x \in I$ and $y \in I]$ is impossible. Similarly, if $[x, y] \in E^3$, then $[x \in K$ and $y \in K]$ is impossible. Represent these constraints by a set of boolean equations: For each vertex $x$ in $V$ define a boolean variable $X$, where $X$ will be true iff $x \in K$. Hence, the set of constraints can be rewritten as a set of boolean equations:

$$\begin{aligned}(\overline{X} \vee \overline{Y}) \text{ for every } [x, y] \in E^3\\(X \vee Y) \text{ for every } [x, y] \in E^1\end{aligned} \qquad (1)$$

**Lemma 4.** *There exists a split sandwich iff system (1) is consistent.*

*Proof.* If a split sandwich exists, then the truth assignment $t(X) = TRUE$ iff $x \in K$ satisfies the system (1). Conversely, if system (1) is consistent, then there is a truth assignment $t$ which satisfies it. Define $x \in K$ iff $t(X) = TRUE$. A sandwich graph $G = (V, E)$ will contain all the clique edges on $K$, none of the edges between vertices in $I$, all $E^1$ edges between $K$ and $I$ in addition to any subset of $E^0$-edges between $K$ and $I$.

We need to show that for every $x, y \in V$ if $x \in K$, $y \in K$ then $[x, y] \in E^1 \cup E^0$, but this is implied by the condition $[x, y] \in E^3 \Rightarrow (\overline{X} \vee \overline{Y})$. Similarly, if $x \in I$ and $y \in I$, then $[x, y] \in E^0 \cup E^3$ follows from the condition $[x, y] \in E^1 \Rightarrow (X \vee Y)$. $\qquad\square$

**Theorem 5.** *The split sandwich problem is solvable in $O(|V| + |E^1| + |E^3|)$ time.*

*Proof.* The transformation of the problem into the set of equations requires $O(|E^1| + |E^3|)$ steps. By Lemma 4, it then suffices to solve the system. Since this system is an instance of 2-Satisfiability, it is solvable in $O(|E^1| + |E^3|)$ time [1]. $\qquad\square$

**3.2 Threshold Graphs:** A graph $G = (V, E)$ is a *threshold graph* if one can assign a non-negative integer value $a(v)$ to each vertex $v$ such that $S \subseteq V$ is independent iff $\sum_{v \in S} a(v) < t$ for some 'threshold' integer value $t$. Threshold graphs were introduced by Chvátal and Hammer [6].

The threshold sandwich problem is polynomial. Hammer et al ([17, Section 4]) have given an $O(|V|^3)$ algorithm for the problem. Below we present a linear algorithm for the problem. Define $Adj(x)$ to be the set of neighbors of a vertex $x$ in a graph. We need the following characterization of threshold graphs:

**Lemma 6 [6].** $G = (V, E)$ *is a threshold graph if and only if for each subset $X \subseteq V$ there exists a vertex $x \in X$ such that $Adj(x) \cap X = \emptyset$ or $Adj(x) \cap X = X - \{x\}$.* □

Hence, if $G$ is a threshold graph and $v$ is a new vertex then $G + v$ and $G \cup v$ are threshold graphs.

**Lemma 7.** *If a threshold sandwich for $(V, E^1, E^3)$ exists, then there is an isolated vertex in $G^1$ or an isolated vertex in $G^3$.*

*Proof.* Suppose $G^s$ is a threshold sandwich for $(V, E^1, E^3)$. According to Lemma 6 it contains a vertex $v$ which is either isolated or adjacent to all others. If $v$ is isolated, since $G^s$ is a sandwich graph, $v$ must be isolated in $G^1$, whereas if $v$ is adjacent to all other vertices it must be isolated in $G^3$. □

**Proposition 8.** *Let $(V, E^1, E^3)$ be a threshold sandwich instance and let $v \in V$ be an isolated vertex in $G^1$ or $G^3$. There is a threshold sandwich for $(V, E^1, E^3)$ iff there is a threshold sandwich for $(V, E^1, E^3)_{V - \{v\}}$.*

*Proof.* The 'only-if' direction is obvious since being a threshold graph is an hereditary property. To prove the converse, suppose there is a threshold sandwich $G^s$ for $(V, E^1, E^3)_{V - \{v\}}$. If $v$ is isolated in $G^1$ then $G^s \cup v$ is a sandwich graph for $(V, E^1, E^3)$ and according to the remark after 6 it is a threshold graph. Similarly, if $v$ is isolated in $G^3$ then $G^s + v$ is a threshold sandwich for $(V, E^1, E^3)$. □

The algorithm for solving the threshold sandwich problem repeatedly reduces the sandwich instance by applying the following simple procedure: Look for an isolated vertex $v$ in $G^1$ or $G^3$, and delete $v$ and all the edges incident to it from the sandwich instance. According to Lemma 7 and Proposition 8, if this procedure ends with an empty sandwich instance there is a threshold sandwich graph and otherwise there is no threshold sandwich.

**Theorem 9.** *The threshold sandwich problem is solvable in $O(|V| + |E^1| + |E^3|)$ steps.*

*Proof.* (Sketch) Validity of the above procedure follows from the previous discussion. The complexity follows by maintaining the sets of active vertices isolated in $G^1$ and in $G^2$. □

**3.3  Cographs:**  A graph is called a complement reducible graph, or a *cograph*, if it does not contain a $P_4$ (a path with 4 vertices) as an induced subgraph [7]. In this section we shall give a polynomial algorithm for the cograph sandwich problem. We shall use two known characterizations for cographs (cf. [7]):

**Theorem 10.** *For a graph $G$, the following statements are equivalent:*
*(1) $G$ is a cograph.*
*(2) $G$ belongs to the set of graphs which can be defined recursively as follows:*
- *A single vertex is a cograph.*
- *If $G_1, G_2, \ldots, G_k$ are cographs, then so is their union $G_1 \cup G_2 \cup \ldots \cup G_k$.*
- *If $G$ is a cograph then so is its complement $\overline{G}$.*

*(3) The complement of any nontrivial connected induced subgraph of $G$ is disconnected.*  □

From characterization (2), using the relation $\overline{G_1 + G_2 + \cdots + G_k} = \overline{G_1} \cup \overline{G_2} \cdots \cup \overline{G_k}$, we obtain:

*Remark.* If $G_1, \ldots, G_k$ are disjoint cographs then $G_1 + G_2 + \cdots + G_k$ is a cograph.

Let $(V, E^1, E^3)$ be a cograph sandwich instance where $|V| > 1$. Using characterization (3) we obtain:

**Lemma 11.** *If $G^1$ and $G^3$ are both connected then there is no cograph sandwich for $(V, E^1, E^3)$.*  □

**Lemma 12.** *Suppose $G^1$ is connected and $G^3$ is disconnected, and let $V_1, \ldots, V_k$ be the vertex sets of the distinct connected components of $G^3$. If $G_1^s, \ldots, G_k^s$ are cograph sandwiches for $(V, E^1, E^3)_{V_1}, \ldots, (V, E^1, E^3)_{V_k}$ respectively, then $G^s = G_1^s + \ldots + G_k^s$ is a cograph sandwich for $(V, E^1, E^3)$.*

*Proof.* By the remark above, $G^s$ is a cograph. But $G^s$ is a sandwich graph for $(V, E^1, E^3)$, since all the edges between different components $\{[u, v] \in V_i \times V_j \,|\, i \neq j\}$ are in $E^2$.  □

**Lemma 13.** *Suppose $G^3$ is connected, $G^1$ is disconnected, and let $V_1, \ldots, V_k$ be the vertex sets of the distinct connected components of $G^1$. If $G_1^s, \ldots, G_k^s$ are cograph sandwiches for $(V, E^1, E^3)_{V_1}, \ldots, (V, E^1, E^3)_{V_k}$ respectively, then $G_1^s \cup \ldots \cup G_k^s$ is a cograph sandwich for $G(V, E^1 \ E^3)$.*

*Proof.* Apply Proposition 2 to Lemma 12.  □

**Lemma 14.** *There exists a cograph sandwich for instance $(V, E^1, E^3)$ iff for every $X \subseteq V$ there exists a cograph sandwich for $(V, E^1, E^3)_X$.*

*Proof.* The 'if' part is obvious. The 'only if' follows since the property of being a cograph is hereditary.  □

We can now describe an algorithm for the cograph sandwich problem: Partition the vertex set into connected components in $G^1$. By Lemma 12, if there is a cograph sandwich for each component, then one can take the union of the sandwiches as the overall solution. By Lemma 14, if there is no cograph sandwich for some component, then there is none for the original problem.

Next, for each vertex set comprising such a connected component in $G^1$, examine the subgraph of $G^3$ induced by this set. If it is connected (and not a singleton), then by Lemma 11 there is no cograph sandwich induced on it, and by Lemma 14 there is no cograph sandwich for the original instance. If it is disconnected, then it suffices to find a cograph sandwich for each component, since by Lemma 13 and the remark following Theorem 10 the join of those sandwiches is a sandwich and a cograph. For each new cograph sandwich instance the algorithm can now be applied recursively.

**Theorem 15.** *The cograph sandwich problem in solvable $O(|V|(|V| + |E^1| + |E^3|))$ steps.*

*Proof.* (Sketch) Validity was discussed above. The complexity follows by observing that there are at most $|V|$ decompositions into connected components of graphs each of which is not larger than the original $G^1$ or $G^3$. □

## 4  NP-complete Sandwich Problems

### 4.1  Comparability Graphs:

A directed graph $D = (V, F)$ is *transitive* if for every $i, j, k \in V$, if $(i, j) \in F$ and $(j, k) \in F$ then $(i, k) \in F$. An undirected graph is a *comparability graph* (or *transitively orientable*, TRO) if one can assign an orientation to each of its edges, so that the resulting directed graph is transitive (see [8], [12].) A comparability graph which has exactly two transitive orientations is called *uniquely transitively orientable* (UTRO). (Clearly, the orientation obtained by reversing all arcs in a transitive orientation is also transitive, so orientations come in pairs.) In this section we prove that the sandwich problem is NP-complete for both of these graph families. The reduction uses as a 'gadget' the TRO sandwich problem in figure 1(a).

**Lemma 16.** *For the sandwich problem in figure 1(a) there are exactly three comparability sandwich graphs, all of which are isomorphic to the graph $G^\sigma$ in figure 1(c).* □

The proof uses the characterization of TRO graphs in terms of implication classes [8], [12].

**Theorem 17.** *The comparability sandwich problem is NP-complete.*

*Proof.* (Sketch) The problem is in NP since recognizing TRO graphs is in P (cf. [12]). We give a reduction from NOT-ALL-EQUAL 3-SATISFIABILITY (NAE-3SAT): Given a 3CNF-formula $\Phi$ with variables $X_1, \ldots, X_n$ and clauses $C_1, \ldots, C_m$, is there a truth assignment for which either one or two literals are true in each clause? This

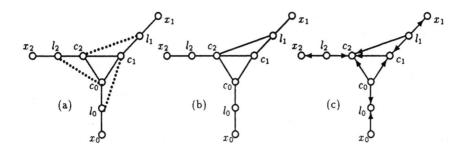

**Fig. 1.** (a) A sandwich problem instance $(W, H^1, H^0)$. $H^1$: Solid edges (mandatory). $H^0$: Dotted edges (optional). $H^2$ is $H^1 \cup H^0$. (b) A comparability sandwich graph $G^\sigma = (W, H^\sigma)$. (c) A transitive orientation for $G^\sigma$.

problem was shown to be NP-complete by Schaefer [24]. Given an instance of NAE-3SAT, construct a TRO sandwich instance $(V, E^1, E^0)$ with $n + 6m$ vertices, $n + 9m$ mandatory edges and $3m$ optional edges, as follows:

The vertex set is $V = \{x_i, \overline{x}_i | i = 1, \ldots, n \} \cup \{l_j^i, c_j^i | i = 1, \ldots, m, j = 0, 1, 2\}$. $x_i$ and $\overline{x}_i$ are called *the vertices of literals* $X_i$ and $\overline{X}_i$, respectively. The six vertices $l_j^i, c_j^i, j = 0, 1, 2$ are called the *private vertices* of clause $C_i$.

To define the edge sets, let $\hat{x}_{i_j}, j = 0, 1, 2$ be the vertices corresponding to the literals in clause $C_i$. Connect the nine vertices $\hat{x}_{i_0}, \hat{x}_{i_1}, \hat{x}_{i_2}, l_0^i, l_1^i, l_2^i, c_0^i, c_1^i, c_2^i$, according to the scheme in in figure 1(a). ($x_k$ is replaced by $\hat{x}_{i_k}$ and $l_k, c_k$ are replaced by $l_k^i, c_k^i$, respectively.) The three edges $[l_j^i, c_{j+1(mod3)}^i], j = 0, 1, 2$, are in $E^0$, and all the other edges are in $E^1$. We call the subgraph induced by these nine vertices the $i$-th *clause subgraph*. Finally, add an edge $[x_i, \overline{x}_i] \in E^1$ for $i = 1, \ldots, n$. All edges between private vertices of different clauses are forbidden.

The reduction is clearly polynomial. The validity proof uses Lemma 16 and the fact that in any transitive orientation of a comparability sandwich induced on a clause subgraph, out of the three edges $[x_i, l_i], i = 0, 1, 2$, either one or two must be oriented towards $x_i$. The orientation of the edges $[x_i, \overline{x}_i]$ is used to induce truth values on the $X_i$-s, and vice versa. $\square$

Suppose the set of clauses in the NAE-SAT instance cannot be partitioned into two nonempty subsets containing disjoint sets of variables. The NAE-SAT problem restricted to such instances clearly remains in NPC. For such instances the graph $G = (V, E^1)$ in the reduction in Theorem 17 is connected. Since the graph sandwich induced on each clause subgraph must be UTRO, the connectedness of $G^s$ implies that all the edges of $G^s$ are $\Gamma^*$-related (cf. [12]) and thus $G^s$ is UTRO. Hence:

**Theorem 18.** *The UTRO sandwich problem is NP-complete.* $\square$

## 4.2 Permutation Graphs:

A *matching diagram* of a permutation $\pi$ on the numbers $1, \ldots, n$ can be described by writing $n$ points on a straight horizontal line and marking them $1, \ldots, n$ in sequence, writing $n$ points on another straight line parallel to the first and marking them $\pi(1), \ldots, \pi(n)$ in sequence, and adding $n$ segments connecting point $i$ above to point $i$ (which is in position $\pi^{-1}(i)$) below. The connecting segments will be called *chords*. A graph is a *permutation graph* iff it is the intersection graph of the chords of a matching diagram (see figure 2(a) for an example). One can define a partial order on the chords of a matching diagram by $a \prec b$ iff the chord $a$ is completely to the left of chord $b$. This defines a partial order on the complement of a permutation graph.

**Theorem 19.** *The permutation sandwich problem is NP-complete.*

*Proof.* (Sketch) Reduction from BETWEENNESS: Given is a set of elements $S = \{a_1, \ldots, a_n\}$ and a set $T = \{T_1, \ldots, T_m\}$ of ordered triplets of elements from $S$, where $T_i = (a_{i_1}, a_{i_2}, a_{i_3})$. Is there a linear order of the elements in $S$ so that for every triplet $T_i$, $a_{i_2}$ is between $a_{i_2}$ and $a_{i_3}$? This problem was shown to be NP-complete by Opatrny [20].

Given an instance of BETWEENNESS, form an instance $(V, E^1, E^3)$ of the permutation sandwich problem as follows: Define a vertex $v_i$ for each element $a_i$, and two vertices $x_j^1$ and $x_j^2$ for triplet $T_j$. The vertex set is $V = \{v_1, \ldots, v_n\} \cup \{x_j^1, x_j^2 | j = 1, \ldots, m\}$. Define the edge sets $E^1$ and $E^3$ such that to each triplet corresponds a 5-chain as shown in figure 2(a) and put all the edges $[v_i, v_j], i \neq j$ in $E^3$ forcing the $v_i$-s to form an independent set in the sandwich graph. All other edges are in $E^0$.

The reduction is clearly polynomial. The key to the validity proof is that in any matching diagram corresponding to a 5-chain $(a, x, b, y, c)$, either $a \prec b \prec c$ or $c \prec b \prec a$ (see figure 2(b)). The freedom in using the $E^0$-edges allows extending an ordering of the elements in $S$ to a placement of the chords and vice versa. $\qquad\Box$

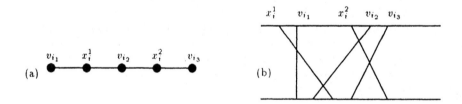

(a)

(b)

**Fig. 2.** (a) The subgraph corresponding to triplet $T_i$. (All edges are $E^1$-edges. Non-edges are in $E^3$.)
(b) A matching diagram of a permutation graph of the 5-chain $(v_{i_1}, x_i^1, v_{i_2}, x_i^2, v_{i_3})$.

Golumbic et al. [15] characterized the cocomparability graphs as the intersection graph of a function diagram. Using this characterization, by a similar reduction from betweenness one can prove that the co-comparability sandwich problem is NP-complete. Using Proposition 2 this gives an alternative proof for Theorem 17. However, this proof does not seem to generalize to UTRO graphs.

**4.3 Circle Graphs:** An undirected graph $G$ is called a *circle graph* if there exists a set of chords $C$ on a circle and a one-to-one correspondence between vertices of $G$ and chords of $C$ such that two distinct vertices are adjacent if and only if their corresponding chords intersect. The set of chords $C$ is a *circle representation* of $G$. The class of circle graphs includes the permutation graphs. (In the matching diagram of a permutation graph identify the right and left endpoints of the two horizontal lines).

**Lemma 20.** *Let $G$ be an arbitrary graph and let $p$ be a graph comprising of a single vertex. $G$ is a permutation graph iff $G + p$ is a circle graph.* □

**Theorem 21.** *The circle graph sandwich problem is NP-complete.*

*Proof.* Reduce permutation-SP to circle-SP, as follows: Given an instance $(V, E^1, E^0)$ for the permutation-SP, add to $V$ a vertex $p$ and take $(\tilde{V}, \tilde{E}^1, \tilde{E}^0) = (V \cup \{p\}, E^1 \cup \{[p,v] | v \in V\}, E^0)$ as the corresponding instance for the circle-SP. According to Lemma 20, $(V, E^1, E^0)$ has a permutation graph sandwich iff $(\tilde{V}, \tilde{E}^1, \tilde{E}^0)$ has a circle graph sandwich. □

**4.4 Interval Graphs:** The intersection graph of a family of intervals on the real line is called an *interval graph*. Golumbic and Shamir [16] recently proved that the interval sandwich problem is NP-complete. A simpler proof for the same result can be obtained by arguments similar those used in Section 4.2.

An interval graph which has an interval representation in which all intervals have unit length is called *unit interval graph*. By a careful modification of the reduction in [16], one can prove:

**Theorem 22.** *The unit interval sandwich problem is NP-complete.* □

**4.5 Circular-Arc and Circular-Permutation Graphs:** The intersection graph of a family of arcs on a circle is called a *circular arc graph* (cf. [26]). If all the arcs have unit length the graph is called *unit circular arc* and if no arc includes another the graph is called *proper circular arc*. Note that these definitions do not coincide. The complexity of circular arc-SP is immediately implied by the complexity of the interval-SP. by the following observation.

**Lemma 23.** *Let $G = (V, E)$ be an arbitrary graph and let $v$ the graph comprising of a single vertex. $G$ is an interval graph iff $G \cup v$ is a circular arc graph.* □

**Theorem 24.** *The circular arc sandwich problem is NP-complete.*

*Proof.* Reduce the interval-SP to the circular arc-SP, as follows: Given an instance $(V, E^1, E^2)$ for the interval-SP, add to $V$ an isolated vertex $v'$ and take $(\tilde{V}, \tilde{E}^1, \tilde{E}^2) = (V \cup \{v'\}, E^1, E^2)$ as the corresponding instance for the circular arc-SP. The reduction is clearly polynomial. By Lemma 23, $(V, E^1, E^2)$ has an interval graph sandwich iff $(\tilde{V}, \tilde{E}^1, \tilde{E}^2)$ has a circular arc graph sandwich. □

Note that $G$ is a unit interval graph iff $G \cup v$ is a unit and proper circular arc graph. Thus by a similar reduction from the unit interval-SP one can prove:

**Corollary 25.** *The unit circular arc-SP and the proper circular arc-SP are NP-complete.* ☐

A larger class containing permutation graphs is the class of circular permutation graphs [23]. By a reduction from permutation sandwich, using the construction used in Theorem 24, one can prove:

**Theorem 26.** *The circular permutation sandwich problem is NP-complete.* ☐

**4.6  Path Graphs:**  A *directed rooted tree* is a directed tree with a distinguished vertex from which there is a directed path to every other vertex. The intersection graph of a family of directed paths on a directed rooted tree is called a *directed path graph* [10]. The intersection graph of a family of paths on an undirected tree is called a *path graph* [11]. Clearly, every interval graph is also a directed path graph and every directed path graph is a path graph.

**Lemma 27.** *Let $G = (V, E)$ be an arbitrary graph and let $G' = G + p$ be the join of $G$ with a singleton $p$. The following statements are equivalent: (1) $G$ is an interval graph. (2) $G'$ is a directed path graph. (3) $G'$ is a path graph.*

*Proof.* (Sketch) Clearly $(1) => (2)$ and $(2) => (3)$. To show that $(3) => (1)$, take a paths representation of $G'$ and use the intersection of each path with $p$'s path together with the Helly property (cf. [12]) to obtain an interval representation of $G$. ☐

Using this lemma one can reduce interval-SP to the path-SP and to the directed path-SP by a construction identical to the one in Theorem 21 and obtain:

**Theorem 28.** *The path sandwich and the directed path sandwich problems are NP-complete.* ☐

## 5  Summary

We have studied the complexity of the sandwich problems for some families of graphs. Figure 3 summarizes many of the results, and indicates the outstanding open problems for subfamilies of perfect graphs.

## References

1. B. Apsvall, M. F. Plass, and R. Tarjan. A linear-time algorithm for testing the truth of certain quantified boolean formulas. *Information Processing letters*, 8(3):121–123, 1979.
2. H. L. Bodlaender, M. R. Fellows, and T. J. Warnow. Two strikes against perfect phylogeny. In W. Kuich, editor, *Proc. 19th ICALP*, pages 273–283, Berlin, 1992. Springer. Lecture Notes in Computer Science, Vol. 623.
3. A. Brandstädt. Special graph classes - a survey. Technical Report SM-DU-199, Universität Duisburg, 1991.

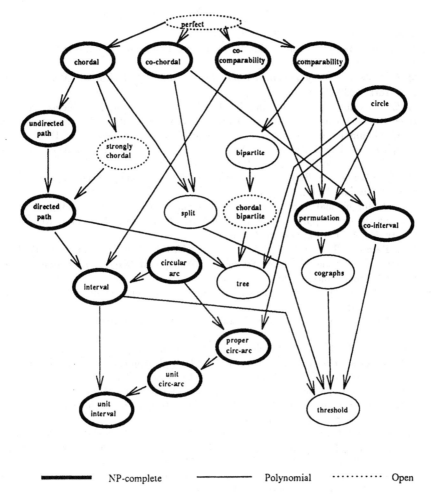

**Fig. 3.** The complexity status of the sandwich problem for some graph classes. An arrow from A to B indicates that class A contains class B.

4. P. Buneman. A characterization of rigid circuit graphs. *Discrete Math.*, 9:205–212, 1974.

5. A. V. Carrano. Establishing the order of human chromosome-specific DNA fragments. In A. D. Woodhead and B. J. Barnhart, editors, *Biotechnology and the Human Genome*, pages 37–50. Plenum Press, 1988.

6. V. Chvátal and P. L. Hammer. Aggregation of inequalities for integer programming. *Ann. Discrete Math.*, 1:145–162, 1977.

7. D. G. Corneil, H. Lerchs, and L. Stewart Burlingham. Complement reducible graphs. *Discrete Applied mathematics*, 3:163–174, 1981.

8. T. Gallai. Transitiv orientierbare graphen. *Acta Math. Acad. Sci. Hungar.*, 18:25–66, 1967.

9. M. R. Garey and D. S. Johnson. *Computers and Intractability: A Guide to the Theory of NP-Completeness*. W. H. Freeman and co., San Francisco, 1979.

10. F. Gavril. A recognition algorithm for the intersection graphs of directed paths in directed trees. *Discrete Math.*, 13:237–249, 1975.

11. F. Gavril. A recognition algorithm for the intersection graphs of paths in trees. *Discrete Math.*, 23:211–227, 1978.

12. M. C. Golumbic. *Algorithmic Graph Theory and Perfect Graphs.* Academic Press, New York, 1980.

13. M. C. Golumbic, H. Kaplan, and R. Shamir. Graph sandwich problems. Technical Report 270-92, Computer Science Dept., Tel Aviv University, 1992. submitted.

14. M. C. Golumbic, H. Kaplan. and R. Shamir. On the complexity of DNA physical mapping. Technical Report 271-93, Computer Science Dept., Tel Aviv University, 1993. To appear in *Advances in Applied Mathematics.*

15. M. C. Golumbic, D. Rotem, and J. Urrutia. Comparability graphs and intersection graphs. *Discrete Math.*, 43:37–46, 1983.

16. M. C. Golumbic and R. Shamir. Complexity and algorithms for reasoning about time: A graph-theoretic approach. Technical Report 91-54, DIMACS Center, Rutgers University, NJ, 1991. to appear in *JACM.*

17. P. L. Hammer, T. Ibaraki, and U. N. Peled. Threshold numbers and threshold completions. In P. Hansen, editor, *Studies on Graphs and Discrete Programming*, pages 125–145. North-Holland, 1981.

18. D. S. Johnson. The NP-completeness column: an ongoing guide. *J. of Algorithms*, 6:434–451, 1985.

19. S. K. Kannan and T. J. Warnow. Triangulating 3-colored graphs. *SIAM J. of Discrete Math.*, 5(2):249–258, 1992.

20. J. Opatrny. Total ordering problems. *SIAM J. Computing*, 8(1):111–114, 1979.

21. F. S. Roberts. *Discrete Mathematical Models, with Applications to Social Biological and Environmental Problems.* Prentice-Hall, Englewood Cliffs, New Jersey, 1976.

22. J. D. Rose. A graph-theoretic study of the numerical solution of sparse positive definite systems of linear equations. In R. C. Reed, editor, *Graph Theory and Computing*, pages 183–217. Academic Press. N.Y., 1972.

23. D. Rotem and J. Urrutia. Circular permutation graphs. *Networks*, 11:429–437, 1982.

24. T. J. Schaefer. The complexity of satisfiability problems. In *Proc. 10th Annual ACM Symp. on Theory of Computing*, pages 216–226, 1978.

25. M. Steel. The complexity of reconstructing trees from qualitative characters and subtrees. *J. of Classification*, 9:91–116, 1992.

26. A. C. Tucker. Characterizing circular arc graphs. *Bull. Amer. Math. Soc.*, 76:1257–1260, 1970.

27. M. Yannakakis. Computing the minimum fill-in is NP-complete. *SIAM J. Alg. Disc. Meth.*, 2, 1981.

# On-line Graph Algorithms for Incremental Compilation [*]

Alberto Marchetti-Spaccamela[1], Umberto Nanni[2], Hans Rohnert[3]

[1] Dipartimento di Informatica e Sistemistica, University of Rome "La Sapienza", Italy
[2] Dipartimento di Matematica Pura ed Applicata, University of L'Aquila, Italy. Part of this work was done while the author was visiting ICSI, Berkeley.
nanni@vxscaq.aquila.infn.it
[3] Most of this report was written while the author was a postdoc at ICSI, Berkeley, on leave from Siemens AG, Munich, Germany. rohnert@ztivax.zfe.siemens.de.

**Abstract.** Compilers usually construct various data structures which often vary only slightly from compilation run to compilation run. This paper describes in a compact and uniform way solutions to several problems arising in order to quickly update these data structures instead of building them from scratch each time. All the considered problems can be reduced to graph problems. Specifically, we give algorithms for the dynamic problems of loop detection, topological order, reachability from the start routine, and transitive closure.

As far as the dynamic maintenance of a topological order is concerned, for which no previous solution is known to the authors, simple algorithms and data structures are provided, and the achieved upper bound is $O(n)$ amortized time per update in a sequence of $O(m)$ edge insertions, which favourably compares to the trivial $O(n+m)$ worst case time bound (applying the off-line algorithm). The additional space requirement, besides the space to represent the graph itself, is $O(n)$. We also discuss by an example the harder *fully dynamic* version of topological order.

## 1 Introduction

One way to speed up compilation times is to incorporate mechanisms for incremental update of important data structures built by the compiler. In fact, the first compilation run of a program is generally followed by several other compilation runs, each time with a slightly changed program. Therefore it is desirable that data structures built during the first compilation run are maintained dynamically without building them from scratch during every subsequent run. Namely, after having made these data structures persistent on disk during the first compilation run, subsequent compilation runs read in these data structures again, modify them as much as necessary because of changes in the input program text, and write them back to disk storage. Increasing memory availability and improving memory technology may make it even possible to hold these data structures in fast random access memory. The

---

[*] Work supported by the ESPRIT II Basic Research Action Program of the European Community under contract No.7141 "Algorithms and Complexity II" and by the Italian MURST 40% project "Algoritmi, Modelli di Calcolo e Strutture Informative".

data structures we have in mind (e.g., the call graph and, notably in object-oriented programs, the inheritance graph) can be modelled by directed graphs. Also, some consistency checks may be reduced to graph problems.

Specifically, we consider the following problems on directed graphs: loop detection, topological order, reachability from the start or "main" vertex, and transitive closure. These problems are considered in their static and dynamic variations: the dynamic operations considered are insertions and/or deletions of edges/vertices of the graph. The emphasis here is on "practical" approaches, leading to algorithms and data structures that are simple and easy to implement.

We will often suggest several solutions to a problem and the best solution for the problem at hand depends on its size and some more specific requirements. In particular, different solutions for the same problem will have different space/time requirements. The arising tradeoffs must be resolved on the basis of the available resources and the size of the input. In fact the size of the graphs is sometimes fairly small or medium (the inheritance graph is seldom really large), but they can also reach a considerable size (the call graph of large systems may have many thousands of vertices). Another issue is the tradeoff between the costs for updating and querying the data structures. We will see that some of the solutions will have a high cost for updating and a small cost for querying; in this case the choice of the best solution will be influenced by the relative frequency of these operations.

Finally, another intriguing issue, when dealing with incremental algorithms, is the evaluation of time performances. The time cost of a dynamic algorithm may be expressed in terms of *worst case* cost (i.e., an upper bound to the time spent for a single operation performed on the data structure), or *amortized* cost [11], which provides the average time spent per operation over the worst possible *sequence* of operations.

As far as the topological order is concerned, no solution for the dynamic version of this problem is known to the authors.

A *topological order* of a partially ordered set $(V, \Omega)$ is any total order *ord* of $V$ such that if $(x, y) \in \Omega$, then $(x, y) \in ord$. It is straightforward to represent a partially ordered set (*poset*) $(V, \Omega)$ by using a directed acyclic graph (*dag*) $G = (V, E)$ such that $(x, y) \in E$ if and only if $(x, y) \in \Omega$. A topological order *ord* of the vertices of a dag with $n$ vertices maps each vertex $v$ to an integer $ord(v)$ between 1 and $n$ which fulfills the following property: for each $(x, y) \in E$ we have $ord(x) < ord(y)$. A simple algorithm to find a topological sorting in a dag, and requiring $O(n + m)$ time, can be found, for example, in [7].

We propose algorithms and data structures to maintain a topological order in a dag while performing a sequence of edge insertions and deletions. In the worst case, the cost per operation is not greater than recomputing a new topological order from scratch, that is $O(n + m)$ [7]. However, we will show that the total time spent to perform a sequence of $m$ edge insertions is $O(mn)$, thus improving on the $O(mn+m^2)$ bound obtained by performing $m$ topological orderings of the graph. Equivalently, the cost of an insertion amortized over a sequence of $O(m)$ insertions is only $O(n)$ (see [11] for examples of amortized analysis). The space requirements are favourable: the algorithm uses $O(n+m)$ space for the representation of the graph and $O(n)$ space to store the topological order; $O(n)$ additional working space is required during the execution of an update operation.

The rest of the paper is organized as follows. In Section 2 we introduce the terminology and give preliminary definitions. In Sections 3 and 4 we will consider the problems of performing loop detection and topological sort in a graph that can change under sequences of insert and delete operations. Finally in Sections 5 and 6 we will consider the problems of reachability from a start vertex and of maintaining transitive closure.

Since one of the purposes of this document is to serve as a collection of optimization hints for compiler writers, some of the presented solutions and techniques are already known in the literature [3, 4, 7, 8, 2] and are described here just for the sake of completeness. We present original contributions in Section 4, with the dynamic maintenance of topological order, and in Section 6, with a parallel implementation of an algorithm for the dynamic transitive closure.

Furthermore some well-known techniques (or similar algorithms matching existing bounds) are presented in a way emphasizing practical and simple implementation.

## 2   Preliminaries

In the following we assume the standard graph theoretical terminology as contained for example in [1, 10]. A *directed graph* $G = (V, E)$ (also referred to as a *digraph* or *dag*) consists of a finite set $V$ of *vertices* and a finite set of ordered pairs $E \subseteq V \times V$ of *edges*. A *path* from $v_1$ to $v_k$ is a sequence $\langle v_1, v_2, \ldots, v_k \rangle$ of vertices such that any adjacent pair $(v_h, v_{h+1})$ is an edge in $E$. A *cycle* is a nonempty path from a vertex to itself. A digraph with no cycles is called a *directed acyclic graph* (in short *dag*). A vertex $y$ is said to be *reachable* from a vertex $x$ if there exists a path from $x$ to $y$. We will denote as $Reach(v)$ the set of vertices reachable from a vertex $v$ (including $v$ itself). Inversely, $Reach_{\text{INV}}(v)$ is the set of vertices that can reach $v$.

If $G = (V, E)$ is a digraph the *transitive closure* of $G$ is defined by the digraph $G^+ = (V, E^+)$ such that there is an edge from $(x, y) \in E^+$ if and only if there exists a nonempty path from $x$ to $y$ in $G$.

In this paper we assume that graphs are represented by storing the edges in adjacency lists, that is using space $O(n + m)$.

We consider several problems on directed acyclic graphs both in the static and the dynamic version. In a static problem we find a solution $S$ for a given problem for a specified graph $G$. If the graph is updated to a new graph $G'$, the old solution $S$ is no longer valid and we must recompute the new solution $S'$. In a typical dynamic situation, we want to maintain a data structure under a sequence of two kinds of operations: *query* and *update*. The update operation will modify the graph by inserting or deleting an edge or a vertex. The query operation requires to find an answer to the particular problem. The trivial approach to this kind of problems consists in using the static algorithm either when a query has to be answered, or when an update occurs.

In this paper we study algorithms and data structures that allow to obtain the new solution $S'$ without recomputing it from scratch. The time and space requirements to maintain a solution to a given problem depend on the kind of updates allowed, for example whether we can only add edges to the graph, or only delete edges, or if we consider a fully dynamic situation in which insertions and deletions are arbitrarily performed. In the former cases we can often gain advantage by the natural monotonicity of the solution.

In the rest of the paper we will concentrate our attention on the insertions and deletions of edges because this is the most interesting problem. In fact we will see that the insertion/deletion of an isolated vertex (i.e., a vertex with no ingoing or outgoing edges) does not pose any problem. Therefore the deletion of a vertex can be performed by executing a sequence of edge deletions followed by an update operation that deletes a vertex with no edges. The case of vertex insertion can be dealt with analogously. Furthermore we do not consider the problem of updating the representation of a graph after insertion and deletion of edges, but only the updates to the structures handled by our algorithms.

## 3   Loop Detection

The first problem that we consider is *loop detection* which consists of deciding whether a given graph is acyclic or not. An example where it is necessary to check a graph for the existence of loops is the inheritance graph of programs written in object-oriented languages, which has to be a directed acyclic graph. The introduction and the deletion of new classes and new inheritance relationships between classes give raise to the problem of repeatedly checking the acyclicity of a dynamically changing graph.

Given a graph $G = (V, E)$ with vertex set $V$ and edge set $E$ it is well-known how to test $G$ for cycles: Simply run a topological sort [7]. If the algorithm terminates and assigns a topological number to every vertex then the graph is acyclic. If it aborts prematurely because in the remaining graph there are no more vertices with no ingoing edges then the graph contains cycles. The time complexity of this procedure is $O(n + m)$ where $n = |V|$ and $m = |E|$. The space complexity is also $O(n + m)$, just for storing $G$.

*Dynamic Loop Detection*

Given an acyclic graph $G = (V, E)$ we deal now with the problem of performing an intermixed sequence of operations of the following kinds:

a) updates to the structure of the graph, consisting of *insertion* or *deletion* of edges;
b) *queries* concerning the existence of loops;
c) *test* whether the insertion of an edge $(v, w)$ would introduce a loop.

Observe that the deletion of an edge $e$ from $G = (V, E)$ is trivial since this operation can not introduce cycles. Furthermore insertions or deletions of isolated vertices require just to handle entries in the tables representing the graph. Therefore, in the sequel of this section, we concentrate on the operation of inserting an edge $e = (v, w)$, with $v, w \in V$.

Trivial solutions to dynamic loop detection consist of applying the $O(m + n)$ algorithm to compute a topological sort of the vertex set either when a query or test is requested, or when an update occurs.

Another simple solution, usually better than performing a topological sort, is to perform a depth first search from vertex $w$ [7]. If it reaches $v$ then adding $e$ introduces a cycle into $G$, otherwise is does not. The time complexity for $dfs(w)$ is $O(|G_{Reach(w)}|)$ where $Reach(w)$ is the set of vertices reachable from $w$ and $G_{Reach(w)}$ is the subgraph induced by $Reach(w)$. In the worst case this means time $O(m)$. The space complexity of $dfs(w)$ is $O(n)$ for storing a marker array and the call stack of the recursive $dfs$ procedure, cf. [7] for details.

As far as a really dynamic solution for the loop detection problem is concerned, it can be reduced to the dynamic maintenance of a topological order, that is considered in Section 4. As shown there, it will require $O(m \cdot \min\{n, k\})$ total time to perform any sequence of $k$ consecutive edge insertions (that leave the graph acyclic) followed by a sequence of $h$ edge deletions. If we start from an empty graph the insertion of $m$ edges requires $O(n)$ amortized time per update, which is a considerable improvement over the $O(n + m)$ static algorithm, especially for long sequences of insertions in a dense graph. In the worst case a single insert operation may require $O(m)$ time. This solution is space efficient since it requires $O(n)$ additional space besides the space necessary for storing the graph.

Using the above solution, to perform a $test(u,v)$ operation (testing whether a new possible edge $(u, v)$ leaves the graph acyclic or not) requires $O(m)$ worst case time (by using $dfs$, for example), since the cost can not be amortized over a sequence of operations.

It is possible to improve the time requirements of the test operation at the expense of space requirements. Testing in constant time is indeed possible if we maintain the transitive closure $G^+$ of the graph $G$. The insertion of the edge $(u, v)$ leaves the graph acyclic if and only if there is no connection from $v$ to $u$ in $G^+$, and this can be checked in constant time. This solution requires more space to be maintained, since we have to store the whole $n \times n$ matrix describing $G^+$. This leads to $O(n^2) \geq O(m)$ space for a simple matrix implementation, that can be reduced for sparse graphs by storing only the meaningful (non-empty) items of $G^+$ by using, for example, perfect hashing [2].

Nevertheless the maintenance of transitive closure might be preferred in some context, and it is considered in Section 6.

## 4 Maintaining a Topological Order

Sometimes a partial information about the ordering of the vertices of a directed acyclic graph is sufficient. As shown in the previous section, finding (or maintaining) a topological order may provide information about the acyclicity of the graph. A description of a simple static solution of topological sorting can be found, for example, in [1, 6, 7].

A dag $G = (V, E)$ defines a partial order among the elements in the set $V$, as the *reflexive transitive closure* of its edges. In the following we will denote as $\preceq^G$ such a binary relation among nodes; $x \preceq^G y$ ($x$ *precedes* $y$) if one of the following cases occurs:

a) $x = y$

b) $(x, y) \in E$

c) there exists a node $z$ such that $x \preceq^G z$, and $(z, y) \in E$.

A *topological ordering* of the set $V$ is any total ordering *ord* such that:

$$\text{if } x \preceq^G y \text{ then } x \preceq^{ord} y.$$

We also denote as *ord* the (bijective) function mapping each of the $n$ nodes in the set $V$ into one of the integers $1, 2, \ldots, n$; using this notation, relationship $ord(x) \leq ord(y)$ means that $x$ precedes $y$ in *ord*.

Given a graph $G$ let *ord* be a topological ordering of its vertices. Deleting an edge does not require to update *ord*, since it is still a valid topological order for the new graph.

Edge insertion is computationally more complicated; if we want to insert the edge $(u, v)$ let $G' = (V, E \cup \{(u, v)\})$ where $(u, v) \notin E$. If $u$ precedes $v$, i.e., $ord(u) < ord(v)$, no update has to be performed, since the new graph complies to the existing topological order. On the other hand, if $u$ follows $v$ in the topological order, then the new edge introduces a path from $u$ to $v$ and hence the topological order needs to be modified. The trivial way to update *ord* after an edge insertion is to check first whether the new edge does not introduce a cycle and then to recompute the topological order from scratch. In the worst case the running time is $O(n+m)$, where $n$ is the number of nodes, and $m$ (with $m = O(n^2)$) is the number of edges.

Our approach provides a solution requiring $O(nm)$ total time for any sequence of edge insertions, which means $O(n)$ amortized time per update in a sequence of $m$ insertions. The case of sequences of updates alternating insertions and deletions is discussed at the end of the section.

As remarked previously, insertion and deletion of vertices can be performed via operations on edges and isolated nodes.

The next lemma allows us to minimize the number of nodes that are to be considered to update a topological order of a dag $G$ while a new edge is inserted.

**Lemma 1.** *Given a graph $G = (V, E)$ and a topological order ord of its vertices, let $G' = (V, E \cup \{(u, v)\})$ be the graph obtained from $G$ by inserting edge $(u, v)$. If $G'$ is acyclic, then there is a topological order ord' of $G'$ such that $ord'(w) = ord(w)$ for all $w$ with $ord(w) < ord(v)$ or $ord(w) > ord(u)$.*

*Proof.* Without loss of generality we can assume that $u$ follows $v$ in *ord* (otherwise the lemma is trivially true since *ord* is a topological order of $G'$).

Let $w$ be a vertex that precedes $v$ (follows $u$) in the topological order *ord*. To prove the lemma it is sufficient to show that the set of predecessors (successors) of $w$ is not modified by the insertion of the new edge.

If $w$ is a vertex such that $ord(w) < ord(v)$ then there is no path from $v$ to $w$ in $G$. Hence the insertion of the edge $(u, v)$ does not introduce new paths from any vertex of the graph to $w$ and, therefore, the set of predecessors of $w$ will not be modified by the inserted edge.

Analogously, if $ord(w) > ord(u)$ there is no path in $G$ from $w$ to $u$. Hence the insertion of edge $(u, v)$ does not introduce new paths from $w$ to any other vertex of the graph and therefore the set of successors of $w$ will not be modified by the edge insertion.

In both cases, we can conclude that if $G'$ is acyclic then there must be a topological ordering $ord'$ such that $ord'(w) = ord(w)$.

As a consequence of the lemma it is sufficient to focus on vertices $w$ such that $ord(v) \leq ord(w) \leq ord(u)$.

**Definition 2.** Given a graph $G = (V, E)$ and a topological order $ord$ of its vertices, if $(u, v)$ does not belong to $E$ let us define the following set of *unstable nodes* for the pair $(u, v)$:

$$\mathcal{U}_{ord}(u, v) = \{w \mid ord(v) \leq ord(w) \quad \text{and} \quad ord(w) \leq ord(u)\}.$$

By Lemma 1 and the above definition the only vertices that may eventually have to change their position in $ord$ while inserting an edge $(u, v)$ belong to $\mathcal{U}_{ord}(u, v)$.

We assume that the topological order is stored in array $A$ and we show how to maintain $A$ under insertions and deletions of edges.

Let $G = (V, E)$ be a directed acyclic graph and let $A$ be an array representing a topological order of the vertices of $G$ in the obvious way: $A[i] = v$ if $ord(v) = i$ (i.e., vertex $v$ is the $i$-th vertex in the topological order). Furthermore, we store array $ord$, indexed by the nodes in $V$. According to the notation introduced in the previous section, we say that vertex $y$ follows vertex $x$ in $ord$ if $ord(y) > ord(x)$; equivalently, we also say that $x$ precedes $y$ in $ord$.

We also remark that in this paper we do not explicitly deal with the updates to the structure of a graph after insertion and deletion of edges, but only with the updates to the data structures handled by our algorithms.

In order to obtain a new topological order $ord'$ after the insertion of a new edge $(u, v)$, it is necessary to modify the position of these vertices in such a way that $v$ and all vertices in $Reach(v)$ follow $u$ and all vertices in $Reach_{INV}(u)$ in $ord'$. These modifications must be done without violating any order relationship in the graph before the insertion of edge $(u, v)$. In order to perform this task we perform a depth first search $dfs$ starting from $v$. If $ord(u) > ord(v)$ then $v$ and all its successors, i.e., all vertices in $Reach(v)$, have to be moved behind $u$ in the new topological order (if they are not yet already behind $u$). Those vertices which were in the topological order between $v$ and $u$ (inclusive $u$) and are not successors of $v$, have to be shifted ahead while preserving their relative ordering to make room for $v$ and its successors. Due to Lemma 1 we can consider the unstable vertices only, i.e., the set $\mathcal{U}_{ord}(u, v)$. The details follow.

The algorithm, reported as procedure *Insert* in Figure 1, consists of two stages. In the first stage it detects whether the new edge introduces a cycle or not and marks those vertices in $\mathcal{U}_{ord}(u, v)$ that in the old topological ordering $ord$ are before $u$ and must follow $u$ in the new topological ordering $ord'$. The second stage is performed only if no cycle was detected and computes the new arrangement.

The first stage consists of a $dfs$ starting from $v$ to explore $Reach(v)$. The individual $dfs$ invocations stop when hitting a sink with no outgoing edges or when hitting

**Procedure** *Insert*$((u, v) :$ **edge**$)$;

1. **begin**

    **Stage 1**

2.           Perform a *dfs* starting from $v$, and
               when an unmarked vertex $w$ is found:
3.          if $ord(w) > ord(u)$    {i.e., if $w \notin \mathcal{U}_{ord}(u,v)$ }
4.            then prune the search
5.            else mark node $w$;
6.       if vertex $u$ has been marked during Stage 1
7.            then HALT;    {a cycle has been detected}

    **Stage 2**

8.        $shift := 0$;
9.        let $L$ be an empty list;
10.       for $i := A[ord(v)]$ to $A[ord(u)]$
11.          do if $A[i]$ has been marked during Stage 1
12.            then begin
13.                append $A[i]$ to the end of $L$;
14.                $shift := shift + 1$;
15.            end
16.            else begin
17.                $ord(A[i]) := i - shift$;
18.                $A[i - shift] := A[i]$;
19.            end;
20.       insert vertices in list $L$ into array $A$
            using positions $A[i - shift + 1]$ to $A[i]$ and update their *ord* values;
21. **end**;

**Fig. 1.** Maintaining a Topological sort while inserting edges

a vertex $x \notin \mathcal{U}_{ord}(u,v)$, i.e., $ord(x) > ord(u)$. Such vertices are already behind $u$ in the topological order, and therefore, by Lemma 1, they and any of their successors do not have to be considered since they already are behind $u$ in *ord*. During the depth first search the algorithm marks the entries of the array $A$ that correspond to the set $Reach(v) \cap \mathcal{U}_{ord}(u,v)$. Assume $ord(u) > ord(v)$. (Otherwise, the *dfs* already stops at its startpoint in line 4 with $w = v$ and the *for* loop in stage 2 is empty.) The set of marked vertices includes $v$ and corresponds to the set that must follow $u$ in the new topological ordering *ord'* in contrast to the old topological ordering *ord*. If (and only if) the depth first search hits vertex $u$, then edge $(u,v)$ introduces a loop and the procedure stops, since there is no way to define a topological order.

If no cycle was detected, the second stage of the algorithm updates array $A$. The new order *ord'* is obtained from *ord* in such a way that $v$ and all its successors will follow $u$ without modifying their relative order.

The second stage is implemented by scanning array $A$ from $A[ord(v)]$ to $A[ord(u)]$. We use a counter *shift* initialized with 0. When the algorithm considers entry $i$ of array $A$ it distinguishes between two possibilities. If $A[i]$ has been marked during the first stage then the counter *shift* is increased by one: the corresponding vertex, $A[i]$, is one of the vertices to be moved behind $u$ and is inserted into the ordered list $L$.

If $A[i]$ has not been marked during the first stage then the algorithm moves the corresponding vertex in the topological order by *shift* many places ahead by decreasing its position by *shift*, i.e., $A[i-shift] := A[i]$. When the algorithm considers the entry corresponding to vertex $u$, i.e., $A[ord(u)]$, then the counter *shift* is equal to the number of vertices marked during the first stage. Therefore after moving vertex $u$ ahead there is room in the array for placing the marked vertices. These vertices are inserted into the array preserving their relative ordering which was consistent with a full topological ordering of the graph before the insertion of $(u, v)$.

Notice that if we maintain the topological order using a doubly linked list $A$, then no additional space is necessary (beside the stack used by *dfs*). In fact, during stage 1, the visited vertices can be eliminated from the original list $A$ and enqueued to form a new list $L$. Stage 2 consists of inserting the new list $L$ into $A$ just after vertex $u$. So, the decision is whether to use an array of size $n$ plus a variably sized list (of length up to $n - 1$) or two doubly-linked lists of total length $n$.

**Lemma 3.** *Given a graph $G = (V, E)$ and a topological order of its vertices. If adding edge $(u, v)$ does not introduce a cycle then procedure Insert finds a new topological order.*

*Proof.* Let $ord'$ be the order constructed from $ord$ by procedure *Insert*. By Lemma 1 it is sufficient to show that vertices in $\mathcal{U}_{ord}(u, v)$ are topologically sorted in order to show that $ord'$ is a topological order. This is equivalent to show that for each pair of vertices $(x, y)$ such that $(x, y) \in E$, with $x, y \in \mathcal{U}_{ord}(u, v)$, we have $ord'(x) < ord'(y)$. For such vertices $x, y$, one of three possible cases arises:

1. $x \in Reach(v)$
   In this case both $x$ and $y$ are unmarked during stage 1 of the algorithm. Then during stage 2 their relative order is not modified and, hence, $x$ precedes $y$ in $ord'$.
2. $x, y \notin Reach(v)$
   In this case neither $x$ nor $y$ are marked during stage 1 of the algorithm. Therefore, during stage 2, they are moved ahead in the topological order. Notice that this shift is performed without changing the relative order of the marked vertices. Hence $x$ precedes $y$ in $ord'$.
3. $x \notin Reach(v)$ and $y \in Reach(v)$
   In this case $y$ is marked during stage 1 of the algorithm while $x$ is not marked. Therefore, during stage 2, $x$ is moved towards the beginning in the topological order while $y$ moves in the other direction. Hence $x$ precedes $y$ in $ord'$.

This completes the correctness proof.

As far as the worst case time complexity of the insertion algorithm is concerned we observe that the running time of the second stage is proportional to the size of the set of marked vertices, i.e., to the set $Reach(v) \cap \mathcal{U}_{ord}(u, v)$; therefore, the worst case cost of the second stage is $O(n)$. On the other hand, the time complexity of the first stage is proportional to the size of the *dfs*: due to the pruning strategy, the number of scanned edges is equal to the total outdegree of $v$ and of all its successors that are also in the set $\mathcal{U}_{ord}(u, v)$; this can be $O(m)$ in the worst case. Therefore $O(n + m)$ is the worst case running time of procedure *Insert* in order to restore the topological order after a single edge insertion.

We now show that the total time spent to perform any sequence of $k$ edge insertions is $O(m \cdot \min\{n, k\})$ where $m$ is the total number of edges in the graph after the insertions have been performed.

**Theorem 4.** *There exists an algorithm and data structures to maintain a topological order in a dag $G$ under a sequence of $k$ edge insertions which require $O(m \cdot \min\{n, k\})$ total running time.*

*Proof.* The correctness of procedure *Insert* is proved in Lemma 3. As to the time complexity observe that the worst case running time of procedure *Insert* is $O(m)$ for the first stage and $O(n)$ for the second stage; therefore, to prove the theorem it is sufficient to show that the running time of the *dfs* performed in stage 1 during a sequence of $m$ edge insertions is $O(mn)$. In order to prove this claim we will show that, during any sequence of edge insertions, each edge of the graph is scanned at most $n$ times.

If a node $w$ is marked in stage 1 of algorithm *Insert* due to the insertion of an edge $(u, v)$ (with $v$ possibly coincident with $w$) then a path from $u$ to $w$ has been introduced. Hence, from now on, $ord(u) < ord(w)$, and no further edge insertion can reverse the relative position of $u$ and $w$: this means that node $w$ will be never marked again due to the insertion of an edge $(u, v')$, for any $v'$. This leads to conclude that any node is marked at most $n$ times in any sequence of edge insertions.

Since an edge $(w, y)$ is scanned by *dfs* in stage 1 of procedure *Insert* only when the node $w$ is marked (otherwise the search is pruned at $w$), it turns out that any edge is scanned at most $n$ times for any sequence of edge insertions.

Note that the above theorem is not true if some edge deletion occurs during the sequence of insertions. This is due to the fact that we can not bound anymore the number of times that an edge is scanned, since the deletion may cut the path from $u$ to $w$, where $w$ is any node marked after the insertion of an edge $(u, v)$. In other words our algorithm, although working correctly also in case of an interleaved sequence of both edge insertions and deletions, is not *fully dynamic*, since the amortized time bound only holds for any subsequence of edge insertions. On the other hand, the fully dynamic version of the problem is intrinsically harder, as shown in the following example.

Let $G_i = (V_i, E_i)$, for $i = 1, 2$, be two dags and $s_i, p_i \in V_i$ be vertices such that for any $x \in V_i$, we have $x \in Reach(s_i)$, and $x \in Reach_{INV}(p_i)$. Let now be $G = (V_1 \cup V_2, E_1 \cup E_2)$, and consider the following sequence of updates of $G$ that alternates insertion and deletion of edges $(p_1, s_2)$ and $(p_2, s_1)$:

$Insert(p_1, s_2), Delete(p_1, s_2), Insert(p_2, s_1), Delete(p_2, s_1), Insert(p_1, s_2), Delete(p_1, s_2), .$

It is easy to see that the above sequence maintains the graph acyclic and that the proposed algorithm for edge insertions searches the whole graph for any two successive *Insert* operations.

This example also shows the intrinsic hardness of this problem. In fact if a topological order has to be stored in an array or, equivalently, if the complexity is computed in terms of the number of elements which change their position, a sequence of $k$ updates alternating insertion and deletion of edges requires $\Omega(kn)$ operations, while a sequence of $k$ edge insertions requires at most $O(n)$ updates in the array (it would be sufficient to guess a topological order conformal to the final dag).

The same example shows that, if we consider the number of pair inversions required to restore a topological order, the number of such inversions can be as large as $\Omega(kn^2)$ in a sequence of $k$ updates alternating insertions and deletions.

Considering the interest in minimizing the updates to a topological order of a dag, since any exchange requires some update to the data structure, we suggest the following problem, consisting in dealing with a *competitive* approach to this problem: find a strategy $A$ to update a topological order of the nodes of a dag $G$ under a sequence $\sigma$ of edge insertions and deletions in such a way that the total number of pair inversions $C_A(\sigma)$, for any sequence $\sigma$ of updates, is bounded by the following relationship:

$$C_A(\sigma) \leq c \cdot C_{\text{OPT}}(\sigma)$$

where $c$ is a suitable constant, and $C_{\text{OPT}}(\sigma)$ is the *optimal* number of required inversions (which can be computed when the whole sequence of updates is known).

## 5 Maintaining Reachability from Start Vertex

Another question often checked in compilers is whether a specific routine is reachable via routine calls from the start routine $M$ or which routines overall are reachable from $M$. The same problem arises in garbage collectors which try to identify data blocks no more reachable by a chain of pointers from one or at least one of several starting pointers. This problem will be referred to as the *single source reachability* or, in short, *reachability* problem.

A generalization of this problem is the *transitive closure* of a directed graph, that is the problem of maintaining the reachability between any pair of vertices in the graph. This is considered in Section 6.

The evident static solution to the reachability problem is to perform a graph search such as *dfs* starting in start vertex $s$ and to check which vertices are reached. The time complexity of this approach is obviously $O(|G_{Reach(s)}|) \leq O(|G|)$.

### Dynamic Reachability

A very efficient dynamic solution for this problem is possible in case of either sequences of edge insertions or sequences of edge deletions in a dag. The solution proposed here is similar to the one proposed in [8], and has the same time and space complexity as [3, 4, 5].

The performances of the algorithms are the following:

- $O(m)$ time for any sequence of edge insertions, that is constant amortized time per edge insertion starting from an empty graph;
- $O(m)$ time for any sequence of edge deletions, that is constant amortized time per edge deletion finishing with the empty graph;
- constant time to test whether a given vertex $v$ is reachable from the "main" vertex $M$.

The extra space required by the algorithm presented here is $O(n)$, namely a counter $C[h]$ for each vertex $h \in V$. The procedures to perform insertion and deletion of an edge $(i, j)$ are pseudo-coded in Figures 2 and 3 respectively.

```
        Procedure INS(i, j : vertex);
1.    begin
2.        if C[i] > 0
3.            then begin
4.                    Set-queue (Q, {j});
5.                    while Q not empty
6.                        do begin
7.                                dequeue (Q, h);
8.                                increment C[h];
9.                                if C[h] = 1
10.                                   then for each (h, y) ∈ OUT-LIST[h]
11.                                            do enqueue(Q, y);
12.                           end;
13.                 end;
14.   end;
```

**Fig. 2.** Maintaining $Reach(M)$ while inserting edges

The counter $C[h]$ for vertex $h$ is equal to the number of edges entering in $h$ and coming from a vertex which is in $Reach(M)$:

$$C[h] = |\{(x, h) \mid x \in Reach(M)\}|$$

The graph is supposed to be stored by using adjacency lists for each vertex. In the algorithms the adjacency list of vertex $h$ is referred to as OUT-LIST[h].

The initialization required before using the algorithms and starting from the empty graph is the following:

$$C[M] = 1$$

$$C[h] = 0 \quad \text{for any} \quad h \neq M.$$

The algorithms are to be used in order to update the counters $C[h]$ when an edge insertion or deletion occurs.

The behavior of algorithm INS is the following. When an edge $(i, j)$ is inserted, if the vertex $i$ is not reachable from $M$, no further work is required (line 2). Otherwise,

**Procedure** DEL($i, j$ : **vertex**);
```
1.  begin
2.      if C[i] > 0
3.          then begin
4.                  Set-queue (Q, {j});
5.                  while Q not empty
6.                      do begin
7.                              dequeue (Q, h);
8.                              decrement C[h];
9.                              if C[h] = 0
10.                                 then for each (h, y) ∈ OUT-LIST[h]
11.                                     do enqueue(Q, y);
12.                          end;
13.             end;
14. end;
```

**Fig. 3.** Maintaining *Reach*($M$) while deleting edges

vertex $j$ is inserted into the queue $Q$ (line 4). Thereafter a vertex $y$ is enqueued (lines 10, 11) if and only if there exists an edge $(h, y)$ such that $h$ has become reachable from $M$ due to the insertion of the new edge $(i, j)$ (lines 8, 9). On the other hand, any time a vertex $h$ is dequeued (line 7), the counter $C[h]$ is increased by one (line 8). The correctness of the algorithm can be proved by induction on the number of the iteration of the *while* loop.

As far as the time complexity is concerned, we observe that for any enqueued vertex we perform a constant number of operations (lines 7–9 are charged to the vertex $h$ being dequeued and lines 10–11 are charged to the vertex $y$ being enqueued). Since any time a vertex is enqueued its counter is increased by one, the time spent by the procedure in a sequence of edge insertions is equal to the total sum of the counters, and this is bounded by the number of edges in the graph. The additional space required, as noted before, is $O(n)$.

The procedure DEL is symmetrical to the previous one and its correctness and time complexity are not discussed. Though, it is worth to notice that procedure INS retains its validity while inserting edges in any directed graph, while DEL can be used only in acyclic graphs.

Efficient deletion of edges from a cyclic directed graph is indeed a hard problem that still does not have a practical solution. The problem is addressed in [5].

More details about the proof of correctness and complexity of the proposed algorithms can be found in [8].

# 6  Dynamic Transitive Closure

In this section we shortly describe the implementation of a simple and fast algorithm to maintain the transitive closure $G^+$ of a graph $G$ during insertion and deletion of edges (in case of deletions we only deal with directed *acyclic* graphs). The graph $G^+$ has an edge $(x, y)$ if and only if there is a path from $x$ to $y$ in $G$. This problem may be interesting per se and furthermore, as mentioned in Section 3, maintaining explicitly the transitive closure of a graph, allows us to test in constant time whether the insertion of a given edge $(i, j)$ would leave the graph acyclic. If this is a dominant operation in a given application, it might be worth spending some additional space for the implementation of this algorithm. Dynamic algorithms for this problem have been considered, e.g., in [3, 4, 5, 12].

It is a straightforward generalization of algorithms and data structures described in the previous section. In the following we present the algorithm also in a simple parallel version which has the advantage to be perfectly suited for easy adaptation to any number of processors from 1 to $n$, where $n$ is the number of vertices in $G$. Note that this implementation can not be considered an *efficient* parallel algorithm, since it requires a polynomial instead of a polylogarithmic running time.

The main data structure presented here requires additional space $O(n^2)$ to store the $n \times n$ matrix of counters, which can be reduced to $O(|E^+|)$, that is to the number of edges in $G^+$ by representing only the non-zero values of the matrix. It is worth notice, for the parallel implementation, that any processor will have access to a separate set of elements of this matrix (namely a row). So there is no need of either memory sharing or interprocessor communication.

As to the time required to update the data structures during insertions and deletions the sequential implementation achieves the following bounds (the complexity of the parallel implementation is discussed below):

- $O(nm)$ worst case total time for any sequence of edge insertions, that is $O(n)$ per insertion starting from the initial graph;
- $O(nm)$ worst case total time for any sequence of edge deletions in a directed *acyclic* graph, that is $O(n)$ per deletion in a sequence that leaves the graph empty.
- testing whether there is a path from any vertex $x$ to any vertex $y$ (equivalent to testing whether a new edge $(y, x)$ would introduce a cycle in the graph) requires constant time by means of table look-up.

The pseudocode of the two procedures for insertion and deletion of an edge $(i, j)$ is given in Figures 4 and 5 respectively. From now on $C$ will denote a matrix such that, for any $x, y \in V$, $C[x, y]$ is equal to the number of edges entering in $y$ and coming from a vertex which is reachable from $x$ (including the vertex $x$ itself):

$$C[x, y] \stackrel{.}{=} |\{(z, y) \mid z \in Reach(x)\}|$$

Procedure PAR-INS, handling the insertion of an edge, uses a procedure CLOS requiring two parameters: the first one is the vertex $k$ whose closure has to be propagated, and the second one is the first vertex to be inserted in the queue (namely the endpoint of the new inserted edge).

If the $n$ iterations of the **pardo** in line 3 are implemented on $n$ different processors, any processor $P_k$ is in charge to update the set $Reach(k)$, i.e., to update all the counters $C[k, x]$ for any $x$. Using $p$ processors, any of them would be in charge

```
        Procedure PAR-INS(i, j : vertex);
1.      begin
2.          for each {vertex} k ∈ V pardo
3.              if C[k, i] > 0 then CLOS(k, j);
4.      end;

        Procedure CLOS(k, j : vertex);
1.      begin
2.          Set-queue (Q_k, {j});
3.          while Q_k not empty
4.              do begin
5.                      dequeue (Q_k, h);
6.                      increment C[k, h];
7.                      if C[k, h] = 1
8.                      then for each (h, y) ∈ OUT-LIST[h]
9.                              do enqueue(Q_k, y);
10.             end;
11. end;
```

Fig. 4. Parallel transitive closure: insertion of an edge.

```
        Procedure PAR-DEL(i, j : vertex);
1.      begin
2.          for each {vertex} k ∈ N pardo
3.              if C[k, i] > 0 then DECLOS(k, j);
4.      end;

        Procedure DECLOS(k, j : vertex);
1.      begin
2.          Set-queue (Q_k, {j});
3.          while Q_k not empty
4.              do begin
5.                      dequeue (Q_k, h);
6.                      decrement C[k, h];
7.                      if C[k, h] = 0
8.                      then for each (h, y) ∈ OUT-LIST[h]
9.                              do enqueue(Q_k, y);
10.             end;
11. end;
```

Fig. 5. Parallel transitive closure: deletion of an edge.

to update the closure of $\lceil n/p \rceil$ or $\lfloor n/p \rfloor$ vertices. With $p = 1$ we get the sequential implementation.

Using $n$ or $k$ processors the work load for any of them in a sequence of edge insertions or edge deletions is $O(m)$ or $O(m \cdot \lceil n/k \rceil)$, respectively. As far as the time bound per update is concerned, for any edge insertion we have to wait for the last processor which could take $O(m)$ time. But also in the worst case the time for any single update is always bounded by $O(|G_{Reach(j)}|)$.

# 7 Conclusions

In this paper we have studied the problem of dynamically maintaining data structures that are built during compilation runs. The proposed solutions show that in several cases it is possible to maintain these data structures under a sequence of update operations without building them from scratch, thus speeding up successive compilation runs. We have presented in a compact and uniform way data structures and algorithms simple and easy to implement. We have seen that for some problems several solutions are possible that have different time/memory requirements. The choice of the particular solution has to be based on the available resources and their cost. A new result, consisting in the dynamic maintenance of a topological order of a dag, is included in the paper, achieving $O(n)$ amortized time per update in any sequence of edge insertions, where $n$ is the number of nodes in the graph.

Many more questions on several graph problems need to be further investigated, arising in many applications. Static (and often optimal) solutions for these problems are known but, when we turn to dynamic cases, many basic questions surprisingly remain with no satisfying answer, at least in general cases, especially when we consider fully dynamic situations (i.e., when both insertions and deletions are allowed): transitive closure and shortest path are noticeable examples. The considerable gap between the complexity of known algorithms and often trivial lower bounds need to be filled up, either by showing the intrinsic complexity of these problems or by devising more powerful techniques, possibly to amortize over updates and queries, or to get algorithms with a competitive behavior.

## Acknowledgements

The above problems were originally suggested by Stephen Omohundro in the context of defining the new version 1.0 of the object-oriented language Sather and designing its compiler [9].

## References

1. A. V. Aho, J. E. Hopcroft, and J. D. Ullman. *The Design and Analysis of Computer Algorithms*. Addison-Wesley, Reading, MA, 1974.
2. M. Dietzfelbinger, A. Karlin, K. Mehlhorn, F. Meyer auf der Heide, H. Rohnert, and R.E. Tarjan. Dynamic perfect hashing: Upper and lower bounds. Technical Report Bericht Nr. 77, Fachbereich Mathematik-Informatik, Universität-Gesamthochschule Paderborn, 4790 Paderborn, Germany, January 1991, also presented at FOCS '88.

3. G. F. Italiano. Amortized efficiency of a path retrieval data structure. *Theoret. Comput. Sci.*, 48:273–281, 1986.
4. G. F. Italiano. Finding paths and deleting edges in directed acyclic graphs. *Inform. Process. Lett.*, 28:5–11, 1988.
5. J. A. La Poutré and J. van Leeuwen. Maintenance of transitive closure and transitive reduction of graphs. In *Workshop on Graph-Theoretic Concepts in Computer Science*, Lecture Notes in Computer Science, 314, pages 106–120. Springer-Verlag, 1988.
6. J. A. McHugh. *Algorithmic Graph Theory*. Prentice Hall, 1990.
7. K. Mehlhorn. *Graph Algorithms and NP-Completeness*, volume 2 of *Data Structures and Algorithms*. Springer-Verlag, Berlin Heidelberg New York Tokyo, 1984.
8. U. Nanni and P. Terrevoli. A fully dynamic data structure for path expressions on dags. *R.A.I.R.O. Theorical Informatics and Applications* 25, 5, 1991.
9. S. M. Omohundro, C. Lim, and J. Bilmes. The sather language compiler/debugger implementation. Technical Report TR-92-017, International Computer Science Institute, Berkeley, Ca., 1992.
10. R. E. Tarjan. *Data structures and network algorithms*, volume 44 of *CBMS-NSF Regional Conference Series in Applied Mathematics*. SIAM, 1983.
11. R. E. Tarjan. Amortized computational complexity. *SIAM J. Alg. Disc. Meth.*, 6:306–318, 1985.
12. D. M. Yellin. Speeding up dynamic transitive closure for bounded degree graphs. *Acta Informatica*, to appear.

# Average Case Analysis of Fully Dynamic Connectivity for Directed Graphs *

Paola Alimonti[1] **, Stefano Leonardi[1], Alberto Marchetti-Spacccamela[1] and
Xavier Messeguer[2]

[1] Dipartimento di Informatica e Sistemistica, Università di Roma "La Sapienza",
via Salaria 113, 00198 Roma, Italia
[2] Departament de Llenguatges i Sistemes Informátics, Universitat Politécnica de
Catalunya, Pau Gargallo 5, 08028 Barcelona, España

**Abstract.** In this paper we consider the problem of maintaining the transitive closure in a directed graph under both edge insertions and deletions from the point of view of average case analysis. Say $n$ the number of nodes and $m$ the number of edges. We present a data structure that supports the report of a path between two nodes in $O(n \cdot \log n)$ expected time and $O(1)$ amortized time per update, and connectivity queries in a dense graph in $O(1)$ expected time and $O(n \cdot \log n)$ expected amortized time per update. If $m > n^{4/3}$ then connectivity queries can be performed in $O(1)$ expected time and $O(\log^3 n)$ expected amortized time per update. These bounds compares favorably with the best bounds known using worst case analysis. Moreover we consider an intermediate model beetween worst case analysis and average case analysis, the semi-random adversary introduced in [2].

## 1 Introduction

Significant progress has been recently made in the design of algorithms and data structures for dynamic graphs [1, 4, 5, 9, 6, 10, 11, 17, 14, 15, 18, 19, 22]. These data structures support insertions and deletions of edges and/or nodes in a graph, in addition to several types of queries. The goal is to compute the new solution in the modified graph without having to recompute it from scratch. Usually, the sequence of insertions/deletions of edges is not known in advance and each operation must be completed before the next operation is known. If the data structure supports only insertions or only deletions then it is said *partially dynamic*, while a data structure is said *fully dynamic* if it supports both insertions and deletions.

The problem of dynamic maintenance of connected components in undirected graphs has received several efficient solution both for the partially dynamic and fully dynamic version [10, 16, 22]. In this case the main problem is the insertion/deletion of edges. In fact the insertion/deletion of vertices can be reduced to the analougous operations on edges [11].

* This work was partly supported by the ESPRIT Basic Research Action No. 7141 (AL-COM II) and by the italian projects "Algoritmi, Modelli di Calcolo e Strutture Informative", Ministero dell'Università e della Ricerca Scientifica e Tecnologica, and "Progetto Finalizzato Trasporti II", Consiglio Nazionale delle Ricerche.
** Supported by a grant from Consiglio Nazionale delle Ricerche, Italia.

If we consider directed graphs, the problem of maintaining the transitive closure appears much more difficult. If we consider graphs with $n$ vertices and $m$ edges then, for an arbitrary sequence of insertions and connectivity queries between a pair of vertices, the update amortized time for directed graphs is $O(n)$ instead of $O(\alpha(n,n))$ for undirected graphs [1, 14, 21]. If we consider deletion of arcs there are solutions for special classes of graphs such as directed acyclic graphs [15]. There are also solutions for the fully dynamic problem [9, 18, 17] but, to the best of our knowledge, no fully dynamic data structure exists for general directed graphs that, in the worst case, achieves a bound of $o(m)$ for connectivity queries and update operations.

In this work we consider the problem of on-line maintaining the transitive closure in a fully dynamic directed graph from the point of view of average case analysis instead of worst case analysis. This kind of analysis has been already applied to undirected graphs: in [19] a data structure that supports fully dynamic operations in $O(\log n)$ amortized expected time is presented. In [19] the authors have introduced the concept of stochastic graph process to model the evolution of a random graph under a sequence of random insertions/deletions of edges. We extend their approach to directed graph and we consider queries of two different kinds: a) *report-path* queries that report a path that connects two nodes if it exists; b) *connect* queries that report the information on connectivity between two nodes.

We propose algorithms and data structures that support report path queries in $O(n \cdot \log n)$ expected time and $O(1)$ amortized time for update, and queries on connectivity in a dense graph ($m > \Lambda \cdot n \cdot \log n$) in $O(1)$ expected time and $O(n \cdot \log n)$ expected amortized time for update. For a graph with $m > n^{3/2}$ the expected cost for report path queries is $O(n^{1/2})$ and $O(1)$ amortized time for update, while for a graph with $m > n^{4/3}$ we allow queries on connectivity in $O(1)$ expected time and $O(\log^3 n)$ expected amortized time for update.

Our algorithms perform favorably with the best known algorithms for computing the transitive closure in random graphs [12, 13]. In fact Karp has proposed an algorithm for computing the transitive closure of a dense directed graph in $O(n)$ expected time and linear space has been given [12]. Notice that this algorithm answers only connectivity queries but does not allow the report of a path.

Finally in section 4 we consider an intermediate case between worst case analysis and average case analysis, the so-called semi-random graph model [2]. In this model the starting graph and the sequence of insertions and deletions is created by an adversary each of whose decisions is reversed with some small probability $p$. Our data structure supports report path queries in $O(n \cdot \log^2 n)$ for low reverse probability $p = n^{-1/2}$.

## 2 Preliminaries

In this section we first recall some standard notations. The model used for average case analysis is the standard random directed graph model [3]. Let $D = (V, E)$ be a directed graph with $n = |V|$ nodes and $m = |E|$ edges. Let $D_{n,m}$ be the set of all the directed graphs with $n$ nodes and $m$ edges in which all the graphs have the same probability, and let $D_{n,p}$ be the set of all the directed graphs with $n$ nodes, in which the edges are chosen indipendently to occour with probability $p$. In order to analyze randomly changing random graphs a *stochastic graph process* has been introduced [19]. We extend this notion to capture randomly changing random directed graphs:

**Definition 2.1** *A stochastic digraph process (sdp) on a set of vertices $V = \{1, 2, ..., n\}$ is a Markov chain $D^* = \{D_t\}_0^\infty$ whose states are directed graphs on $V$. The process starts with $D_0$ being some directed graph on $V$.*

Let $D = (V, E_t)$ be the state of a sdp at time $t$; then we indicate by $E_t^c$ all the possible edges not in $E_t$, where $D = (V, E_t)$ is the state of the process at time $t$.

**Definition 2.2** *A stochastic digraph process on $V = \{1, 2, ..., n\}$ is called fair (fsdp) if*

1. *$D_0$ is the empty graph.*
2. *There is a $t_1 > 0$ such that $\forall t \leq t_1$, $D_t$ is obtained from $D_t - 1$ by an addition of an edge uniformly at random among all edges in $E_t^c$ (Up to $t_1$, an edge is added at each $t \leq t_1$).*
3. *$\forall t > t_1$, $D_t$ is obtained from $D_{t-1}$ by either an adddition of one new edge which happens with probability 1/2 (and all edges are equiprobable), or by the deletion of one existing edge which happens with probability 1/2 (and all existing edges are equiprobable).*

In the following we extend foundamental results of random graphs theory to random directed graphs. The proofs of these results are omitted because they are simple extensions of the original results presented in random graphs.

The first lemma is a straightfoward extension to stochastic digraph process of a lemma given in [19] for stochastic graph process.

**Lemma 2.1** *Let $D^*$ be an fsdp on $\{1, 2, .., n\}$. Let $D_t = (V, E_t)$ be the state of the process at time $t$, then $D_t$ is a random digraph from $D_{n,m}$.*

In this paper we are mainly concerned with connectivity, which is monotone property (i.e. a property that is maintained while new edges are inserted in the graph). The following lemma extends to random directed graphs a result given in [3] for random graphs.

**Lemma 2.2** *If a monotone property holds for the model $D_{n,m}$ then it also holds for the the corresponding model $D_{n,p}$, with $p = \frac{m}{n^2}$.*

The next lemma, derived from a result by Karp [12], shows that certain known results about random graphs obtained through any standard sequential algorithms, such as breadth-first search or depth-first search, can be converted directly to results on random digraphs.

**Lemma 2.3** *Let $G$ be drawn from $G_{n,p}$ and $D$ be from $D_{n,p}$. Then the random variables representing the number of vertices in the connected component of $G$ containing vertex 1 and the number of vertices reachable from vertex 1 in $D$ are identically distributed.*

The last result is a well known foundamental theorem proved by Erdös and Rényi [8] (see also [3]).

**Lemma 2.4** *Let $c \in R$ be fixed, and $p = \frac{\log n + c + o(1)}{n}$. Then*

$$Prob(G_{n,p} \text{ is connected}) \to e^{-e^{-c}}.$$

If $c$ is not a constant then from [3] it is easy to derive the following bound.

**Lemma 2.5** *Let $p = \Lambda \cdot \frac{\log n}{n}$ and $\Lambda > 1$. Then*

$$Prob(G_{n,p} \text{ is not connected}) < n^{-\Lambda+1}.$$

Using the previous lemma, we are able to extend the result on threshold connectivity from random graphs to random directed graphs.

**Lemma 2.6** *Let $D$ be drawn from $D_{n,p}$. If $p > \Lambda \cdot \frac{\log n}{n}$, with $\Lambda > 2$, then the digraph $D$ is almost surely strongly connected with probability greater than $1 - n^{-\Lambda+2}$.*

*Proof.* Let $p = \Lambda \cdot \frac{\log n}{n}$. By lemmas 2.3 and 2.5 we have:

$$Prob(\text{vertex 1 does not reach all vertices in } D_{n,p}) =$$

$$Prob(G_{n,p} \text{ is not connected}) < n^{-\Lambda+1}$$

Therefore

$$Prob(D_{n,p} \text{ is strongly connected}) =$$

$$1 - Prob(\exists \text{ a vertex that does not reach all verteces}) > 1 - n^{-\Lambda+2}$$

Finally, observe that the following property holds for the expected degree of a node: the expected number of edges leaving or entering a node is $\frac{m}{n}$.

## 3 Fully dynamic directed graphs connectivity

In this paper we consider two kinds of operation on directed graphs: updates and queries. Namely we have:

- *report-path(i, j)*: returns a path from node $i$ to node $j$ if such a path exists;
- *connect(i,j)*: returns the information on connectivity from node $i$ to node $j$;
- *insert(i,j)*: inserts the edge from node $i$ to node $j$;
- *delete(i,j)*: deletes the edge from node $i$ to node $j$.

Now we introduce a data structure that allows to perform efficient connectivity queries while edges are inserted and deleted.

The underlying idea is to perform connectivity queries in a subgraph randomly drawn from the original graph. The subgraph is, the same as the original graph, with high probability strongly connected.

More precisely, we randomly extract from the digraph $D = (V, E)$ a subgraph $D_K = (V_K, E_K)$ where:

- $V_K$ is a set of $k = \frac{c}{p} \cdot \log n$ independent nodes $V_K = \{v_1, .., v_k\}$, randomly chosen from $V = \{1, ..., n\}$, where $c$ is an appropriate constant (see section 3.1);
- $E_K = \{(g, h) \in E : g, h \in V_K\}$.

We refer to the graph $D_K = (V_K, E_K)$ as the *black graph*, and we will use it in place of the original graph $D$ to perform connectivity queries. Moreover we call the set of vertices $V_K$, the set of edges $E_K$, and the set of remaining vertices $V - V_K$. the sets of *black nodes*, *black edges*, and *white nodes* respectively.

Finally we associate to each white node $i \in V - V_K$:

- a double linked list $E_i = \{(g, i) \in E - E_K | g \in V_K\}$ of entering edges whose starting node is black,
- a double linked list $L_i = \{(i, g) \in E - E_K | g \in V_K\}$ of outgoing edges whose ending node is black.

With each edge we associate a pointer to the relative position in a list.

The time complexity of building the data structure is $O(m)$, and it takes $O(m)$ space.

In the following of this section we will study the expected time complexity of a fully dynamic algorithm for report-path query, and the expected time complexity of a fully dynamic algorithm for connect query.

## 3.1 Report-path query

The algorithm for report-path query first looks for a path formed only by edges in the black graph $D_K$ possibly except the first and the last edge.

Namely assume that both $i$ and $j$ are white edges. In this case the algorithm checks if in $L_i$ ($E_j$) there is at least one black vertex $g$ (respectively $h$) connected to $i$ (respectively to $j$). In this case if there exists a path $P$ in $D_K$ form $g$ to $h$ then the algorithm returns the path $(i, g), P, (h, j)$. If either $L_i$ is empty or $E_j$ is empty or there is no path in $D_K$ from $g$ to $h$ then the algorithm searches the whole graph for a path from $i$ to $j$. As we will see this happens with very low probability. The cases in which either $i$ or $j$ are not black vertices are analougous.

The algorithm for report-path query is as follows:

**Algorithm Report-path$(i, j)$**

```
begin
   case
      i, j ∈ V − V_K
         if L_i <> empty and E_j <> empty
            then begin
               (i, g) ← first element of L_i
               (h, j) ← first element of E_j
               if there exists a path P from g to h in D_K
                  then returns <(i, g), P, (h, j)>
                  else searches the whole graph D
            end
            else searches the whole graph D;
```

$i \in V - V_K$, $j \in V_K$
    **if** $L_i <>$ *empty*
        **then begin**
            $(i, g) \leftarrow$ first element of $L_i$
            **if** there exists a path $P$ from $g$ to $j$ in $D_K$
                **then returns** $<(i,g),P>$
                **else** searches the whole graph $D$
        **end**
        **else** searches the whole graph $D$;
$i \in V_K$, $j \in V - V_K$
    **if** $E_j <>$ *empty*
        **then begin**
            $(h, j) \leftarrow$ first element of $E_j$
            **if** there exists a path $P$ from $i$ to $h$ in $D_K$
                **then returns** $<P,(h,j)>$
                **else** searches the whole graph $D$
        **end**
        **else** searches the whole graph $D$;
$i, j \in V_K$
    **if** there exists a path $P$ from $i$ to $j$ in $D_K$
        **then returns** $<P>$
        **else** searches the whole graph $D$
  **endcase**
**end.**

To study the expected computational cost for report-path query we need the following preliminary lemmas.

**Lemma 3.1** *Let $i$ be a white node. The probability that either $E_i$ or $L_i$ is empty is smaller than $2 \cdot n^{-c}$.*

*Proof.* Consider the set $E_i$. The probability that $E_i$ is empty is equal to the probability that does not exist any edge from $i$ to a black node, that is

$$(1-p)^k \leq e^{-k \cdot p}.$$

Assuming $k = \frac{c}{p} \cdot \log n$, follows that $E_i$ is empty with probability smaller than $e^{c \cdot \log n} = n^{-c}$. The same holds for $L_i$.

Moreover, we must assure that with high probability there exists a path between any pair of black nodes.

**Lemma 3.2** *If $c > \Lambda$ then the graph $D_K$ is strongly connected with probability greater than $1 - n^{-\Lambda+2}$.*

*Proof.* By Lemma 2.6 it is sufficient to show that $p_K$, the edge probability of $D_K$ satisfies $p_K > \Lambda \cdot \frac{\log k}{k}$, where it easy to show that $p_K = p$, the edge probability in $D$. The inequality is satisfied for $k > \Lambda \cdot \frac{\log k}{p}$. Since we choose $k = \frac{c}{p} \cdot \log n$, we must take $c > \Lambda$.

Then, the following lemma gives the computational cost for report-path query.

**Lemma 3.3** *If $c > \Lambda > 4$ then the expected cost for report-path query in a directed graph with $m > \Lambda \cdot n \cdot \log n$ edges is $O(\frac{n^2}{m} \cdot (\log^2 n))$.*

*Proof.* We say $A$ the event "the algorithm searches the whole graph". The expected cost $x$ of a query can be bounded as follows:

$$E[x] \leq E[x|\neg A] + E[x|A] \cdot \text{Prob}(A).$$

$E[x|\neg A]$ is $O(\frac{n^2}{m} \cdot (\log^2 n))$. In fact in this case the running time of the algorithm is dominated by the expected number of edges in $D_K$. The average degree of a node in $D_K$ is $p \cdot n \cdot \frac{k}{n} = p \cdot k$. Hence, the average number of edges in $D_K$ is given by $O(k^2 \cdot n)$, that is $O(\frac{n^2}{m} \cdot (\log^2 n))$, since $p = \frac{m}{n^2}$.

On the other side by lemmata 3.1 and 3.2 $\text{Prob}(A) < n^{-\Lambda+2} + 2 \cdot n^{-c}$; moreover in this case the running time of the algorithm is $O(m)$. Hence if $c > \Lambda > 4$ we have $E[x|A] \cdot \text{Prob}(A) = O(1)$.

The data structure must be updated while edges are inserted and deleted. If an edge whose extreme nodes are both black is inserted or deleted then the edge is inserted or deleted in the black graph $D_K$. If an edge from a white node $i$ to a black node is inserted then the edge is inserted in $L_i$, *viceversa* if the edge is deleted, it is deleted from $L_i$. Analogously for an insertion or a deletion of an edge from a black node to a white node. All these updates are performed in constant time since we use doubled linked list and we mantain a pointer from any edge to its position in a list.

Moreover, the structure must be rebuilt when the number of edges is either excessively increased or decreased. In the former case, by choosing a smaller number of black nodes, a lower query time is possible, while in the latter case the structure no longer satisfies the requirements stated in lemmata 3.1 and 3.2.

Now let $m^*$ be the number of edges in the graph when the structure is built. The same structure is used until $m$ is within a range beetween $\frac{m^*}{2}$ and $2 \cdot m^*$. To satisfy the condition of lemma 3.1, we take $k = 2 \cdot \frac{c}{p} \cdot \log n$, and we rebuild the data structure from scratch when $m$ is outside this range. Then, since the structure is used for at least $O(m)$ updates, and the cost to build the data structure is $O(m)$, the amortized cost for update is $O(1)$.

Then we can state the following theorem:

**Theorem 3.4** *There exists a data structure such that the expected cost for report-path query in a directed graph with $m > \Lambda \cdot n \cdot \log n$ is $O(\frac{n^2}{n} \cdot \log^2 n)$, and that can be update in $O(1)$ amortized time for each insertion or deletion.*

*Proof.* The analysis of the running time is given by lemma 3.3 and by the considerations following the lemma.

As far as the correctness of the procedure report-path$(i,j)$ suppose that both $i$ and $j$ are white nodes (i.e. $i, j \in V - V_K$). In this case, if $(i, g) \in E$, $(h, j) \in E$, and there is a path in $D_K$ from $g$ to $h$ then, clearly, $j$ is reachable from $i$ and the algorithm report a path. Otherwise, if the above conditions are not satisfied then the algorithm applies a search procedure to find all vertices reachable from $i$. The correctness of the algorithm in the remaining cases is analougous and it is omitted.

Notice that for a directed graph with number of edges $m > n^{3/2}$ the expected computational cost for report-path query is $O(n^{1/2})$.

Since our analysis applies only to graphs with number of edges greater than $\Lambda \cdot n \cdot \log n$, the expected computational cost for report-path query is $O(n \cdot \log n)$. Clearly, also for $m \leq \Lambda \cdot n \cdot \log n$ the computational cost for a report-path query is $O(n \cdot \log n)$.

**Corollary 3.5** *There exists a data structure such that the expected cost for report-path query in a directed graph is $O(n \cdot \log n)$, and that can be update in $O(1)$ amortized time for each insertion or deletion.*

## 3.2 Connect query

In the following of this section we present an algorithm to perform connect queries beetween any pair of nodes in a dense directed graph (i.e. $m > \Lambda \cdot n \cdot \log n$). In this case the report of a path that connects the pair of nodes is no longer required.

Our algorithm is based on a previous result by Karp: in [12] is given an algorithm to perform the transitive closure in dense directed graphs in $O(n)$ expected time. The transitive closure is represented in a compact form with $O(n)$ expected space requirement. This data structure allows connect queries in $O(1)$ expected time, but, it does not support report-path queries.

The data structure proposed for report-path query is slightly modified by representing in compact form the transitive closure of the black graph $D_K$. The algorithm proposed for report-path queries is then modified for connect queries. The query connect$(i, j)$ returns *yes* if there exists a path from node $i$ to node $j$ in $D_K$ possibly except the first and the last edge, otherwise it searches the whole graph $D$.

As we have shown in the previous section this is very unlikely to happen if $c$ and $\Lambda$ are large enaugh.

**Algorithm Connect $(i, j)$**

```
begin
  case
    i, j ∈ V − V_K
      if L_i <> empty and E_j <> empty
        then begin
          (i, g) ← first element of L_i
          (h, j) ← first element of E_j
          if g is connected to h in D_K
            then returns yes:
            else check if i is connected to j in D
        end
      else check if i is connected to j in D
    i ∈ V − V_K, j ∈ V_K
      if L_i <> empty
        then begin
          (i, g) ← first element of L_i
          if g is connected to j in D_K
            then returns yes
```

```
                else check if i is connected to j in D
            end
        else check if i is connected to j in D
    i ∈ V_K, j ∈ V − V_K
        if E_j <> empty
        then begin
            (h, j) ← first element of E_j
            if i is connected to h in D_K
               then returns yes
               else check if i is connected to j in D
            end
        else check if i is connected to j in D
    i, j ∈ V_K
        if i is connected to j in D_K
        then returns yes
        else check if i is connected to j in D
    endcase
end.
```

A straightforward analysis of this algorithm gives the following result:

**Theorem 3.6** *If $c > \Lambda > 4$ then the expected cost for connect query in a directed graph with $m > \Lambda \cdot n \cdot \log n$ is $O(1)$.*

Next, we consider the expected cost for updating the data structure when edges are inserted or deleted. The main problem concerns updates that modify the black graph $D_K$. In fact the transitive closure of the black graph must be updated and we are forced to rebuild it from scratch.

**Theorem 3.7** *The amortized expected cost to update the data structure for each insertion or deletion is $O(\frac{n^4}{m^3} \cdot \log^3 n)$.*

*Proof.* Since the probability that a random edge is black is $\frac{k}{n} \cdot \frac{k}{n}$, and the expected cost for computing the transitive closure is $O(k)$, the expected computational cost for computing from scratch the transitive closure after any insertion or deletion is $O(\frac{k^3}{n^2}) = O(\frac{n^4}{m^3} \cdot \log^3 n)$. Moreover the amortized cost for update is $O(1)$ since the data structure is used for at least $O(m)$ insertions or deletions and the cost to re-build it from scratch is $O(m)$.

**Corollary 3.8** *There exists a data structure such that the expected cost for connect query in a directed graph with $m > \Lambda \cdot n \cdot \log n$ is $O(1)$, and that can be update in $O(n \cdot \log n)$ amortized expected time for each insertion or deletion.*

For a directed graph with $m \geq n^{4/3}$ we state the following corollary.

**Corollary 3.9** *There exists a data structure such that the expected cost for connect query in a directed graph with $m \geq n^{4/3}$ edges is $O(1)$, and that can be update in $O(\log^3 n)$ amortized expected time for each insertion or deletion.*

# 4 Semi-random directed graphs

In this section we consider the problem of performing efficient connectivity queries in a semi-random graph. In this case we consider an intermediate model beetween worst case analysis and average case analysis.

Namely, we consider a model in which each decision about the graph is taken not by a worst case adversary, but by an adversary each of whose decisions is reversed with probability $p$. This type of adversary is derived from the *semi-random* source of Santha and Vazirani [20] and has been applied for approximate coloring of 3-chromatic graphs by Avrim Blum [2].

The decisions of the adversary regarding whether or not to include an edge in the starting graph, and the inclusion of any insertion or deletion of an edge in the sequence of updates, are accepted with probability $1-p$ and rejected with probability $p$ by the algorithm.

Notice that the adversary would propose the repeated deletion of the same edge. In this way if $p$ is the probability of rejection we have that after $k$ trials the probability that an edge is in the graph is negligible. Then, before repeating in a sequence the deletion (insertion) of the same edge, the adversary must be care to re-insert (re-delete) the edge. However, the insertion of an edge already existing or the deletion of an edge not existing does not modify the graph at all.

The semi-random graph generated by the previous procedure has the following property:

**Lemma 4.1** *Each edge has probability greater than $p - p^2$ to appear in a graph created by an adversary each of whose decisions is reversed with probability $p$.*

*Proof.* The property holds for the starting graph since each edge included by the adversary is in the graph with probability $1 - p$, and each edge not included by the adversary is in the graph with probability $p$. Moreover, insertions and deletions of the same edge are alternate in the sequence of updates given by the adversary. The probability that an edge appears in the graph after the $i^{th}$ deletion is $P_i = p \cdot ((1 - p) + p \cdot P_{i-1})$, where $p \cdot (1 - p)$ is the probability that an edge is in the graph after the $i^{th}$ deletion if the $i - 1^{th}$ deletion successes, while $p^2 \cdot P_{i-1}$ is the probability that the edge is in the graph after the $i^{th}$ deletion if the $i - 1^{th}$ insertion fails. Moreover, $P_1 = p \cdot (1 - p)$ if the edge is in the starting graph, and $P_1 = p$ if the edge is not in the starting graph assuming that the adversary has imposed a deletion to exclude the edge from the starting graph. It can be shown by induction that the probability that each edge appears in the graph is $p$ for an edge not included in the starting graph and $p \cdot (1 - p^i)$ with $i \geq 1$ for an edge included in the starting graph.

The above lemma state that we can always consider the presence of a random directed graph $D_{n,p-p^2}$ overlapping the semi-random graph.

The presence of a random directed graph hidden in the semi-random directed graph can be used to apply the same data structure developed in section 3.1 for report-path queries. In particular we randomly choose $k = \frac{c}{p-p^2} \cdot \log n$ nodes to build the data structure where the constant $c$ is large enough to take into account the constraints given in lemmata 3.1 and 3.2. Observe that the number of edges in a semi-random direcetd graph never falls below $O(p \cdot n^2)$.

Report-path queries in semi-random directed graphs are performed by the same algorithm given in section 3.1, and the expected computational cost of a query is given by the number of edges in the black graph, that is $O(k^2)$ in the worst case.

**Theorem 4.2** *The expected computational cost for report-path query in a semi-random directed graph is* $O(\frac{1}{p^2} \cdot \log^2 n)$.

Notice that for low reverse probability such as $p = n^{-\frac{1}{2}}$ [2] the expected computational cost for report-path query is $O(n \cdot \log^2 n)$.

With regard to the update of the structure, observe that this is built only once since the set of $k$ nodes randomly choosen depends only on the initial choice of $p$. Then the cost for building the structure can be considered as preprocessing, while the structure can be update in $O(1)$ worst case time bound for each insertion or deletion.

The data structure for connect query given in section 3.2 does not easily extend to semi-random directed graphs since the algorithm given in [12] for computing the transitive closure in random directed graphs does not apply to semi-random directed graphs.

## 5    Conclusions and open problems

In this paper we have shown efficient solutions for answering connectivity queries in stochastic random digraphs. The obvious open problem is to reduce the time requirements for connectivity queries in the case of graphs with $m \leq n^{4/3}$. In particular, it would be interesting to show that connectivity in a stochastic digraph process has the same complexity of the problem of maintaining connectivity in stochastic undirected graph process.

Another interesting open problem is to investigate the complexity of the shortest path problem in a stochastic digraph process.

## References

1. G. Ausiello, G. F. Italiano, A. Marchetti-Spaccamela, U. Nanni, "Incremental algorithms for minimal length paths", *J. of Algorithms*, 12 (1991), 615-638.
2. A. Blum, "Some tools for approximate 3-coloring", *Proceedings of the 31st IEEE Symposium on Foundations of Computer Science*, 1990.
3. B. Bollobas, *"Random graphs"*, Academic Press, 1985.
4. G. Di Battista and R.Tamassia, "Incremental planarity testing", *Proc. 30th annual Symp. on Foundations of Computer Science*, 1989.
5. G. Di Battista and R.Tamassia, "On-line graph algorithms with SPQR-trees", *Proc. 17th Int. Coll. on Automata, Languages and Programming*, Lect. Not. in Comp. Sci., Springer-Verlag, 1990.
6. D. Eppstein, Z.Galil, G. F. Italiano, A.Nissenzweig, "Sparsification - a technique for speeding up dynamic graph algorithms", *Proc. 33rd Symp. on Found. of Computer Sci.* 1992.
7. D. Eppstein, G. F. Italiano, R. Tamassia, R. E. Tarjan, J. Westbrook, M. Young, "Maintenance of a minimum spanning forest in a dynamic planar graph", *Proc. 1st ACM-SIAM Symp. on Discrete Algorithms*, S.Francisco, 1990.

8. P. Erdős, A. Rènyi, "On random graphs I", *Publ. Math. Debrecen*, 6, 290-297.

9. S. Even and H. Gazit, "Updating distances in dynamic graphs", *Methods of Operations Research*, 49, 1985.

10. Z. Galil, G.F. Italiano, "Fully dynamic algorithms for edge-connectivity problems", *Proc. 23rd ACM Symp. on Theory of Comp.*, (1991) 317-327.

11. Z. Galil, G.F. Italiano, "Reducing edge connectivity to vertex connectivity", *SIGACT News*, 22 (1) (1991) 57-61.

12. R.M. Karp, "The transitive closure of a random digraph", Technical Report-89-047, International Computer Science Institute (ICSI), August 1989.

13. R.M. Karp, R.E. Tarjan, "Linear expected-time algorithms for connectivity problems", *Proc. of the 11th. annual ACM Symp. on Theory of Computing*, 368-377, 1980.

14. G.F. Italiano, "Amortized efficiency of a path retrieval data structure", *Theoret. Comp. Sci.*, 48, 1986, 273-281.

15. G.F. Italiano, "Finding paths and deleting edges in directed acyclic graphs", *Inf. Proc. Lett.*, 28, 1988, 5-11.

16. J.A. La Poutré, "Maintenance of triconnected components of graphs", *Proc. 19th Int. Coll. Automata Languages and Programming, Lect. Not. in Computer Sci.*, Springer Verlag, (1992), 354-365.

17. J.A. La Poutré, J. van Leeuwen, "Maintenance of transitive closure and transitive reduction of graphs", *Proc Work. on Graph Theoretic concepts in Comp. Sci., Lect. Not. in Computer Sci.*, vol. 314, Springer Verlag, Berlin, 1985, 106-120.

18. H. Rohnert, "A dynamization of the all-pairs least cost path problem", *Proc. of the 2nd Symp. on Theoretical Aspects of Computer Science*, Lect. Not. in Comp. Sci., vol. 182, Springer-Verlag, 1990.

19. J.H. Reif, P.G. Spirakis, M. Yung, "Fully dynamic graph connectivity in logarithmic expected time", Alcom Technical Report, 1992.

20. M. Santha, U.V. Vazirani, "Generating quasi-random sequences from semi-random sources", *Journal of Computer and Systems Science*, 33:75-87,1986.

21. R.E. Tarjan, Jan van Leeuwen, "Worst case analysis of set union algorithms", *Journal of Assoc. Comput. Mach.*, 31 (1984), 245-281.

22. J. Westbrook, "Algorithms and data structures for dynamic graph problems", Ph.D. Dissertation, Tech. Rep. CS-TR-229-89, Dept. of Computer Science, Princeton University, 1989.

# Fully Dynamic Maintenance of Vertex Cover[1]

Zoran Ivković                    Errol L. Lloyd

{ivkovich,elloyd}@cis.udel.edu

Department of Computer and Information Sciences, University of Delaware
Newark, DE, 19716

**Abstract.** The problem of maintaining an approximate solution for **vertex cover** when edges may be inserted and deleted dynamically is studied. We present a fully dynamic algorithm $A_1$ that, in an amortized fashion, efficiently accommodates such changes. We further provide for a generalization of this method and present a family of algorithms $A_k$, $k \geq 1$. The amortized running time of each $A_k$ is $\Theta((v+e)^{\frac{1+\sqrt{1+4(k+1)(2k+3)}}{2(2k+3)}})$ per *Insert/Delete* operation, where $e$ denotes the number of edges of the graph $G$ at the time that the operation is initiated. It follows that this amortized running time may be made arbitrarily close to $\Theta((v+e)^{\frac{\sqrt{2}}{2}})$. Each of the algorithms given here is 2-competitive, thereby matching the competitive ratio of the best existing off–line approximation algorithms for vertex cover.

## 1   Introduction

Recall that *vertex cover* is a classic problem in combinatorial optimization. The study of vertex cover in computer science has been diverse: it was one of the original *NP*–complete problems [9]; it is often used as a technical tool in performing reductions [6]; various approximation algorithms for vertex cover have been proposed over the past two decades [1, 2, 7, 12]; and, recently, some results in the area of structural complexity theory have deepened our understanding of the difficulties in designing better approximations [11].

Recall also that *fully dynamic* algorithms are aimed at situations where the problem instance is changing (slowly) over time. This situation occurs often in interactive design processes, and fully dynamic algorithms incorporate these incremental changes without any knowledge of the existence and nature of future changes. The objective of course is to develop fully dynamic algorithms that are "competitive" with existing off–line algorithms.

Although the bulk of the existing work on fully dynamic algorithms has been directed toward problems known to be in *P*, e.g. [3, 4, 5], some recent attention has been paid to fully dynamic *approximation* algorithms for problems that are *NP*-complete [8, 10]. In this case, being competitive with off–line algorithms means that the quality of the approximation produced by the fully dynamic approximation algorithm should be as good as that produced by the off–line

---

[1]Partially supported by the National Science Foundation under Grant CCR–9120731.

algorithms. Further, the running time per operation (i.e. change) of the fully dynamic algorithm should be as small as possible.

Thus, in this paper, we consider *fully dynamic approximation algorithms for vertex cover*, where edges are inserted into, and deleted from, the graph in a dynamic fashion, and the vertex cover may be adjusted to accommodate the changes to the instance via insertions and deletions of edges (i.e. vertices may be added to and removed from the vertex cover).

## 1.1 Vertex cover

In the vertex cover problem, an undirected graph $G = (V, E)$, $v = |V|$ and $e = |E|$, with no multiple edges, and no self–loops is given. The goal is to find a minimum cardinality *vertex cover* $V'$. That is, a subset of $V$ that covers all edges in $E$ in the following sense: for each edge $uw \in E$, at least one of $u$ and $w$ belongs to $V'$.

Vertex cover is well-known as one of the earliest *NP*–complete problems [9]. And, as mentioned earlier, vertex cover has a number of polynomial time off–line *approximation algorithms* [1, 2, 7, 12]. For such an approximation algorithm $A$, it will be convenient to define the **quotient** $Q(A, G) = \frac{A(G)}{OPT(G)}$, where $A(G)$ and $OPT(G)$ denote the cardinality of a vertex cover of a graph $G$ produced by $A$, and the cardinality of a optimal vertex cover of $G$. Then, the usual measure of the quality of a solution produced by a vertex cover approximation algorithm $A$ is its **competitive ratio** $R(A)$ defined as $R(A) = \sup\{Q(A, G)\}$. In this case, $A$ is said to be **R(A)–competitive**.

All of the off–line approximation algorithms for vertex cover referenced above are 2–competitive , and no algorithm developed to date has a smaller competitive ratio. Since it is known that vertex cover is approximable in polynomial time only up to some fixed competitive ratio [11], it could be the case that 2 is the best possible competitive ratio (unless, of course $P = NP$). Finally, we note that in regard to running times, the fastest of the above algorithms, e.g. [7, 12], attain the competitive ratio of 2 in time $\Theta(v + e)$.

## 1.2 Fully dynamic algorithms for vertex cover

Fully dynamic approximation algorithms for vertex cover process a sequence of *Inserts* and *Deletes* of edges[2]. This is a truly on–line situation, in that the algorithms have no advance knowledge of what future changes (if any) there may be to graph $G$. As a consequence, in the course of processing *Inserts* and *Deletes*, it may be necessary to change the current vertex cover both in order to maintain a cover, and to insure that the cover is of the appropriate quality. Thus, as the algorithm proceeds, the status of a vertex may change as it is either made part of the cover, or is removed from the cover. Further, such changes may occur arbitrarily often as edges are inserted and deleted.

---

[2]Note that the size of the vertex set, $v = |V|$, is fixed.

An important consideration that arises in the fully dynamic context, is just what internal form a "solution" may take. In this sense, it is natural to require that, in addition to processing *Insert* and *Delete* operations, fully dynamic approximation algorithms for vertex cover should also handle "lookup" queries of the following form: *size*, which returns in $\mathcal{O}(1)$ the number of vertices in the current vertex cover, *cover*, which returns a list of the vertices in the current vertex cover, in time linear in the size of that cover, and *incover(v)*, which returns in $\mathcal{O}(1)$ the value *true* if $v$ is in the current vertex cover, and *false* otherwise. These queries may be interspersed in the *Insert/Delete* sequence in any fashion. Of course, since this is an on–line situation, the algorithm has no advance knowledge of where such queries may occur.

Next we discuss the notion of competitiveness in the context of developing fully dynamic approximation algorithms. We begin by noting that with respect to the definitions of *quotient* and *competitive ratio*, there is no need to make a distinction between fully dynamic and off–line algorithms. In each case, these measures reflect the size of the vertex covers produced by the algorithm relative to the size of optimal covers.

In this framework, let $A$ be an (off–line) approximation algorithm for vertex cover, and let $B$ be a fully dynamic approximation algorithm for vertex cover. Then, $B$ is **approximation–competitive** with $A$ if $R(B) \leq R(A)$. Thus, our goal in this paper is to develop fully dynamic approximation algorithms for vertex cover that are 2-competitive.

In regard to running time, the goal is to develop such a $B$ whose running time for any instance $G$ of vertex cover, and any (valid) change of $G$, call it $\delta$, is $T_B(G, \delta) = o(T_A(\delta(G))$, where $T_B(G, \delta)$ denotes the running time required by $B$ to perform the change $\delta$ on $G$, and $T_A(\delta(G))$ denotes the running time of $A$ on $\delta(G)$, the modified instance.

In the case of vertex cover, this translates into developing approximation–competitive fully dynamic algorithms having a running time that is $o(v + e)$ per operation. We say that a fully dynamic approximation algorithm $B$ for vertex cover has **running time** $\mathcal{O}(f(v, e))$ if the time taken by $B$ to process a change to $G$ on $v$ vertices and $e$ edges is $\mathcal{O}(f(v, e))$. If $\mathcal{O}(f(v, e))$ is a *worst case* time bound, then $B$ is **uniform**. If $\mathcal{O}(f(v, e))$ is an *amortized* time bound, then $B$ is **amortized**.

With the preliminaries concluded, the remainder of the paper is organized as follows: In the next section we provide a simple algorithm $A_0$ for fully dynamic vertex cover. This algorithm is a straightforward extension of a maximal matching based off–line approximation algorithm for vertex cover. However, as we indicate in section 2, $A_0$ may not offer any savings over total recomputation for very sparse graphs. We remedy this situation in sections 3 and 4, where we present our main results. We begin in section 3 by giving a fully dynamic 2– competitive algorithm $A_1$ having an $\Theta((v + e)^{\frac{1+\sqrt{41}}{10}})$ amortized running time per operation. This algorithm features a specific technique that has two essential ingredients. First, we utilize two different methods (**Clean** and **Dirty**) for the processing of *Insert* and *Delete* operations. Second, we partition the processing of operations into *stages*, with a certain number of insertions/deletions handled per stage. Within each stage, all of the operations are handled in an identical fashion (either **Clean** or **Dirty**). The description of $A_1$ and the techniques used therein lead to the result of section 4. There we generalize $A_1$ and present a family of fully dynamic approximation algorithms $A_k$, $k \geq 1$ for vertex cover. For this family of algorithms, we show that, at the expense of some additional effort (for $A_k$, $k + 1$ "tests"), the asymptotic running time can be made arbitrarily close to $\Theta((v + e)^{\frac{\sqrt{2}}{2}})$ amortized, per *Insert* or *Delete*. Each of these algorithms is 2-competitive, thereby matching the competitive ratio of the best existing off–line algorithms for vertex cover.

# 2   A Simple Algorithm for Incremental Vertex Cover

Motivated by the notion of approximation-competitiveness introduced in the preceding section, a natural approach to the development of fully dynamic vertex cover algorithms is to try to adapt off–line methods to a fully dynamic context. Thus, we begin by doing just that.

Specifically, we consider a 2–competitive approximation algorithm for vertex cover that runs in $\Theta(v + e)$ time, and is based on **maximal matching** [7]. This algorithm simply computes a maximal matching in the graph G, and then takes as a vertex cover all of the vertices that are incident to the edges in the matching.

A conversion of this off–line algorithm into a fully dynamic one is given below. The conversion is relatively straightforward, and not particularly interesting. However, the ideas involved in the design of this algorithm, as well as an appreciation of the situations where it performs rather slowly, are important factors that guide the development of subsequent algorithms.

In what follows we describe both the *Insert* and *Delete* operations. In each case, it is assumed that, immediately before an *Insert/Delete* is processed, there is an existing maximal matching $M$ of the current $G$ (i.e. not including the effect of this operation), and that the graph $G$ is represented with adjacency lists.

- *Insert(a)* – to insert an edge $a$ into $G$, check whether either of its endpoints is already an endpoint in $M$. If so, then the set of vertices incident to the edges of $M$ is also a vertex cover of $G + a$, and nothing further needs to be done in this regard. If neither of the endpoints of $a$ is an endpoint of an edge in $M$, then $a$ is added to $M$. Clearly $M + a$ is a maximal matching of $G + a$. In either case, we conclude by updating the adjacency lists of both endpoints of $a$ (to reflect the inclusion of $a$ in $G$). All of these operations can be performed, with a proper implementation, in $\mathcal{O}(1)$ time.

- *Delete(a)* – to delete an edge $a$ from $G$, check whether $a$ is in $M$. If it is not, then the set of vertices incident to the edges of $M$ is a vertex cover of $G - a$, and nothing further needs to be done in this regard. If $a$ **is** in $M$, then, by deleting $a$ from $G$, we may "uncover" some edges. These "uncovered" edges of $G$ are the ones whose endpoints are **not** adjacent to any endpoints of edges in $M$, other than $a$. To cover all such edges in $G - a$, the following will be done for each endpoint of $a$:

1. Let $u$ be an endpoint of $a$. Then the adjacency list of $u$ is scanned until finding a vertex $w$ that is **not** an endpoint of an edge in $M$, or until the end of the list is encountered.

2. If such a $w$ is found, add edge $uw$ to $M$. Otherwise, do nothing, since all of the edges incident on $u$ in $G - a$ are covered by endpoints of edges in $M - a$.

Clearly, the matching that results from this processing is maximal for $G-a$.

In either case, the adjacency lists of both endpoints of $a$ are updated to reflect the deletion of $a$.

Finally, note that the deletion of a non–matching edge can be performed in $\mathcal{O}(1)$ time, and the deletion of a matching edge may involve scanning an entire adjacency list, and hence takes $\Theta(v)$ time.

Hence, we have the following theorem:

**Theorem 1** $A_0$ is a fully dynamic approximation algorithm for vertex cover, and it is approximation–competitive with the standard off–line algorithm for computing a maximal matching (that is, $A_0$ is 2–competitive). Further, Insert operations, as well as Delete operations of non–matching edges, take $\mathcal{O}(1)$ uniform running time, and Delete operations of matching edges take $\Theta(v)$ uniform running time.

Let us remark on some facts about this simple algorithm: its uniform running time is bounded by $\mathcal{O}(v)$, incurred only while deleting matching edges. The operations are fast in all the other cases, taking only constant time per operation in the worst case. A natural question that arises here is whether there is a way to somehow avoid the undesirably high running time of some *Delete* operations by a careful manipulation of the dynamically changing information on $G$, or perhaps perform an amortized analysis of $A_0$, and attempt to charge some of the cost of a deletion of a matching edge to some other operation, some vertex or edge. Unfortunately, a situation where an edge (whose endpoints have high degree) "toggles" into, and out of, $G$ and $M$, a number of times, seems to defeat these ideas.

We do note that for graphs $G$ for which $v = o(e)$, even the use of $A_0$ is a definite improvement over a total recomputation of a maximal matching after each *Insert* and *Delete*. Unfortunately, many important classes of graphs do not belong to this family, e.g. planar graphs.

An obvious question is whether it is possible to improve upon $A_0$, and design an algorithm that would guarantee, for each operation performed by that algorithms, savings over recomputation for any graph.

In the next section we present an algorithm that accomplishes just that, albeit with an *amortized* running time.

# 3   An Improved, Amortized Algorithm $A_1$

We begin this section by describing an algorithm $B_1$ for fully dynamic vertex cover whose amortized running time is $\Theta((v+e)^{\frac{3}{4}})$ per *Insert/Delete* operation. In our discussion of algorithm $B_1$, we introduce the technique of **Clean and Dirty stages**. The ideas involved in designing $B_1$ are then used to obtain another algorithm $A_1$, with a slightly sharper amortized running time bound of $\Theta((v+e)^{\frac{1+\sqrt{41}}{10}})$ (roughly $\Theta((v+e)^{0.74})$).

## 3.1   What is Clean and what is Dirty?

Here we give an intuitive introduction to our technique. Based on the concepts developed here, we sketch $B_1$. In later sections we build on the ideas from this section, developing more and more efficient algorithms.

The algorithm $B_1$ is once again based on the computation of a maximal matching to provide a vertex cover. For the purposes of fully dynamic approximation of vertex cover, this method is appealing because whether or not an edge is in the matching depends only on the "local properties" of that edge, i.e. on the *status* of the endpoints of the edge under consideration. Here, the *status* of a vertex is either *matched*, i.e. being an endpoint of some matching edge, or *non–matched*, i.e. not being an endpoint of some matching edge in the current maximal matching.

We proceed by reiterating the difficulties encountered in attempting to speed up $A_0$. The weak spot of $A_0$ is that a *Delete* of a matching edge may involve scanning as many as $\Theta(v)$ elements of the adjacency lists of the two endpoints of that edge, searching for an unmatched vertex. This being the case, it would seem reasonable to do the following: For each vertex $u$, separate the adjacency list of $u$ into two sublists, with one sublist containing the matched neighbors of $u$, and the other sublist containing the non-matched neighbors of $u$. Such a partition would make it trivial to locate an unmatched neighbor of $u$. Unfortunately, the maintenance of such a partition is quite costly. In particular, the sequence of *Inserts* and *Deletes* may be such that a vertex of high degree "toggles" between being and not being an endpoint of a matching edge. This forces each neighbor $w$ of that vertex to continually register these changes by moving the vertex between the two parts of the adjacency list of $w$. If such vertices have as many as $\Theta(v)$ neighbors, there are $\Theta(v)$ changes to be performed per operation.

### 3.1.1   The Dirty strategy

Despite the difficulties just discussed, the idea of separating adjacency lists into two sublists in order to facilitate the location of unmatched neighbors is an appealing one, and is one that we utilize, in a modified form, in all of the remaining algorithms of this paper. To attempt to deal with the "toggling" problem discussed above, we relax the stringent condition that *all* of the changes of status be recorded immediately. The idea is to proceed with the computation, with the awareness that some of the information about the status of the neighbors of a vertex may be false. Thus, the algorithm tolerates some "inaccuracy" in the sublists. To make this work, there needs to be an efficient mechanism for testing whether the information read from the adjacency list (two sublists) is actually correct. This can be easily done in $\mathcal{O}(1)$ uniform running time: in a record associated with each vertex, one bit of information will suffice to record whether or not that vertex is an endpoint in the current maximal matching; and, each entry of an adjacency list (two sublists) may have an additional pointer that points to the record associated with the respective vertex.

It is also critical that there be a bound on the number of operations that the algorithm performs while tolerating inaccuracy. After this number of operations, call it $s$, the algorithm "cleans up" all of the inaccurate information, and then again proceeds in the described fashion until the next "clean–up", and so on. The "clean-up" may take longer than $\Theta(v + e)$, but if it is performed rather rarely, the cost may be amortized by charging to the operations performed between the two successive "clean-ups". *Insert* and *Delete* operations between two successive "clean-ups" may require $\Theta(s)$ time in order to deal with $\mathcal{O}(s)$ inaccurate information that may be present.

This is the essence of the **Dirty** strategy. As will be seen later, if $G$ is sparse, the **Dirty** strategy provides significant savings; whereas, if $G$ is dense, then the **Dirty** strategy will not be beneficial.

### 3.1.2   The Clean strategy

The **Clean** strategy performs *Insert* and *Delete* operations in precisely the manner described in the discussion of $A_0$, along with the additional requirement that, for each vertex, both adjacency sublists are maintained correctly from operation to operation. As noted before, $A_0$ provides generous savings if G *is* dense.

## 3.2   Incorporating both strategies

Recall that our goal is to offer savings over recomputation, regardless of whether $G$ is dense or sparse. And, as noted earlier, the density of $G$ may change in the course of performing many operations. Thus performing only one of the two strategies will not do, since a dense graph may well become sparse and vice versa. Therefore, in the algorithms that follow, we incorporate a certain level of supervision, such that, from time to time, the algorithm will pause and determine whether the next several operations will be performed in the **Clean** or **Dirty** way. This algorithm operates in stages, processing a stream of on–line requests for *Inserts/Deletes* and queries (*size*, *cover*, *incover(u)*).

We now proceed with a description of $B_1$. Here, we define a **stage** in the execution of $B_1$ as:

- the execution of two tests that will determine whether all of the *Insert* and *Delete* operations in that stage will be done in the **Clean** or **Dirty** manner;

- the execution of $v^a$, $a = \frac{3}{4}$ *Insert/Delete* operations, in an on–line fashion;

- in case the chosen strategy was **Dirty**, the "clean–up" of the inaccurate information in the adjacency sublists.

Next, we give the details of the tests performed at the beginning of each stage, and give a thorough description of the **Clean** and **Dirty** modes of operation.

Tests – two tests are performed. If either of the two tests succeeds, then the **Clean** strategy will be used. Conversely, if both tests fail, then the **Dirty** strategy will be used.

1. The first test asks whether there are at least $v^{b_2}$, $b_2 = \frac{2}{3}$ vertices of degree at least $v^{c_2}$, $c_2 = \frac{2}{3}$. If so, the **Clean** strategy should be used throughout the stage (i.e. for the processing of $v^a$, $a = \frac{3}{4}$ *Insert/Delete* operations). While performing this test, label the vertices meeting the degree requirement as *high*, and the other vertices as *low*. These labels will, in case both tests fail and the **Dirty** strategy is adopted, provide the information needed to properly execute the slightly inaccurate operations and the "clean–up".

2. The second test asks whether there are at least $v^{b_1}$, $b_1 = \frac{1}{2}$ vertices of degree at least $v^{c_1}$, $c_1 = \frac{5}{6}$. If so, then the **Clean** strategy should be used throughout the stage.

These tests may be implemented in time $\Theta(v)$. This does not increase the running time of the algorithm in either the **Clean** or **Dirty** mode of operation.

**Clean** – as noted before, we perform the operations in a manner almost identical to that of $A_0$, with one exception: since each vertex now has two adjacency sublists, recording the neighbors currently in the matching and not in the matching respectively, this information needs to be maintained rigorously, and upon the completion of each operation it should be correct. Note that the running time is $\Theta(v)$ per *Insert/Delete* operation, but in the case of **Clean**, this is permissible: $\Theta(v) = \Theta(e^{\frac{3}{4}})$. In addition, the relatively small number of operations in the stage cannot significantly change the fact that $G$ contains a relatively high number of edges: in case the first test succeeded, $v^a \ll v^{b_2+c_2}$, i.e. $\Theta(v^{b_2+c_2} \pm v^a) = \Theta(v^{b_2+c_2})$ (similar reasoning for the second test). The remaining issue of accounting for the running time required to perform the test(s) is resolved by assigning the entire cost ($\Theta(v)$) to the first operation in the stage.

**Dirty** – the very fact that **Dirty** is being executed implies that both tests failed. This provides some useful information. In particular, consider the following classification of vertices of $G$ at the beginning of such a stage:

- Vertices of degree at least $v^{c_1}$. We call the set of such vertices XL (extra large), and note that there are fewer than $v^{b_1}$ such vertices at the beginning of the stage. While performing the amortized analysis we conservatively assume that at the end of the stage, all XL vertices are of degree $v$.

- Vertices with degree at least $v^{c_2}$, but less than $v^{c_1}$. We call the set of such vertices L (large), and note that there are less than $v^{b_2}$ such vertices[3] at the beginning of the stage. While performing the amortized analysis we will conservatively assume that at the end of this stage, all L vertices are of degree $v^{c_1} + v^a = \Theta(v^{c_1})$. Note that the XL and L vertices are precisely the vertices that are labeled as *high* while performing the first test.

- All other vertices. We call the set of such vertices S (small), and note that these are vertices with degree less than $v^{c_2}$ at the beginning of the stage, thus labeled as *low*. Note that the degree of such vertices may grow within the stage to at most $\mathcal{O}(v^a)$.

With these preliminaries concluded, we proceed with a description of a **Dirty** stage. As noted, in a **Dirty** stage, $v^a$ *Inserts/Deletes* are performed. Both *Inserts* and *Deletes* depend upon the information about the status of the endpoints of an edge. That status (recall: matched/non–matched) is always maintained correctly in the record associated with each of the endpoints of the edge. The aforementioned "inaccuracy" pertains to the fact that some elements might belong to the wrong adjacency sublist.

To simplify our description of a **Dirty** stage, we divide *Inserts* and *Deletes* into two categories:

*Operations that do not change the status of any vertices* – this category includes both the *Inserts* of edges having at least one endpoint in the current maximal matching, and *Deletes* of non–matching edges. Since our goal is simply to maintain a maximal matching, in these cases no vertices should change their status. It follows that we need only add (for *Inserts*) or remove (for *Deletes*) one endpoint, call it $u$, to/from the other endpoint's, call it $w$, adjacency sublists and vice versa. In the case of insertion, $u$ should be added into the appropriate sublist according to the status of $u$. This needs to be done in a specific way: if $u$ is a *high* vertex, the algorithm insists that $u$ should be added at the front of the appropriate sublist (matched/non–matched); if $u$ is a *low* vertex, and it needs to be added to the non–matched sublist of $w$, then $u$ is added at the front of the sublist; whereas if it needs to be added to the matched sublist of $w$, then the algorithm insists that $u$ should be added at the end of the sublist.

In addition, we maintain an *update list* (doubly linked), for each vertex $x$, that identifies all the occurrences of $x$ in the adjacency lists throughout $G$. So, before completing the *Inserts* and *Deletes* in this category, the algorithm will add the newly created adjacency list entries to (remove the deleted adjacency list entries from) $u$'s and $v$'s respective update lists.

All of these actions may be performed in constant time.

*Operations that change the status of one or more vertices (at most two vertices per operation)* – this category includes both the *Inserts* of edges neither of whose endpoints are in the current maximal matching, and *Deletes* of matching edges.

---

[3] Note that this a very loose bound, since we did not exclude the XL vertices. It will turn out that this generosity does not affect the analysis, and we chose to keep matters simple.

First, we consider the *Insert* of an edge $h$ neither of whose endpoints ($u$ and $w$) is in the current maximal matching. Hence, $h$ needs to be added to the matching. To reflect the inclusion of $h$ in $G$, $u$ and $w$ are added into each other's adjacency sublists, and update lists. We do not however *necessarily* update the entries for $u$ and $w$ in their neighbor's sublists (to reflect that $h$ is now in the matching). This is the essence of the **Dirty** technique: first for $u$ and then for $w$, the algorithm checks whether $u$ ($w$) is labeled *low*. If so, the entire update list associated with $u$ ($w$) is scanned, and the status information at all of the neighbors' adjacency sublists is updated. If, on the other hand, $u$ ($w$) is labeled *high*, the sublists of the neighbors of $u$ ($w$) will not be updated. This inaccuracy in some adjacency sublists is the essential ingredient that leads to a reduced running time per operation. Finally, it follows from the vertex degrees that an *Insert* requires $\Theta(v^a)$ time.

Next, we consider a *Delete* of a matching edge $f$. Similarly to above, the adjacency sublists, and the update lists of $u$ and $w$ (the endpoints of $f$) are updated to reflect the deletion of $f$ from $G$. In addition, in order to maintain the matching, up to two edges may need to be added to the matching, and, of course, $f$ needs to be removed. To do this we first check if any of the remaining edges incident to $u$ and $w$ can be added to the matching. Checking for such edges is somewhat involved, since adjacency sublists may contain some inaccurate information. Thus, let $x$ be first $u$ and then $w$. The algorithm scans the adjacency sublists of $x$. Since the goal is to locate a non–matched vertex adjacent to $x$, the sublist of $x$ containing non–matched vertices is scanned first. This scan proceeds until a vertex $y$ is found that both claims to be, and truthfully is, a non–matched vertex. Note that when scanning $x$'s non–matched sublist, the only "liars" are *high* vertices, because *low* vertices are updated after each operation. It follows that one of two cases will occur: either a truthfully non–matched neighbor of $x$ will be found after scanning $\mathcal{O}(v^a)$ elements of the non–matched sublist of $x$, or the scanning of the entire non–matched sublist of $x$ will not yield a non–matched neighbor. Note that, in the latter case, the sublist is rather short ($\mathcal{O}(v^a)$). However, this does **not** necessarily mean that there is not a non–matched neighbor of $x$. There could be such a neighbor who is sitting in the matched sublist of $x$, but is a "liar". Here, our discipline for inserting *high* and *low* vertices into matched adjacency sublists comes to the rescue: *high* vertices are at the front of the matched sublist. Thus, the algorithm proceeds by scanning the matched sublist until either a "liar" is found, or a *low* vertex is encountered, or the entire sublist is scanned. It follows that no more than $\mathcal{O}(v^a)$ elements in both sublists need to be scanned until either a non–matched vertex has been found, or until it becomes certain that there is no non–matched neighbor of $x$.

If a non–matched neighbor $y$ of $x$ is located, then the edge $xy$ needs to be added to the matching by updating the status of $x$ and $y$. This is done precisely as described above for dealing with the insertion of an edge, neither of whose endpoints is in the matching. This means that if $x$ ($y$) is a *low* vertex, then a full update of the appropriate sublists is performed, and if $x$ ($y$) is a *high* vertex, then nothing further is done. Therefore, the total running time for operations of this kind is bounded by $\Theta(v^a)$.

We conclude our description of the processing in a **Dirty** stage, by considering the "clean–up" to be done at the end of the stage. This "clean–up" is simple: for each *high* vertex $u$, it suffices to scan the update list of $u$, and using the entries on that list, to place $u$ in the appropriate sublist of each of its neighbors. The time required is proportional to the degree of $u$. The overall time for this "clean–up" is thus $\mathcal{O}(v^{2a})$.

There is yet another "clean–up" that needs to be performed. Namely, in the process of executing insertions and deletions within a stage, a certain number of vertices ($\mathcal{O}(v^a)$) may, although initially labeled as *low*, actually become of high degree, and should be labeled as *high*, and vice versa. Thus, this needs to be rectified by "cleaning–up", i.e. placing the transient (*low* → *high*) vertices at the appropriate end of all the matched sublists where they occur. There can be $\mathcal{O}(v^a)$ such vertices. The overall time required for this "clean–up" in a **Dirty** stage is again $\mathcal{O}(v^{2a})$.

Note that we can charge the time required to perform the tests by simply adding it to the time for "clean–up", since $2a > 1$, i.e. $\mathcal{O}(v^{2a} + v) = \mathcal{O}(v^{2a})$.

The cost of the "clean–up", i.e. the incurred running time, can be charged to the preceding $v^a$ *Insert/Delete* operations, allocating a charge of $\mathcal{O}(v^a)$ to each operation, so the amortized running time of each operation in a **Dirty** stage of $B_1$ is $\mathcal{O}(v^a)$.

Summarizing, each operation in a **Clean** stage requires $\Theta(v)$ time, which is, due to the large number of edges, $\Omega(e^a)$, $a = \frac{3}{4}$. Also, each operation in a **Dirty** stage requires amortized time $\Theta(v^a)$, $a = \frac{3}{4}$. Hence, operations performed in both **Clean** and **Dirty** fashion achieve significant savings over recomputation. The same will be true of all the subsequent algorithms. Thus we have the following theorem:

**Theorem 2** $B_1$ *is a 2-competitive fully dynamic approximation algorithm for vertex cover (thus, it is approximation–competitive with the standard off–line algorithm). Further, both Insert and Delete operations require* $\Theta((v + e)^{\frac{3}{4}})$ *amortized running time.*

## 3.3   A slight improvement – Algorithm $A_1$

Our description of $B_1$ involved the constants $a$, $b_1$, $c_1$, $b_2$, and $c_2$. The values of these constants were carefully chosen to produce a time bound of $\Theta(v^a)$. In this section we describe algorithm $A_1$, a variation of $B_1$, that achieves a slightly improved running time through a different selection of the values of those constants. Further, we can show that this particular choice of the values of the constants is "the best possible"[4].

Due to space constraints, we furnish without proof:

---

[4] Note that this does not at all suggest that $A_1$ is the best among all algorithms. In the next section, we present a family of algorithms $A_k$, $k \geq 1$, all of which, except of course $A_1$ itself, are superior to $A_1$.

**Theorem 3** $A_1$ *is a 2-competitive fully dynamic approximation algorithm for vertex cover (thus, it is approximation-competitive with the standard off-line algorithm). Further, Insert operations, as well as Delete operations require* $\Theta((v+e)^{\frac{1+\sqrt{41}}{10}})$ *amortized running time. The choice of constants in $A_1$, i.e.* $a = \frac{1+\sqrt{41}}{10}$, $b_1 = 2a-1$, $c_1 = \frac{a+1}{2}$, $b_2 = \frac{3a-1}{2}$, *and $c_2 = a$, is optimal in the sense that no other choice of $a$, ... , $c_2$ can improve the running time.*

# 4  Further Improvements – Algorithms $A_k$

In this section we develop a family of fully dynamic algorithms $A_k$, $k \geq 1$, each of which improves upon the running time of the basic algorithm $A_1$.

The difference between $A_1$ and any $A_k$, $k > 1$, is only in the number of tests performed prior to the execution of a stage. In general, $A_k$ will perform $k+1$ tests of the form, $k+1 \geq i \geq 1$:

Are there at least $v^{b_i}$ vertices of degree at least $v^{c_i}$?

As before, an affirmative answer to any of the $k+1$ questions implies the execution of a **Clean** stage, whereas a negative answer to all of the questions implies the execution of a **Dirty** stage. As in $A_1$, during the computation of the first test, each vertex is labeled *high/low*. Aside from the use of the additional tests, each of these algorithms proceeds in a manner identical to that of $A_1$.

Our results are summarized in the following theorem, furnished without proof:

**Theorem 4** *For each integer $k > 0$, $A_k$ is a 2-competitive fully dynamic approximation algorithm for vertex cover (thus, it is approximation-competitive with the standard off-line algorithm). Further, Insert operations, as well as Delete operations performed by $A_k$ require* $\Theta((v+e)^{\frac{1+\sqrt{1+4(k+1)(2k+3)}}{2(2k+3)}})$ *amortized running time. The choice of constants in $A_k$, is optimal in the sense that no other choice of $a$, ... , $c_{k+1}$ can improve the running time. Here, the constants are $a = \frac{1+\sqrt{1+4(k+1)(2k+3)}}{2(2k+3)}$, $b_1$, ... , $c_{k+1}$, where $b_i$'s and $c_i$'s satisfy the following equations:*

$$b_1 + c_1 = \ldots = b_{k+1} + c_{k+1} = \frac{1}{a}, \qquad b_{k+1} + c_k = b_k + c_{k-1} = \ldots = b_1 + 1 = 2a,$$

$$and \quad c_{k+1} = a.$$

**Corollary 1** *For any $\alpha$, $\frac{\sqrt{2}}{2} < \alpha$ there is an amortized fully dynamic approximation algorithm $C_\alpha$ for vertex cover that is 2-competitive. Further, Insert and Delete operations performed by $C_\alpha$ require* $\mathcal{O}((v+e)^\alpha)$ *amortized running time.*

# References

[1] R. Bar–Yehuda and S. Even. (1985). A Local–Ratio Theorem for Approximating the Weighted Vertex Cover Problem. *Annals of Discrete Mathematics* **25**, pp. 27–46.

[2] K. L. Clarkson. (1983). A Modification of the Greedy Algorithm for Vertex Cover. *Information Processing Letters* **16**(1), pp. 23–25.

[3] D. Eppstein, Z. Galil, G. F. Italiano, and A. Nissenzweig. (1992). Sparsification – A Technique for Speeding up Dynamic Graph Algorithms. *Proceedings of the 33rd IEEE Symposium on Foundations of Computer Science*, pp. 60–69.

[4] G. Frederickson. (1985). Data Structures for On–Line Updating of Minimum Spanning Trees, with Applications. *SIAM Journal on Computing* **14**(4), pp. 781–798.

[5] Z. Galil, G. F. Italiano, and N. Sarnak. (1992). Fully Dynamic Planarity Testing. *Proceedings of the 24th ACM Symposium on Theory of Computing*, pp. 495–506.

[6] M. R. Garey and D. S. Johnson. (1979). *Computers and Intractability: A Guide to the Theory of NP–Completeness*. Freeman, San Francisco.

[7] F. Gavril. (1974). See [6, pp. 134]

[8] Z. Ivković and E. L. Lloyd. (1993). Fully Dynamic Algorithms for Bin Packing: Being (Mostly) Myopic Helps. *Proceedings of the 1st European Symposium on Algorithms*.

[9] R. M. Karp. (1972). Reducibility among Combinatorial Problems. In *Complexity of Computations* (R. E. Miller and J. W. Thatcher, Eds.), pp. 85–103. Plenum, New York.

[10] P. N. Klein and S. Sairam. (1993). Fully Dynamic Approximation Schemes for Shortest Path Problems in Planar Graphs. Manuscript.

[11] C. H. Papadimitriou and M. Yannakakis. (1991). Optimization, Approximation, and Complexity Classes. *Journal of Computer and System Sciences* **43**, pp. 425–440.

[12] C. Savage. (1982). Depth–First Search and the Vertex Cover Problem. *Information Processing Letters* **14**(5), pp. 233–235.

# Dynamic Algorithms for Graphs with Treewidth 2[*]

Hans L. Bodlaender

Department of Computer Science, Utrecht University
P.O. Box 80.089, 3508 TB Utrecht, the Netherlands

**Abstract.** In this paper, we consider algorithms for maintaining tree-decompositions with constant bounded treewidth under edge and vertex insertions and deletions for graphs with treewidth at most 2 (also called: partial 2-trees, or series-parallel graphs), and for almost trees with parameter $k$. Each operation can be performed in $O(\log n)$ time. For a large number of graph decision, optimization and counting problems, information can be maintained using $O(\log n)$ time per update, such that queries can be resolved in $O(\log n)$ or $O(1)$ time. Similar results hold for the classes of almost trees with parameter $k$, for fixed $k$.

## 1  Introduction

Two recently popular areas of investigations in graph algorithms are dynamic graph algorithms, and algorithms for graphs with small treewidth. In this paper, we consider dynamic algorithms for graphs with treewidth at most 2, (also known as *partial 2-trees* or *series-parallel graphs*.) These contain all outerplanar graphs.

Many problems become linear time solvable for graphs, given together with a tree-decomposition with treewidth bounded by some constant $k$. (Such a tree-decomposition can be found in $O(n)$ time [4] (see also [13, 15]).) These problems include many well known NP-complete problems, like Hamiltonian Circuit, Independent Set, Graph Coloring, etc., counting problems like *How many Hamiltonian circuits does $G$ have?* and some PSPACE-complete problems. Also, these problems when restricted to graphs with bounded treewidth belong to NC [3, 7, 12].

We consider the problem of solving these problems on graphs with treewidth at most 2 that change dynamically. We allow the following operations: insertion and deletion of isolated vertices, insertion of edges that do not result in a graph with treewidth larger than 2, and deletion of edges. One can check (using the results of this paper) in $O(\log n)$ time whether a desired edge insertion would yield a graph with treewidth at least 3. Also, when considering problems on weighted or labeled graphs, we allow operations that change the label or weight of a vertex or edge.

For a large class of graph decision, optimization, and counting problems, we show that data structures can be maintained, such that each such operation and queries to the problem can be performed in $O(\log n)$ time. These problems include almost all problems known to be linear time solvable on graphs with bounded treewidth.

Besides graphs with treewidth at most two, similar results also hold for the classes of graphs of *almost trees with parameter $k$*, for some constant $k$.

---

[*] This work was partially supported by the ESPRIT Basic Research Actions of the EC under contract 7141 (project ALCOM II).

*Related results.* In a recent paper, Cohen et al [8] designed algorithms for the maintenance of graphs with treewidth at most 2 or 3. For graphs with treewidth at most 2, insertions and deletions can be done in $O(\log^2 n)$ time, while queries cost $O(\log n)$ time. For graphs with treewidth at most 3, their data structure allows insertions (no deletions) to be performed in $O(\log n)$ amortized time. The class of graph problems that can be queried with their approach is much smaller than the class, dealt with in the present paper. The techniques used in [8] and this paper are quite different.

Frederickson [10, 11] found independently similar results for trees, forests, and $k$-terminal trees under several operations, including label changes. The technique in this paper for maintaining trees and forests is very similar to the technique used in [10, 11].

Fernandez-Baca and Slutzki [9] considered parametrized algorithms on graphs with bounded treewidth.

*Overview of this paper.* In section 2, definitions and some preliminary results are reviewed. In section 3, we discuss the class of query problems we can deal with. In section 4, we use 'parallel tree-contraction' to come to the data structure and algorithms that maintain suitable tree-decompositions of binary forests. In section 5, we discuss how the result of section 4 can be used to deal with larger classes of graphs, including the graphs with treewidth at most 2, and the almost trees with parameter $k$. Consequences of these results, open problems, and some other final remarks can be found in section 6.

## 2   Definitions and preliminary results

The notion of tree-width and tree-decomposition were introduced by Robertson and Seymour in their series of fundamental papers on graph minors[16].

**Definition 1.** A *tree-decomposition* of a graph $G = (V, E)$ is a pair $(\{X_i \mid i \in I\}, T = (I, F))$ with $\{X_i \mid i \in I\}$ a collection of subsets of vertices, and $T$ a tree, such that

- $\bigcup_{i \in I} X_i = V$.
- $\forall (v, w) \in E : \exists i \in I : v, w \in X_i$.
- $\forall v \in V : \{i \in I \mid v \in X_i\}$ induced a connected subtree of $T$.

The treewidth of tree-decomposition $(\{X_i \mid i \in I\}, T = (I, F))$ is $\max_{i \in I} |X_i| - 1$. The treewidth of a graph $G$ is the minimum treewidth over all possible tree-decompositions of $G$.

The third condition can be equivalently be replaced by: for all $i, j, k \in I$, if $j$ lies on the path from $i$ to $k$ in $T$, then $X_i \cap X_k \subseteq X_j$. There are several notions, equivalent to treewidth, e.g. a graph is a 'partial $k$-tree', iff its treewidth is at most $k$ (see e.g. [17]). We say a tree-decomposition is *nice*, if $T$ is a rooted binary tree, with for root node $r$: $X_r = \emptyset$. It is easy to transform each tree-decomposition into a nice one with the same treewidth.

A $\leq k$-boundary graph $G = (V, E, B)$ is a 3-tuple, with $(V, E)$ an undirected graph, and $B$ a set of at most $k$ vertices in $V$. Consider a nice tree-decomposition

$(\{X_i \mid i \in I\}, T = (I, F))$ of treewidth $k - 1$. For $i \in I$, let $Y_i = \{v \in X_j \mid j = i$ or $j$ is a descendant of $i\}$. Write $G[Y_i] = (Y_i, E_i)$. For all $i \in I$, $(Y_i, E_i, X_i)$ is a $\leq k$-terminal graph.

Many linear time algorithms that solve problems on graphs with a given (nice) tree-decomposition can be expressed in the following way (after 'cosmetical changes'):

For each node $i \in I$, some information about $(Y_i, E_i, X_i)$ is computed. Denote this information by $\alpha(i)$. Each value $\alpha(i)$ can be computed in $O(1)$ time when given all values $\alpha(j)$ for all children $j$ of $i$. Thus, the values can be computed in $O(n)$ time in total, working bottom-up in the decomposition tree. From $\alpha(r)$, the answer to the problem that is to be solved can be determined in $O(1)$ time. In case each $\alpha(i)$ contains only a constant bounded number of bits, we call the problem *finite state* (after Abrahamson and Fellows [1].)

Next, we review some useful results on the structure of graphs with treewidth at most 2. A graph has treewidth at most 1, if and only if it is a forest.

Consider a graph $G = (V, E)$ with treewidth at most 2. Let $H = (V, E')$ be the graph, obtained by adding to $G$ all edges $(v, w) \notin E$ for all pairs $v, w \in V$, $v \neq w$, for which there are at least three vertex disjoint paths between $v$ and $w$ in $G$. $H$ has also treewidth $\leq 2$, and is called the *cell completion* of $G$. $H$ is the cell completion of itself. We say a graph $G$ is completed, if it is the cell completion of a graph with treewidth at most 2.

A completed graph $H$ has a nice, and relatively easy structure: each biconnected component either consists of a single edge, or can be made in the following way: start with a simple cycle, and then iteratively add zero or more cycles, each new cycle sharing exactly one edge with the part of the component made so far. (This characterization is due to Ton Kloks. See e.g. [5].) We say an edge $(v, w)$ in a completed graph $H$ is a *base edge*, if, besides the edge $(v, w)$, there are two other vertex disjoint paths from $v$ to $w$ in $H$. (It is an edge where cycles were 'glue-ed together'.)

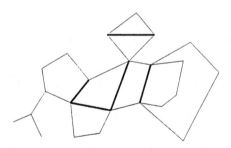

**Fig. 1.** A completed graph with base edges

A graph $G = (V, E)$ is an *almost tree with parameter* $k$, if $G$ has a spanning forest $T = (V, F)$, such that each biconnected component of $G$ contains at most $k$ edges not in $T$. A graph is a *cactus*, if it is an almost tree with parameter 1. Equivalently, if no edge belongs to more than one simple cycle. The treewidth of an almost tree with parameter $k$ is at most $k + 1$.

# 3   Td-open problems

Suppose we have a dynamically changing graph $G = (V, E)$ with a tree-decomposition that may change dynamically too. When solving problems (like Hamiltonian Circuit, Independent Set) on $G$, it is desirable that only few values $\alpha(i)$ (see section 2) need to be recomputed during a change of $G$ and its tree-decomposition. We will see that for many problems, the update time can made linear in the number of *affected* nodes, where *affected* is defined as follows:

- In case of a change of a weight or a label of a vertex $v \in V$, or an edge $(w, x) \in E$, let $i \in I$ be the highest node in the tree containing $v$, or containing both $w$ and $x$, respectively. All predecessors of $i$ (not $i$ itself) are *affected*.
- In case of a change of $G = (V, E)$ to $G' = (V', E')$ and of its tree-decomposition $(\{X_i \mid i \in I\}, T = (I, F))$ to tree-decomposition $(\{X_i' \mid i \in I'\}, T' = (I', F'))$ of $G'$, the following nodes in $I'$ are *affected*:
  - All nodes in $I'$, not in $I$.
  - All nodes $i \in I \cap I'$ whose subtree of $T$, rooted at $i$ is not identical to the subtree of $T'$, rooted at $i$. This includes the case that for some descendant $j \in I \cap I'$ of $i$, $X_j \neq X_j'$.
  - All nodes $i \in I'$, for which there are vertices $v, w \in V \cap V'$ with $(v, w) \in E - E' \cup E' - E$ and $i$ is a predecessor of the highest node $j$ in $T$ with $v, w \in X_j$.

Note that any predecessor of an affected node is also affected. Graph problems are described as functions, that map graphs $G = (V, E)$ with vertex and edge labelings $lab_V : V \to L_V$, $lab_E : E \to L_E$ to some set of answers.

**Definition 2.** A graph problem $P$ is *td-open*, if there exists a function $\alpha$, that maps nodes in nice tree-decompositions of the input graph to a value in some set $S$, such that

- for each node $i$, $\alpha(i)$ can be computed in $O(1)$ time, given all values $\alpha(j)$ for all children $j$ of $i$.
- the desired answer $P(G, lab_V, lab_E)$ can be determined in $O(1)$ time from $\alpha(r)$ (where $r$ is the root node of the decomposition tree).
- when $G$, $lab_V$, $lab_E$ and/or the tree-decomposition is changed, then $\alpha(i)$ is not changed for nodes that are not affected by the change.

If $P$ is td-open, then maintaining the necessary information to solve $P$ can be done in time, proportional to the number of affected nodes.

A powerful language to express many linear time solvable problems on graphs with given tree-decomposition of bounded treewidth is the Monadic Second Order Logic (MSOL) and its extensions. See e.g. [2, 6].

**Theorem 3.** *(i) If a problem $P$ is finite state, then $P$ is td-open.*
*(ii) If a problem $P$ is expressible in extended monadic second order logic, then $P$ is td-open.*

*Proof.* (i) The basic idea is: $\alpha(i)$ contains all states (as in the original algorithm) for all possible labelings of $X_i$, $E[X_i]$, and all possible sets of edges in $X_i \times X_i$. This information is still an element from a finite set, can be computed in $O(1)$ time from the information from the children of $i$, and does not change under updates that do not affect $i$.
(ii) By simple modification of the algorithm in [6]. □

Also, counting versions of finite state or MSOL problems (like counting the number of Hamiltonian Circuits) are td-open. Most other problems, solvable in linear time on graphs with given tree-decomposition of bounded treewidth can also shown to be td-open. Many td-open problems are listed in section 6. Clearly, problems that can have output of non-constant size (like: output a set of vertices that is a maximum independent set) cannot be td-open.

## 4 Maintenance of tree-decompositions of binary trees with logarithmic depth

**Definition 4.** Let $\mathcal{G}$ be a class of graphs, and let $op$ be a set of operations $\mathcal{G} \to \mathcal{G}$. We say that $(\mathcal{G}, op)$ can be *strongly td-maintained* with cost $O(f(n))$, if data-structures and algorithms exist, which maintain a nice tree-decomposition of a dynamically changing (by operations in $op$) graph $G \in \mathcal{G}$, such that

- the depth of the decomposition tree is $O(\log n)$.
- there is a uniform upper bound on the treewidth of the tree-decompositions.
- each operation in $op$ can be executed in $O(f(n))$ time, such that there are $O(f(n))$ affected nodes.
- for each vertex $v \in V$, there are at most $O(f(n))$ nodes $i$ with $v \in X_i$ or $i$ a predecessor of a node $j$ with $v \in X_j$.

If the last condition does not hold, we say that $(\mathcal{G}, op)$ is *td-maintained* with cost $O(f(n))$.

**Theorem 5.** *Suppose $(\mathcal{G}, op)$ can be td-maintained with cost $O(f(n))$. For each finite set of td-open problems on (labeled) graphs in $\mathcal{G}$, a data-structure exists, that allows operations in $op$, label changes of a vertex and/or edge, and queries to the problems to be executed in $O(f(n))$ time.*

Usually, queries cost even only $O(1)$ time. In this section we show that binary forests with deletions, and insertions of and isolated vertices (and several other 'local' operations) can be td-maintained with cost $O(\log n)$..

We use the notations $del_{\mathcal{G}}$, $ins_{\mathcal{G}}$, $delv_{\mathcal{G}}$, $insv_{\mathcal{G}}$, $compr_{\mathcal{G}}$, $contr_{\mathcal{G}}$, $subdiv_{\mathcal{G}}$ for the operations: delete an edge, insert an edge, delete an isolated vertex, insert an isolated vertex, compress over a vertex of degree 2 (remove the vertex and connect its two old neighbors), contract over an edge, subdivide an edge, in each case provided that the resulting graph belongs to $\mathcal{G}$. For a class of graphs $\mathcal{G}$, denote $OP_{\mathcal{G}}^- = \{deleg,$ $inseg, delv_{\mathcal{G}}, insv_{\mathcal{G}}, compr_{\mathcal{G}}, subdiv_{\mathcal{G}}\}$, and denote $OP_{\mathcal{G}} = OP_{\mathcal{G}}^- \cup \{contr_{\mathcal{G}}\}$.

We base our approach on applying parallel tree-contraction, as introduced by Miller and Reif [14]. In [3] tree-contraction was used in the first proof of membership in NC of many problems on graphs with constant bounded treewidth. (Later results, especially those of Lagergren [12] gave very large savings in the number of used processors.) We use a version of tree-contraction that is most suitable for our purposes.

Define the following operations on forest $T$ with maximum vertex degree 3.

- RAKE. Remove a leaf node $v$, and its adjacent edge $(v, w)$. We say that $v$ and $(v, w)$ are *involved* in this operation and that $w$ represents $v$ and $(v, w)$ after the rake.
- COMPRESS. Take a node $v$ with degree 2. Let $w$ and $x$ be the neighbors of $v$. Replace $v$, $(v, w)$, and $(v, x)$ by an edge $(w, x)$. We say that $v$, $(v, w)$, and $(v, x)$ are *involved* in the operation and that $(w, x)$ *represents* $v$, $(v, w)$, $(v, x)$.
- 0-REMOVE. Remove an isolated vertex $v$ from $T$. $v$ is *involved* in this operation.

A set of rake, compress and 0-remove operations is said to be a *good RC-set*, if no vertex or edge is involved in more than one operation, and it is maximal, i.e., every vertex of degree 0 is 0-removed, and every vertex of degree 1 or 2 is adjacent to an edge that is involved.

A contraction series of a forest $T = (V, E)$ is a sequence $(T_0,\ S_0, T_1,\ S_1, \ldots, T_{r-1},$ $S_{r-1},\ T_r)$ with $T_0 = T$, $T_r$ the empty graph, and each $T_i$ $(i \geq 1)$ is the forest, obtained by applying the good RC-set $S_{i-1}$ to forest $T_{i-1}$. Write $T_i = (V_i, E_i)$.

**Lemma 6.** *The length $r$ of a contraction series $(T_0,\ S_0,\ T_1, \ldots, T_r)$ of a forest $T = (V, F)$ is $O(\log |V|)$.*

*Proof.* Similar as in [14]. $\qquad\qquad\qquad\qquad\qquad\qquad\qquad\qquad\qquad\qquad\qquad\qquad$ $\square$

Given a contraction series $(T_0,\ S_0,\ T_1,\ \ldots,\ T_r)$ of a binary tree $T = T_0$, we can build a tree-decomposition $(\{X_i \mid i \in I\}, \mathcal{T} = (I, F))$ of $T$ with treewidth 2 with $\mathcal{T}$ a tree of depth $r$ with degree at most 6, in the following way. Let $I$ be the disjoint union of $V_0$, $V_1, \ldots, V_r$, $E_0, E_1, \ldots, E_r$. (We use superscripts to denote to what $V_j$, $E_j$ a vertex or edge belongs.) If a vertex $v^j \in V_j$ (edge $(v, w)^j \in E_j$) is not involved in an operation in $S_j$, then $v^{j+1}$ $((v, w)^{j+1})$ has one child, namely $v^j$ $((v, w)^{j+1})$, and $X_{v^{j+1}} = \{v\}$ $(X_{(v,w)^{j+1}} = \{v, w\})$. If a rake on $v^j$, $(v, w)^j$ is done in $S_j$, then the children of $w^{j+1}$ are those vertices and edges in $V_j \cup E_j$ that are represented by $w$ (at most two vertices and two edges, if both children of $w$ are raked). $X_{w^{j+1}}$ contains $w$ and $v$, and possibly the other child of $w$ if that is also raked in $S_j$. If a compress on $v$, $(v, w)$, $(v, x)$ is done, then $(w, x)$ has three children: $v^j$, $(v, w)^j$, and $(v, x)^{j+1}$.

Using the modification described in section 2, $\mathcal{T}$ can be made binary, while its depth increases with only a constant factor. Extending this technique to binary forests can be done without much problems by building for each tree a tree-decomposition as above, and then using extra nodes $i$ with $X_i = \emptyset$, which are at the top of the resulting tree and together have logarithmic depth.

Let $\mathcal{F}_3$ denote the set of all forests with maximum vertex degree 3. The reason why the construction described above is useful for dynamic algorithms is the following theorem, which can be shown by an extensive case analysis.

**Theorem 7.** *There exists a finite set $X$ of operations $\mathcal{O}$, each involving a constant bounded number of vertices and edges, which contains all operations in $OP_{\mathcal{F}_3}$, such that for every forest $T_0 = (V_0, E_0)$, for every forest $T_0' = (V_0', E_0')$ obtained by applying an operation in $\mathcal{O}$ to $T_0$, and for every forest $T_1 = (V_1, F_1)$ that is obtained by applying a good RC-set $S$ to $T_0$, there exists a good RC-set $S'$ on $T_0'$, such that the forest $T_1'$ obtained by applying $S'$ to $T_0'$ can also be obtained by applying one operation from $\mathcal{O}$ to $T_1$.*

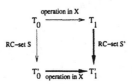

**Fig. 2.** Graphical representation of theorem 7

**Theorem 8.** $(\mathcal{F}_3, OP_{\mathcal{F}_3})$ *can be strongly td-maintained with cost $O(\log n)$.*

*Proof.* Maintain a contraction series of the forest, making repetive use of theorem 7. Changes in the contraction series are directly reflected in the corresponding tree-decompositions. A 'small change' in the forest will affect only a constant bounded vertices and edges per forest in the contraction series, hence $O(\log n)$ nodes will be affection in total with an operation. Strongness follows as each vertex belongs to $O(1)$ nodes per level of the decomposition tree. □

## 5 Larger classes of graphs

In this section, we show for larger classes of graphs that they can be td-maintained (with the usual operations) with $O(\log n)$ cost.

First, we consider the class $\mathcal{CAC}_3$ of the cactus graphs with maximum vertex degree 3. We maintain a maximal spanning forest $\mathbf{TI} = (V, E')$ of cactus $G = (V, E)$. If we have a nice tree-decomposition $(\{X_i \mid i \in I\}, \mathcal{T} = (I, F))$ of $\mathbf{T}$ with treewidth 2, we can make a nice tree-decomposition of $G$ in the following way: for each edge not in the spanning forest $(v, w) \in E - E'$, add either $v$ or $w$ to all nodes in $I$ on the path in $\mathcal{T}$ between the highest node that contains $v$ and the highest node that contains $w$ (inclusive). One can show that this gives a tree-decomposition of $G$ with treewidth at most 5. Simple analysis of the different cases show that each insertion or deletion can be done in $O(\log n)$ time, such that also $O(\log n)$ nodes are affected. As each other operation in $OP_{\mathcal{CAC}_3}$ can be expressed in $O(1)$ insertions and/or deletions, we have:

**Lemma 9.** $(\mathcal{CAC}_3, OP_{\mathcal{CAC}_3})$ *can be td-maintained with cost* $O(\log n)$.

The main technique to come to larger classes of graphs is 'interpreting' a graph into another graph from a 'simpler' class of graphs.

**Definition 10.** An *interpretation* of a graph $G = (V_G, E_G)$ into a graph $H = (V_H, E_H)$ is a function $f : V_G \to \mathcal{P}(V_H)$, mapping each vertex $v \in V_G$ to a set of vertices $\subseteq V_H$, such that for all $v \in V$, $f(v)$ induces a connected subgraph in $H$, and for all $(v, w) \in E_G$: there exist $x \in f(v)$, $y \in f(w)$ with $x = y$ or $(x, y) \in E_H$.

The *inverse* of interpretation $f$ is a pair $(f^{-1,V}, f^{-1,E})$ with $f^{-1,V} : V_H \to \mathcal{P}(V_G)$ defined by $f^{-1,V}(v) = \{w \in V_G \mid v \in f(w)\}$, and $f^{-1,E} : E_H \to \mathcal{P}(E_G)$ defined by $f^{-1,E}((x,y)) = \{(v, w) \in E_G \mid x \in f(v) \text{ and } y \in f(w)\}$.

The *width* of interpretation $f$ is $w(f) = \max_{x \in V_H} |f^{-1,V}(x)|$. The *breadth* of interpretation $f$ is $b(f) = \max_{v \in V_G} |f(v)|$.

**Lemma 11.** *Let* $(\{X_i \mid i \in I\}, T = (I, F))$ *be a tree-decomposition of* $H = (V_H, E_H)$ *with treewidth* $k$, *and let* $f : V_G \to V_H$ *be an interpretation of* $G = (V_G, E_G)$ *into* $H$. *Then* $(\{Y_i \mid i \in I\}, T = (I, F))$ *with* $Y_i = \bigcup_{x \in X_i} f^{-1,V}(x)$ *is a tree-decomposition of* $G$ *with treewidth at most* $w(f) \cdot (k + 1) - 1$.

We say that a class of graphs $\mathcal{G}$ with set of operations $\mathcal{G} \to \mathcal{G}$, $op_\mathcal{G}$ can be *interpreted* with width $c$ (and breadth $b$) into a class of graphs $\mathcal{H}$ with set of operations $\mathcal{H} \to \mathcal{H}$, $op_\mathcal{H}$, if for each $G \in \mathcal{G}$, we have a non-empty collection $C_G$ of pairs $(H, f)$, with $H \in \mathcal{H}$, and $f$ an interpretation of $G$ into $H$ with width at most $c$ (and breadth at most $b$), such that for all $G \in \mathcal{G}$, $(H, f) \in C_G$, and $G' \in \mathcal{G}$ obtained by applying one operation from $op_{calG}$ to $G$, there exists a sequence $opseq$ of operations in $op_\mathcal{H}$ with its length bounded by some constant, such that when the operations in $opseq$ are sequentially applied to $H$, a graph $H'$ results, and there exists an interpretation $f'$ of $G'$ into $H$ with $(H', f') \in C_{G'}$. Moreover $f^{-1,V}$ and $(f')^{-1,V}$ must be identical for all vertices in $H \cap H'$ that are not involved in one or more operations in $opseq$, and the time, needed to find the sequence $opseq$ must be at most $O(\log n)$, plus the time needed for solving $O(1)$ td-open problems on $G$ and/or on $H$. (As tree-decompositions of $G$ and $H$ are maintained, the time for solving these td-open problems will be also $O(\log n)$.)

**Lemma 12.** *Let* $\mathcal{G}, \mathcal{H}$ *be classes of graphs, and* $op_\mathcal{G}$, $op_\mathcal{H}$ *be finite sets of operations on graphs in* $\mathcal{G}, \mathcal{H}$, *respectively, where each operation in these classes never involves more than a constant bounded number of vertices and edges.*
*(i) If for constants* $c, b \in N$, $(\mathcal{G}, op_\mathcal{G})$ *can be interpreted into* $(\mathcal{H}, op_\mathcal{H})$ *with width* $c$ *and breadth* $b$, *and* $(\mathcal{H}, op_\mathcal{H})$ *can be strongly td-maintained with cost* $O(\log n)$, *then* $(\mathcal{G}, qp_\mathcal{G})$ *can be strongly td-maintained with cost* $O(\log n)$.
*(ii) If for constant* $c \in N$, $(\mathcal{G}, op_\mathcal{G})$ *can be interpreted into* $(\mathcal{H}, op_\mathcal{H})$ *with width* $c$, *and* $(\mathcal{H}, op_\mathcal{H})$ *can be strongly td-maintained with cost* $O(\log n)$, *then* $(\mathcal{G}, op_\mathcal{G})$ *can be td-maintained with cost* $O(\log n)$.

*Proof.* $C_G$ denotes the possible (graph - interpretation) pairs for $G$. We maintain one such pair. Then the tree-decomposition of $G$ is made from the tree-decomposition of $H$ as in lemma 5.2. One operation in $G$ translates to $O(1)$ operations to its 'interpreted graph' and interpretation, resulting in $O(\log n)$ changes to the tree-decompositions of $G$ and $H$. We maintain also the '$\alpha$-information' of the td-open problems on $G$ and $H$, necessary to find the operation sequences *opseq*. □

Let $\mathcal{F}$ ($\mathcal{F}_c$) denote the class of forests (with maximum vertex degree $c$); let $\mathcal{CAC}$ ($\mathcal{CAC}_c$) denote the class of cactus graphs (with maximum vertex degree $c$). The following lemma illustrates our technique.

**Lemma 13.** *($\mathcal{F}_4, OP_{\mathcal{F}_4}$) can be interpreted into ($\mathcal{F}_3, OP_{\mathcal{F}_3}$) with width 1 and breadth 2.*

*Proof.* Each forest $T = (V, F) \in \mathcal{F}_4$ is interpreted into forests $T'$, where every vertex of degree 4 in $T$ is replaced by two adjacent vertices of degree 3. See for an example figure 3. Adding an edge $(v, w)$ to $T$ can be done as follows: if the degree of $v$ is

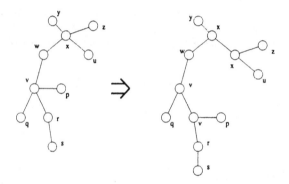

**Fig. 3.** Interpreting a tree with degree 4 into a tree with degree 3

3, then subdivide one edge adjacent to $f(v)$, put the new vertex also in $f(v)$, and attach the edge to be added to the new vertex. Similar for $f(w)$. Thus, the operation sequence consists of at most two subdivisions and one edge addition.

When deleting an edge $(v, w)$ from $T$, first delete its counterpart, say $(x, y)$ from $T'$. Check whether $x$ now has degree 2 and is adjacent to a vertex $z$ with $f^{-1,V}(x) = f^{-1,V}(z)$. If so, compress over $x$. Do the same for $y$.

The other operations are straight-forward. □

It follows that $(\mathcal{F}_4, OP_{\mathcal{F}_4})$ can be strongly td-maintained. We give many similar results without proof, or with only an indication of the used interpretations.

**Lemma 14.** *(i) For all $c \geq 4$, ($\mathcal{F}_c, OP_{\mathcal{F}_c}$) can be strongly interpreted into ($\mathcal{F}_3, OP_{\mathcal{F}_3}$) with width 1 and breadth $c - 2$.*
*(ii) ($\mathcal{F}, OP_{\mathcal{F}} - \{contr_{\mathcal{F}}\}$) can be interpreted into ($\mathcal{F}_3, OP_{\mathcal{F}_c}$) with width 1.*
*(iii) For all $c \geq 4$, ($\mathcal{CAC}_c, OP_{\mathcal{CAC}_c}$) can be strongly interpreted into ($\mathcal{CAC}_3, OP_{\mathcal{CAC}_3}$) with width 1 and breadth $c - 2$.*
*(iv) ($\mathcal{CAC}, OP_{\mathcal{CAC}}^-$) can be interpreted into ($\mathcal{CAC}_3, OP_{\mathcal{CAC}_3}$) with width 1.*

Let $COM$ ($COM_c$) be the class of the completed graphs (with maximum vertex degree $c$). We say a completed graph is *cactaceous*, if no two base edges share a vertex. $COMC_c$ denotes the cactaceous completed graphs with maximum vertex degree $c$. $TW_2$ denotes the class of graphs with treewidth at most 2.

**Lemma 15.** $(COMC_4, OP_{COMC_4})$ can be interpreted into $(CAC_6, OP_{CAC_6})$ with width 2 and breadth 1.

*Proof.* If we contract each edge in a cactaceous completed graph, then we obtain a cactus. □

**Lemma 16.** (i) $(COM_4, OP_{COM_4})$ can be interpreted into $(COMC_4, OP_{COMC_4})$ with width 1 and breadth 2.
(ii) $(COM, OP_{COM}^-)$ can be interpreted into $(COM_4, OP_{COM_4})$ with width 1.

*Proof.* (i) Split a vertex which is adjacent to two base edges. No vertex can be adjacent to more than two base edges in a graph in $COM_4$. See figure 5.
   (ii) Each base edge is replaced by a ladder, as shown in figure 5. Vertices that are adjacent to more than one biconnected component may be split, similar as in lemma 13. □

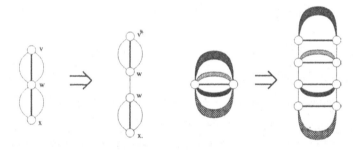

**Fig. 4.** Interpretations of lemma 16

**Lemma 17.** (i) $(TW_2, OP_{TW_2}^-)$ can be interpreted into $(COM, OP_{COM}^-)$ with width 1 and breadth 1.
(ii) $(TW_2, OP_{TW_2}^-)$ can be interpreted into $(COM_4, OP_{COM_4})$ with width 1.

*Proof.* (i) We interprete a graph $G$ with treewidth $\leq 2$ into its completed graph $H$. E.g., when adding an edge to $G$, at most one base edge may have to be added to the completed graph. The problem to determine if a new base edge must be added, and if so, to find the endpoints of such a new base edge can be seen to be td-open, using a formulation of the problem in monadic second order logic. Similarly, when an edge is deleted, it may be possible that one extra edge (a base edge) disappears from the completed graph. Again, this edge can be found if necessary by solving a td-open problem, (hence in $O(\log n)$ time).

(ii) By combining techniques used in lemma 16(ii) and part (i) of this theorem.

□

**Corollary 18.** $(\mathcal{TW}_2, OP^-_{\mathcal{TW}_2})$ *can be td-maintained with cost $O(\log n)$.*

The treewidth of the resulting tree-decompositions is at most 11. Using similar methods, one can also obtain the following result:

**Theorem 19.** *Let $ALMOST_k$ denote the class of almost trees with parameter $k$. $(ALMOST_k, OP^-_{ALMOST_k})$ can be td-maintained with cost $O(\log n)$.*

# 6 Conclusions

By combining the results of the earlier sections, we get the following result. In all cases, directed and mixed graphs can be handeled as undirected graphs with a direction labeling on the edges.

**Theorem 20.** *For each problem $P$ from table 1, there exists a data-structure that allows the following operations on graphs with treewidth $\leq 2$:*

- *delete an isolated vertex*
- *add an isolated vertex*
- *delete an edge*
- *add an edge such that the treewidth of the resulting graph is at most 2*
- *check for a given pair of vertices whether adding this edge would increase the treewidth to larger than 2*
- *change the label or weight of a vertex or edge (in case $P$ is a problem on labeled or weighted graphs)*
- *query (solve $P$ on current graph)*

*such that each operation can be carried out in $O(\log n)$ time. Queries cost $O(1)$ or $O(\log n)$ time, as described in the table. When given a graph $G$ with treewidth $\leq 2$, the data-structure can be build in $O(n)$ time.*

The same result holds, when we use almost trees with parameter $k$ for some fixed $k$, instead of graphs with treewidth at most 2.

An interesting open problem is to extend these results to graphs with treewidth at most $k$, for constant $k$. Using techniques from this paper (especially those in sections 3 and 4) and some older results, one can build a data-structure, that supports queries to a td-open problem $P$, and has the following time bounds:

- deleting an isolated vertex: $O(\log n)$ amortized, $O(n)$ worst case
- adding an isolated vertex: $O(\log n)$
- deleting an edge: $O(\log n)$
- adding an edge such that the treewidth of the resulting graph is at most $k$: $O(n)$ (just rebuild the data structure from scratch ...)

**Problems with $O(1)$ query time:**
Vertex cover, dominating set. domatic number, chromatic number (graph coloring), achromatic number for fixed $k$, monochromatic triangle, feedback vertex set, feedback edge set, feedback arc set, partial feedback edge set, minimum maximal matching, partition into triangles, partition into isomorphic subgraphs for fixed $H$, partition into Hamiltonian subgraphs, partition into forests, partition into cliques, partition into perfect matchings for fixed $k$, covering by cliques, covering by complete bipartite subgraphs, clique, independent set, induced subgraph with property $P$ (for monadic second order properties $P$), induced connected subgraph with property $P$ (for monadic second order properties $P$), balanced complete bipartite subgraph, bipartite subgraph, degree-bounded connected subgraph for fixed $k$, planar subgraph, transitive subgraph, uniconnected subgraph, Hamiltonian completion, Hamiltonian path, Hamiltonian circuit, directed Hamiltonian circuit (path), subgraph isomorphism for fixed $H$, induced subgraph isomorphism for fixed $H$, path with forbidden pairs for fixed $n$, multiple choice matching for fixed $J$, kernel, $k$-closure, path distinguishers, degree constraint spanning tree, maximum leaf spanning tree, bounded diameter spanning tree for fixed $d$, $k$th best spanning tree for fixed $k$, bounded component spanning forest for fixed $k$, multiple choice branching for fixed $m$, Steiner tree in graphs, maximum cut, minimum cut into bounded sets, Chinese postman for mixed graphs, Stacker-crane, rural postman, longest circuit, chordal graph completion for fixed $k$, chromatic index for fixed $k$, spanning tree parity problem, distance $d$ chromatic number for fixed $d$ and $k$, thickness $\leq k$ for fixed $k$. membership for each class $C$ of graphs that is closed under minor taking, vertex generalized geography, maximum matching, minimum spanning tree, outerplanarity, connectivity, biconnectivity, strong connectivity, triangulating colored graphs with 3 colors, (and counting variants of many of the problems described above) ...

**Problems with $O(\log n)$ query time:** *(endpoint of paths etc. can be specified with the query):*
longest path, shortest path, Hamiltonian path between specified endpoints, k'th shortest path for fixed $k$, disjoint connected paths for fixed $k$, maximum length-bounded disjoint paths for fixed $J$, maximum fixed length disjoint paths for fixed $J$, minimum cut between given endpoints, do $s$ and $t$ belong to the same: connected component, biconnected component, ... (and counting variants of many of the problems described above) ...

(See among others [2, 6].)

**Table 1.** List of problems

- checking for a given pair of vertices whether adding this edge would increase the treewidth to larger than $k$: $O(\log n)$
- changing the label or weight of a vertex or edge: $O(\log n)$
- queries: $O(1)$
- building time for given graph $G$ with treewidth $\leq k$: $O(n)$ (with use of result in [4] and tree-contraction)

Possible improvements may well be possible here. Most challenging seems to bring down the edge insertion time. Restrictions to planar graphs, or graphs with bounded vertex degree (while still assuming an upper bound on the treewidth) seem also interesting, and may be easier to solve.

# References

1. K. R. Abrahamson and M. R. Fellows. Finite automata, bounded treewidth and well-quasiordering. In *Graph Structure Theory, Contemporary Mathematics vol. 147*, pages 539–564. American Mathematical Society, 1993.
2. S. Arnborg, J. Lagergren, and D. Seese. Easy problems for tree-decomposable graphs. *J. Algorithms*, 12:308–340, 1991.
3. H. L. Bodlaender. NC-algorithms for graphs with small treewidth. In J. van Leeuwen, editor, *Proc. Workshop on Graph-Theoretic Concepts in Computer Science WG'88*, pages 1–10. Springer Verlag, Lecture Notes in Computer Science, vol. 344, 1988.
4. H. L. Bodlaender. A linear time algorithm for finding tree-decompositions of small treewidth. In *Proceedings of the 25th Annual Symposium on Theory of Computing*, pages 226–234. ACM Press, 1993.
5. H. L. Bodlaender and T. Kloks. A simple linear time algorithm for triangulating three-colored graphs. *J. Algorithms*, 15:160–172, 1993.
6. R. B. Borie, R. G. Parker, and C. A. Tovey. Automatic generation of linear-time algorithms from predicate calculus descriptions of problems on recursively constructed graph families. *Algorithmica*, 7:555–582, 1992.
7. N. Chandrasekharan. *Fast Parallel Algorithms and Enumeration Techniques for Partial k-Trees*. PhD thesis, Clemson University, 1990.
8. R. F. Cohen, S. Sairam, R. Tamassia, and J. S. Vitter. Dynamic algorithms for bounded tree-width graphs. Technical Report CS-92-19, Department of Computer Science, Brown University, 1992.
9. D. Fernández-Baca and G. Slutzki. Parametic problems on graphs of bounded treewidth. In O. Nurmi and E. Ukkonen, editors, *Proceedings 3rd Scandinavian Workshop on Algorithm Theory*, pages 304–316. Springer Verlag, Lecture Notes in Computer Science, vol. 621, 1992.
10. G. N. Frederickson. A data structure for dynamically maintaining rooted trees. In *Proceedings of the 4th ACM-SIAM Symposium on Discrete Algorithms*, pages 175–184, 1993.
11. G. N. Frederickson. Maintaining regular properties dynamically in k-terminal graphs. Manuscript, 1993.
12. J. Lagergren. Efficient parallel algorithms for tree-decomposition and related problems. In *Proceedings of the 31rd Annual Symposium on Foundations of Computer Science*, pages 173–182, 1990.
13. J. Matoušek and R. Thomas. Algorithms finding tree-decompositions of graphs. *J. Algorithms*, 12:1–22, 1991.
14. G. L. Miller and J. Reif. Parallel tree contraction and its application. In *Proceedings of the 26th Annual Symposium on Foundations of Computer Science*, pages 478–489, 1985.
15. B. Reed. Finding approximate separators and computing tree-width quickly. In *Proceedings of the 24th Annual Symposium on Theory of Computing*, pages 221–228, 1992.
16. N. Robertson and P. D. Seymour. Graph minors. II. Algorithmic aspects of tree-width. *J. Algorithms*, 7:309–322, 1986.
17. J. van Leeuwen. Graph algorithms. In *Handbook of Theoretical Computer Science, A: Algorithms and Complexity Theory*, pages 527–631, Amsterdam, 1990. North Holland Publ. Comp.

# Short disjoint cycles in graphs with degree constraints

Andreas Brandstädt
FB Mathematik/FG Informatik
Universität -GH- Duisburg
D 47048 Duisburg Germany

Heinz-Jürgen Voss
Abteilung Mathematik
TU Dresden
D 01062 Dresden

### Abstract

We show that each finite undirected graph $G = (V, E), |V| = n, |E| = m$ with minimum degree $\delta(G) \geq 3$ and maximum degree $\Delta = \Delta(G)$ contains at least $\frac{n}{4(\Delta-1)log2n}$ pairwise vertex-disjoint cycles of length at most $4(\Delta - 1) \cdot log2n$. Furthermore collections of such cycles can be determined within $O(n \cdot (n + m))$ steps. For constant $\Delta$ this means $\Omega(n/logn)$ cycles of length $O(logn)$. This bound is also an optimum.

A similar approach yields similar bounds for subgraphs with more edges than vertices instead of cycles. Furthermore also collections of many small pairwise disjoint induced subgraphs of this type can be determined within $O(n \cdot (n + m))$ steps similarly as for cycles.

## 1  Introduction

In connection with the bisection problem of transputer networks (cf. e.g. [4]) the problem arises whether cubic graphs have many pairwise disjoint short cycles and small "useful" (i.e. more edges than vertices) subgraphs. In order to give a more precise bound on the bisection width of transputer networks it is important to show that in cubic graphs there are many short disjoint useful subgraphs.

In [1] it has been shown that every cubic bridgeless graphs with n vertices contains $\Omega(n^{1/7})$ pairwise disjoint cycles of length $O(n^{6/7})$ and $\Omega(n^{1/14})$ pairwise disjoint useful subgraphs with $O(n^{13/14})$ vertices. The proof given in [1] bases on the theorem of Petersen about the 1– and 2–factor partition of cubic bridgeless graphs and is long and complicated.

Here we improve and generalize the results of [1] and obtain a considerably simpler proof on the basis of repeated application of Breadth–First Search (BFS):
The improvements and generalizations consist in the following:

1. The bounds are much better than in [1].

2. The bounds are shown not only for cubic graphs but much more general for graphs with $\delta(G) \geq 3$.

3. The approach is constructive i.e. it is not only shown that many short pairwise disjoint cycles exist but also collections of those cycles can be constructed within $O(n \cdot (n + m))$ steps.

4. These improvements are carried over also to the case of many small pairwise disjoint subgraphs with more edges than vertices instead of many short pairwise disjoint cycles.

## 2  The construction of many short disjoint cycles

Throughout this paper all graphs $G = (V, E)$ are finite, undirected and may contain loops and multiple edges. We use standard notations. Let $|V| = n(G) = n, |E| = m$. Let $C$ be a cycle in $G$. $|C|$ denotes the *length* (number of vertices) of $C$.

The *girth* $g(G)$ of $G$ is

$$g(G) = \begin{cases} min\{|C| : C \text{ is a cycle in } G\} & \text{if there is a cycle in } G \\ \infty & \text{otherwise} \end{cases}$$

Let $\delta(G)$ denote the minimum degree and $\Delta = \Delta(G)$ denote the maximum degree of vertices of G. As usual a loop incident to $v$ increases the degree of $v$ by 2.
The following property is well-known (see e.g. [3]):

**Lemma 2.1** *Let $G = (V, E)$ be a graph with $\delta(G) \geq 3$. Then $g(G) \leq 2log2n$.*

(log always denotes the logarithm with basis 2).
This can be seen by using BFS on G. It is easy to see that each vertex $v \in V$ of a graph with $\delta(G) \geq 3$ taken as start vertex of BFS is contained in a cycle of length $\leq 2 \cdot log2n$ which can also be determined by BFS. Then the repeated application of BFS for each vertex as start vertex allows the construction of minimum length cycles.
A cycle C of G is called *short* if $|C| \leq 4(\Delta - 1) \cdot \log 2n$. Our aim is to construct "many" short pairwise disjoint cycles in graphs G with $\delta(G) \geq 3$.
For constant $\Delta$ the best one could hope for (cf. also the end of section 2) are $O(n/logn)$ such cycles. Surprisingly this bound can be reached as we will show by the following construction which bases on the repeated deletion of a shortest cycle in a graph $G_i$ starting with $G_1 = G$. In order to apply repeatedly BFS to determine a cycle of length $\leq 2 \cdot log2n$ one has to ensure that each graph $G_i$ has minimum degree at least 3.
The deletion of a shortest cycle C in $G_i$ together with all edges incident to C leads in general to a graph $G'_i$ whose vertices can have degree less than 3. Thus the construction of $G_{i+1}$ is caused by the deletion of a shortest cycle in $G_i$ and has to ensure that

(i) all vertices of $G_{i+1}$ have at least degree 3

(ii) cycles in $G_{i+1}$ represent cycles in the original graph G

Observe that the cycles of $G_{i+1}$, however, can have a greater length in G than in $G_{i+1}$. Note also that $G_{i+1}$ may be the empty graph.
We first study the deletion of an induced subgraph $H = (V', F)$ in a graph $G = (V, E)$ with $\delta(G) \geq 3$. Let Q be the set of edges between $V'$ and $V \setminus V'$ and let $G'$ be the graph $G' = (V \setminus V', E \setminus (F \cup Q))$. $G'$ may contain also nodes of degree 0, 1 and 2. In order to transform $G'$ to a graph $G''$ which fulfills the conditions (i), (ii) above we repeatedly omit nodes of degree less than 2 until there are no such nodes in the graph and afterwards contract the subgraphs defined by degree 2 nodes (these are paths and/or cycles – cf. Figure 1) as follows:

**Algorithm 1** (The deletion of subgraph H and the reconstruction of minimum degree 3)

**Input:** A graph $G = (V, E)$ with $\delta(G) \geq 3$ and a subset $V' \subseteq V$
Let $H = (V', F)$ denote the induced subgraph of $V'$ in G and let Q denote the set of edges between $V'$ and $V \setminus V'$
**Output:** A graph $G''$ fulfilling the conditions (i), (ii) above

(1)  Construct $G' = (V \setminus V', E \setminus (F \cup Q))$;
(2)  **repeat**
         omit all nodes of degree $\leq 1$ and the edges incident to them
      **until** there are only nodes of degree $\geq 2$;
(3)  Omit all nodes of degree 2 as follows:

(3.1)     Omit all nodes of degree 2 which form a cycle which is not incident to a node of degree at least 3, and omit the edges incident to them;

(3.2)     For all paths of nodes $v_1, ..., v_k$ of degree 2 with end-nodes $v_1, v_k$ incident to nodes $v_0, v_{k+1}$ of degree at least 3, resp., omit $v_1, ..., v_k$ and the edges incident to them and replace the path by an edge $v_0 v_{k+1}$ labeling this edge with the path $v_1, ..., v_k$.
          (Note that also $v_0 = v_{k+1}$ is possible – in this case $v_0$ gets a self–loop).

Observe that Algorithm 1 can be done in linear time $O(n + m), n = |V|, m = |E|$. Let $D_{i,j}$ denote the set of nodes having degree $j, j \in \{0, 1, 2\}$ after the $i$-th iteration of (2) and let $d_{i,j} = |D_{i,j}|$. ($D_{0,j}$ are the corresponding sets before (2) is carried out the first time.) Furthermore let $D$ be the set of all nodes deleted by Algorithm 1 in steps (2) and (3) and let $d = |D|$ and $q = |Q|$. Then the following estimation holds:

**Lemma 2.2** $d \leq q$.

**Proof:**
Obviously $q \geq 3d_{0,0} + 2d_{0,1} + d_{0,2}$ holds. Let $I = d_{0,0} + 2d_{0,1} + d_{0,2}$.
If a node of degree 0 is deleted then $I$ is diminished by 1. The deletion of a node $v$ of degree 1 decreases $I$ by 2 if the valency of the only neighbour of $v$ is at least 4, and by 1 if this valency is 3, 2 or 1.
If a cycle all of its $k$ nodes have valency 2 is omitted according to (3.1) or a path of nodes $v_1, ..., v_k$ of degree 2 with end–nodes $v_1, v_k$ incident with nodes $v_0, v_{k+1}$ of degree at least 3, resp., is replaced by the edge $v_0 v_{k+1}$ according to (3.2) then $I$ is diminished by $k$.
Consequently, the number of nodes deleted according to Algorithm 1 is at most $I$. Hence

$$d \leq I = d_{0,0} + 2d_{0,1} + d_{0,2} \leq q.$$

$\square$

We now describe the construction of $G_{i+1} = (V_{i+1}, E_{i+1})$ from $G_i = (V_i, E_i)$ in more detail:

**Algorithm 2** (The construction of many short pairwise disjoint cycles in $G$)

**Input:**   A graph $G = (V, E)$ with $\delta(G) \geq 3$.
**Output:** A collection $\mathcal{C} \subseteq \{C_1, ..., C_s\}$ of short pairwise disjoint cycles of $G$.

(1)     $i := 1; G_i := G; \mathcal{C} := \emptyset;$
(2)     **repeat**
(2.1)       with BFS determine a cycle $C'_i$ in $G_i$ of length $\leq 2 \cdot log2n$;
(2.2)       determine the cycle $C_i$ in $G$ which corresponds to the cycle $C'_i$ in $G_i$;
(2.3)       **if** $|C_i| \leq 4 \cdot (\Delta - 1) \cdot log2n$ **then** $\mathcal{C} := \mathcal{C} \cup \{C_i\};$
(2.4)       with input $H = C'_i$ and $G = G_i$ carry out Algorithm 1;
(2.5)       $i := i + 1; G_i :=$ output of Algorithm 1
        **until** $G_i$ contains no further cycle

Let $G_i = (V_i, E_i)$.

**Lemma 2.3** $|V_i| - |V_{i+1}| \leq |C'_i| \cdot (\Delta - 1)$ *for all* $i \geq 1$.

**Proof:**
According to Lemma 2.2 $d \leq |C'_i| \cdot (\Delta - 2)$ holds. Thus Lemma 2.3 is an immediate consequence of Lemma 2.2.

$\square$

**Theorem 2.4** *For each graph $G = (V, E)$ with minimum degree $\delta(G) \geq 3$ and maximum degree $\Delta = \Delta(G)$ Algorithm 2 constructs at least $\frac{n(G)}{4(\Delta-1)log2n}$ pairwise vertex-disjoint cycles of length at most $4(\Delta - 1)log2n$ in $G$.*
*Furthermore the time bound of Algorithm 2 is $O(n \cdot (n + m))$.*

**Proof:**
Recall that $C_1', ..., C_s'$ is the set of cycles constructed by Algorithm 2 and let
$n(G_i) = |V_i|, i = 1, ..., s$.
The resulting cycles $C_1', ..., C_s'$ are obviously pairwise node-disjoint cycles in $G_1, ..., G_s$ and the corresponding $C_1, ..., C_s$ are pairwise node-disjoint cycles in $G$.
Since $n(G) \leq n(G_1) - n(G_2) + n(G_2) - n(G_3) + n(G_3) - ... - n(G_{s+1})$, $n(G_{s+1}) = 0$, and because of Lemma 2.3 we have

$$n(G) \leq s \cdot (\Delta - 1) \cdot 2log2n$$

i.e.

$$s \geq \frac{n(G)}{(\Delta - 1)2log2n}.$$

At least one half of these cycles $C_1, ..., C_s$ contains not more than $4(\Delta-1)log2n$ vertices: Assume to the contrary that more than one half of these cycles contains more than $4(\Delta - 1)log2n$ vertices. Then $G$ contains more than

$$s/2 \cdot 4(\Delta - 1)log2n \geq \frac{n(G) \cdot 4(\Delta - 1)log2n}{2(\Delta - 1)2log2n} \geq n(G)$$

vertices – a contradiction.

The time bound for determining such a collection of cycles follows from the following observation:

The *repeat*–loop of Algorithm 2 is carried out at most n times since $G_{i+1}$ has less vertices than $G_i$. It remains to show that each of the steps (2.1)–(2.5) can be done in linear time $O(n + m)$. This is clear for (2.1). For (2.2) it can be done using the edge labels for edges in $G_i$ which are the result of a contraction in step (3.2) of Algorithm 1. In this way also the length of the cycle $C_i$ in $G$ can be determined. For (2.3) the assertion is clear. For (2.4) it is the linear time bound of Algorithm 1, and for (2.5) nothing remains to show. $\square$

**Corollary 2.5** *If $\Delta$ is bounded by a constant and $\delta(G) \geq 3$ then $G$ contains $\Omega(n/logn)$ pairwise vertex-disjoint cycles whose length is bounded by $O(logn)$.*

For $r$–regular graphs one has an even stronger property. Here a lower bound for the number of cycles can be given which does not depend on $r$:

**Theorem 2.6** *Let $n, r$ be integers with $n \geq r + 1 \geq 4$. Each $r$–regular graph with n vertices has at least $\frac{1}{24} \cdot \frac{n}{log(2n+4)}$ pairwise vertex-disjoint cycles of length at most $24 \cdot log(2n + 4)$.*

**Proof:**
Let $G$ be a $r$–regular graph with $k$ pairwise disjoint cycles containing no $k+1$ pairwise disjoint cycles. Let $R$ be a smallest set of vertices representing all cycles of $G$ i.e. $G \setminus R$ has no cycle. Hence $G \setminus R$ is a forest and has at most $|V(G \setminus R)| - 1$ edges.
Since $G$ is $r$–regular the sum $\Sigma_1$ of valencies of $G \setminus R$ is
$\Sigma_1 = r \cdot |V(G \setminus R)|$ and the sum $\Sigma_2$ of valencies of $R$ is $\Sigma_2 = r \cdot |R|$. Since $G \setminus R$ has at most $|V(G \setminus R)| - 1$ edges at least $\Sigma_1 - 2(|V(G \setminus R)| - 1)$ edges join $G \setminus R$ and R. Hence

$$r \cdot |V(G \setminus R)| - 2|V(G \setminus R)| + 1 \leq r \cdot |R|.$$

Consequently,

$$|V(G \setminus R)| \leq \frac{r}{r-2} \cdot |R|.$$

With $\frac{r}{r-2} \leq 3$ the order is

$$n(G) = |V(G \setminus R)| + |R| \leq 4|R|.$$

In [5] Theorem 4.2, page 166 the following result is obtained:

$$|R| \leq 4k \cdot log(2k+4).$$

With $k \leq n$ we have

$$n \leq 4|R| \leq 16k \cdot log(2k+4) \leq 16k \cdot log(2n+4).$$

Therefore

$$k \geq \frac{1}{16} \cdot \frac{n}{log(2n+4)}.$$

Next we prove
(*) $G$ contains at least $\frac{k}{2}$ pairwise disjoint cycles of length $\leq 32log(2n+4)$:

Assume that $G$ has more than $\frac{k}{2}$ pairwise disjoint cycles of length greater than $32 \cdot log(2n+4)$. Hence

$$n(G) > \frac{k}{2} \cdot 32log(2n+4) \geq \frac{1}{32} \cdot \frac{n}{log(2n+4)} \cdot 32 \cdot log(2n+4) = n.$$

This contradiction proves (*). $\qquad \square$

Now we still show that the bounds we have reached are optimal up to constant factors:

Let $G(r,t)$ denote a $r$–regular graph of girth $t$ with minimum number of vertices.

**Theorem 2.7** $g(G(3,t)) > log\, n(G(3,t)) - 1.$

**Proof:**
P. Erdős and H. Sachs [2] proved for r-regular graphs

$$n(G(r,t)) \leq 4\left(\frac{(r-1)^{t-1} - 1}{r-2} - 1\right).$$

Hence for r=3

$$n(G(3,t)) \leq 4(2^{t-1} - 2) < 4 \cdot 2^{t-1} = 2^{t+1}.$$

Therefore,

$$g(G(3,t)) > log(n(G(3,t))) - 1.$$

$\qquad \square$

**Theorem 2.8** *There is an infinite sequence of integers* $n_0, n_1, n_2, \ldots$ *such that for each index* $i$ *there is a cubic graph* $G_i$ *of order* $n_i$ *having only cycles of lengths* $\geq g(G_i) > log n_i - 1.$

**Corollary 2.9** *The maximum number of vertex-disjoint cycles of* $G_i$ *is less than* $\frac{n}{log n - 1}.$

# 3 The construction of many small disjoint useful subgraphs

For this section we introduce the following notions:

A subgraph $H = (U, F)$ of $G = (V, E), U \subseteq V, F \subseteq E$ is *useful* if $H$ is connected and $|F| \geq |U| + 1$ (i.e. $H$ contains more edges than vertices).

A subgraph $H = (U, F)$ is *small* if $|U| \leq 8(\Delta - 1) \cdot \log 2n$.

We want to show that not only a collection of many short pairwise disjoint cycles but also of many small pairwise disjoint useful subgraphs can be constructed. We first show an analogon of Lemma 2.1:

**Lemma 3.1** *Let $G = (V, E)$ be a graph with $\delta(G) \geq 3$. Then $G$ contains a useful subgraph with at most $4 \cdot \log 2n$ vertices.*

**Proof:**

By Lemma 2.1 the graph $G$ contains a cycle $C$ of length $\leq 2 \log 2n$.

Contract all edges of $C$. Hence $C$ is contracted into a single node $w$. In the obtained graph $G'$ the node $w$ has at least degree 1 and all other nodes have degree $\geq 3$. By using BFS on $G'$ with start node $w$ it can easily be seen that $G'$ contains a subgraph $L$ of size $\leq 2 \log 2n$ consisting of a cycle $C'$ and a path joining $C'$ and $w$ (called a *lasso* with start node $w$). (Note that the path of the lasso can have length 0). $C$ and the edges of $L$ induce a useful subgraph of $G$ of size $\leq 4 \log 2n$. □

The construction of many small useful subgraphs can be done in an analogous way as for cycles by repeatedly carrying out the following steps starting with $G_1 = G$:

(1) Choose a small useful subgraph $C$ in $G_i$ and construct $G'_i$ by deleting $C$.

(2) Construct $G_{i+1}$ in steps (2.1)–(2.5) analogously to the case of cycles.

A little difference appears only in the analogon of Lemma 2.3 – the upper bound for $|V_i| - |V_{i+1}|$ is a little bit worse than in the case of cycles:

**Lemma 3.2** $|V_i| - |V_{i+1}| \leq |C_i| \cdot (\Delta - 1) \leq 4(\Delta - 1) \log 2n$.

The proof is analogous to the proof of Lemma 2.3.

**Theorem 3.3** *Every graph $G = (V, E)$ with $\delta(G) \geq 3$ and maximum degree $\Delta = \Delta(G)$ contains at least $\frac{n}{8(\Delta-1)\log 2n}$ pairwise disjoint useful subgraphs which have at most $8(\Delta - 1)\log 2n$ vertices.*

*Furthermore such a collection of useful subgraphs can be determined within $O(n^2 \cdot m)$ steps.*

The proof is analogous to the proof of Theorem 2.4.

Finally a generalization is given for graphs which do not fulfill the condition $\delta(G) \geq 3$:

**Corollary 3.4** *Let $G = (V, E)$ be a graph with $n$ vertices of degree 2 or 3 and at most $c \cdot n$ vertices $v$ have degree $\deg(v) \leq 2$, $0 \leq c < 1$. Then the following holds:*

a) *$G$ contains a cycle of length $\leq \frac{8}{1-c} \log 2n$ (i.e. $g(G) \leq \frac{8}{1-c} \cdot \log 2n$).*

b) *$G$ contains a useful subgraph of size $\leq \frac{16}{1-c} \log 2n$.*

**Proof:**

For given $G$ a graph $G' = (V', E')$ is constructed by contracting paths with vertices of degree 2 to one edge. $G'$ has at least $(1 - c) \cdot n$ vertices: $n' = |V'| \geq (1 - c) \cdot n$.

The theorems imply that $G'$ contains

a) at least $\frac{1}{8} \cdot \frac{n'}{\log 2n'}$ pairwise disjoint cycles of length $\leq 8 \log 2n'$

b) at least $\frac{1}{16} \cdot \frac{n'}{\log 2n'}$ pairwise disjoint useful subgraphs with at most $16 \log 2n'$ vertices.

Now by a pigeon hole argument the assertion follows. □

If $\Delta$ is unbounded then there are graphs $G$ with minimum degree $\geq 3$ and with large maximum degree having no two disjoint cycles. Such a graph is the wheel $W_n$, $n \geq 4$, consisting of an induced cycle of length $n$ with an additional central node adjacent to all cycle nodes.

Recently we found many short disjoint cycles in graphs without small cycles. The following theorem presents one of these results.

**Theorem 3.5** *Let $G$ be a graph with minimum degree $\geq 3$ and girth $\geq 7$. Then $G$ contains at least $\frac{\sqrt{2}}{8} \cdot \frac{\sqrt{n}}{\log n}$ disjoint cycles of lengths $\leq 12 \cdot \log n$.*

**Figure 1**

Subgraphs induced by degree 2 nodes    Contraction to:

a) paths:                                                        b) cycles :

no new edge

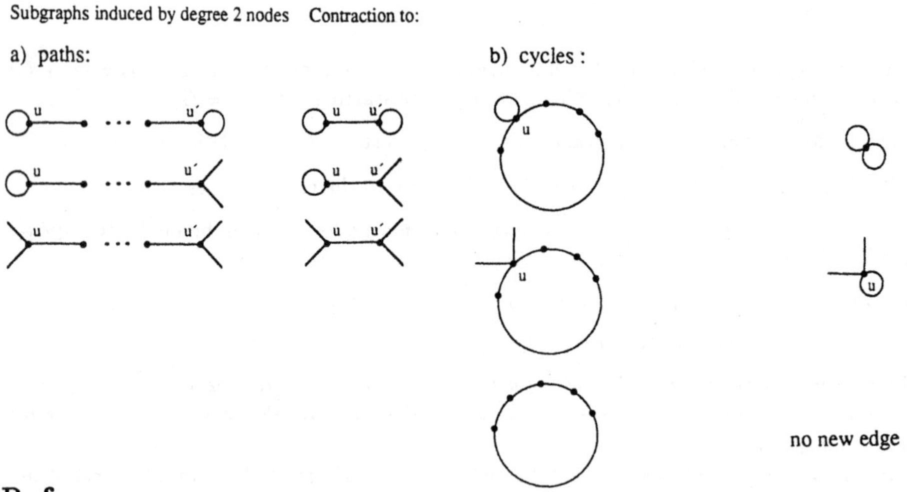

# References

[1] A. BRANDSTÄDT, Short disjoint cycles in cubic bridgeless graphs, Proc. of the 17th Intern. Workshop WG'91 - Graph Theoretic Concepts in Computer Science, *Lecture Notes in Computer Science*, 570, 239-249, 1992

[2] P. ERDÖS and H. SACHS, Reguläre Graphen gegebener Taillenweite mit minimaler Knotenzahl, *Wiss. Z. Univ. Halle–Wittenberg*, Math.-Nat. R. 12 (1963), 251-258

[3] R. HALIN, Graphentheorie, *Akademie–Verlag*, Berlin 1989

[4] J. HROMKOVIC and B. MONIEN, The Bisection Problem for Graphs of Degree 4 (Configuring Transputer Systems), MFCS 1991, 211–220, *Lecture Notes in Computer Science*

[5] H. WALTHER and H.–J. VOSS, Über Kreise in Graphen, *Deutscher Verlag der Wissenschaften*, Berlin 1977

# Efficient Algorithms for Tripartitioning Triconnected Graphs and 3-Edge-Connected Graphs

Koichi Wada* and Kimio Kawaguchi

Nagoya Institute of Technology
Gokiso-cho, Syowa-ku, Nagoya 466, JAPAN

**Abstract.** The extended $k$-partition problem is defined as follows. For the following inputs (1)an undirected graph $G = (V, E)(n = |V|, m = |E|)$, (2)a vertex subset $V'(\subseteq V)$, (3)distinct vertices $a_i \in V'(1 \leq i \leq k)$ and (4)natural numbers $n_i(1 \leq i \leq k)(n_1 \leq \ldots \leq n_k)$ such that $n_1 + \ldots + n_k = n' = |V'|$, we compute a partition $V_1 \cup \ldots \cup V_k$ of $V$ and a partition $V_1' \cup \ldots \cup V_k'$ of $V'$ such that (a)each $V_i'$ is included in $V_i$, (b)each $V_i'$ contains the specified vertex $a_i$, (c)$|V_i'| = n_i$ and (d)each $V_i$ induces a connected subgraph. If $V' = V$, then the problem is called the $k$-partition problem. In this paper, we show that if the input graph is triconnected the extended tripartition problem can be solved in $O(m + (n - n_3) \cdot n)$ time and that the algorithm solves the original tripartition problem in $O(m + (n_1 + n_2) \cdot n)$ time. Furthermore, we show that for a $k$-edge-connected graph $G = (V, E)$ there exists a partition $V_1 \cup \ldots \cup V_k$ of $V$ such that each $V_i$ contains the specified vertex $a_i$, $|V_i| = n_i$ and $k$ subgraphs $G_1, \ldots, G_k$ are mutually edge disjoint and each of $G_i$ contains all of elements in $V_i(1 \leq i \leq k)$ and the case in which $k = 3$ can be solved in $O(n^2)$ time.

## 1 Introduction

*The $k$-vertex-partition($k$-partition for short) problem* is described as follows.
**Input:**
(1) an undirected graph $G = (V, E)$ with $n = |V|$ vertices and $m = |E|$ edges;
(2) $k$ distinct vertices $a_i(1 \leq i \leq k) \in V$, $a_i \neq a_j(1 \leq i < j \leq k)$; and
(3) $k$ natural numbers $n_1, n_2, \ldots, n_k$ such that $\sum_{i=1}^{k} n_i = n$.
**Output:** a partition $V_1 \cup V_2 \cup \ldots \cup V_k$ of vertex set $V$ such that for each $i(1 \leq i \leq k)$
(a) $a_i \in V_i$;
(b) $|V_i| = n_i$; and
(c) each $V_i$ induces a connected subgraph of $G$.

* Partially supported by the Grant-in-Aid for Scientific Research of the Ministry of Education, Science and Culture of Japan under Grant: (C)05680271.

We extend the $k$-partition problem in the following and we call it *the vertex-subset $k$-partition problem.*

**Input:**

(1) an undirected graph $G = (V, E)$ with $n = |V|$ vertices and $m = |E|$ edges;

(2) a vertex subset $V'(\subseteq V)$ with $n' = |V'| \geq k$;

(3) $k$ distinct vertices $a_i(1 \leq i \leq k) \in V'$, $a_i \neq a_j(1 \leq i < j \leq k)$; and

(4) $k$ natural numbers $n_1, n_2, \ldots, n_k$ such that $\sum_{i=1}^{k} n_i = n'$.

**Output:** a partition $V_1 \cup V_2 \cup \ldots \cup V_k$ of vertex set $V$ and a partition $V'_1 \cup V'_2 \cup \ldots \cup V'_k$ of vertex set $V'$ such that for each $i(1 \leq i \leq k)$

(a) $a_i \in V'_i$;

(b) $|V'_i| = n_i$;

(c) $V'_i \subseteq V_i$ and

(d) each $V_i$ induces a connected subgraph of $G$.

This problem is extended in the sense that a specified vertex subset is partitioned. If $V' = V$ then it coincides with the original $k$-partition problem. Fig. 1 shows an instance of the vertex-subset tripartition problem and a solution.

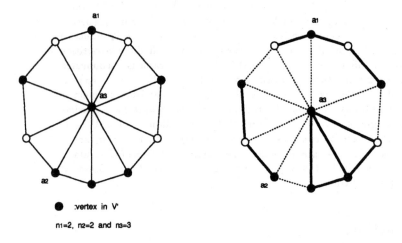

**vertex in $V'$**

$n_1=2$, $n_2=2$ and $n_3=3$

**Fig. 1.** An instance of the vertex-subset tripartition problem and a solution.

The $k$-partition problem is NP-hard in general even if $k$ is limited to 2 [2]. Győri and Lovász independently showed that the $k$-partition problem has a solution if the input graph is $k$-connected [5, 9]. The bipartition problem($k = 2$) can be solved in $O(m)$ time [11] and the tripartition problem($k = 3$) can be solved in $O(n^2)$ time [12]. Throughout this paper, $n$ and $m$ denote the number of vertices in $G$ and the number of edges in $G$, respectively. It is an open problem whether or not we can solve the $k$-partition problem for $k \geq 4$ in polynomial time if the input graph is $k$-connected [12].

In this paper, we show that the vertex-subset $k$-partition problem has a solution if the input graph is $k$-connected for any $k(\geq 2)$. This is a generalization of Győri and Lovász's result. We also present an algorithm to solve the vertex-subset tripartition problem for triconnected graphs. Our algorithm utilizes a nonseparating ear decomposition which characterizes triconnected graphs[1] and it runs in $O(m + (n - max(n_1, n_2, n_3)) \cdot n)$ time. Our algorithm is not only completely different from that of [12], which works only for the tripatition problem, but also is more efficient for the tripartition problem. It solves the tripartition problem for triconnected graphs in $O(m + (min_{1 \leq i < j \leq 3}(n_i + n_j)) \cdot n)$ time.

Furthermore, we characterize $k$-edge-connected graphs by using a vertex-partition. In general the $k$-(vertex-)partition problem does not have a solution for $k$-edge-connected graphs. However by using Győri and Lovász's result the following edge-partition can be done[5]:

Given a $k$-edge-connected graph $G = (V, E)$, $k$ edges $e_1, e_2, \ldots, e_k \in E$ and $k$ positive integers $m_1, m_2, \ldots, m_k$ such that $\sum_{i=1}^{k} m_i = |E|$, there exists an edge-partition $E = E_1 \cup E_2 \cup \ldots \cup E_k$ such that $e_i \in E_i$, $|E_i| = m_i$ and $G_i = (V(E_i), E_i)$ is connected for each $i(1 \leq i \leq k)$, where $V(E')$ denotes the set of vertices incident with at least one member of $E'$.

In this paper, we propose the following $k$-partition problem which is obtained by replacing the condition (c) of the $k$-partition problem with the condition (c-v) or (c-e).

**Input:**
(1) an undirected graph $G = (V, E)$ with $n = |V|$ vertices and $m = |E|$ edges;
(2) $k$ distinct vertices $a_i(1 \leq i \leq k) \in V$, $a_i \neq a_j(1 \leq i < j \leq k)$; and
(3) $k$ natural numbers $n_1, n_2, \ldots, n_k$ such that $\sum_{i=1}^{k} n_i = n$.
**Output:** a partition $V_1 \cup V_2 \cup \ldots \cup V_k$ of vertex set $V$ such that for each $i(1 \leq i \leq k)$
  (a) $a_i \in V_i$;
  (b) $|V_i| = n_i$; and
  (c-v)((c-e)) For any pair $(v_i, u_i)$ of vertices in any $V_i$ and any pair $(v_j, u_j)$ of vertices in any $V_j(1 \leq i < j \leq k)$, there exist a path $P_i$ between $v_i$ and $u_i$ and a path $P_j$ between $v_j$ and $u_j$ such that $P_i$ and $P_j$ are vertex-disjoint(edge-disjoint).

We refer to this problem as *the k-partition problem with respect to vertex-disjointness (edge-disjointness)*. Since a solution for the $k$-partition problem is also a solution for the $k$-partition problem with respect to vertex-disjointness from the definition, the $k$-partition problem with respect to vertex-disjointness has a solution if the input graph is $k$-connected.

We show that we can solve the $k$-partition problem with respect to edge-disjointness for $k$-edge-connected graphs by using the fact that we can solve the vertex-subset $k$-partition problem for $k$-connected graphs. Moreover, by using the result and the algorithms shown here, we show the cases in which $k = 2$ and $k = 3$ can be solved in $O(m)$ and $O(n^2)$ time, respectively.

## 2 Preliminary

We deal with an undirected graph $G = (V, E)$ with vertex set $V$ and edge set $E$. For a graph $G$, the vertex set is denoted by $V(G)$. A graph $G$ is $k$-

*connected(k-edge-connected)* if there exist $k$ node-disjoint(edge-disjoint) paths between every pair of distinct nodes in $G$. Usually 2-connected graphs are called *biconnected graphs* and 3-connected graphs are called *triconnected graphs*. The *distance* between nodes $x$ and $y$ in $G$ is the length of the shortest path between $x$ and $y$ and is denoted by $dis_G(x, y)$. For a graph $G = (V, E)$ and a vertex subset $V'$, the induced subgraph is denoted by $G[V']$. For two graphs $G = (V, E)$ and $G' = (V', E')$, the graph $(V \cup V', E \cup E')$ is denoted by $G \cup G'$.

# 3 Vertex-Subset $k$-partition Problem

**Theorem 1.** [14] *For any integer $k(\geq 2)$, the vertex-subset $k$-partition problem has a solution for $k$-connected graphs.*

This theorem is a generalization of Györi and Lovász's result and can be proved by extending the method used in proving the original problem[5]. The detail of the proof is shown in [14].

# 4 Tripartition of Triconnected Graphs

In order to solve the tripartition problem efficiently, we utilize the concepts of a nonseparating ear decomposition which characterizes triconnected graphs and an s-t numbering for biconnected graphs.

## 4.1 Nonseparating Ear Decomposition

An *ear decomposition* of a biconnected graph $G = (V, E)$ is a decomposition $G = P_0 \cup P_1 \cup \ldots \cup P_k$, where $P_0$ is a cycle and $P_i (1 \leq i \leq k)$ is a path whose end vertices are distinct and the vertices in common with $P_0 \cup \ldots \cup P_{i-1}$ are the end vertices. Each $P_i$ is called an open ear. Note that $P_0 \cup P_1 \cup \ldots \cup P_i$ is biconnected for each $i(1 \leq i \leq k)$. An ear which has length greater than one is called *nontrivial* and an ear which consists of an edge is called *trivial*. We can assume that for an ear decomposition $P_0 \cup P_1 \cup \ldots \cup P_k$, there exists $j$ such that each ear $P_i(1 \leq i \leq j)$ is nontrivial and the other ears $P_i(j+1 \leq i \leq k)$ are trivial, and we also call $P_0 \cup P_1 \cup \ldots \cup P_j$ an ear decomposition of $G$.

Given an ear decomposition $P_0 \cup P_1 \cup \ldots \cup P_k$ of $G$, let $V_i = V(P_0) \cup \ldots \cup V(P_i)$, let $G_i = G[V_i]$ and let $\overline{G_i} = G[V - V_i]$ for each $i(1 \leq i \leq k)$.

We say that $G = P_0 \cup P_1 \cup \ldots \cup P_k$ is an *ear decomposition through edge* $(a, b)$ *and avoiding vertex* $c$, if the cycle $P_0$ contains the edge $(a, b)$ and the last nontrivial ear, say $P_q$, is of length 2 and has $c$ as its only internal vertex.

An ear decomposition $P_0 \cup P_1 \cup \ldots \cup P_q$ of a graph $G$ through edge $(a, b)$ and avoiding vertex $c$ is *nonseparating* if for all $i(1 \leq i < q)$, each graph $\overline{G_i}$ is connected and each internal vertex of the ear $P_i$ has a neighbour in $\overline{G_i}$. Fig. 2 shows an example of a nonseparating ear decoomposition through $(a, b)$ and avoiding $c$.

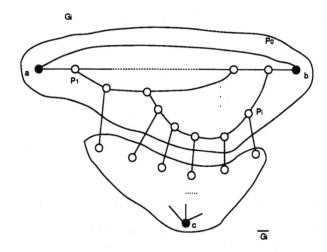

**Fig. 2.** A nonseparating ear decomposition $P_0 \cup \ldots \cup P_i \cup \ldots \cup P_q$ through edge $(a, b)$ and avoiding vertex $c$.

**Proposition 2.** [1] *For a triconnected graph $G = (V, E)$, any edge $(a, b) \in E$ and any vertex $c(\neq a, b) \in V$, a nonseparating ear decomposition $P_0 \cup P_1 \cup \ldots \cup P_q$ through edge $(a, b)$ and avoiding vertex $c$ can be constructed in $O(|V| \cdot |E|)$, where the path $P_q$ has the vertex $c$ as its only internal one. In particular, the cycle $P_0$ and each path $P_i (1 \leq i \leq q)$ can be constructed in $O(|E|)$.*

## 4.2 s-t Numbering

Given an edge $(s, t)$ of a biconnected graph $G = (V, E)$, a bijective function $g : V \to \{1, 2, \ldots, |V| = n\}$ is called an *s-t numbering* if the following conditions are satisfied:

- $g(s) = 1$, $g(t) = n$ and
- Every node $v \in V - \{s, t\}$ has two adjacent nodes $u$ and $w$ such that $g(u) < g(v) < g(w)$.

**Proposition 3.** [3] *Let $G = (V, E)$ be a biconnected graph. For any edge $(s, t) \in E$, an s-t numbering can be computed in $O(|E|)$ time.*

The following lemma holds from the definition of the s-t numbering.

**Lemma 4.** *Let $g$ be an s-t numbering for a biconnected graph $G = (V, E)$ and an edge $(s, t)$. For any $i(1 \leq i \leq |V|)$, the two induced subgraphs $G[\{g^{-1}(j)|1 \leq i \leq j\}]$ and $G[\{g^{-1}(j)|i + 1 \leq j \leq |V|\}]$ are connected, where $g^{-1}$ is the inverse function of $g$.*

The following theorem is easily shown from Lemma 4 and Proposition 3.

**Theorem 5.** *The vertex-subset bipartition problem can be solved in $O(m)$.*

*Proof.* Let $G = (V, E)$ be a biconnected graph and $G$, $V'(\subseteq V)$, $a_1$ and $a_2$ and $n_1$ and $n_2$ $(n_1 + n_2 = n' = |V'|)$ be inputs for the problem. If $(a_1, a_2) \notin E$ then let $G = (V, E \cup \{(a_1, a_2)\})$. Let $g$ be an $a_1$-$a_2$ numbering for $G$. the node set $V$ is partitioned into two sets $V_1 = \{g^{-1}(j)|1 \leq i \leq j\}$ and $V_2 = \{g^{-1}(j)|i + 1 \leq j \leq n\}$ such that $|V_1 \cap V'| = n_1$ and $|V_2 \cap V'| = n_2$. Clearly, $a_1 \in V_1$ and $a_2 \in V_2$. The two induced subgraphs $G[V_1]$ and $G[V_2]$ are connected from Lemma 4 and contain $n_1$ and $n_2$ vertices of $V'$, respectively. This bipartition can be done in $O(m)$ time from Proposition 3.

## 4.3  Algorithm for Vertex-Subset Tripatition Problem

In this section, we present an efficient algorithm PART3 for solving the vertex-subset tripartition problem for a triconnected graph $G$. We may assume without loss of generality that $n_1 \leq n_2 \leq n_3$ for a given input.

The algorithm PART3 is based on the following idea. Let $P_0 \cup \ldots \cup P_q$ be a nonseparating ear decomposition through $(a_1, a_2)$ and avoiding $a_3$. Let $i$ be the least value such that $|(V(P_0) \cup V(P_1) \ldots \cup V(P_i)) \cap V'| \geq n_1 + n_2$. If the equality holds, the desired tripartion can be done. Because the graph $G_i$ is biconnected and the graph $\overline{G_i}$ containing $a_3$ and $n_3$ vertices in $V'$ is connected. Since $G_i$ contains $a_1$ and $a_2$, it can be bipartitioned into desired subgraphs. If the equality does not hold, the ear $P_i$ is tripartitioned. Since the graph $G_{i-1}$ is biconnected and contains $a_1$ and $a_2$, it is bipartitioned into two connected subgraphs. These subgraphs does not contain the specified number of vertices and they are adjusted to $n_1$ and $n_2$ by using the ear $P_i$ with preserving connectivity of each subgraph. Since the remaining vertices(say $W$) of $P_i$ have neighbours in $\overline{G_i}$, the induced subgraph $G[V(\overline{G_i}) \cup W]$ is still connected. Thus, the desired tripartition can be done. The algorithm is as follows:

**Algorithm PART3**$(G = (V, E); V'; a_1, a_2, a_3; n_1, n_2, n_3)$

> **begin**
> **if** $(a_1, a_2) \notin E$ **then** $E \leftarrow E \cup \{(a_1, a_2)\}$;
> Let $P_0 \cup P_1 \cup \ldots \cup P_q$ be a nonseparating ear decomposition
> through edge $(a_1, a_2)$ and avoiding vertex $a_3$ for the graph $G$;
> {Note that this algorithm does not construct all paths of the ear
> decomposition.}
> $i \leftarrow 1$;
> **while** $|V(G_i) \cap V'| < n_1 + n_2$ **do** $i \leftarrow i + 1$;
> **if** $|V(G_i) \cap V'| = n_1 + n_2$ **then**
> > **begin**
> > Let $g$ be an $a_1$-$a_2$ numbering of $G_i$ and
> > let $n'_1$ satisfy $n_1 = |\{g^{-1}(j)|1 \leq j \leq n'_1\} \cap V'|$ and $g^{-1}(n'_1) \in V'$;
> > $V_1 \leftarrow \{g^{-1}(j)|1 \leq j \leq n'_1\}$;
> > $V_2 \leftarrow \{g^{-1}(j)||G_i| = n_1 + n_2 \geq j \geq n'_1 + 1\}$;

```
        V₃ ← V(Ḡᵢ);
        return(V₁, V₂, V₃)
    end
else {|V(Gᵢ) ∩ V'| > n₁ + n₂ > |V(Gᵢ₋₁) ∩ V'|}
    begin
        Let g be an a₁-a₂ numbering of Gᵢ₋₁,
        let Pᵢ = (x₀, ..., xᵣ) such that g(x₀) < g(xᵣ),
        let n₁' satisfy n₁ = |{g⁻¹(j)|1 ≤ j ≤ n₁'} ∩ V'| and g⁻¹(n₁') ∈ V',
        let n₂' satisfy n₂ = |{g⁻¹(j)||V(Gᵢ₋₁)| ≥ j ≥ |V(Gᵢ₋₁)| − n₂'+1}∩
        V'| and g⁻¹(|V(Gᵢ₋₁)| − n₂' + 1) ∈ V';
        U₁ ← {g⁻¹(j)|1 ≤ j ≤ n₁'};
        U₂ ← {g⁻¹(j)||V(Gᵢ₋₁)| ≥ j ≥ |V(Gᵢ₋₁)| − n₂' + 1};
        I ← U₁ ∩ U₂; { Fig. 3 }
        if x₀ ∈ U₁ − I then { Fig. 4 }
            begin
                Let j = |I ∩ V'| and
                let j' satisfy j = |{x₁, ..., xⱼ'} ∩ V'| and xⱼ' ∈ V';
                V₁ ← (U₁ − I) ∪ {x₁, ..., xⱼ'};
                V₂ ← U₂;
                V₃ ← V(Ḡᵢ) ∪ {xⱼ'₊₁, ..., xᵣ₋₁};
                return(V₁, V₂, V₃)
            end
        else if xᵣ ∈ U₂ − I then { Similar to Fig. 4 }
            begin
                Let j = |I ∩ V'| and
                let j' satisfy j = |{xᵣ₋ⱼ', ..., xᵣ₋₁} ∩ V'| and xᵣ₋ⱼ' ∈ V';
                V₁ ← U₁;
                V₂ ← (U₂ − I) ∪ {xᵣ₋ⱼ', ..., xᵣ₋₁};
                V₃ ← V(Ḡᵢ) ∪ {x₁, ..., xᵣ₋ⱼ'₋₁};
                return(V₁, V₂, V₃)
            end
        else {x₀, xᵣ ∈ I} { Fig. 5 }
            begin
                Let I = {z₁, ..., z|I|} and
                let x₀ = zₛ and xᵣ = zₜ (1 ≤ s < t ≤ |I|),
                let j₁ = |{zₛ₊₁, ..., z|I|} ∩ V'|,
                let j₂ = |{z₁, ..., zₛ} ∩ V'|,
                let j₁' satisfy j₁ = |{x₁, ..., xⱼ₁'} ∩ V'| and xⱼ₁' ∈ V' and
                let j₂' satisfy j₂ = |{xᵣ₋ⱼ₂', ..., xᵣ₋₁} ∩ V'| and xᵣ₋ⱼ₂' ∈ V';
                V₁ ← (U₁ − I) ∪ {z₁, ..., zₛ} ∪ {x₁, ..., xⱼ₁'};
                V₂ ← (U₂ − I) ∪ {zₛ₊₁, ..., z|I|} ∪ {xᵣ₋₁, ..., xᵣ₋ⱼ₂'};
                V₃ ← V(Ḡᵢ) ∪ {xⱼ₁'₊₁, ..., xᵣ₋ⱼ₂'₋₁};
                return(V₁, V₂, V₃)
            end
    end
end
```

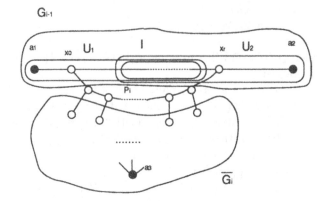

**Fig. 3.** The case in which $|V(G_i) \cap V'| > n_1 + n_2 > |V(G_{i-1}) \cap V'|\}$.

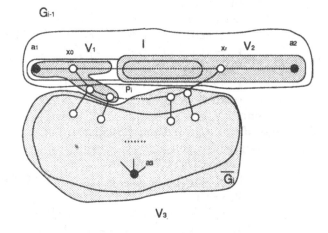

**Fig. 4.** The case in which $x_0 \in U_1 - I$.

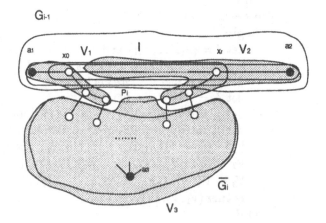

**Fig. 5.** The case in which $x_0, x_r \in I$.

**Theorem 6.** *The vertex-subset tripartition problem for triconnected graphs can be solved in $O(m + i \cdot n)$, where $i$ denotes the number of ears constructed in PART3.*

*Proof.* The correctness of PART3 is easily derived from the definition of the nonseparating ear decomposition and Lemma 4. Since a spanning triconnected subgraph $G' = (V, E')$ with $|E'| = O(|V|)$ can be computed in $O(|E|)$ time for a triconnected graph $G = (V, E)$ [10, 12], Since the cycle $P_0$ and each ear $P_i$ is computed in $O(|E'|) = O(n)$ time from Proposition 2 and s-t numbering is comouted in $O(n)$ time from Proposition 3, the time complexity of the theorem is obtained.

The next corollaries are easily obtained from the construction of the algorithm. Since the graph $\overline{G_i}$ has at least $n_3$ vertices, the number of ears constructed in PART3 is at most $n - n_3$. Moreover, for the tripartition problem it holds that $n - n_3 = n_1 + n_2$.

**Corollary 7.** *The vertex-subset tripartition problem for triconnected graphs can be solved in $O(m + (n - n_3) \cdot n)$.*

**Corollary 8.** *The tripartition problem for triconnected graphs can be solved in $O(m + (n_1 + n_2) \cdot n)$, where $n_1 \leq n_2 \leq n_3$ for the given input.*

**Corollary 9.** *If $n_1 + n_2 = O(1)$ or $dis_G(a_1, a_2) \geq n_1 + n_2$ then the tripartition problem for triconnected graphs can be solved in linear time.*

## 5   The $k$-partition with respect to Edge-disjointness

In this section, in order to solve the vertex-subset $k$-partition problem with respect to edge-disjointness (in which a vertex-subset $V'$ is specified instead of $V$ like the vertex-subset $k$-partition problem), we compute mutually $k$ edge-disjoint connected subgraphs $G_i = (U_i, E_i)(1 \leq i \leq k)$ such that
   (a) $a_i \in V_i'$,
   (b) $V_i' \subseteq U_i$ and
   (c) $|V_i'| = n_i$.
If we can find the mutually $k$ edge-disjoint subgraphs stated above, we can easily show that the vertex-partition $V' = V_1' \cup V_2' \cup \ldots \cup V_k'$ is a solution for the vertex-subset $k$-partition problem with respect to edge-disjointness.

We utilize the method transforming $k$-edge-connected graphs into $k$-connected graphs [4].

Let $k \geq 2$. Given a graph $G = (V, E)$, define the graph $\varphi_k(G) = (\varphi(V), \varphi(E))$ as follows. For every vertex $v \in V$, there are $k-2$ vertices $\varphi(v_1), \varphi(v_2), \ldots, \varphi(v_{k-2})$ in $\varphi(V)$. These vertices are called *node-vertices* of $\varphi_k(G)$. For every edge $e \in E$, there is a vertex $\varphi(e)$ in $\varphi(V)$. This vertex is called *arc-vertex* of $\varphi_k(G)$.

The edge set $\varphi(E)$ is defined as follows: Let $v$ be any vertex in $V$ and $u_0, u_1, \ldots, u_{d-1}$ be the vertices adjacent to $v$. Let $e_i = (v, u_i)(0 \leq i \leq d-1)$.

Then there are edges $(\varphi(e_i), \varphi(e_{(i+1) mod\ d}))(0 \le i \le d-1)$ and $(\varphi(e_i), \varphi(v_j))(0 \le i \le d-1, 1 \le j \le k-2)$ in $\varphi(E)$. Note that if $d = 2$, there is an edge $(\varphi(e_0), \varphi(e_1))$ in $\varphi(E)$.

From the definition $\varphi_k(G)$ has $(k - 2)|V| + |E|$ vertices and $O(k|E|)$ edges and it can be computed in $O(k(|V| + |E|))$.

Note that if $k = 2$, there is no node-vertex in $\varphi_2(G)$. For a biconnected graph $G = (V, E)$, an injective function $I : V \to E$ can be defined and it is computed in $O(|E|)$ time by using a spanning tree of $G$. Since $\varphi_2(G)$ has no node-vertex, an arc-vertex $\varphi(I(v))$ plays the role of node-vertex for $v$ in $\varphi_2(G)$.

**Proposition 10.** [4] *For any* $k(\ge 2)$, $G$ *is* $k$-*edge-connected if and only if* $\varphi_k(G)$ *is* $k$-*connected.*

Let $U$ be a vertex subset of $\varphi_k(G)$. The subgraph $\varphi^{-1}(U) = (\varphi_V^{-1}(U), \varphi_E^{-1}(U))$ of $G = (V, E)$ is defined to be
$\varphi_V^{-1}(U) = \{v | \varphi(v_i) \in U \text{ and } v \in V\} \cup \{ \text{endpoints of } e | \varphi(e) \in U \text{ and } e \in E\}$ and
$\varphi_E^{-1}(U) = \{e | \varphi(e) \in U \text{ and } e \in E\}$.
The next lemma can be easily shown.

**Lemma 11.** *If the induced subgraph* $\varphi_k(G)[U]$ *is connected, then the graph* $\varphi^{-1}(U)$ *is connected.*

From the Proposition 10 and Lemma 11, the following lemma can be proved.

**Lemma 12.** *If the vertex-subset* $k$-*partition problem can be solved, the vertex-subset* $k$-*partition problem with respect to edge-disjointness can be solved.*

*Proof.* Let $G = (V, E)$ be a $k$-connected graph and let $G, V'(\subseteq V), a_i (\in V')(1 \le i \le k)$ and $n_i(1 \le i \le k)(\Sigma_{i=1}^k n_i = |V'|)$ be the inputs for the vertex-subset $k$-partition problem with respect to edge-disjointness. We transform $G$ into $\varphi_k(G)$. Since $\varphi_k(G)$ is $k$-connected from Proposition 10, the vertex-subset $k$-partition problem is applied for $\varphi_k(G)$ as follows:

1. the input graph : $\varphi_k(G)$,
2. the vertex subset : if $k \ge 3$ then $\{\varphi(v_1) | v \in V'\}$ else $(k = 2)$ $\{\varphi(I(v)) | v \in V'\}$,
3. the specified vertices : if $k \ge 3$ then $\varphi((a_i)_1)(1 \le i \le k)$ else $(k = 2)$ $\varphi(I(a_i))(1 \le i \le k)$,
4. the specified numbers : $n_i(1 \le i \le k)$.

Let $U_i(1 \le i \le k)$ be the solution for the problem. The graphs $\varphi^{-1}(U_i)$ are connected from Lemma 11 and they are mutually edge-disjoint since for any arc-vertex $\alpha$ in $\varphi(V)$, there exists a unique edge $e$ such that $\varphi(e) = \alpha$. It is easily verified that they satisfy the specified conditions.

The next theorem is derived from Lemma 12 and Theorem 1.

**Theorem 13.** *For any integer* $k(\ge 2)$, *the* $k$-*partition problem with respect to edge-disjointness has a solution for* $k$-*edge-connected graphs.*

By using Theorem 5 and Corollary 7, the cases in which $k = 2, 3$ can be computed efficiently.

**Theorem 14.** *The vertex-subset bipartition problem with respect to edge-disjointness can be solved in $O(m)$ time.*

**Theorem 15.** *The vertex-subset tripartition problem with respect to edge-disjointness can be solved in $O(n^2)$ time.*

## 6 Concluding Remarks

We present an algorithm which solves the vertex-subset tripartition problem for triconnected graphs. This algorithm solves the tripartition problem for triconnected graphs efficiently. Compared with the previous result, the worst case of our algorithm is the same as that in [12], since there is a case that $n_1 + n_2 = \Omega(n)$. However, we have applications which satisfies $n_1 + n_2 = o(n)$. For example, the tripartition is necessary in order to define efficient fault-tolerant routings for triconnected graphs and in that case it satisfies that $n_1 + n_2 = O(1)$ or $n_1 + n_2 = O(\log n)$ [6, 7]. Thus, the tripartition can be solved in $O(m)$ or $O(m + n \log n)$ time.

The $k$-partition problem with respect to vertex-disjoint and/or edge-disjointness is related to fault-tolerant routings for $k$-connected graphs and/or $k$-edge-connected graphs [6, 8, 13].

It still remains as a further study to solve the $k$-partition problem($k \geq 4$) for $k$-connected graphs in polynomial time. However, a restricted $k$-edge-partition problem can be solved in polynomial time. Like the $k$-partition problem for $k$-connected graphs, it is not known whether or not we can solve the $k$-edge-partition problem for $k \geq 4$ in polynomial time even if the input graph is $k$-edge-connected. (The answers are 'yes' for the cases in which $k = 2$ and $k = 3$ from the result shown in this paper.) In [15]it is shown that we can solve a restricted $k$-edge-partition problem without specified $k$ edges in polynomial time even if the input graph is not $k$-edge-connected. This problem is called the $k$-edge-partition problem without specified edges. The $k$-edge-partition problem without specified edges($k \geq 4$) can be solved for 4-edge-connected graphs in $O(n^{1.5}\sqrt{\log n} + m)$ time and the 3-edge-partition problem without specified edges can be solved for 2-edge-connected graphs in $O(n^2)$ [15]. It is also interesting whether or not the $k$-partition problem without specified vertices can be solved in polynomial time.

## References

1. J. Cherian and S.N. Maheshwari : Finding nonseparating induced cycles and independent spanning trees in 3-connected graphs, Journal of Algorithms,**9**(1988)507–537

2. M.E Dyer and A.M. Frieze : On the complexity of partitioning graph into connected subgraphs, Discrete Applied Mathematics, **10**(1985)139–153

3. S.Even: Graph algorithms, Computer Science Press, Potomac, MD(1979)

4. Z. Galil and G.F. Italiano: Reducing edge connectivity to vertex connectivity, SIGACT NEWS, **22**,1(1991)57-61

5. E. Győri: On division of connected subgraphs, in: Combinatorics(Proc. 5th Hungarian Combinational Colloquy, 1976, Keszthely) North-Holland, Amsterdam(1978)485–494

6. M.Imase and Y.Manabe: Fault tolerant routings in a $\kappa$-connected network, Information Processing Letters, **28**,4(1988)171-175

7. K.Kawaguchi and K.Wada: New results in graph routing, Information and Computation(1993)(to appear)

8. K.Kawaguchi, K.Wada and T.Sugiura: Improvement of the sufficient conditions and the computational complexity in defining optimal graph routings, Trans. of IEICE,**J76-D-I**, 6(1993)247–259

9. L. Lovász: A homology theory for spanning trees of a graph, Acta mathematica Academic Science of Hungary,**30**(1977)241–251

10. H.Nagamochi and T.Ibaraki : A linear-time algorithm for finding a sparse $k$-connected spanning subgraph of a $k$-connected graph, Algorithmica, **7**,5/6(1992)583–596

11. H.Suzuki, N.Takahashi and T.Nishizeki: A linear algorithm for bipartition of biconnected graphs, Information Processing Letters, **33**,5(1990)227–232

12. H.Suzuki, N.Takahashi T.Nishizeki, H. Miyano and S.Ueno: An algorithm for tripartitioning 3-connected graphs, Information Processing Society of Japan, **31**,5(1990)584–592

13. K.Wada, T.Shibuya, E.Shamoto and K.Kawaguchi: A linear time $(L, k)$-edge-partition algorithm for connected graphs and fault-tolerant routings for $k$-edge-connected graphs, Trans. of IEICE, **J75-D-I**,11(1992)993-1004

14. K.Wada, K.Kawaguchi and N.Yokoyama: On a generalized $k$-partition problem for graphs, Proc. the 6th Karuizawa Workshop on Circuit and Systems(1993)243–248

15. K.Wada, A.Takagi and K.Kawaguchi: Efficient algorithms for $k$-edge-partition problem without specified edges, Kawaguchi Lab. Technical Report of ECE in Nagoya Institute of Technology,TR-01-93(1993)

# Towards a Solution of the Holyer's Problem

Zbigniew Lonc

Institute of Mathematics
Warsaw University of Technology
00-661 Warsaw, Poland

**Abstract.** Let $H$ be a fixed graph. We say that a graph $G$ admits an $H$-decomposition if the set of edges of $G$ can be partitioned into subsets generating graphs isomorphic to $H$. Denote by $\mathcal{P}_H$ the problem of exsitence of an $H$-decomposition of a graph. The Holyer's problem is to classify the problems $\mathcal{P}_H$ according to their computational complexities. In this paper we outline the proof of polynomiality of the problem $\mathcal{P}_H$ for $H$ being the union of $s$ disjoint 2-edge paths. This case is believed to bear the main difficulties among so far uncovered cases.

## 1 Introduction

In this paper we deal with so-called edge decompositions of graphs. A set $\{G_1, ..., G_s\}$ is called a *decomposition* of $G$ if $E(G_1) \cup ... \cup E(G_s) = E(G)$ and $E(G_i) \cap E(G_j) = \emptyset$, for $i \neq j$. Let $H$ be a graph. An $H$-decomposition is a decomposition $\{G_1, G_2, ..., G_s\}$ of $G$ such that each $G_i$ is isomorphic to $H$. Denote by $K_p$ the complete graph of order $p$ and by $K_{p,q}$ the complete bipartite graph with $p$- and $q$-element vertex classes. For vertex disjoint graphs $G$ and $H$ let $G \uplus H$ be their union and let $pH$ be the disjoint union of $p$ copies of $H$.

Edge decompositions have become a fast growing area of research in graph theory for the last two decades (see the survey papers by Bermond and Sotteau [2], Chung and Graham [6] and a more recent monograph by Bosak [4]) motivated by many connections to other branches of combinatorics and applications in operations research.

Among others much effort has been made towards solving computational complexity problems in this area. The central question, we also consider in this paper, is to establish the complexity status of the problem $\mathcal{P}_H$ of existence of an $H$-decomposition of a graph $G$, where $H$ is a fixed graph (not a part of the instance).

Holyer [11] proved the NP-completeness of the problem $\mathcal{P}_H$ for complete graphs of order at least 3, for paths of order at least 4 and for cycles. When $H$ is a path of order 2 the problem is trivially polynomial and for $H$ being a path of order 3 its polynomiality is not hard to show.

Holyer [11] conjectured that the problem $\mathcal{P}_H$ is NP-complete if and only if $H$ has at least 3 edges. His conjecture turned out to be false. Bialostocki and Roditty [3] proved polynomiality of the problem $\mathcal{P}_H$ for $H = 3K_2$. Their result was generalized by Alon [1] who gave a polynomial algorithm solving $\mathcal{P}_H$ when $H = sK_2$ for any fixed positive integer $s$. Similar results were also obtained independently by Brouwer and Wilson [5].

Other positive results were proved by Favaron, Lonc and Truszczyński [9] who gave a polynomial characterization of the class of graphs admitting an $H$-decomposition for $H = K_{1,2} \uplus K_2$. Their result was extended by Priesler and Tarsi who showed that $\mathcal{P}_H$ is polynomial when $H = K_{1,2} \uplus tK_2$, for any fixed positive integer $t$.

Negative (i.e. NP-completeness) results were also obtained by several authors (see Cohen and Tarsi [7], Masuyama and Hakimi [12]). Recently Dor and Tarsi [8] proved a strong result that the problem $\mathcal{P}_H$ is NP-complete whenever $H$ contains a connected component with at least 3 edges. They conjecture that $\mathcal{P}_H$ is polynomial in the remaining cases, i.e. when $H = sK_{1,2} \uplus tK_2$. In view of the above mentioned results, to solve completely the problem of complexity of $\mathcal{P}_H$, we have to establish it for $H = sK_{1,2} \uplus tK_2$ and $s \geq 2$.

Here is the main result of this paper.

**Main Theorem** *The problem $\mathcal{P}_H$, where $H = sK_{1,2}$, is polynomial.*

This case is believed to bear the main difficulties among so far uncovered cases. We suppose that the methods developed in this paper should lead to the complete classification of the problems $\mathcal{P}_H$ according to their computational complexity.

It is perhaps surprising that an analogous problem for vertex partitions is much easier and was solved completely by Hell and Kirkpatrick [10]. By a *strong $H$-factor* of a graph $G$ we mean (c.f. [10]) a set $\{G_1, ..., G_s\}$ of induced subgraphs of $G$ such that the sets $V(G_1), ..., V(G_s)$ form a partition of $V(G)$ and each $G_i$ is isomorphic to $H$. Let S-FACT($H$) denote the problem of existence of a strong $H$-factor for a given graph $G$. Hell and Kirkpatrick [10] have shown that the problem S-FACT($H$) is NP-complete if and only if $H$ has at least 3 vertices. Otherwise it is polynomial.

## 2 Outline of the Method

We call a $K_{1,2}$-decomposition $\pi$ of $G$ *k-good with respect to* $X \subseteq V(G)$ if for every $x \in X$, $|\pi(x)| \leq k$, where $\pi(x)$ is the set of elements of $\pi$ whose vertex sets contain $x$. By a *k-good* decomposition of $G$ we mean a $k$-good decomposition with respect to $V(G)$.

Let us sketch the main steps of the proof of polynomiality of the problem of existence of an $sK_{1,2}$-decomposition of a graph. Note that trivial necessary conditions for existence of an $sK_{1,2}$-decomposition of a graph $G$ are that $e(G)$

is divisible by $2s$ and $G$ admits an $e(G)/2s$-good decomposition. It is shown that, for a fixed $s$, the problem to decide whether a graph $G$ admits an $e(G)/2s$-good decomposition is polynomial. Moreover we prove that, in a sense, the above necessary conditions are "almost" sufficient. More precisely, we show that if $e(G)$ is large enough and divisible by $2s$, $G$ admits an $e(G)/2s$-good decomposition, $e(G) \equiv 0 \pmod{2s}$ and $G \notin \mathcal{G}$ (for some specific class of graphs $\mathcal{G}$) then $G$ has an $sK_{1,2}$-decomposition. Both recognition of the class $\mathcal{G}$ and existence of an $sK_{1,2}$-decomposition of a graph in $\mathcal{G}$ are polynomial problems. These facts together prove that the problem of existence of an $sK_{1,2}$-decomposition is polynomial. The algorithm would proceed as follows:

1. Check whether $G$ has an $e(G)/2s$-good decomposition. If the answer is NO then STOP ($G$ does not admit an $sK_{1,2}$-decomposition) otherwise go to 2.

2. Check whether $G \notin \mathcal{G}$. If the answer is YES then STOP ($G$ admits an $sK_{1,2}$-decomposition) otherwise go to 3.

3. Check whether $G \in \mathcal{G}$ admits an $sK_{1,2}$-decomposition.

## 3 Preliminary Lemmas

By $|G|$ and $e(G)$ we denote the order and the size of the graph $G$, respectively.

Let $Y$ be a finite set of vertices and $\pi$ a collection of graphs isomorphic to $K_{1,2}$. Then denote

$$\pi_i^Y = \{f \in \pi : f \text{ has its central vertex in } Y \text{ and } |V(f) \cap Y| = i\}.$$

Moreover define $\pi^Y = \pi_1^Y \cup \pi_2^Y \cup \pi_1^Y$, $\pi_i^Y(x) = \pi(x) \cap \pi_i^Y$ and $\pi^Y(x) = \pi(x) \cap \pi^Y$.

A collection $\pi$ of graphs isomorphic to $K_{1,2}$ is called a *star* if there is a vertex $x$ (the *center*) such that $\forall f, g \in \pi, V(f) \cap V(g) = \{x\}$. Note that for every $Y$ such that $x \in Y$, $\pi_1^Y(x)$ is a star. By a $K_{1,2}$-*matching* we mean a collection $\pi$ of graphs isomorphic to $K_{1,2}$ such that $\forall f, g \in \pi, V(f) \cap V(g) = \emptyset$.

A collection $\pi$ of graphs isomorphic to $K_{1,2}$ is called *fair with respect to* $X \subseteq V$ if

(1) $$\forall x \in X, \quad \pi_2^{V-X}(x) \text{ is a star,}$$

(2) $$|\pi_1^{V-X}| \le |X| - 1 \quad \text{and}$$

(3) $$|\pi_2^{V-X}| \le o(G - X) + 2(|X| - 1).$$

(For a graph $F$, $o(F)$ denotes the number of components of odd order in $F$.)
Let $\delta_G(X) = e(G - X) - o(G - X)$.

For a collection $\pi$ of graphs isomorphic to $K_{1,2}$ denote by $G(\pi)$ the graph with the vertex set $\bigcup_{f \in \pi} V(f)$ and the edge set $\bigcup_{f \in \pi} E(f)$.

Here we are some preliminary lemmas.

**Lemma 1** *If $\pi$ is a $k$-good fair decomposition of a graph $G = (V, E)$ with respect to $X \subseteq V$ then*

$$2 \sum_{x \in X} |\pi(x)| \leq e(G) - \delta_G(X) + \frac{1}{2}|X|(|X| + 3).$$

**Lemma 2** *Suppose a graph $G$ admits a $K_{1,2}$-decomposition $\pi$. If $r \leq \frac{e(G)}{6\Delta(G)}$ then $\pi$ contains a $K_{1,2}$-matching of size $r$.*

**Proof.** Let $\pi'$ be is the largest $K_{1,2}$-matching contained in $\pi$ and suppose that $|\pi'| < r$. Note that at most $|\pi'|(1 + 2(\Delta(G) - 1) + \Delta(G) - 2) = 3|\pi'|(\Delta(G) - 1)$ members of $\pi$ have a common vertex with some member of $\pi'$. Since

$$|\pi| - 3|\pi'|(\Delta(G) - 1) \geq \frac{1}{2}e(G) - 3(r - 1)(\Delta(G) - 1)$$

$$\geq 3r\Delta(G) - 3(r - 1)(\Delta(G) - 1) \geq 1,$$

$\pi'$ can be extended into a larger $K_{1,2}$-matching. ■

**Lemma 3** *Let a bipartite graph $G = (X, Y; E)$ be a forest. If all vertices of degree at most 1 belong to $X$ then $|G| \leq 2|X| - 1$.*

The proof is routine.

**Definition** Let $\pi_1, ..., \pi_k$, $|\pi_1| \geq ... \geq |\pi_k|$, be stars with centers $x_1, ..., x_k$, respectively. If $\pi_1, ..., \pi_k$, are vertex-disjoint then $\pi = \pi_1 \cup ... \cup \pi_k$ is called a *constellation* of type $(|\pi_1|, ..., |\pi_k|)$, with the base $\{x_1, ..., x_k\}$. If $x_i \notin V(\pi_j)$ and $E(\pi_i) \cap E(\pi_j) = \emptyset$, for $i \neq j$ then $\pi = \pi_1 \cup ... \cup \pi_k$ is called a *pseudoconstellation* with the base $\{x_1, ..., x_k\}$.

**Lemma 4** *Let $t$ be an integer and $\pi$ a pseudoconstellation with the base $X = \{x_1, ..., x_s\}$. If, for $i = 1, ..., s$, $|\pi(x_i)| \geq 2ts$ then there is a constellation $\nu \subseteq \pi$ with base $X$ of type $(t, t, ..., t)$, where $t$ is repeated $s$ times.*

## 4 Characterization of $k$-good Decompositions

Let $G = (V, E)$ be a graph and $X \subseteq V$. Define $Y$ to be the set of components of $G - X$ and $Z$ the set of edges with both endvertices in $X$. Let $B(G)$ be a bipartite graph with vertex classes $X$ and $Y \cup Z$ such that $xy$ is an edge in $B(G)$ if either $x \in X$, $y \in Y$ and there is an edge in $G$ with one endvertex $x$ and the other one in $y$ or $x \in X$, $y \in Z$ and $x$ is an endvertex of $y$ in $G$.

**Theorem 1** *The following conditions are equivalent*

(a) *G admits a k-good decomposition with respect to $X$,*
(b) *G admits a k-good fair decomposition with respect to $X$,*
(c) *the graph $B(G)$ has a subgraph $F$ such that*

    (i) $\forall x \in X, \quad \deg_F x \leq 2k - \deg_G x,$
    (ii) $\forall x \in X, \quad \deg_F x \equiv \deg_G x \pmod 2,$
    (iii) $\forall y \in Y, \quad \deg_F y \equiv e(y) \pmod 2,$
    (iv) $\forall y \in Z, \quad \deg_F y = 1.$

To give the reader a flavor of the methods used in the proof of this theorem we shall show the hardest part:

    (c) $\Rightarrow$ (b).

Notice that we can assume without loss of generality that $F$ is a forest for if it contains a cycle then we can delete its edges and the resulting graph will still satisfy (i)-(iv). We shall split the graph $G$ into several parts admitting $K_{1,2}$-decompositions.

For every edge $xy$ in $F$, $x \in X$, $y \in Y$, choose an edge $e_x^y$ in $G$ with one endvertex at $x$ and the other one in $y$. Such an edge exists by the definition of $B(G)$. Let, for $y \in Y$, $G_y$ be the connected graph induced by the union of the edges of $y$ and the set $\{e_x^y : xy \in E(F)\}$. Each vertex of $X$ belonging to $G_y$ has the degree equal to 1 in $G_y$. Clearly, by (iii)

$$e(G_y) = e(y) + \deg_F y \equiv 0 \pmod 2.$$

Thus $G_y$ admits a $K_{1,2}$-decomposition. Denote it by $\pi_y$. All central vertices of elements of $\pi_y$ belong to $y$.

Let, for $x \in X$, $G_x$ be the connected graph induced by the set of the edges with one endvertex at $x$ and the other one not in $X$ which do not belong to any graph $G_y$, $y \in Y$ and the edges $z \in Z$ with one endvertex at $x$ such that $x$ is not joined to $z$ in $F$. Clearly each graph $G_x$ is a star.

Notice that by (ii)

$$e(G_x) = \deg_G x - \deg_F x \equiv 0 \pmod 2$$

so $G_x$ admits a $K_{1,2}$-decomposition. Denote it by $\pi_x$.

It is routine to check (using (iv)) that $\{G_y : y \in Y\} \cup \{G_x : x \in X\}$ is a decomposition of $G$. Define $\pi = \bigcup_{w \in X \cup Y} \pi_w$. Clearly $\pi$ is a $K_{1,2}$-decomposition of $G$. Since, for any $x \in X$, $G_x$ is a star with the center at $x$, we get by (i)

$$|\pi(x)| \leq \frac{1}{2}e(G_x) + \deg_G x - e(G_x) = \deg_G x - \frac{1}{2}e(G_x) = \frac{1}{2}(\deg_G x + \deg_F x) \leq k.$$

Thus $\pi$ is $k$-good with respect to $X$.

Consider any $x \in X$ and $a, b \in \pi_2^{V-X}(x)$, $a \neq b$, and suppose that there is $v \in (V(a) \cap V(b)) - \{x\}$. Assume first that $v \in y$ for some component $y \in Y$. Then, by the definition of $\pi$, $a, b \in \pi_y$. Thus the central vertices of $a$ and $b$ belong to $y$ and consequently $x$ has degree at least 2 in $G_y$, a contradiction.

Now suppose that $v \in X$. Since $a, b \in \pi^{V-X}$, the central vertices of $a$ and $b$ belong to some components $y_a$ and $y_b$ of $V - X$, respectively. If $y_a = y_b$ then $x$ has again the degree at least 2 in $G_{y_a}$. Hence $y_a \neq y_b$. By the definition of $\pi$, $xy_a, xy_b, vy_a, vy_b \in E(F)$. These edges form, however, a cycle in $F$ contrary to the assumption that $F$ is a forest. Thus (1) is satisfied.

To show (2) notice that by the definition of $\pi$

$$(4) \qquad |\pi_1^{V-X}| = \sum_{y \in Y} |\pi_1^y| \leq \sum_{y \in Y} \lfloor \tfrac{1}{2} \deg_F y \rfloor \leq \frac{1}{2} \sum_{y \in Y'} \deg_F y,$$

where $Y' \subseteq Y$ is the set of vertices of degree at least 2 in $F$. The subgraph F' of F induced by $X \cup Y'$ has all its vertices of degree less then 2 in X. Thus by Lemma 3

$$(5) \qquad \sum_{y \in Y'} \deg_F y = e(F') \leq |F'| - 1 \leq 2|X| - 2.$$

We get (2) by (4) and (5).

Finally to show (3), by the definition of $\pi$ we get

$$(6) \qquad |\pi_2^{V-X}| = e(F) = |Y''| + e(F'),$$

where $Y'' \subseteq Y$ is the set of vertices of degree 1 in $Y$. By (iii), for every $y \in Y''$, $e(y) \equiv \deg_F y = 1 \pmod 2$. Thus

$$(7) \qquad |Y''| \leq o(G - X).$$

By (5), (6) and (7)

$$|\pi_2^{V-X}| \leq o(G - X) + 2(|X| - 1).$$

Thus $\pi$ is fair with respect to $X$. ∎

Let $t = |X|$ be a fixed number not depending on the size of the graph $G$. We shall show that then the condition (c) in Theorem 1 can be checked in polynomial number of steps with respect to the order of the graph $G$.

First notice that without loss of generality we can assume that the subgraph $F$ in (c) is a forest (see the first paragraph of the part (c)⇒(b) in the proof of Theorem 1).

We shall show that there are only polynomially many possibilities to check while looking for a subgraph $F$ of $B(G)$ being a forest and satisfying (iii) and (iv).

By Lemma 3 all but at most $2t - 1$ vertices in $F$ are vertices of degree 1 belonging to $Y \cup Z$. Indeed, after deletion of these degree 1 vertices from $F$ we get a forest whose all degree 1 vertices are in $X$. Now we use Lemma 3.

Trying to construct a forest $F$ satisfying (iii) and (iv) we first choose edges whose no endvertex is a degree 1 vertex in $F$ belonging to $Y \cup Z$. Since there are at most $2t - 2$ such edges, we have $O(e(B(G))^{2t-2}) \leq O(m^{2t-2})$ possibilities to choose from (where $m = e(G)$). It remains to add the edges with degree 1 endvertices in $Y \cup Z$.

Divide the set $Y_1 \cup Z$, where $Y_1 = \{y \in Y : e(y) \equiv 1 \pmod{2}\}$, into $2^t - 1$ classes $Y_A = \{y \in Y_1 \cup Z : A$ is the set of neighbors of $y$ in $B(G)\}$, where $A$ are nonempty subsets of $X$. To explore all possibilities of constructing $F$, we divide each class $Y_A$ into $|A| + 1$ subclasses $Y_A^x = \{y \in Y_A : xy$ is an edge in $F\}$, where $x \in X$ and $Y_A^\bullet = Y_A' - \bigcup_{x \in X} Y_A^x$. Note that all vertices in $Y_A$ have the same neighbors in $B(G)$ so the number of possibilities to check here is equal to the number of partitions of the number $|Y_A|$ into $|A| + 1$ parts. Since $|Y_A| \leq |Y| + |Z| \leq n + \binom{t}{2}$ (where $n = |G|$) and $|A| \leq t$, the number is not greater than $O(n^t)$. Summarizing, while constructing a subgraph $F$ of $B(G)$ being a forest and satisfying (iii) and (iv), we have to check not more than $O(m^{2t-2} \cdot (2^t - 1) \cdot n^t) \leq O(n^{5t-2})$ possibilities.

For each such constructed $F$ we check if the conditions (i) and (ii) are satisfied. This, clearly, can be done in polynomially many steps too.

As we noted at the beginning of this paper, a necessary condition for existence of an $sK_{1,2}$-decomposition of a graph $G$ is existence of an $e(G)/2s$-good decomposition of $G$. We shall show that existence of an $e(G)/2s$-good decomposition of $G$ is equivalent to existence of an $e(G)/2s$-good decomposition of $G$ with respect to the set $X$ of $4s$ vertices in $G$ with the highest degrees. Indeed, let $\pi$ be an $e(G)/2s$-good decomposition of $G$ with respect to $X$ and $x_0 \notin X$. Then $\deg_G x_0 + \sum_{x \in X} \deg_G x \leq 2e(G)$, so

$$|\pi(x_0)| \leq \deg_G x_0 \leq \frac{2e(G)}{4s+1} < \frac{e(G)}{2s}$$

and consequently $\pi$ is $e(G)/2s$-good.

Thus we have the theorem.

**Theorem 2** *Let $s$ be fixed. Deciding whether a graph $G$ has an $e(G)/2s$-good decomposition is a polynomial problem.*

## 5 Polynomiality of the $sK_{1,2}$-decomposition Problem

The next theorem shows that for "most" graphs existence of an $e(G)/2s$-good decomposition implies existence of an $sK_{1,2}$-decomposition.

**Theorem 3** *Let $G$ be a graph such that $(\forall Y \subseteq V(G), |Y| \leq s + 1)$ $\delta_G(Y) \geq R$, where $R = R(s) = 32 \cdot 10^8 \cdot s^{10}$. Then $G$ admits an $sK_{1,2}$-decomposition if and only if $e(G) \equiv 0 \pmod{2s}$ and $G$ admits an $e(G)/2s$-good decomposition.*

(Recall that $\delta_G(Y) = e(G - Y) - o(G - Y)$, where $o(G - Y)$ denotes the number of components of odd order in $G - Y$.)

This is a crucial theorem so we shall describe a plan of its proof. First we find a large constellation in $G$ using Lemma 4 and delete it for a while. For the resulting graph $G'$ we apply the following lemma.

**Lemma 5** *If $G$ admits an $e(G)/2s$-good decomposition then $G$ can be decomposed into graphs $G_1$ and $G_2$ such that $G_1$ admits an $sK_{1,2}$-decomposition, $G_2$ admits an $e(G)/2s$-good decomposition and $e(G_2) \leq 500s^3$.*

Finally we decompose into the graphs $sK_{1,2}$ the union of the deleted large constellation and the (small) graph $G_2$.

Let $\mathcal{G}$ be the set of graphs $G$ such that the size of the set $Y_0$ of vertices of degree at least $2R$ in $G$ is equal to either $s$ or $s + 1$. Call this set $Y_0$ the *kernel* of $G \in \mathcal{G}$. We shall show a corollary from Theorem 3.

**Theorem 4** *Let $e(G) \geq 2s(5R-2)$. If $G$ admits an $e(G)/2s$-good decomposition, $e(G) \equiv 0 \pmod{2s}$ and $G \notin \mathcal{G}$ then $G$ admits an $sK_{1,2}$-decomposition.*

**Proof.** Suppose first that there are at least $s + 2$ vertices of degree at least $2R$ in $G$. Let $Y$ be any set of vertices of cardinality at most $s + 1$ and let $x_0$ be a vertex of degree at least $2R$ not in $Y$. Then the component of $G - Y$ containing $x_0$ has at least $2R - (s+1) \geq R+1$ edges. Thus $e(G-Y) \geq o(G-Y)-1+R+1$ so $\delta_G(Y) \geq R$ and we are done by Theorem 2.

Now suppose that there are at most $s - 1$ vertices of degree at least $2R$ in $G$. Let $Y$ be any set of vertices of cardinality at most $s + 1$ and let $\pi$ be any $e(G)/2s$-good with respect to $Y$ decomposition of $G$. Since $|\pi(x)| \leq e(G)/2s$, there are at most $\frac{e(G)}{2s}(s-1)+2(2R-1)$ members of $\pi$ containing a vertex from $Y$. Thus at least $\frac{e(G)}{2} - \frac{e(G)}{2s}(s-1) - 2(2R-1) = \frac{e(G)}{2s} - 2(2R-1)$ of them do not contain any vertex from $Y$. Thus

$$e(G - Y) - o(G - Y) \geq \frac{e(G)}{2s} - 2(2R - 1) \geq R$$

so we are done by Theorem 2 again. ∎

It is clear that the recognition of the class $\mathcal{G}$ is polynomial. So to prove our Main Theorem it suffices to show that, for a fixed $s$, deciding whether $G \in \mathcal{G}$ has an $sK_{1,2}$-decomposition is a polynomial problem. It follows from Lemma 6 and Theorem 5.

**Lemma 6** *If $G \in \mathcal{G}$, $e(G) \geq 13Rs$ and $\delta_G(Y_0) \leq 5R$, where $Y_0$ is the kernel of $G$ then $G$ has an $sK_{1,2}$-decomposition if and only if it can be decomposed into graphs $G_1$ and $G_2$ admitting $sK_{1,2}$-decompositions and such that $e(G_2) \leq 12Rs$, $e(G_1) \geq Rs$ and $G_1$ is a pseudoconstellation with the base $Y_0$.*

**Theorem 5** *If $G \in \mathcal{G}$ is a pseudoconstellation with the base $Y_0$ and $e(G) \geq Rs$ then $G$ has an $sK_{1,2}$-decomposition if and only if $e(G) \equiv 0 \pmod{2s}$ and $G$ has an $e(G)/2s$-good decomposition.*

One can easily check that if $\delta_G(Y_0) > 5R$, where $Y_0$ is the kernel of $G \in \mathcal{G}$ then $\delta_G(Y) > R$, for every subset $Y \subseteq V(G)$ of cardinality at most $s + 1$. So by Theorems 2 and 3 deciding whether $G$ admits an $sK_{1,2}$-decomposition is polynomial. Thus Lemma 6 covers the only uncovered case.

Note that checking whether a graph $G$ can be decomposed into subgraphs $G_1$ and $G_2$ described in Lemma 6 can be done in polynomially many steps with respect to the size of $G$. To see it notice that the size of $G_2$ is a constant with respect to the size of $G$. Thus there are only polynomially many decompositions of $G$ into subgraphs $G_1$ and $G_2$ such that $e(G_2) \leq 12Rs$. Deciding whether $G_1$ has an $sK_{1,2}$-decomposition is also a polynomial problem which follows from Theorems 2 and 5. This completes the proof of polynomiality of the Main Theorem.

# References

1. N. Alon: A note on the decomposition of graphs into isomorphic matchings, Acta Math. Acad. Sci. Hung. 42 (1983), 221-223
2. J.C. Bermond, D. Sotteau: Graph decomposition and $G$-designs, Proc. of the 5th British combinatorial conference, Aberdeen 1975
3. A. Bialostocki, Y. Roditty: $3K_2$-decomposition of a graph, Acta Math. Acad. Sci. Hung. 40 (1982), 201-208
4. J. Bosak: Graph Decompositions, Springer-Verlag 1990
5. A.E. Brouwer, R.M. Wilson: The decomposition of graphs into ladder graphs, Stiching Mathematisch Centrum (zn 97/80), Amsterdam 1980
6. F.R.K. Chung, R.L. Graham: Recent results in graph decompositions, In Combinatorics (H.N.V. Temperley, ed.), London Math. Soc., Lecture Notes Series 52 (1981), 103-124
7. E. Cohen, M. Tarsi: NP-completeness of graph decomposition problem, Journal of Complexity 7 (1991), 200- 212
8. D. Dor, M. Tarsi: Graph decomposition is NPC - A complete proof of Holyer's conjecture, 1992, preprint
9. O. Favaron, Z. Lonc, M. Truszczyński: Decomposition of graphs into graphs with three edges, Ars Combinatoria 20 (1985), 125-146
10. P. Hell, D.G. Kirkpatrick: On the complexity of general graph factor problems, SIAM Journal of Computing 12(1983), 601-609
11. I. Holyer: The NP-completeness of some edge partition problems, SIAM J. of Comp. 10 (1981), 713-717
12. Masuyama, Hakimi: Edge packing in graphs, preprint

# Graphs, Hypergraphs and Hashing

George Havas[1], Bohdan S. Majewski[1], Nicholas C. Wormald[2]
and Zbigniew J. Czech[3]

[1] Key Centre for Software Technology, Department of Computer Science, University
of Queensland, Queensland 4072, Australia
[2] Department of Mathematics, University of Melbourne, Parkville, Victoria 3052,
Australia
[3] Institutes of Computer Science, Silesia University of Technology and Polish
Academy of Sciences, 44–100 Gliwice, Poland

**Abstract.** Minimal perfect hash functions are used for memory efficient storage and fast retrieval of items from static sets. We present an infinite family of efficient and practical algorithms for generating minimal perfect hash functions which allow an arbitrary order to be specified for the keys. We show that almost all members of the family are space and time optimal, and we identify the one with minimum constants. Members of the family generate a minimal perfect hash function in two steps. First a special kind of function into an $r$−graph is computed probabilistically. Then this function is refined deterministically to a minimal perfect hash function. We give strong practical and theoretical evidence that the first step uses linear random time. The second step runs in linear deterministic time. The family not only has theoretical importance, but also offers the fastest known method for generating perfect hash functions.

## 1 Introduction

Consider a set $W$ of $m$ keys, where $W$ is a subset of a universe $U = \{0, \ldots, u-1\}$. For simplicity we assume that the keys in $W$ are either integers or strings of characters. In the latter case the keys can either be treated as numbers base $|\Sigma|$ where $\Sigma$ is the alphabet in which the keys were encoded, or as sequences of characters over $\Sigma$. For convenience we assume that $u$ is a prime.

A *hash function* is a function $h : W \to I$ that maps the set of keys $W$ into some given interval of integers $I$, say $[0, k-1]$, where $k \geq m$. The hash function, given a key, computes an address (an integer from $I$) for the storage or retrieval of that item. The storage area used to store items is known as a *hash table*. Keys for which the same address is computed are called *synonyms*. Due to the existence of synonyms a situation called *collision* may arise in which two keys are mapped into the same address. Several schemes for resolving collisions are known. A perfect hash function is an injection $h : W \to I$, where $W$ and $I$ are sets as defined above, $k \geq m$. If $k = m$, then we say that $h$ is minimal perfect hash function. As the definition implies, a perfect hash function transforms each key of $W$ into a unique address in the hash table. Since no collisions occur each item can be retrieved from the table in a single probe.

Minimal perfect hash functions are used for memory efficient storage and fast retrieval of items from a static set, such as reserved words in programming languages, command names in operating systems, commonly used words in natural languages, etc. Overviews of perfect hashing are given in Gonnet and Baeza-Yates (1991, §3.3.16) and Meyer auf der Heide (1990), the area is surveyed in Lewis and Cook (1988) and Havas and Majewski (1992), and some recent independent developments appear in Fox, Chen, Daoud and Heath (1991) and Fox, Heath, Chen and Daoud (1992).

Various other algorithms with different time complexities have been presented for constructing perfect or minimal perfect hash functions. They fall into four broad categories.

1. Number theoretical methods. These involve the determination of a small number of numeric parameters, using methods based on results in number theory. They include: Sprugnoli (1977); Jaeschke (1981); Chang (1984); Chang and Lee (1986); Chang and Chang (1988); Winters (1990b).

2. Perfect hash functions with segmentation. Keys are initially distributed into smaller sets, kept in buckets, by an ordinary, first-stage hash function. For all keys in a bucket a perfect hash function is computed. Algorithms falling into this category are: Du, Hsieh, Jea and Shieh (1983); Fredman, Komlós and Szemerédi (1984); Slot and van Emde Boas (1984); Cormack, Horspool and Kaiserwerth (1985); Yang and Du (1985); Jackobs and van Emde Boas (1986); Dietzfelbinger and Meyer auf der Heide (1990); Schmidt and Siegel (1990b); Winters (1990a); Dietzfelbinger, Gil, Matias and Pippenger (1992).

3. Algorithms based on restricting the search space. These methods usually use some kind of backtracking procedures to search through the space of all possible functions, in order to find a perfect hash function. To limit the search space, and in effect to speed up the search, an ordering heuristic is applied to keys before the search begins. Solutions belonging to this category include: Cichelli (1980); Cook and Oldehoeft (1982); Cercone, Boates and Krause (1985); Sager (1985); Haggard and Karplus (1986); Brain and Tharp (1989); Gori and Soda (1989); Fox, Heath, Chen and Daoud (1992); Fox, Chen and Heath (1992); Czech and Majewski (1992).

4. Algorithms based on sparse matrix packing. The main idea behind these solutions is to map $m$ keys uniformly into an $m \times m$ matrix. Then, using a matrix packing algorithm (Mehlhorn 1984, p. 108–118), compress the two-dimensional array into linear space. This type of approach is adopted by Tarjan and Yao (1979), Brain and Tharp (1990) and Chang and Wu (1991). A modification of this approach, which leads to a more compact hash function, is presented by Chang, Chen and Jan (1991).

The algorithms in each of the categories provide distinct solutions, using similar ideas but different methods to approach them. Significant advances are made in Fox, Heath, Chen and Daoud (1992), Fox, Chen and Heath (1992) and Czech, Havas and Majewski (1992) where algorithms using linear space and time are described.

We present a related family of algorithms based on generalized random graphs for finding order preserving minimal perfect hash functions of the form

$$h(w) = g\big(f_1(w)\big) \diamond g\big(f_2(w)\big) \diamond \cdots \diamond g\big(f_r(w)\big)$$

where $\diamond$ is a binary operation. For simplicity we choose $\diamond$ to be addition modulo $m$. (Alternatively we could choose exclusive or, giving benefits in speed and avoiding overflow for large $m$.) We show that each member of the family, for a suitable choice of parameters, constructs a minimal perfect hash function for $W$ in $O(m)$ expected time and requires $O(m \log m)$ space. (The theoretical derivation is based on a reasonable assumption about uniformity of the graphs involved in our algorithms.) These algorithms are both efficient and practical.

Throughout this paper $\log(n)$ means $\log_2(n)$, and $y^{\underline{i}}$ denotes the falling factorial $y$, $y^{\underline{i}} = y(y-1) \cdots (y-i+1)$. A generalized graph, called an $r$–graph, is a graph in which each edge is a subset of $V$ containing precisely $r$ elements, $r \geq 1$.

## 2  Graphs and Hashing

The first method which used graphs (implicitly) was published by Fredman, Komlós and Szemerédi (1984). Although the authors did not cast their algorithm in graph theoretical terms it is easy to do so. Thus we can regard their first step as mapping a set of $m$ keys into $m$ primary vertices. For each group of keys that have been mapped into the same primary vertex, a second mapping takes each key in the group into a unique vertex. The result is an graph which is a union of star shaped trees. Fredman, Komlós and Szemerédi provide a structure that efficiently encodes the resulting graph and prove that the number of the vertices, which determines the size of the structure, does not exceed $11m$. Since the $m$ primary vertices also need to store parameters for the second mapping, the total size of the structure is $13m$. (Fredman, Komlós and Szemerédi (1984) mention a complicated refinement that allows some reduction in the size of the encoding.)

The graph created in this scheme is a union of stars. Such graphs are not very common, which is why the number of vertices must be much greater than the number of edges (which correspond to the keys). This reduces the practicality of the method, requiring more space than desirable. Our techniques rely upon mapping the keys into any acyclic graph. This leads to two advantages: much smaller constants and the ability to generate perfect hash functions that allow arbitrary arrangement of keys in the hash table. The space requirements for our major data structures can be as little as $1.23m$.

## 3  The Family

In order to generate a minimal perfect hash function we first compute a special kind of function from the $m$ keys into an $r$–graph with $m$ edges and $n$ vertices, where $n$, depending on $m$ and $r$, is determined in Section 5. The special feature is that the edges of the resulting $r$–graph must be independent, a notion we define later. We achieve edge independence probabilistically. Then deterministically we refine this function (from the keys into an $r$–graph) to a minimal perfect hash function. The expected time for finding the hash function is $O(rm + n)$. This type of approach works for any $r > 0$. As the family of $r$–graphs, for $r > 0$ is infinite, we have an infinite family of algorithms for generating minimal perfect hash functions.

Consider the following assignment problem. For a given $r$–graph $G = (V, E)$, $|E| = m$, $|V| = n$, where each $e \in E$ is an $r$–subset of $V$, find a function $g : V \rightarrow [0, m - 1]$ such that the function $h : E \rightarrow [0, m - 1]$ defined as

$$h(e = \{v_1, v_2, \ldots, v_r\} \in E) = (g(v_1) + g(v_2) + \cdots + g(v_r)) \bmod m$$

is a bijection. In other words we are looking for an assignment of values to vertices so that for each edge the sum of values associated with its vertices, modulo the number of edges, is a unique integer in the range $[0, m - 1]$.

This problem does not always have a solution for arbitrary graphs. However, if the graph $G$ fulfills an edge independence criterion, a simple procedure can be used to find values for each vertex.

**Definition** *Edges of an $r$–graph $G = (V, E)$ are independent if and only if repeated deletion of edges containing vertices of degree 1 results in a graph with no edges in it.*

Observe that the definition is equivalent to the requirement that the $r$–graph does not contain a subgraph with minimum degree 2. Such subgraphs, discussed by Duke (1985, p. 407), are a natural generalization of cycles in 2–graphs.

To solve the assignment problem we proceed as follows. Associate with each edge a unique number $h(e) \in [0, m - 1]$ in any order. Consider the edges in reverse order to the order of deletion during a test of independence, and assign values to each as yet unassigned vertex in that edge. As the definition implies, each edge (at the time it is considered) will have at least one unique vertex to which no value has yet been assigned. Let the set of unassigned vertices for edge $e$ be $\{v_1, v_2, \ldots, v_j\}$. For edge $e$ assign 0 to $g(v_1)$, $g(v_2)$, $\ldots$, $g(v_{j-1})$ and set $g(v_j) = (h(e) - \sum_{v \in e, v \neq v_j} g(v)) \bmod m$.

To prove the correctness of the method it is sufficient to show that the value of the function $g$ is computed exactly once for each vertex, and for each edge we have at least one unassigned vertex by the time it is considered. This property is clearly fulfilled if the edges of $G$ are independent and they are processed in the reverse order to that imposed by the test for independence.

Unfortunately, the independence test suggested directly by the definition is not fast enough. For $r$–graphs, with $r > 1$ it is easy to find examples for which it requires $O(m^2)$ time. Consequently we must find a better method to test and arrange edges of an independent $r$–graph, for $r > 0$. One solution follows.

Initially mark all the edges of the $r$–graph as not removed. Then scan all vertices, each vertex only once. If vertex $v$ has degree 1 (that is, belongs to only one edge) then remove the edge $e \ni v$ from the $r$–graph. As soon as edge $e$ is removed check if any other of its vertices has now degree 1. If yes, then for each such a vertex remove the unique edge to which this vertex belongs. Repeat this recursively until no further deletions are possible. After all vertices has been scanned, check if the $r$-graph contains edges. If so, the $r$-graph has failed the independence test. If not, the edges are independent and the reverse order to that in which they were removed is one we are looking for. This method can be implemented so that the running time is $O(rm + n)$.

The solution to this assignment problem becomes the second part of our algorithms for generating minimal perfect hash functions. Now we are ready to present an algorithm for generating a minimal perfect hash function. The algorithm comprises two steps: mapping and assignment. In the mapping step the input set in mapped into an $r$-graph $G = (V, E)$, where $V = \{0, \ldots, n-1\}$, $E = \{\{f_1(w), f_2(w), \ldots, f_r(w)\} : w \in W\}$, and $f_i : U \rightarrow \{0, \ldots, n-1\}$. The step is repeated until graph $G$ passes the test for edge independence. Once this has been achieved the assignment step is executed. Generating a minimal perfect hash function is reduced to the assignment problem as follows. As each edge $e = \{v_1, v_2, \ldots, v_r\} \in E$ corresponds uniquely to some key $w$, such that $f_i(w) = v_i$, $1 \le i \le r$, the search for the desired function is straightforward. We simply set $h(e = \{f_1(w), f_2(w), \ldots f_r(w)\}) = i - 1$ if $w$ is the $i$th word of $W$, yielding the order preserving property. Then values of function $g$ for each $v \in V$ are computed by the assignment step (which solves the assignment problem for $G$). The function $h$ is an order preserving minimal perfect hash function for $W$.

To complete the description of the algorithm we need to define the mapping functions $f_i$. Ideally the $f_i$ functions should map any key $w \in W$ randomly into the range $[0, n-1]$. Total randomness is not efficiently computable, however the situation is far from hopeless. Limited randomness is often as good as total randomness (Carter and Wegman 1979b; Karlin and Upfal 1986; Schmidt and Siegel 1989; Schmidt and Siegel 1990a).

A suitable solution comes from the field originated by Carter and Wegman (1977) and called universal hashing. A class of universal hash functions $\mathcal{H}$ is a collection of generally good hash functions from which we can easily select one at random. A class is called $k$ universal if any member of it maps $k$ or less keys randomly and independent of each other. Carter and Wegman (1979a) suggested the use of polynomials as hash functions. They prove that polynomials of degree $d$ constitute a class of $(d+1)$ universal hash functions. Dietzfelbinger and Meyer auf der Heide (1990) proposed another class of universal hash functions. Their class, based on polynomials of a constant degree $d$ and tables of random numbers of size $m^\delta$, $0 < \delta < 1$, is $(d+1)$ universal but shares many properties of truly random functions (see Dietzfelbinger and Meyer auf der Heide (1990) and Meyer auf der Heide (1990) for details). Another class was suggested by Siegel (1989) which unfortunately requires a considerable amount of space. Finally, Dietzfelbinger, Gil, Matias and Pippenger (1992) proved that polynomials of degree $d \ge 3$ are reliable, meaning that they perform well with high probability. An advantage that this class offers is a compact representation of functions, as each requires only $O(d \log u)$ bits of space. Any of the above specified classes can be used for our purposes. Our experimental results show that polynomials of degree 3 or the class defined by Dietzfelbinger and Meyer auf der Heide (1990) are good choices, and in practice polynomials of degree one work and are faster.

The above suggested classes perform quite well for integer keys. Character keys however are more naturally treated as sequences of characters. For that reason we define one more class of universal hash functions, $C_n$, designed specially for character keys. (This class has been used by others including Fox, Heath, Chen and Daoud (1992).) We denote the length of the key $w$ by $|w|$ and its $i$-th character by $w[i]$. A member of this class, a function $f_i : \Sigma^* \to \{0, \ldots, n-1\}$ is defined as $f_i(w) = \left(\sum_{j=1}^{|w|} T_i(j, w[j])\right) \bmod n$, where $T_i$ is a table of random integers modulo $n$ for each character and for each position of a character in a word. Selecting a member of the class is done by selecting (at random) the mapping table $T_i$. We can prove the following theorem:

**Theorem 1.** *The expected number of keys mapped independently by a member of class $C_n$ is $O(|\Sigma| \log L)$.*

*Proof (outline).* Consider the following painting problem. We are given an urn containing $b$ white balls. At each step we take one ball at random, paint it red and return it to the urn. What is the expected number of white balls in the urn after $k$ steps? The urn corresponds to one column of the mapping table, and $b = |\Sigma|$. Choosing a white ball represents taking an unused entry of the column, hence the key for which such an entry is selected will be mapped independently of other keys. Solving the painting problem for $L$ independent urns proves the above theorem. □

The above defined class allows us to treat character keys in the most natural way, as sequences of characters from a finite alphabet $\Sigma$. However this approach has an unpleasant theoretical consequence. For any fixed maximum key length $L$, the total number of keys cannot exceed $\sum_{i=1}^{L} |\Sigma|^i \sim |\Sigma|^L$ keys. Thus either $L$ cannot be treated as a constant and $L \geq \log_{|\Sigma|} m = \Omega(\log m)$ or, for a fixed $L$, there is an upper limit on the number of keys. In the former case, strictly speaking, processing a key character by character takes nonconstant time. In practice, however, it is often faster and more convenient to use this method than to treat a character key as a binary string. Other hashing schemes use this approach, asserting that the maximum key length is bounded (for example: Sager 1985; Haggard and Karplus 1986; Pearson 1990; Fox, Heath, Chen and Daoud 1992). This is an abuse of the RAM model (Aho, Hopcroft and Ullman 1974, pp. 5–14), however it is a purely *practical* abuse. (In the RAM model with uniform cost measure it is assumed that each operation on a numeric key takes constant time, which corresponds to being able to process a character key character by character in constant time.)

## 4  Some Benefits of Arbitrary Key Arrangements

Unlike traditional hashing functions, our method allows the keys to be arranged in any specified order in the hash table. Furthermore, we can modify the perfect assignment problem so that for any predefined function $h$ into the cardinal numbers, not necessarily a bijection, we are able to find a suitable function $g$ in linear time. This can be easily extended to functions into integers, rational numbers or character strings in natural ways. It can then provide an effective method for evaluation of various kinds of discrete functions.

This property offers some advantages. Consider a dictionary for character keys. In a standard application the hash value is an index into an array of pointers that point to the beginning of the hashed strings. This means that on top of the space required for the hashing function we need $m$ pointers. With an arbitrary order perfect hash function we can make each value point directly to the beginning of each string, saving the space required for the pointers. Another simple example is when keys form disjoint classes. Instead of storing with each key the name of a class to which it belongs, we simply assign to all keys from a given class the same hash value. Our hashing scheme also facilitates implementing a total ordering not otherwise directly computable from the data elements. These are just a few examples where hashing with arbitrary hash value selection can help.

## 5   Complexity Analysis

We give strong theoretical evidence that the expected time complexity of the algorithm is $O(rm+n)$. For $r > 1$, $n$ is $O(m)$ and thus the method runs in $O(m)$ time for suitably chosen $n$.

The second step of the algorithm, assignment as described above, runs in $O(rm + n)$ time. We now show that each iteration of the mapping step takes $O(rm+n)$ time, and that we can choose $n$ suitably so that the expected number of iterations is bounded above by a constant as $m$ increases. For this we use the assumption that the edges in our graphs appear at random independently of each other. We call this the "uniformity assumption" since it implies that all graphs have the same probability of occurring.

In each iteration of the mapping step, the following operations are executed: (i) selection of a set of $r$ hash functions from some class of universal hash functions; (ii) computation of values of auxiliary functions for each key in a set; (iii) testing if edges of the generated graph $G$ are independent.

Operation (i) takes no more than $O(m+n)$ time (for class $C_n$ the time depends on the maximum length of a word in the set $W$ times size of alphabet $\Sigma$ times $r$. For a particular set and predefined alphabet this may be considered to be $O(r)$). Operations (ii) and (iii) need $O(rm)$ and $O(rm + n)$ time, respectively. Hence, the complexity of a single iteration is $O(rm + n)$.

The expected number of iterations in the mapping step can be made constant by a suitable choice of $n$. Let $p$ denote the probability of generating in one mapping step an $r$–graph with $m$ independent edges and $n$ vertices. Then the expected number of iterations is $\sum_{i>0} ip(1 - p)^{i-1} = 1/p$.

To obtain a high probability of generating an $r$–graph with independent edges we use very sparse graphs. We choose $n = cm$, for some $c$. In the following subsections we present three theorems which estimate $c$'s for each $r > 0$, such that as $m$ goes to infinity the associated probability $p^\infty$ is a nonzero constant. For $r > 2$, $p^\infty = 1$. (For detailed proofs and models see Czech, Havas and Majewski (1992) and Majewski, Wormald, Havas and Czech (1992)).

## 5.1 Case 1; 1–graphs

**Theorem 2.** *The probability that a random 1–graph with $n = cm$ vertices and $m$ edges has independent edges is a non-zero constant iff $c = \Omega(m)$.*

*Proof.* The result follows easily from the solution to the occupancy problem. To prove the above result in the case of limited randomness we may use Fredman, Komlós and Szemerédi (1984, Corollary 2) or Dietzfelbinger, Gil, Matias and Pippenger (1992, Fact 3.2). □

The solution for 1-graphs is not acceptable, primarily because of its space requirements. It also requires $O(m^2)$ time to build the hash function.

## 5.2 Case 2; 2–graphs

This case is described in detail in Czech, Havas and Majewski (1992), including pseudocode for the algorithms.

**Theorem 3.** *Let $G$ be a random graph with $n$ vertices and $m$ edges obtained by choosing $m$ random edges with repetitions. Then if $n = cm$ holds with $c > 2$ the probability that $G$ has independent edges, for $n \to \infty$, is*

$$p = e^{1/c}\sqrt{\frac{c-2}{c}}$$

*Proof.* For 2–graphs edge independence is equivalent to acyclicity. By the well known result of Erdős and Rényi (1960) the probability that a random graph has no cycles, as $m$ tends to infinity, is $\exp(1/c + 1/c^2)\sqrt{\frac{c-2}{c}}$. As our graphs may have multiple edges, but no loops, the probability that the graph generated in the mapping step is acyclic is equal to the probability that there are no multiple edges times the probability that there are no cycles, conditional upon there being no multiple edges. The $j$-th edge is unique with probability $(\binom{n}{2} - j + 1)/\binom{n}{2}$ conditional on the earlier edges being distinct. Thus the probability that all $m$ edges are unique is $\binom{n}{2}^{\underline{m}}\binom{n}{2}^{-m} \sim \exp(-1/c^2 + o(1))$. Multiplying probabilities proves the theorem. □

For limited randomness we may use Fredman, Komlós and Szemerédi (1984, Corollary 4) to prove that the probability of having no multiple edges tends to a nonzero constant, and differs only slightly form the result obtained for unlimited randomness. For longer cycles we rely on the uniformity assumption.

As a consequence of the above theorem, if $c = 2 + \epsilon$ the algorithm constructs a minimal perfect hash function in $\Theta(m)$ random time. For 2-graphs we can speed up the detection of cycles. One practical method is to use a set union algorithm.

## 5.3 Case 3; $r$–graphs for $r > 2$

Edge independence is equivalent to the requirement that the $r$–graph does not contain a subgraph with minimum degree at least 2. For $r > 2$ the analysis is much more complicated than that for $r \leq 2$. We only show that there exists a constant, $c_{\text{inv}}$, such that if $m \leq c_{\text{inv}} n$ then the expected number of edge-minimal subgraphs on $i$ edges, $E(Y_i)$, of minimum degree at least 2 tends to 0 for all $i \leq m$ as $m$ goes to infinity. We determine the minimum possible $c \leq 1/c_{\text{inv}}$ experimentally.

**Theorem 4.** *For any $r$–graph there exists a constant $c_{\text{inv}}$ depending only on $r$ such that if $m \leq c_{\text{inv}} n$ the probability that a random $r$–graph has independent edges tends to 1.*

*Proof (outline).* Here also we use the uniformity assumption. To prove the above theorem we need to estimate the number of subgraphs of minimum degree at least 2 in a random $r$–graph with $n$ vertices and $m$ edges. This number can be shown to be

$$E(X_i) = O(m^{\frac{i}{2}}) \left( \frac{i^{r-1} r^r}{e^{r-1} n^r} \right)^{i\, ir/2} \sum_{k=r}^{} \left( \frac{\alpha^\alpha}{\rho^{1-\alpha}} \right)^{ir} \binom{n}{k}$$

where $\alpha = k/(ir)$ and $\rho = \rho(\alpha)$ is defined by $\dfrac{e^\rho - 1 - \rho}{\rho} = \alpha$. Then, using the fact that for edge-minimal subgraphs removing any edge must reduce the degree of at least one vertex in the subgraph to 1, we deduce that any such subgraph with $i$ edges must have at least $i/2$ vertices. Now, exploiting different approximation techniques, we can show that for each $r$ there exists a constant $c_{\text{inv}}$ depending only on $r$ such that if $m \leq c_{\text{inv}} n$ the expected number of edge-minimal subgraphs tends to 0. □

## 6 Experimental Results

Four algorithms for $r \in \{2, 3, 4, 5\}$, without any specific improvements, were implemented in the C language. All experiments were carried out on a Sun SPARC station 2, running under the SunOS™ operating system. For $r \in \{2, 3\}$ and character keys the results are summarized in Table 1. The values shown under *iterations, mapping, assignment* and *total* are the average number of iterations in the mapping step, time for the mapping step, time for the assignment step and total time for the algorithm, respectively. All times are in seconds.

For integer keys and $r = 3$ the results are presented in Table 2. The mapping functions $f_i$, $1 \leq i \leq 3$ were selected from class $\mathcal{H}_n^3$ and keys were chosen from the universe $U = \{0, \ldots, 2^{31} - 2\}$. Each row in the table represents the average taken over 200 experiments. (Notice that computing three polynomials of degree 3 takes about twice as much time as evaluating three functions from class $C_n$. In practice polynomials of degree one suffice and are faster.) Again, as for character

**Table 1.** Experimental results for character keys

| $m$ | $r = 2, c = 2.1$ | | | | $r = 3, c = 1.23$ | | | |
|---|---|---|---|---|---|---|---|---|
| | iterations | mapping | assignment | total | iterations | mapping | assignment | total |
| 1024 | 2.248 | 0.068 | 0.016 | 0.085 | 2.188 | 0.105 | 0.007 | 0.112 |
| 2048 | 2.540 | 0.134 | 0.030 | 0.164 | 1.928 | 0.166 | 0.013 | 0.179 |
| 4096 | 2.536 | 0.246 | 0.056 | 0.302 | 1.664 | 0.271 | 0.027 | 0.298 |
| 8192 | 2.828 | 0.526 | 0.123 | 0.650 | 1.332 | 0.444 | 0.053 | 0.497 |
| 16384 | 2.620 | 0.972 | 0.255 | 1.227 | 1.108 | 0.783 | 0.111 | 0.895 |
| 24692 | 2.880 | 1.565 | 0.392 | 1.958 | 1.061 | 1.099 | 0.144 | 1.243 |
| 32768 | 2.660 | 2.109 | 0.529 | 2.638 | 1.010 | 1.584 | 0.236 | 1.820 |
| 65536 | 2.700 | 4.189 | 1.067 | 5.256 | 1.000 | 3.169 | 0.495 | 3.664 |
| 131072 | 2.824 | 8.582 | 2.148 | 10.730 | 1.000 | 6.368 | 1.025 | 7.393 |
| 262144 | 2.868 | 18.022 | 4.620 | 22.642 | 1.000 | 12.176 | 2.086 | 14.262 |
| 524288 | 2.756 | 33.448 | 8.563 | 42.011 | 1.000 | 24.855 | 4.201 | 29.056 |

**Table 2.** Experimental results for integer keys

| $m$ | $r = 3, c_3 = 1.23$ | | | |
|---|---|---|---|---|
| | iterations | mapping | assignment | total |
| 1000 | 2.290 | 0.222 | 0.007 | 0.229 |
| 2000 | 1.655 | 0.328 | 0.014 | 0.343 |
| 4000 | 1.325 | 0.534 | 0.026 | 0.560 |
| 8000 | 1.215 | 0.978 | 0.053 | 1.031 |
| 16000 | 1.100 | 1.799 | 0.109 | 1.908 |
| 32000 | 1.010 | 3.357 | 0.231 | 3.588 |
| 64000 | 1.005 | 6.693 | 0.486 | 7.179 |
| 128000 | 1.000 | 13.277 | 1.003 | 14.280 |
| 256000 | 1.000 | 26.447 | 2.034 | 28.481 |
| 512000 | 1.000 | 52.676 | 4.102 | 56.778 |

keys, for $m > 64000$ the number of iterations stabilized at 1. Hence the average is also the worst case behavior for sufficiently large $m$. For $r \in \{3, 4, 5\}$ we experimentally determined constants $c_r$, such that if $n = c_r m$ then the expected number of iterations in the mapping step is a nonincreasing function of $m$. These values are: $c_3 = 1.23$, $c_4 = 1.29$, $c_5 = 1.41$. For these constants and for increasing $m$, the observed average number of iterations in the mapping step approached 1. Thus the generation algorithm for 3–graphs outperforms the algorithm for 2–graphs as the number of keys goes up, because the expected number of iterations for the mapping step for $r = 3$ goes down, while the expected number is constant for $r = 2$. New theoretical results for random hypergraphs agree very closely with the values of $c_3$, $c_4$ and $c_5$ given above (see Spencer and Wormald (1993)).

The experimental results fully back the theoretical considerations. Also, the time requirements of the new algorithm are very low. Likewise the mapping, assignment and total times grow approximately linearly with $m$.

# 7 Discussion

The method presented is a special case of solving a set of $m$ linearly independent integer congruences with a larger number of unknowns. These unknowns are the entries of array $g$. We generate the set of congruences probabilistically in $O(m)$ time. We require that the congruences are consistent and that there exists a sequence of them such that 'solving' $i - 1$ congruences by assignment of values to unknowns leaves at least one unassigned unknown in the $i$th congruence. We find the congruences in our mapping step and such a solution sequence in our independence test. It is conceivable that there are other ways to generate a suitable set of congruences, with at least $m$ unknowns, possibly deterministically. It may be that memory requirements for such a method would be smaller than for the given method. However, any space saving can only be by a constant factor, since $O(m \log m)$ space is required for order preserving minimal perfect hash functions (see Havas and Majewski (1992)). Further, it remains to be seen if the solution (such values for array $g$ that the resulting function is minimal and perfect) can then be found in linear time.

## Acknowledgements

The first and third authors were supported in part by the Australian Research Council.

# References

A.V. Aho, J.E. Hopcroft, and J.D. Ullman. *The Design and Analysis of Computer Algorithms.* Addison-Wesley, 1974.

M.D. Brain and A.L. Tharp. Near-perfect hashing of large word sets. *Software — Practice and Experience* **19** (1989) 967–978.

M.D. Brain and A.L. Tharp. Perfect hashing using sparse matrix packing. *Information Systems,* **15** (1990) 281–290.

J.L. Carter and M.N. Wegman. Universal classes of hash functions. In *9th Annual ACM Symposium on Theory of Computing – STOC'77,* 106–112, 1977.

J.L. Carter and M.N. Wegman. New classes and applications of hash functions. In *20th Annual Symposium on Foundations of Computer Science — FOCS'79,* 175–182, 1979a.

J.L. Carter and M.N. Wegman. Universal classes of hash functions. *Journal of Computer and System Sciences,* **18** (1979b) 143–154.

N. Cercone, J. Boates, and M. Krause. An interactive system for finding perfect hash functions. *IEEE Software,* **2** (1985) 38–53.

C-C. Chang and T-C. Wu. Letter oriented perfect hashing scheme based upon sparse table compression. *Software — Practice and Experience,* **21** (1991) 35–49.

C.C. Chang. The study of an ordered minimal perfect hashing scheme. *Communications of the ACM,* **27** (1984) 384–387.

C.C. Chang and C.H. Chang. An ordered minimal perfect hashing scheme with single parameter. *Information Processing Letters,* **27** (1988) 79–83.

C.C. Chang, C.Y. Chen, and J.K. Jan. On the design of a machine independent perfect hashing. *The Computer Journal*, **34** (1991) 469–474.

C.C. Chang and R.C.T. Lee. A letter-oriented minimal perfect hashing scheme. *The Computer Journal*, **29** (1986) 277–281.

R.J. Cichelli. Minimal perfect hash functions made simple. *Communications of the ACM*, **23** (1980) 17–19.

C.R. Cook and R.R. Oldehoeft. A letter oriented minimal perfect hashing function. *SIGPLAN Notices*, **17** (1982) 18–27.

G.V. Cormack, R.N.S. Horspool, and M. Kaiserwerth. Practical perfect hashing. *The Computer Journal*, **28** (1985) 54–55.

Z.J. Czech, G. Havas, and B.S. Majewski. An optimal algorithm for generating minimal perfect hash functions. *Information Processing Letters*, **43** (1992) 257–264.

Z.J. Czech and B.S. Majewski. Generating a minimal perfect hashing function in $O(m^2)$ time. *Archiwum Informatyki*, **4** (1992) 3–20.

M. Dietzfelbinger, J. Gil, Y. Matias, and N. Pippenger. Polynomial hash functions are reliable. In *19th International Colloquium on Automata, Languages and Programming – ICALP'92*, 235–246, Vienna, Austria, 1992. LNCS 623.

M. Dietzfelbinger and F. Meyer auf der Heide. A new universal class of hash functions, and dynamic hashing in real time. In *17th International Colloquium on Automata, Languages and Programming – ICALP'90*, 6–19, Warwick University, England, 1990. LNCS 443.

M.W. Du, T.M. Hsieh, K.F. Jea, and D.W Shieh. The study of a new perfect hash scheme. *IEEE Transactions on Software Engineering*, **9** (1983) 305–313.

R. Duke. Types of cycles in hypergraphs. *Ann. Discrete Math.*, **27** (1985) 399–418.

P. Erdős and A. Rényi. On the evolution of random graphs. *Publ. Math. Inst. Hung. Acad. Sci.*, **5** (1960) 17–61. Reprinted in J.H. Spencer, editor, *The Art of Counting: Selected Writings*, Mathematicians of Our Time, 574–617. Cambridge, Mass.: MIT Press, 1973.

E. Fox, Q.F. Chen, A. Daoud, and L. Heath. Order preserving minimal perfect hash functions and information retrieval. *ACM Transactions on Information Systems*, **9** (1991) 281–308.

E. Fox, Q.F. Chen, and L. Heath. LEND and faster algorithms for constructing minimal perfect hash functions. In *Proc. 15th Annual International ACM SIGIR Conference on Research and Development in Information Retrieval – SIGIR'92*, Copenhagen, Denmark, 266–273, 1992.

E.A. Fox, L.S. Heath, Q. Chen, and A.M. Daoud. Practical minimal perfect hash functions for large databases. *Communications of the ACM*, **35** (1992) 105–121.

M.L. Fredman, J. Komlós, and E. Szemerédi. Storing a sparse table with $O(1)$ worst case access time. *Journal of the ACM*, **31** (1984) 538–544.

G.H. Gonnet and R. Baeza-Yates. *Handbook of Algorithms and Data Structures.* Addison-Wesley, Reading, Mass., 1991.

M. Gori and G. Soda. An algebraic approach to Cichelli's perfect hashing. *BIT*, **29** (1989) 209–214.

G. Haggard and K. Karplus. Finding minimal perfect hash functions. *ACM SIGCSE Bulletin*, **18** (1986) 191–193.

G. Havas and B.S. Majewski. Optimal algorithms for minimal perfect hashing. Technical Report 234, The University of Queensland, Key Centre for Software Technology, Queensland, 1992.

C.T.M. Jackobs and P. van Emde Boas. Two results on tables. *Information Processing Letters*, **22** (1986) 43–48.

G. Jaeschke. Reciprocal hashing: A method for generating minimal perfect hashing functions. *Communications of the ACM*, **24** (1981) 829–833.

A. Karlin and E. Upfal. Parallel hashing - an efficient implementation of shared memory. In *18th Annual ACM Symposium on Theory of Computing - STOC'86*, 160–168, 1986.

T.G. Lewis and C.R. Cook. Hashing for dynamic and static internal tables. *Computer*, **21** (1988) 45–56.

B.S. Majewski, N.C. Wormald, Z.J. Czech, and G. Havas. A family of generators of minimal perfect hash functions. Technical Report 16, DIMACS, Rutgers University, New Jersey, USA, 1992.

K. Mehlhorn. *Data Structures and Algorithms 1: Sorting and Searching*, volume 1. Springer-Verlag, Berlin Heidelberg, New York, Tokyo, 1984.

F. Meyer auf der Heide. Dynamic hashing strategies. In *15th Symposium on Mathematical Foundations of Computer Science - MFCS'90*, 76–87, Banska Bystrica, Czechoslovakia, 1990.

P.K. Pearson. Fast hashing of variable-length text strings. *Communications of the ACM*, **33** (1990) 677–680.

T.J. Sager. A polynomial time generator for minimal perfect hash functions. *Communications of the ACM*, **28** (1985) 523–532.

J.P. Schmidt and A. Siegel. On aspects of universality and performance for closed hashing. In *21st Annual ACM Symposium on Theory of Computing - STOC'89*, 355–366, Seattle, Washington, 1989.

J.P. Schmidt and A. Siegel. The analysis of closed hashing under limited randomness. In *22st Annual ACM Symposium on Theory of Computing - STOC'90*, 224–234, Baltimore, MD, 1990.

J.P. Schmidt and A. Siegel. The spatial complexity of oblivious $k$-probe hash functions. *SIAM Journal on Computing*, **19** (1990) 775–786.

A. Siegel. On universal classes of fast high performance hash functions, their time-space trade-off, and their applications. In *30th Annual Symposium on Foundations of Computer Science - FOCS'89*, 20–25, 1989.

C. Slot and P. van Emde Boas. On tape versus core: An application of space efficient hash functions to the invariance of space. In *16th Annual ACM Symposium on Theory of Computing - STOC'84*, 391–400, Washington, DC, 1984.

J. Spencer and N.C. Wormald. On the $k$-core of random graphs. Manuscript, 1993.

R. Sprugnoli. Perfect hashing functions: A single probe retrieving method for static sets. *Communications of the ACM*, **20** (1977) 841–850.

R.E. Tarjan and A.C-C Yao. Storing a sparse table. *Communications of the ACM*, **22** (1979) 606–611.

V.G. Winters. Minimal perfect hashing for large sets of data. In *International Conference on Computing and Information - ICCI'90*, 275–284, Niagara Falls, Canada, 1990.

V.G. Winters. Minimal perfect hashing in polynomial time. *BIT*, **30** (1990) 235–244.

W.P. Yang and M.W. Du. A backtracking method for constructing perfect hash functions from a set of mapping functions. *BIT*, **25** (1985) 148–164.

# Coloring $k$-colorable graphs in constant expected parallel time

Luděk Kučera

Charles University,
Prague, Czechoslovakia [*]
and
Max Planck Institute for Computer Science,
Saarbrücken, Germany

### Abstract

*A parallel (CRCW PRAM) algorithm is given to find a k-coloring of a graph randomly drawn from the family of k-colorable graphs with n vertices, where $k = \log^{O(1)} n$. The average running time of the algorithm is constant, and the number of processors is equal to $|V| + |E|$, where $|V|$, $|E|$, resp. is the number of vertices, edges, resp. of the input graph.*

## Introduction

The graph coloring problem is NP-complete, and it has been proved that finding a good approximation solution is as difficult as computing of the optimal one [10]. The complexity classes $R$ and $BPP$, representing practical use of randomization, do not seem to contain NP-complete problems. Therefore the only way to cope with coloring-type problems is presently a probabilistic analysis of behavior of (usually deterministic) polynomial time algorithms, applied to random inputs.

The most common way of generating random input graphs is represented by the random graph $\mathcal{G}_n$, which gives all graphs with $n$ vertices equaly likely, or by a more general model $\mathcal{G}_{n,p}$, where $p$ is a parameter chosen usually so that sparser graphs are more likely, but all graphs with the same number of edges are generated with the same probability (the distribution $\mathcal{G}_n$ is equal to $\mathcal{G}_{n,1/2}$). Throughout the paper, we will suppose for simplicity that $p$ is a constant, but most results can easily be generalized. We will denote $1/(1-p)$ by $b$.

Random graphs $\mathcal{G}_{n,p}$ do not represent an interesting tool for testing heuristic coloring algorithms. It is known that $\chi(\mathcal{G}_{n,p}) = (1 + o(1)) \frac{n}{2 \log_b n}$ almost surely (for the lower bound see [11], for the upper bound [3,18]). On the other hand, it is easy to show that perhaps the simplest coloring algorithm, the greedy one, uses almost surely $(1 + o(1)) \frac{n}{\log_b n}$ colors when applied to $\mathcal{G}_{n,p}$, i.e. the "ratio of optimality" is almost surely $2 + o(1)$. It

[*]This research was partially supported by EC Cooperative Action IC-1000 (project ALTEC: *Algorithms for Future Technologies*)

is interesting that even the most sophisticated among known polynomial graph coloring algorithms seem or are proved to behave in essentially the same way as the greedy one, using almost surely about $2\chi(G)$ colors. (There is however an indication that the problem of (near) optimal coloring of $\mathcal{G}_{n,p}$ might be easier that NP-hard problems (or problems hard "on average" in the sense of Levin [17]), because searching of all independent subsets of size $O(\log n)$ finds a near optimal coloring almost surely in time $n^{O(\log n)}$ ).

The greedy algorithm can also be parallelized so that it achieves essentially the same results in polylogarithmic time [9,7].

All this implies that all known polynomial time algorithms color the random graphs $\mathcal{G}_{n,p}$ using roughly the same number of colors. Therefore this type of analysis, though giving interesting results about random graphs, reveals nothing really useful about algorithms and their computational power. It is much more interesting to study another simple random graph model $\mathcal{G}_{n;k}$, which consists in picking up uniformly a random element of the class of all $k$-colorable graphs with $n$ vertices. It was shown in [22] for $k < (1-\varepsilon)\log_2 n$, and generalized in [15] for $k = o(\sqrt{n/\log n})$, that this distribution can be approximated by the next procedure (using $p = 1/2$):

1. Generate a graph $G$ using the distribution $\mathcal{G}_{n,p}$, (which means that the probability that two vertices are connected by an edge is $p$, and these probabilites are independent for different pairs of vertices),

2. label vertices of $G$ randomly by $k$ colors,

3. remove all edges of $G$ with endpoints labeled by the same label.

The graph distribution defined by this procedure is denoted by $\mathcal{G}_{n,p;k}$. Let us recall that we will suppose that $p$ is a constant, but the results can be generalized (with some restrictions on $p$). Note that the first two phases of the construction are independent and hence the labeling of vertices can be performed first.

It is clear that the labeling constructed in the second step will become a coloring of $G$, and it is possible to prove that for $k = o(\sqrt{n/\log n})$ this labeling is almost surely both the optimal coloring and the unique (up to a permutation of colors) $k$-coloring of $G$.

No fast algorithm is known to give almost surely a good coloring of $\mathcal{G}_{n,p;k}$ for $k = \Omega(\sqrt{n})$. However the case of small $k$ is quite appealing. Though primitive algorithms behave badly unless $k$ is extremely small (e.g. the greedy algorithm uses almost surely $\Omega(n/\log n)$ colors to color $\mathcal{G}_{n,0.5;k}$ even if $k = n^\varepsilon$ for arbitrarily small positive constant $\varepsilon$ [16]), more sophisticated algorithms are able to find the optimal solution almost surely (see [8] for $k$ constant, [22] for $k < (1 - \varepsilon)\log_2 n$, and [15] for $k = o(\sqrt{n/\log n})$ ).

The algorithm of [15] is based on dividing pairs of vertices into two classes using approximation of treshold functions with precision $1/k$. Such an approximation can be done in constant worst case time on CRCW parallel RAM, provided $k = \log^{O(1)} n$ [1,12]. Based on this we will give a processor efficient CRCW PRAM algorithm coloring most $k$-colorable graphs optimally in constant time.

Since the treshold approximation with precision $1/k$ seems to be closely connected to evaluation of treshold functions, known lower bounds to depth of circuits computing symmetric functions [4,19] suggest a conjecture that constant average time algorithms do not exist for coloring of $k$-colorable graphs if $k$ grows faster than any polylogarithmic function.

We first present a constant time CRCW PRAM algorithm with a superpolynomial number of processors in Section 1. Section 2 shows how to find a coloring in constant time using small number of processors. However the algorithm is randomized (its main idea is to apply the algorithm of Section 1 to a small sample of $\log^{O(1)} n$ vertices and then to extent the solution to the whole graph). In Section 3 we show how to derandomize it to behave in the same way on random input graphs. This is based on the fact that the randomized algorithm does not ask for the existence of a large number of edges, and therefore this information can be used as a source of randomness.

# 1 Randomized algorithm with many processors

Let $0 < \vartheta < 1/2$ and $k = O(n^\vartheta)$. In this paragraph we show an algorithm which finds almost surely a $k$-coloring of a random $k$-colorable graph with $n$ vertices. The underlying idea of the algorithm comes from [15].

Given a vertex $v$, let us denote the class of equilabeled vertices obtained in the second step of the generating procedure $\mathcal{G}_{n,p;k}$ and containing $v$ by $C_v$.

We will first show how, given a graph generated by the procedure $\mathcal{G}_{n,p;k}$ and its vertex $v$, we can find almost surely $C_v$:

**Algorithm** $\mathcal{A}_0$.
Input: A $k$-colorable graph $G$ with the vertex set $X$
      and the edge set $E$,
      a vertex $v \in X$.
Output: A color class $C$ of a $k$-coloring of $G$ containing $v$.
Remark: $\mathcal{A}_0$ sometimes gives an incorrect answer.
**begin**
**for all** $w \in X - \{v\}$ **pardo**
    $d(w) := |\{ z \in X \mid \{v, z\}, \{w, z\} \in E \}|$;
**for all** $t := 1$ to $n$ **pardo**
    $\Gamma_t := \{v\} \cup \{ w \in X - \{v\} \mid d(w) \geq t \}$;
$t_v := \min\{ t \mid \Gamma_t$ is an independent set $\}$;
$C := \Gamma_{t_v}$;
**end.**

First we will prove some properties of $\mathcal{G}_{n,p;k}$.

**Lemma 1.1** *Let $A \subset X$. With probability $1 - \exp(-\Omega(|A|/k))$, $\frac{|A|}{2k} < |C_v \cap A| < \frac{3|A|}{2k}$ for each vertex $v$.*

**Proof:** The size of the set $C_v \cap A$ is a random variable with binomial distribution $\mathcal{B}(|A|, \frac{1}{k})$. The Chernoff bound [5,2] implies that

$$\mathbf{Prob} \left( \left| |C_v \cap A| - \frac{|A|}{k} \right| \geq \frac{|A|}{2k} \right) = \exp \left( -\Omega \left( \frac{|A|}{k} \right) \right).$$

♣

**Corrolary 1.2** *With probability* $1 - \exp(-\Omega(n/k))$, $\frac{n}{2k} < |C_v| < \frac{3n}{2k}$ *for each vertex* $v$.

**Lemma 1.3** *The probability that for each* $v$ *the set* $C_v$ *is a proper subset of no independent set of* $\mathcal{G}_{n,p;k}$ *is at least* $1 - nk(1 - p)^{n/2k}$.

**Proof:** Given a fixed set of the form $C_v$ and a fixed $x \notin C_v$, the probability that $x$ has no neighbour in $C_v$ is

$$(1 - p)^{|C_v|} \leq (1 - p)^{n/2k},$$

where we suppose $|C_v| \geq n/2k$, see the previous lemma. ♣

Let us now prove that the algorithm $\mathcal{A}_0$, applied to a vertex $v$ of $\mathcal{G}_{n,p;k}$, returns almost surely $C_v$

**Lemma 1.4** *If* $0 < \vartheta < 1/2$ *is a constant,* $k = O(n^\vartheta)$, *then the probability that* $\Gamma_\tau = C_v$, *where* $\tau = (n - |C_v| - \frac{n}{4k})p^2$, *is* $1 - \exp(-\Omega(n/k^2))$.

**Proof:** Let $w$ be a vertex. Possible neighbours of both $v$ and $w$ are elements of $X - (C_v \cup C_w)$, each with probability $p^2$. Therefore it follows from the Chernoff inequality [5,2] that

$$\left| d(w) - p^2 |X - (C_v \cup C_w)| \right| \geq \frac{n}{4k} p^2 \qquad (1)$$

with probability at most

$$2 \exp\left( -\frac{1}{2} \left( \frac{np^2}{4k} \right)^2 \frac{1}{n - |C_v \cup C_w|} \right) \leq 2 \exp\left( -\frac{p^4}{32} \frac{n}{k^2} \right),$$

Suppose that (1) holds for all $w$, which happens with probability at least

$$1 - 2n \exp\left( -\frac{p^4}{32} \frac{n}{k^2} \right) = 1 - \exp\left( -\Omega\left( \frac{n}{k^2} \right) \right).$$

If $C_v = C_w$, then (1) implies

$$d(w) > (n - |C_v|)p^2 - \frac{n}{4k} p^2 = \tau,$$

if $C_v \neq C_w$, then

$$d(w) < (n - |C_v| - |C_w|)p^2 + \frac{n}{4k} p^2 \leq (n - |C_v| - \frac{n}{2k})p^2 + \frac{n}{4k} p^2 = \tau,$$

which implies $\Gamma_\tau = C_v$. ♣

**Theorem 1.5** *The probability that the algorithm* $\mathcal{A}_0$ *returns* $C_v$ *is* $1 - \exp(-\Omega(n/k^2))$.

**Proof:** Lemma 1.4 implies that $\Gamma_t$ is likely to be $C_v$ for some $t$. In view of Lemma 1.3, $C_v$ is unlikely to be contained in a larger independent set, and therefore $\Gamma_t = C_v$. ♣

It is not surprising that the algorithm is likely to find the color class used in the construction $\mathcal{G}_{n,p;k}$, because it can be proved ([15,22]) that the original coloring is almost surely the unique coloring of $\mathcal{G}_{n,p;k}$ (up to a permutation of colors).

The bounds to the resource requirements of the algorithm will be derived in the following paragraph. Let us just mention that computation of $d(v)$ in constant time requires superpolynomial number of processors. However the algorithms of the next paragraphs will call $\mathcal{A}_0$ to find an independent set of a graph of a polylogarithmic size.

$\mathcal{A}_0$ gives immediately an algorithm to color $\mathcal{G}_{n,p;k}$ almost surely by $k$ colors:

**Algorithm $\mathcal{A}_1$.**
Input: A $k$-colorable graph $G$ with the vertex set $X$
       ordered by a relation $<$ and the edge set $E$.
Output: A $k$-coloring of $G$.
Remark: $\mathcal{A}_1$ sometimes fails without giving an answer.
**begin**
**for all** vertices $v$ **pardo begin**
    find the set $C_v$ using $\mathcal{A}_0$;
    **if** $v = \min C_v$ **then** label $v$ as "selected"
    **else** label $v$ as "unselected";
    **end**;
**for all** selected vertices $v$ **pardo begin**
    $c_v :=$ the number of selected vertices $w$ such that $w < v$;
    color all vertices of $C_v$ by $c_v$;
    **end**;
**for all** edges $e = \{v, w\} \in E$ **pardo**
    **if** $v$ and $w$ are colored by the same color **then**
        the computation failed;
**if** there exists a selected vertex $v$ such that $c_v \geq k$ **then**
    the computation failed;
**end**.

Note that the algorithm $\mathcal{A}_1$ either reports a failure or gives a valid $k$-coloring of the graph by colors $0, 1, \ldots, k - 1$. It is necessary to check the correctness of the result of the computation, because the algorithm $\mathcal{A}_0$ might give an incorrect answer.

It follows from the analysis of $\mathcal{A}_0$ that

**Theorem 1.6** *The algorithm $\mathcal{A}_1$ finds a $k$-coloring of the input graph with probability* $1 - \exp(-\Omega(n/k^2))$.

**Proof:** With probability $1 - n\exp(-\Omega(n/k^2)) = 1 - \exp(-\Omega(n/k^2))$, all sets $C_v$ are computed correctly by $\mathcal{A}_0$. ♣

# 2 Efficient randomized algoritm

In this paragraph we suppose that $k = \log^{O(1)}$. Let $m$ be a number sufficiently larger than $k^2$. It will be quite sufficient to put $m = k^2 \log^a n$ for some constant $a \geq 2$. For reasons that will be clear in the next paragraph, we also choose a fixed set $Q$ of, say, $n/2$ vertices $X$.

We will show that a solution of the coloring problem for a random subset of $m$ vertices of $X - Q$ gives almost surely a solution of the global problem.

**Graph coloring algorithm** $\mathcal{A}_2$.

Input: A $k$-colorable graph $G$ with vertex set $X$ of size $n$
   and edge set $E$, $k = \log^{O(1)} n$,
   random bits $r_{i,j}$, $i = 1, \ldots, s$, $j = 1, \ldots, \lceil \log_2 n \rceil$
   a set $Q$ of $\lfloor n/2 \rfloor$ vertices of $G$,
   a constant $a$.

Output: A coloring of $G$.

Remark: $\mathcal{A}_2$ sometimes fails to find a $k$-coloring.

**begin**

choose a number $m$ such that $m \geq k^2 \log^a n$;

**for** $i = 1$ **to** $m$ **pardo**

   $w_i :=$ the number with binary representation $r_{i,1} \ldots r_{i,t}$,
   where $t = \lceil \log_2 n \rceil$;

$Y := \{w_i \mid 1 \leq i \leq m, \ w_i \in X - Q\}$;

use $\mathcal{A}_1$ to find a $k$ coloring of the graph induced by $G$ on $Y$;

if $\mathcal{A}_1$ fails then halt;

for each vertex $v$ choose an integer $0 \leq \lambda(v) < k$ such that
   there is no $w \in Y$ such that
   $\{v, w\}$ is an edge of $G$ and $w$ is colored by $\lambda(v)$;

if such $\lambda(v)$ does not exit for some vertex $v$ then halt;

if $\lambda$ is not a coloring of $G$ then halt;

color each vertex $v$ of $G$ by $\lambda(v)$;

**end.**

Note that if $v, w \in Q$, the algorithm never asks whether $v$ and $w$ are connected by an edge of $G$.

Randomized algorithms are usually analyzed on assumption that a source of randomness provides independent random bits such that $\mathbf{Prob}(r_i = 1) = 1/2$ for each $i$. The random bits constructed in the next paragraph are independent, but it could only be proved that the probability of being equal to 1 is closed to $1/2$.

We first prove that the size of any set $C_v \cap Y$ is sufficiently large:

**Lemma 2.1** *Let us suppose that random bits $r_{i,j}$, given as an input to $\mathcal{A}_2$, are independent and $\mathbf{Prob}(r_{i,j} = 1) = (1 + O(\log^{-1} n))/2$ for all $i, j$. There exists a constant $c > 0$ such that, with probability $1 - \exp(-\Omega(m/k))$, $|C_v \cap Y| \geq cm/k$ for each vertex $v$.*

**Proof:** Let $\kappa = O(\log^{-1} n)$ be such that $\mathbf{Prob}(r_{i,j} = 1), \mathbf{Prob}(r_{i,j} = 0) \geq (1 - \kappa)/2$ for all $i, j$. There is a constant $\gamma > 0$ such that

$$\mathbf{Prob}(w_j = z) \geq \left(\frac{1 - \kappa}{2}\right)^{\lceil \log_2 n \rceil} \geq \frac{\gamma}{n}$$

for each vertex $z$ and $j \leq m$.

Let $v$ be a fixed vertex of $X - Q$. It follows from Lemma 1.1 that we can suppose that $|C_v - Q| \geq |X - Q|/2k \geq n/4k$, and therefore

$$\mathbf{Prob}(w_j \in (C_v - Q - \{w_\ell | \ell < j\})) \geq \left(\frac{n}{4k} - m\right)\frac{\gamma}{n} \geq \frac{\gamma}{5k}.$$

In view of the Chernoff bound

$$\mathbf{Prob}\left(|C_v \cap Y| < \frac{\gamma m}{6k}\right) \leq \exp\left(-\Omega\left(\frac{m}{k}\right)\right),$$

and the probability that this is true for all $k$ color classes is not more that $k$ times greater.
♣

Note that the Lemma implies that $|Y - Q| \geq cm$ with large probability.

**Theorem 2.2** *Under the same assumptions on bits $r_{i,j}$ as in the preceeding Lemma, if $k = \log^{O(1)} n$, then the probability that the algorithm $A_2$ finds a k-coloring of $G_{n,p;k}$ is at least $1 - \exp(-\Omega(m/k^2)) = 1 - \exp(-\Omega(\log^a n))$.*

**Proof:** Since the algorithm constructs the set $Y$ before it asks for any edge, $Y$ can be determined before the construction $G_{n,p;k}$ is applied. It follows that the probabilistic distribution of the graph induced on $Y$ is the same as $G_{m,p;k}$, and therefore $A_1$ finds the solution given by classes $C_1 \cap Y, \ldots, C_k \cap Y$ with probability at least

$$1 - \exp(-\Omega(m/k^2)) \geq 1 - \exp(-\Omega(\log^a n)),$$

see Theorem 1.6. If this is the case, the choice of $\lambda(v)$ is possible for each $v$ as the color used for $C_v \cap Y$. Now, it is sufficient to prove that no other choice of $\lambda(v)$ is likely for any vertex $v$. In view of the preceeding lemma, the probability that a vertex $v$ is connected to no $w \in C_z \cap Y$, $v \notin C_z$, is at most

$$(1 - p)^{cm/k} = \exp(-\Omega(\log^a n)).$$

♣

We are interested in three resources used by the algorithm $A_0$: the worst case parallel time, the number of processors and the amount of randomness. We will need the following technical lemma, that says that it is possible to add a polylogarithmic number of bits in constant time with small number of processors. It is known that there are families of constant depth circuits that compute the sum of $n$ bits, provided the number of 1's among them is bounded by a polylogarithmic function [1,12]. Unfortunately the proofs are either nonconstructive ([1]), or possess a certain degree of uniformness, but they are complicated. However in our special case there is an easy direct proof, since the most difficult part of the general proof is a compaction of 1's into an array of a polylogarithmic size, which is already done here.

**Lemma 2.3** *Let $0 < \varepsilon, \ell$ be constants. If $i \leq \lfloor \log^\ell n \rfloor$, then the sum of $i$ integers $b_1, \ldots, b_i \in \{0, 1\}$ can be computed in constant time on CRCW PRAM with $O(n^\varepsilon)$ processors.*

**Proof:** In constant time and using $O(AC^{-1}B^C)$ processors, we can reduce computing of the sum of $A$ integers $0 \leq b_1, \ldots, b_A < B$ to computing the sum of at most $A/C$ integers less or equal to $BC$:

Partition numbers into $A/C$ groups of at most $C$ elements each. There are at most $B^C$ different sequences $\sigma$ of $C$ numbers from the interval $[0, B)$. For each of $A/C$ groups of numbers, use $B^C$ families of processors, such that for each sequence $\sigma$ there is exactly one family that checks whether the numbers in the group form a sequence $\sigma$ and, in the affirmative case, outputs the predefined result.

Now put $C = \log^{1/2} n$. If $B \leq A = \log^{O(1)} n$, then

$$AC^{-1}B^C \leq A^{C+1} = \exp((C + 1) \log A) = \exp(O(\log^{1/2} n \log \log n)) =$$

$$= O(\exp(\varepsilon \log n)) = O(n^\varepsilon).$$

Repeating the reduction $2\ell$ times, we can compute the sum of $\log^\ell n$ nembers from $\{0, 1\}$ in constant time with $O(n^\varepsilon)$ processors. ♣

**Theorem 2.4** *Let* $k = \log^{O(1)} n$. *There is an implementation of* $\mathcal{A}_2$ *on CRCW PRAM with* $O(|V| + |E|)$ *processors that runs in constant parallel time and needs* $\log^{O(1)} n$ *random bits.*

**Proof:** $\mathcal{A}_2$ calls $\mathcal{A}_1$ (and through it $\mathcal{A}_0$) on a subgraph of polylogarithmic size. It follows from the preceeding lemma that, given a constant $\varepsilon > 0$, all sums computed by $\mathcal{A}_0$ and $\mathcal{A}_1$ in such a case can be computed in constant parallel time using $O(n^\varepsilon)$ processors.

It is known that the minimum of $\bar{m}$ bits can be computed in constant parallel time on CRCW PRAM with $O(m^2)$ processors [15].

The algorithm $\mathcal{A}_0$ checks if $d(w) \geq t$ for each $t$, $w$. This can be done using $n^2$ computations based on the preceeding Lemma. Remaining part of the computation can easily be done in constant time on CRCW PRAM with polynomially many processors. ♣

# 3 Derandomization

The algorithms given in the previous two paragraphs are randomized. We will now show a way of derandomizing the coloring algorithm $\mathcal{A}_2$.

Choose a fixed vertex $u \in Q$ (for the set $Q$, see Algorithm $\mathcal{A}_2$). Given a vertex $v \in Q - \{u\}$, put $b_v = 1$ if $\{u, v\}$ is an edge, $b_v = 0$ otherwise. The key observation is that $\mathcal{A}_2$ checks the value of $b_v$ for no $v \in Q - \{u\}$. If the input graph is generated by $\mathcal{G}_{n,p;k}$, we can use boolean variables $b_v$, $v \in Q - \{u\}$ as a source of random bits. The probabilities **Prob**$(b_v = 1)$ are independent, but the problem is that they are not equal to $1/2$.

Therefore we will group bits $b_v$ into groups of polylogarithmic size, and we will use the parities of groups instead of the single bits. The parity of $\log^{O(1)} n$ bits can be computed in constant time using $O(n^\varepsilon)$ processors, see Lemma 2.3, bits obtained in this way are independent, and they are almost unbiased:

**Lemma 3.1** *Let* $v_1, \ldots, v_\ell$ *are different elements of* $Q - \{u\}$. *Then*

$$\mathbf{Prob}(b_{v_1} \oplus \cdots \oplus b_{v_\ell} = 1) = \frac{1}{2}(1 + \exp(-\Omega(\ell/k))).$$

**Proof:** The probability that $b_v = 1$ is either 0 if $v \in C_u$ or $p$ if $v \notin C_u$. In view of Lemma 1.1, the probability that $C_u$ contains more than $3\ell/2 k \leq \ell/2$ vertices $v_i$, $1 \leq i \leq \ell$ is $\exp(-\Omega(\ell/k))$.

Suppose now that $w_1, \ldots, w_{\ell/2}$ is a subsequence of $v_1, \ldots, v_\ell$ that does not contain points of $C_u$. Then

$$\mathbf{Prob}\,(b_{w_i} = 1) = \frac{1}{2}(1 + (2p - 1)),$$

$$\mathbf{Prob}\,(b_{w_1} \oplus \cdots \oplus b_{w_\ell/2} = 1) = \frac{1}{2}(1 + (2p - 1)^{\ell/2}),$$

and therefore given any fixed values of $b_{v_j}$ for $v_j$ that do not belong to the sequence $w_i$,

$$\mathbf{Prob}\,(b_{v_1} \oplus \cdots \oplus b_{v_\ell} = 1) = \frac{1}{2}(1 \pm (2p - 1)^{\ell/2}) = \frac{1}{2}(1 + \exp(-\Omega(\ell))).$$

♣

Theorem 2.4 and the preceeding lemma give

**Theorem 3.2** *Let $k = \log^{O(1)} n$, $a$ be a constant. There is a deterministic CRCW PRAM algorithm that runs in constant parallel time and, with probability $1 - n^{\Omega(\log^a n)}$, colors the random graph $\mathcal{G}_{n,p;k}$ with $k$ colors.*

**Proof:** Choose $\ell = k \log^{a+2} n$. It follows from Lemma 3.1 that, with probability at least

$$1 - (\log^{O(1)} n) \exp(-\Omega(\ell/k)) = 1 - n^{\Omega(\log^a n)},$$

a fixed polylogarithmic number of groups of $\ell$ bits $b_v$ give a sequence of random bits that verifies assumption of Theorem 2.4. ♣

# 4 Conclusions

We have proved that $\mathcal{G}_{n,p;k}$, $k = \log^{O(1)} n$, can be colored almost surely with $k$ colors using a constant time CRCW PRAM with $O(|V| + |E|) = O(n^2)$ processors. This means that both time and processor bounds are optimal. Using methods of [8], it is easy to obtain a CRCW PRAM algorithm, which colors $\mathcal{G}_{n,p;k}$ *always* with $k$ colors, such that its *average* running time is constant, and the number of processors equal to the number of vertices and edges of the graphs (it is sufficient to use an $n^{polylog\,n}$ expected time algorithm to color the input graph when $\mathcal{A}_2$ fails). In view of [22,15], the same results are true for a graph drawn randomly (with uniform probability) from the class of all $k$-colorable graphs with $n$ vertices, $k = \log^{O(1)} n$.

Our result are closely connected to the existence of uniform constant - depth probabilistic circuits, approximating the treshold functions. If treshold functions with $n$ inputs can be approximated with multiplicative precision $1/k$, where $k = o(\sqrt{n/\log n})$, then $k$-colorable graphs can be colored almost surely with $k$ colors in constant average time, see Lemma 1.4.

Several problems are left open

- Is it possible to find a constant average time algorithm coloring $\mathcal{G}_{n,p;k}$ by $k$ colors if $k$ is not bounded by a polylogarithmic function?

- What is the best precision bound to constant-depth probabilistic circuits, approximating general treshold functions, and what is the optimal size of such circuits ?

The results show that methods based on taking small random samples of the input information can give very efficient probabilistic approximation algorithms and deterministic algorithms with good behavior on randomly generated inputs. It is likely that this paradigm can be used to solve other problems.

# References

[1] M. Ajtai, and M. Ben-Or, A theorem on probabilistic constant depth computation, *Proceedings of the 16th Symposium on Theory of Computing*, (1984), 471-474.

[2] N. Alon, J. Spencer, and P. Erdős, The probabilistic Method, J. Wiley and Sons, New York, 1992.

[3] B. Bollobás, The chromatic number of random graphs, *Combinatorica* **8**, 49-56.

[4] B. Brustmann, and I. Wegener, The complexity of symmetric functions in bounded depth circuits, *Information Processing Letters* **25** (1987), 217-219.

[5] H. Chernoff, A measure of asymptotic efficiency for tests based on the sum of observations, *Ann. Math. Statist.* **23** (1952), 493-509.

[6] B. Chlebus, K. Diks, T. Hagerup, and T. Radzik, Efficient simulations between CRCW PRAMs, *Proc. 13th Symp. on the Mathematical Foundations of Computer Science*, 1988, 230-239.

[7] D. Coppersmith, P. Raghavan, and M. Tompa, Parallel graph algorithms that are efficient on average, *Proceedings of the 28th Annual IEEE Conference on Foundations of Computer Science*, (1987), 260-269.

[8] M. Dyer, and A. Frieze, The solution of some random NP-hard problems in polynomial expected time, *J.Algorithms* **10** (1989), 451-489.

[9] A. Frieze, and L. Kučera, Parallel colouring of random graphs, in *Random Graphs 87*, M. Karonski, J. Jaworski, A. Rucinski, eds, J. Wiley 1990, 41-52

[10] M. R. Garey, and D. S. Johnson, The complexity of near optimal graph coloring, *J.ACM* **23** (1976), 43-49.

[11] G.R. Grimmett, and C. J. H. McDiarmid, On colouring random graphs, *Math. Proc. Cambridge Phil. Soc.*, **77** (1975), 313-324.

[12] J. Håstad, I. Wegener, N. Wurm, and S.-Z. Yi, Optimal depth, very small size circuits for symmetric functions in $AC^0$, *Tech. Rep. 384*, Univ. Dortmund (1991).

[13] L. Kučera, Expected behavior of graph coloring algorithms, in *FCT'77, M. Karpinski, ed., Lecture Notes in Computer Science* **56** (Springer, Berlin, 1977), 447-451.

[14] L. Kučera, Parallel computation and conflicts in memory access, *Information Processing Letters* **14** (1982), 93-96.

[15] L. Kučera, Graphs with small chromatic numbers are easy to color, *Information Processing Letters* **30** (1989), 233-236.

[16] L. Kučera, The greedy coloring is a bad probabilistic algorithm, *J.Algorithms*, **12** (1991), 674-684.

[17] L. Levin, Average case complete problems, SIAM *J. Computing* **15** (1986), 285-286.

[18] D. Matula, and L. Kučera, An expose-and-merge algorithm and the chromatic number of a random graph, in *Random Graphs' 87, M.Karonski, J.Jaworski, A.Rucinski, eds.*, (J.Wiley and Sons, 1990), 175-187.

[19] S. Moran, Generalized lower bound derived from Håstad's main lemma, *Information Processing Letters* **25** (1987), 383-388.

[20] P. Ragde, The parallel simplicity of compaction and chaining, *ICALP'90, Lecture Notes in Computer Science* **443**, 744-751, 1990.

[21] L. Stockmayer, The complexity of approximate computing, *Proceedings of the 15th Symposium on Theory of Computing* (1983), 118-126.

[22] J. Turner, Almost all $k$-colorable graphs are easy to color, *J. Algorithms* **9** (1988), 63-82.

# DECIDING 3-COLOURABILITY IN LESS THAN $O(1.415^n)$ STEPS

Ingo Schiermeyer
Lehrstuhl C für Mathematik
Technische Hochschule Aachen
D-52056 Aachen, Germany

### Abstract

In this paper we describe and analyze an improved algorithm for deciding the 3-Colourability problem. If $G$ is a simple graph on $n$ vertices then we will show that this algorithm tests a graph for 3-Colourability, i.e. an assignment of three colours to the vertices of $G$ such that two adjacent vertices obtain different colours, in less than $O(1,415^n)$ steps.

**Key words:** Graph, algorithm, k-colouring, complexity

## 1 Introduction

We use Bondy & Murty [1] for terminology and notation not defined here and consider simple graphs only.

An assignment of $k$ colours to the vertices of a graph $G$ is called a *k-colouring* of $G$ if adjacent vertices are coloured by different colours. A graph $G$ is *k-colourable* if there exists an *l-colouring* of $G$, where $l \leq k$. The minimum $k$ for which a graph $G$ is *k-colourable* is called the *chromatic number* of $G$ and is denoted by $\chi(G)$. A graph $G$ is *k-chromatic* if $\chi(G) = k$.

Let *k-Colourability* denote the decision problem whether a given graph $G$ is *k-colourable*. It is well-known that *k-Colourability* is *NP-complete* for $k \geq 3$ and solvable in polynomial time for $k = 1$ and $k = 2$ (cf. [3]). Even the problem of determining the chromatic number of an arbitrary graph to within a given factor $M < 2$ has also been shown to be NP-complete [4]. Moreover, Lund and Yannakakis (cf. [5]) recently proved the following remarkable result for k-Colourability.

**Theorem 1.1** *There is an $\epsilon > 0$ such that no polynomial-time approximation algorithm can have worst-case ratio growing as $O(|V|^\epsilon)$ unless $P = NP$.*

Therefore, analyzing the worst-case running time of exponential-time algorithms is of increasing interest. Lawler [6] showed that determining the chromatic number of an arbitrary graph can be solved by the algorithm of Christofides [2] with worst-case running time of $O(mn \cdot (1 + 3^{1/3})^n)$, where $m$ is the number of edges in the graph and $n$ is the number of vertices, respectively (note: $1 + 3^{1/3} \approx 2,4422$).

The ideas of Christofides algorithm can be described as follows: An independent subset $I$ of vertices is *maximal* if it is not a proper subset of any other independent subset $I'$. It is well-known that if a graph is k-colourable then there is a partition of its vertex set into $k$ independent subsets where at least one of these independent subsets is maximal. Now computing all maximal independent sets of a given graph $G$ and repeating this computation for the remaining graphs the chromatic number of $G$ can be determined.

As suggested by Lawler [6], to test a graph for 3-Colourability, one can generate all maximal independent subsets in time of $O(mn \cdot 3^{n/3})$ and then check the induced subgraph on each complementary set of vertices for bipartiteness. It follows that such a test can be made in $O(mn \cdot 3^{n/3})$ time, where $3^{1/3} \approx 1,4422$. This bound is sharp since graphs on $n$ vertices may have up to $3^{n/3}$ maximal independent sets as has been shown by Moon & Moser [7].

In this paper we present an algorithm which decides 3-Colourability of an arbitrary graph in less than $O(1,415^n)$ steps. Choosing a vertex $u$ of maximum degree $\Delta(G)$ there are at most $3^{(n-1-\Delta(G))/3}$ maximal independent sets containing $u$. Exploiting the structure of the graph and a sophisticated case analysis in the generation process of all maximal independent sets containing $u$ leads to the improvement of the constant to $1,415$. Moreover, in several stages the algorithm can verify that a given graph $G$ is not 3-colourable without generating all maximal independent sets containing $u$.

## 2 The 3-Colour Algorithm

Let $u$ and $v$ be two non-adjacent vertices of a graph $G$. Then in any proper colouring of $G$ they are either assigned the same or different colours. If they obtain the same colour then we may identify (or *contract*) $u$ and $v$ and replace them by a single vertex $w$. We write $G \cdot \{u, v\}$ for this operation, where $N(w) = N(u) \cup N(v)$. More general, let $S \subseteq V(G)$ be a subset of $s \geq 2$ vertices $v_1, \ldots, v_s$ which are assigned the same colour. We write $G \cdot S$ for the *contraction* of $S$ to a single vertex $w$, where $N_{V(G)-S}(w) = N_{V(G)-S}(v_1) \cup \cdots \cup N_{V(G)-S}(v_s)$.

By $\square$ we denote the *empty graph* which contains no vertices and edges. If we k-colour a graph $G$ then let the integers $1, 2, \ldots, k$ represent these $k$ colours. We write $c(v) := i$ if a vertex $v$ is assigned colour $i$ with $1 \leq i \leq k$.

## 3-Colour Algorithm

Input:  *Graph G*
Output: $3 - colourable$, if $G$ is 3-colourable
        not $3 - colourable$, else

Procedure $Col(G)$;

1. if $G$ is empty then $G$ is 3-colourable;

2. $M := \{v \in V(G) \mid d(v) \le 2\}$;

    if $M \neq \emptyset$ then
      $G' := G - M; Col(G')$;
      if $G'$ is 3-colourable then $G$ is 3-colourable
                            else $G$ is not 3-colourable
    (* now $\delta(G) \ge 3$ *)

3. if $G$ is disconnected with components $G_1, G_2, \ldots, G_p$ then
      $Col(G_1); Col(G_2); \ldots; Col(G_p)$;
      if each of $G_1, G_2, \ldots, G_p$ is 3-colourable
        then $G$ is 3-colourable
        else $G$ is not 3-colourable

4. if there is an edge $v_1 v_2$ such that $|N(v_1) \cap N(v_2)| \ge 2$ then
      if $E(G[N(v_1) \cap N(v_2)]) \neq \emptyset$ then $G$ is not 3-colourable
                              else $G' := G \cdot \{N(v_1) \cap N(v_2)\}; Col(G')$;
      if $G'$ is 3-colourable then $G$ is 3-colourable
                           else $G$ is not 3-colourable
    (* now $G$ contains no induced $K_4$ and $K_4 - e$ *)

5. if $G[N(v)]$ is not 2-colourable for a vertex $v \in V(G)$
    then $G$ is not 3-colourable;

6. if there exists a vertex $v \in V(G)$ such that $G[N(v)]$
      contains a component $F$ with $|V(F)| \ge 3$ then
        2-colour $F$ with colour classes $F_1$ and $F_2$;
        $G' := G \cdot F_1; G'' := G' \cdot F_2; Col(G'')$;
        if $G''$ is 3-colourable then $G$ is 3-colourable
                          else $G$ is not 3-colourable
    (* now $\Delta(G[N(v)]) \le 1 \ \forall v \in V(G)$ *)

7. choose a vertex $u \in V(G)$ with maximum degree $d(u) = \Delta(G)$;
   if $\Delta(G) = n - 1$ then $G$ is 3-colourable else
   $c(u) := 1; H := G[V - \{u\} - N(u)]; S := N(u);$
   choose a vertex $w \in V(H)$ with $d_H(w) = \Delta(H)$;

8. if $\Delta(H) = 0$ then $G$ is 3-colourable;

   $I := \{v \in V(H) \mid d_H(v) = 0\};$
   if $I \neq \emptyset$ then
       $c(v) := 1 \, \forall v \in I; \, G' := G - I; \, Col(G');$
       if $G'$ is 3-colourable then $G$ is 3-colourable
                               else $G$ is not 3-colourable
   (* now $\Delta(H) \geq \delta(H) \geq 1$ *)

9. if $\Delta(H) \geq 3$ then
       $G_1 := G \cdot \{u, w\}; Col(G_1);$      (* $d_H(w) = \Delta(H)$ *)
       if $G_1$ is 3-colourable then $G$ is 3-colourable else
           $G_2 := G + uw; \, Col(G_2);$
           if $G_2$ is 3-colourable then $G$ is 3-colourable
                                   else $G$ is not 3-colourable
   (* now $1 \leq \Delta(H) \leq 2$ *)

10. if $\Delta(H) = 1$ or $H$ contains an isolated edge then
        choose an isolated edge $w_1 w_2 \in E(H)$;
        $uncol := true; i := 1;$
        while $i \leq 2$ and $uncol$ do
            $c(w_i) := 1;$
            $G' := G \cdot \{u, w_i\}; G_i := G' \cdot N_S(w_{3-i}); Col(G_i);$
            if $G_i$ is 3-colourable then
                $uncol := false; G$ is 3-colourable;
            else $i := i + 1;$
        if $i = 3$ then $G$ is not 3-colourable
    (* now $\Delta(H) = 2$ *)

11. if $H$ contains a path $P_t = w_1 w_2 \ldots w_t$ with $t \geq 3$ then
        $c(w_1) := 1; G' := G \cdot \{u, w_1\}; G_1 := G' \cdot N_S(w_2); Col(G_1);$
        if $G_1$ is 3-colourable then $G$ is 3-colourable else
            $c(w_2) := 1;$
            if $N_S(w_1) \cap N_S(w_3) \neq \emptyset$ then
                $G' := G \cdot \{u, w_2\}; G'' := G' \cdot \{w_1, w_3\};$
                $G_2 := G'' \cdot (N_S(w_1) \cup N_S(w_3)); Col(G_2);$
            else
                $G' := G \cdot \{u, w_2\}; G'' := G' \cdot N_S(w_1);$
                $G_2 := G'' \cdot N_S(w_3); Col(G_2);$

**if** $G_2$ is 3-colourable **then** $G$ is 3-colourable
**else** $G$ is not 3-colourable
(* now $\delta(H) = 2 = \Delta(H)$ *)

12. **if** $H$ contains a cycle $C_4 = w_1 w_2 w_3 w_4$ **then**
$c(w_1) := 1; c(w_3) := 1;$
**if** $N_S(w_2) \cap N_S(w_4) \neq \emptyset$ **then**
$G' := G \cdot \{u, w_1, w_3\}; G'' := G' \cdot \{w_2, w_4\};$
$G_1 := G'' \cdot (N_S(w_2) \cup N_S(w_4)); Col(G_1);$
**else**
$G' := G \cdot \{u, w_1, w_3\}; G'' := G' \cdot N_S(w_2); G_1 := G'' \cdot N_S(w_4);$
$Col(G_1);$
**if** $G_1$ is 3-colourable **then** $G$ is 3-colourable;
**else**
$c(w_2) := 1; c(w_4) := 1;$
**if** $N_S(w_1) \cap N_S(w_3) \neq \emptyset$ **then**
$G' := G \cdot \{u, w_2, w_4\}; G'' := G' \cdot \{w_1, w_3\};$
$G_2 := G'' \cdot (N_S(w_1) \cup N_S(w_3)); Col(G_2);$
**else**
$G' := G \cdot \{u, w_2, w_4\}; G'' := G' \cdot N_S(w_1); G_2 := G'' \cdot N_S(w_3);$
$Col(G_2);$
**if** $G_2$ is 3-colourable **then** $G$ is 3-colourable
**else** $G$ is not 3-colourable
**if** $H$ contains a cycle $C_t = w_1 w_2 \ldots w_t$ for some $t \geq 5$
with $|N_S(w_2)| \geq 2$ **then**
$uncol := true; i := 1;$
**while** $i \leq 2$ **and** $uncol$ **do**
$c(w_{i+1}) := 1;$
**if** $N_S(w_i) \cap N_S(w_{i+2}) \neq \emptyset$ **then**
$G' := G \cdot \{u, w_{i+1}\}; G'' := G' \cdot \{w_i, w_{i+2}\};$
$G_i := G'' \cdot \{N_S(w_i) \cup N_S(w_{i+2}); Col(G_i);$
**else**
$G' := G \cdot \{u, w_{i+1}\}; G'' := G' \cdot N_S(w_i); G_i := G'' \cdot N_S(w_{i+2});$
$Col(G_i);$
**if** $G_i$ is 3-colourable **then**
$uncol := false; G$ is 3-colourable;
**else** $i := i + 1;$
**if** $N_S(w_2) \cap N_S(w_3) \neq \emptyset$
**then** $G$ is not 3-colourable;
**else**
$c(w_1) := 1; c(w_4) := 1;$
$G' := G \cdot \{u, w_1, w_4\}; G'' := G' \cdot \{w_2 \cup N_S(w_3)\};$
$G_3 := G'' \cdot \{w_3 \cup N_S(w_2)\}; Col(G_3);$
**if** $G_3$ is 3-colourable **then** $G$ is 3-colourable

else $G$ is not 3-colourable

else

  if $H$ contains a cycle $C_3 = w_1w_2w_3$ such that $|N_S(w_1)| = 1$ then

    $G' := G \cdot N_S(w_3); G'' := G' \cdot \{w_3, N_S(w_1)\}; G_1 := G'' - w_1; Col(G_1);$

    if $G_1$ is 3-colourable then $G$ is 3-colourable;

    $G' := G \cdot \{w_1, N_S(w_2)\}; G_2 := G' - \{w_2, w_3\}; Col(G_2);$

    if $G_2$ is 3-colourable then $G$ is 3-colourable

else $G$ is not 3-colourable

  if $|N_S(w_i)| \geq 2$ for $1 \leq i \leq 3$ then

    $uncol := true; j := 1;$

    while $j \leq 3$ and $uncol$ do

        $c(w_j) := 1;$

        $l := j + 1 \bmod 3; p := j + 2 \bmod 3; G' := G \cdot \{u, w_j\};$

        $G'' := G' \cdot \{w_l, N_S(w_p)\}; G_{j+3} := G'' \cdot \{w_p, N_S(w_l)\};$

        $Col(G_{j+3});$

        if $G_{j+3}$ is 3-colourable then

          $uncol := false; G$ is 3-colourable;

        else $j := j + 1;$

    if $j = 4$ then $G$ is not 3-colourable

else

    $F := G[N(u)]; p := |E(F)|; q := 2^{\Delta(G)-p};$

    let $F_1, F_2, \ldots, F_q$ be the $q$ possible 2-colourings of $F$

    with colours 2 and 3;

    $uncol := true; i := 1;$

    while $i \leq q$ and $uncol$ do

        2-colour $F$ according to $F_i;$

        if $F_i$ can be extended to a 3-colouring of $G$ then

          $uncol := false; G$ is 3-colourable;

        else

        $i := i + 1;$

    if $i = q + 1$ then $G$ is not 3-colourable;

In order to show the correctness of the algorithm we prove the following lemma.

**Lemma 2.1** *Let $G$ be a simple graph with a vertex $u, S := N(u), F := G[S]$ and $H := G[V(G) - \{u\} - N(u)]$, where $H$ is the union of cycles with $d_S(w) = 1 \; \forall w \in V(H)$. Let $C$ be a 2-colouring of $F$. Then $C$ can be extended to a 3-colouring of $G$ if and only if $H$ contains no odd cycle $C$ such that $N_S(C)$ is monochromatic.*

**Proof:** If $H$ contains an odd cycle $C$ such that $N_S(C)$ is monochromatic then $C$ cannot be extended to a 3-colouring of $G$ since $\chi(C) = 3$ and $d_S(w) = 1$ for all vertices $w \in V(H)$. Otherwise, suppose $c(u) = 1$ and $S$ is coloured with colours 2 and 3. Let $C_k = w_1 w_2 \ldots w_k$ be a cycle of length $k \geq 3$. Suppose $c(N_S(w_1)) = 2$ (note that $d_S(w_i) = 1$ for $1 \leq i \leq k$). Choose $p$ maximal such that $c(N_S(w_i)) = 2$ for $1 \leq i \leq p$. We then choose $c(w_i) = 3$ for $1 \leq i \leq p$, $i$ odd, and $c(w_i) = 1$ for $1 \leq i \leq p$, $i$ even. If $p < k$ we then exchange colours 2 and 3 and repeat this (alternating) process until all vertices of $C_k$ are coloured (note that $p < k$ if $k$ is odd). Thus $C$ can be extended to a 3-colouring of $G$.

$\square$

**Correctness of the algorithm:** If $G$ is empty then $G$ is 3-colourable. Next the algorithm deletes all vertices of $G$ having degree at most two. Clearly, if the reduced graph $G'$ is 3-colourable then so is $G$, since any proper 3-colouring of $G'$ can be extended to a 3-colouring of $G$ by assigning a color to a deleted vertex which has not been assigned to its (at most two) neighbors. Hence, after the execution of 2., we have $\delta(G) \geq 3$. If $G$ is disconnected with components $G_1, G_2, \ldots, G_p$ then $G$ is 3-colourable if and only if each of $G_1, G_2, \ldots, G_p$ is 3-colourable. If $G$ contains an induced $K_4$ then $G$ is not 3-colourable and if $G$ contains an induced $K_4 - e$ then the two non-adjacent vertices must be assigned the same colour. Hence, after the execution of 4., $G$ contains no induced $K_4$ and no induced $K_4 - e$. Next the algorithm tests whether the induced graph $G[N(v)]$ is 2-colourable for each vertex $v \in V(G)$, which is a necessary condition for $G$ to be 3-colourable. If there exists a vertex $v \in V(G)$ such that $G[N(v)]$ contains a component $F$ with $|V(F)| \geq 3$ then $F$ can be arbitrarily 2-coloured with color classes $F_1$ and $F_2$. Moreover, $F_1$ and $F_2$ can be contracted and the reduced graph $G''$ is 3-colourable if and only if $G$ is 3-colourable. Hence, after the execution of 6., for every vertex $v \in V(G)$ the induced graph $G[N(v)]$ contains only isolated vertices or edges, i.e., $\Delta(G[N(v)]) \leq 1 \ \forall v \in V(G)$. Now the algorithm chooses a vertex $u \in V(G)$ with maximum degree $d(u) = \Delta(G)$. Clearly, if $d(u) = n - 1$, then $G$ is 3-colourable since $G[N(u)]$ is 2-colourable. Hence we may assume that $d(u) < n - 1$. Let $H := G[V(G) - \{u\} - N(u)]$ and $S := N(u)$. If $\Delta(H) = 0$ then $G$ is 3-colourable since $G[N(u)]$ is 2-colourable and $\{u\} \cup V(H)$ is an independent set. By the same argument all isolated vertices of $H$ can be deleted from $G$ and the reduced graph $G'$ is 3-colourable if and only if $G$ is 3-colourable. Hence, after the execution of 8., we have $\Delta(H) \geq \delta(H) \geq 1$. If $\Delta(H) \geq 3$ we branch and generate two subproblems $G_1$ and $G_2$, where $u$ and $w$ are assigned the same or different colours, respectively. Clearly, $G$ is 3-colourable if and only if at least one of $G_1$ and $G_2$ is 3-colourable. Hence we now have $1 \leq \Delta(H) \leq 2$.

In the following the algorithm considers all maximal independent sets containing $u$. Depending on the structure of $H$ it generates up to three new graphs and $G$ is 3-colourable if and only if at least one of the generated graphs is 3-colourable. Let $I$ be an arbitrary maximal independent set containing $u$. If $\Delta(H) = 1$ then $H$ is the union of isolated edges. Let $w_1 w_2$ be an isolated edge of $H$. Then either $w_1 \in I$ or $w_2 \in I$. Hence, after the execution of

10., we have $\Delta(H) = 2$. Next the algorithm tests whether $H$ contains a path $P_t = w_1 w_2 \ldots w_t$ for some $t \geq 3$. Then either $w_1 \in I$ or $w_2 \in I$. Hence, after the execution of 11., we have $\delta(H) = 2 = \Delta(H)$, i.e., $H$ is the union of cycles. Let $C_t = w_1 w_2 \ldots w_t$ be such a cycle. If $t = 4$ then either $w_1, w_3 \in I$ or $w_2, w_4 \in I$. If $t \geq 5$ then either $w_2 \in I$ or $w_3 \in I$ or $w_1, w_4 \in I$. Finally, if $t = 3$, let $C_3 = w_1 w_2 w_3$ be such a cycle. If $|N_S(w_1)| = 1$ then two graphs $G_1$ and $G_2$ are generated. If $|N_S(w_i)| \geq 2$ for $1 \leq i \leq 3$ then three new graphs $G_4, G_5$ and $G_6$ are generated. Otherwise, let $F := G[N(u)]$. For each possible 2-colouring of $F$ the algorithm tests whether this colouring can be extended to a 3-colouring of $G$ using lemma 2.1.

## 3 Complexity-analysis

**Lemma 3.1** *The execution time of Col between two successive calls is bounded by $O(n(m+n))$ when applied to a graph $G$.*

**Proof:** The execution time of statements 1., 2., 3. and 4. can be bounded by $O(1), O(n), O(m+n)$ and $O(mn)$, respectively. Testing a graph for bipartiteness can be performed in $O(m + n)$ time (cf. [6]). Therefore, the execution time of statement 5. is bounded by $O(n(m + n))$. The same bound holds for statement 6.. For statements 7., 8. and 9. we obtain $O(n)$. Since the contraction $G \cdot \{u, v\}$ of two vertices $u, v$ can be performed in $O(n)$ time, the execution time of statements 10., 11. and 12. is bounded by $O(n^2)$. Altogether the execution time can be bounded by $O(n(m + n))$.

□

In order to compute the running time of the 3-Colour Algorithm it remains to compute the number $T(G)$ of recursive calls Col has to perform to test a graph $G$ for 3-Colourability. Define

$$T(n) \quad := \quad max\{T(G) \mid G \text{ is a graph of order } n\}.$$

In order to estimate $T(n)$ we choose a function $F(n, k) = c\alpha^{n-k}\beta^k$, where $\alpha = 1,0294$, $\beta = 1,4147$ and $c$ is a sufficiently large constant.

**Lemma 3.2** *Let $G$ be a graph of order $n$ and $k := n - \Delta(G) - 1$. Then*

$$T(G) \leq F(n, k).$$

**Proof:** 1. If $G$ is the empty graph □ then $T(□) = 1 \leq c$.
2. Let $p := |M|$. If $M \neq \emptyset$ then $p \geq 1$. Let $n' := |V(G')|$ and $k' := n' - \Delta(G') - 1$. Since $\Delta(G') \leq \Delta(G)$ and $n' = n - p$ we have $k \geq k' \geq k - p$. Let $p =: p_1 + p_2$

with $\Delta(G) := \Delta(G') + p_1$ and $k := k' + p_2$. Then

$$
\begin{aligned}
T(G) &\leq 1 + T(G') \\
&\leq 1 + c\alpha^{n'-k'}\beta^{k'} \leq 1 + c\alpha^{(n-p_1-p_2)-(k-p_2)}\beta^{k-p_2} \\
&= 1 + c\alpha^{n-k-p_1}\beta^{k-p_2} < c\alpha^{n-k}\beta^{k}
\end{aligned}
$$

since $\alpha, \beta > 1$ and $p_1 + p_2 \geq 1$.

3. Let $n_i := |V(G_i)|$ with $k_i := n_i - \Delta(G_i) - 1$ for $1 \leq i \leq p$ and $k_{max} := max\{k_i \mid 1 \leq i \leq p\}$. With $\delta(G) \geq 3$ we have $n_i \geq 4$ and thus $k \geq k_i + 4(p-1)$ for $1 \leq i \leq p$. Then

$$
\begin{aligned}
T(G) &\leq 1 + \sum_{i=1}^{p} T(G_i) \\
&\leq 1 + \sum_{i=1}^{p} c\alpha^{n_i-k_i}\beta^{k_i} \leq 1 + c\alpha^{n-k}\sum_{i=1}^{p}\beta^{k_i} \\
&\leq 1 + c\alpha^{n-k}\sum_{i=1}^{p}\beta^{k_{max}} = 1 + c\alpha^{n-k}\beta^{k_{max}}p \\
&\leq 1 + c\alpha^{n-k}\beta^{k_{max}}2^{p}2^{p-2} = 1 + c\alpha^{n-k}\beta^{k_{max}}4^{p-1} \\
&< c\alpha^{n-k}\beta^{k_{max}}\beta^{4(p-1)} \leq c\alpha^{n-k}\beta^{k}
\end{aligned}
$$

since $p \leq 2^p$ for $p \geq 1, \beta^4 > 4,0055 > 4$ and $k \geq k_{max} + 4(p-1)$.

4. Let $G$ be a graph of order $n$ and $k := n - \Delta(G) - 1$. Let $T \subseteq V(G)$ be an arbitrary set with $|T| =: t \geq 2$ and $G' := G \cdot T$ with $n' := |V(G')|$ and $k' := n' - \Delta(G') - 1$. Then $\Delta(G') \geq \Delta(G) - (t-1) \Rightarrow k' \leq k$ since, by the contraction of $T$, $d_{G'}(v) \geq d_G(v) - (t-1) \ \forall v \in V(G')$. Now let $|N(v_1) \cap N(v_2)| =: t \geq 2$. Then

$$
\begin{aligned}
T(G) &\leq 1 + T(G') \\
&\leq 1 + c\alpha^{n'-k'}\beta^{k'} < c\alpha^{n-k}\beta^{k}
\end{aligned}
$$

since $n = n' + (t-1) > n', k \geq k'$ and $\alpha, \beta > 1$.

6. Let $t_i := |F_i|$ for $1 \leq i \leq 2$. Then $(t_1 - 1) + (t_2 - 1) \geq 1$ since $|V(F)| \geq 3$. As in 4. we obtain

$$
\begin{aligned}
T(G) &\leq 1 + T(G'') \\
&\leq 1 + c\alpha^{n'-k'}\beta^{k'} < c\alpha^{n-k}\beta^{k}
\end{aligned}
$$

since $n = n' + (t_1 - 1) + (t_2 - 1) > n', k \geq k'$ and $\alpha, \beta > 1$.

**8.** Let $p := |I|$. Then $\Delta(G') \geq \Delta(G) - p$ and $k' \leq k$. Hence

$$
\begin{aligned}
T(G) &\leq 1 + T(G') \\
&\leq 1 + c\alpha^{n'-k'}\beta^{k'} \leq 1 + c\alpha^{n'-k}\beta^k \\
&= 1 + c\alpha^{n-k}\beta^k \cdot \frac{1}{\alpha^p} < c\alpha^{n-k}\beta^k
\end{aligned}
$$

since $k \geq k', n = n' + p > n'$ and $\alpha > 1$.

**9.** Let $n_i := |V(G_i)|$ and $k_i := n_i - \Delta(G_i) - 1$ for $1 \leq i \leq 2$. Then $n_1 = n - 1, \Delta(G_1) \geq \Delta(G) + 3, k_1 \leq k - 4, n_2 = n, \Delta(G_2) = \Delta(G) + 1$ and $k_2 = k - 1$. Thus

$$
\begin{aligned}
T(G) &\leq 1 + T(G_1) + T(G_2) \\
&\leq 1 + c\alpha^{n_1 - k_1}\beta^{k_1} + c\alpha^{n_2 - k_2}\beta^{k_2} \\
&\leq 1 + c\alpha^{n-k+4-1}\beta^{k-4} + c\alpha^{n-k+1}\beta^{k-1} \\
&= 1 + c\alpha^{n-k}\beta^k\left(\frac{\alpha^3}{\beta^4} + \frac{\alpha}{\beta}\right) < c\alpha^{n-k}\beta^k
\end{aligned}
$$

since $\frac{\alpha^3}{\beta^4} + \frac{\alpha}{\beta} < 0,99998$.

**10.** Let $|N_S(w_1)| =: p \geq 2$ and $N_S(w_2)| =: q \geq 2$. Then

$$
\begin{aligned}
T(G) &\leq 1 + T(G_1) + T(G_2) \\
&\leq 1 + c\alpha^{n-k+2-1-(q-1)}\beta^{k-2} + c\alpha^{n-k+2-1-(p-1)}\beta^{k-2} \\
&\leq 1 + c\alpha^{n-k}\beta^{k-2} + c\alpha^{n-k}\beta^{k-2} \\
&= 1 + c\alpha^{n-k}\beta^k \cdot \frac{2}{\beta^2} < c\alpha^{n-k}\beta^k
\end{aligned}
$$

since $\frac{2}{\beta^2} < 0,9994$.

**11.** Let $p := |N_S(w_1)|, q := |N_S(w_2)|$ and $r := |N_S(w_3)|$ with $p \geq 2$ and $q, r \geq 1$. Then

$$
\begin{aligned}
T(G) &\leq 1 + c\alpha^{n-k+2-1-(q-1)}\beta^{k-2} \\
&\quad + max\{c\alpha^{n-k+3-1-1-max\{p-1,r-1\}}\beta^{k-3}, c\alpha^{n-k+3-1-(p-1)-(r-1)}\beta^{k-3}\} \\
&\leq 1 + c\alpha^{n-k+1}\beta^{k-2} + c\alpha^{n-k+1}\beta^{k-3} \\
&= 1 + c\alpha^{n-k}\beta^k\left(\frac{\alpha}{\beta^2} + \frac{\alpha}{\beta^3}\right) < c\alpha^{n-k}\beta^k
\end{aligned}
$$

since $\frac{\alpha}{\beta^2} + \frac{\alpha}{\beta^3} < 0,878$.

**12.** Let $C_t = w_1 w_2 \ldots w_t$ be a cycle of length $t \geq 3$ with $p := |N_S(w_1)|, q := |N_S(w_2)|, r := |N_S(w_3)|$ and $s := |N_S(w_4)|$. If $t = 4$ then

$$
\begin{aligned}
T(G) \;\leq\;& 1 + T(G_1) + T(G_2) \\
\leq\;& 1 + max\{ca^{n-k+4-2-1-max\{q-1,s-1\}}\beta^{k-4}, ca^{n-k+4-2-(q-1)-(s-1)}\beta^{k-4}\} \\
& + max\{ca^{n-k+4-2-1-max\{p-1,r-1\}}\beta^{k-4}, ca^{n-k+4-2-(p-1)-(r-1)}\beta^{k-4}\} \\
=\;& 1 + ca^{n-k+2}\beta^{k-4} + ca^{n-k+2}\beta^{k-4} \\
=\;& 1 + ca^{n-k}\beta^k \cdot \frac{2\alpha^2}{\beta^4} < ca^{n-k}\beta^k
\end{aligned}
$$

since $\frac{2\alpha^2}{\beta^4} < 0,5292$.

If $t \geq 5$ then $p, r, s \geq 1$ and $q \geq 2$. Hence

$$
\begin{aligned}
T(G) \;\leq\;& 1 + T(G_1) + T(G_2) + T(G_3) \\
\leq\;& 1 + max\{ca^{n-k+3-1-1-max\{p-1,r-1\}}\beta^{k-3}, ca^{n-k+3-1-(p-1)-(r-1)}\beta^{k-3}\} \\
& + max\{ca^{n-k+3-1-1-max\{q-1,s-1\}}\beta^{k-3}, ca^{n-k+3-1-(q-1)-(s-1)}\beta^{k-3}\} \\
& + ca^{n-k+4-2-(q-1+1)-(r-1+1)}\beta^{k-4} \\
=\;& 1 + ca^{n-k+2}\beta^{k-3} + ca^{n-k+1}\beta^{k-3} + ca^{n-k-1}\beta^{k-4} \\
=\;& 1 + ca^{n-k}\beta^k \left( \frac{\alpha^2}{\beta^3} + \frac{\alpha}{\beta^3} + \frac{1}{\alpha\beta^4} \right) < ca^{n-k}\beta^k
\end{aligned}
$$

since $\frac{\alpha^2}{\beta^3} + \frac{\alpha}{\beta^3} + \frac{1}{\alpha\beta^4} < 0,9804$.

If $t = 3$ and $p = 1$ then

$$
\begin{aligned}
T(G) \;\leq\;& 1 + T(G_1) + T(G_2) \\
\leq\;& 1 + ca^{n-k+1-1}\beta^{k-1-1} + ca^{n-k+1-1-(q-1)}\beta^{k-1-2} \\
\leq\;& 1 + ca^{n-k}\beta^k \left( \frac{1}{\beta^2} + \frac{1}{\beta^3} \right) < ca^{n-k}\beta^k
\end{aligned}
$$

since $\frac{1}{\beta^2} + \frac{1}{\beta^3} < 0,8529$.

If $t = 3$ and $p, q, r \geq 2$ then

$$
\begin{aligned}
T(G) \;\leq\;& 1 + T(G_4) + T(G_5) + T(G_6) \\
\leq\;& 1 + ca^{n-k+3-1-(q-1+1)-(r-1+1)}\beta^{k-3} \\
& + ca^{n-k+3-1-(p-1+1)-(r-1+1)}\beta^{k-3} \\
& + ca^{n-k+3-1-(p-1+1)-(q-1+1)}\beta^{k-3} \\
\leq\;& 1 + ca^{n-k}\beta^k \cdot \frac{3}{\alpha^2\beta^3} < ca^{n-k}\beta^k
\end{aligned}
$$

since $\frac{3}{\alpha^2\beta^3} < 0,99991$.

Finally, let $S := S_1 \cup S_2$, where $S_1 = \{v \in V(F) \mid d_F(v) = 1\}$ and $S_2 = \{v \in V(F) \mid d_F(v) = 0\}$. Then $|S_1| = 2p$ since $p := |E(F)|$. Let $e(S, H) := |\{vw \mid v \in S, w \in V(H)\}|$. Since $\delta(G) \geq 3$ and $S = N(u)$ we have $d_H(v) \geq 1 \; \forall v \in S_1$ and $d_H(v) \geq 2 \; \forall v \in S_2$. Thus $e(S, H) \geq 2p + 2(\Delta(G) - 2p) = 2(\Delta(G) - p)$. Since $\delta(H) = \Delta(H) = 2$ and $d_S(w) = 1 \; \forall w \in V(H)$ we have $k = |V(H)| = e(S, H) \geq 2(\Delta(G) - p)$. We may now interpret each 2-colouring of $F$ as a recursive call of $Col$. Then

$$
\begin{aligned}
T(G) &\leq 1 + 2^{\Delta(G)-p} \\
&< c\alpha^{n-k}\beta^{2(\Delta(G)-p)} \leq c\alpha^{n-k}\beta^k
\end{aligned}
$$

since $\alpha > 1, c \geq 1$ and $\beta^2 > 2,0013 > 2$.

□

With lemmata 3.1 and 3.2 we obtain the main result of this chapter.

**Theorem 3.3** *The 3-colour algorithm tests every graph $G$ for 3-Colourability in $O(n(m+n) \cdot 1,4147^n)$ steps.*

**Corollary 3.4** *The 3-colour algorithm tests every graph $G$ for 3-Colourability in less than $O(1,415^n)$ steps.*

# References

[1] J. A. Bondy and U. S. R. Murty, *Graph Theory with Applications* (Macmillan, London and Elsevier, New York, 1976).

[2] N. Christofides, *An Algorithm for the Chromatic Number of a Graph*, Computer J. 14 (1971) 38 - 39.

[3] M. R. Garey and D. S. Johnson, *Computers and Intractability, A Guide to the Theory of NP-Completeness*, W. H. Freeman and Company, New York, 1979.

[4] M. R. Garey and D. S. Johnson, *The complexity of Near-Optimal Graph Coloring*, J. ACM 23 (1976) 43 - 49.

[5] D. S. Johnson, *The NP-Completeness Column: An Ongoing Guide*, J. of Alg. 13 (1992) 502 - 524.

[6] E. L. Lawler, *A Note on the Complexity of the Chromatic Number Problem*, Inform. Process. Lett. 5 (1976) 66 - 67.

[7] J. W. Moon and L. Moser, *On Cliques in Graphs*, Israel J. of Math. 3 (1965) 23 - 28.

# A rainbow about $T$-colorings
# for complete graphs

Klaus Jansen[1]

Fachbereich 11 - Mathematik, FG Informatik, Universität Duisburg,
Postfach 11 05 03, D-47048 Duisburg, Germany

**Abstract.** Given a finite set $T$ of positive integers, with $0 \in T$, a $T$-coloring of a graph $G = (V, E)$ is a function $f : V \rightarrow \mathbb{N}_0$ such that for each $\{x, y\} \in E$, $|f(x) - f(y)| \notin T$. The $T$-span is the difference between the largest and smallest color and the $T$-span of $G$ is the minimum span over all $T$-colorings of $G$. We show that the problem to find the $T$-span for a complete graph is NP-complete.

## 1 Introduction

**Problem definition.** Let $G = (V, E)$ be a graph, and let $T$ be a set of positive integers, with $0 \in T$. A $T$-*coloring* is a function $f$ which assigns to each vertex $x \in V$ a positive integer such that $\{x, y\} \in E$ implies $|f(x) - f(y)| \notin T$.

$T$-colorings arose from the channel assignment problem (see [2, 6]). The *span* of a $T$-coloring is defined by

$$sp_{T,f}(G) = max_{x,y \in V} |f(x) - f(y)|$$

and the $T$-*span of* $G$ is the minimum span of a $T$-coloring of $G$, denoted by $sp_T(G)$. If $T = \{0\}$, then the $T$-coloring is the same as an ordinary coloring. In this case, $sp_T(G) = \chi(G) - 1$ where $\chi(G)$ is the chromatic number of $G$. Therefore, the problem to find $sp_T(G)$ in general is NP-complete.

We study the problem to find $sp_T(K_n)$ for a complete graph $K_n$ with $n$ vertices. The simplest algorithm for a complete graph is the greedy algorithm. In each step, the smallest possible color that will not violate the definition of a $T$-coloring is chosen. The greedy $T$-coloring does not always obtain the optimum span. For $T = \{0, 1, 4, 5\}$ and $n = 3$, the greedy $T$-coloring uses colors $\{0, 2, 8\}$ and the optimum $T$-coloring $\{0, 3, 6\}$ achieves $sp_T(K_3) := 6$ [9]. For other results about $T$-colorings, we refer to [1, 3, 7, 8, 11, 12].

The $T$-coloring problem on complete graphs was mentioned as an open problem by F. Roberts [10]. Using a reduction from CLIQUE and a number theoretical result, A. Gräf [5] proved independently the NP-completeness of $T$-coloring on complete graphs where $T_{all}$, a finite set of positive allowed distances, satisfies $max\ T_{all} = O(|T_{all}|^3)$. In our paper, we give a direct reduction from 3-SAT. We show that the $T$-coloring problem on complete graphs remains NP-complete even if $max\ T_{all} = O(|T_{all}|)$.

## 2 First results

In this section, we give two new results about $T$-colorings for general graphs $G = (V, E)$. Later, we use these results in the proof of our NP-completeness result. Moreover, we give an upper bound for the $T$-span of $G$.

**Lemma 1.** *Let $G = (V, E)$ be a graph, let $T$ be a finite set of positive integers, with $0 \in T$, and let $f$ be a $T$-coloring with $k = min_{x \in V} f(x) > 0$. Then, $f'$ defined by $f'(x) = f(x) - k$ is also a $T$-coloring with $sp_{T,f'}(G) = sp_{T,f}(G)$.*

*Proof.* Let $\{x, y\} \in E$ with $x \neq y$. Since $f$ is a $T$-coloring, we get

$$|f'(x) - f'(y)| = |f(x) - k - f(y) + k| = |f(x) - f(y)| \notin T.$$

In addition, we have $f'(x) = f(x) - k \geq k - k = 0$. Therefore, $f'$ is also a $T$-coloring. Furthermore, the span of $f'$ is equal to the span of $f$. □

We call a $T$-coloring *minimal*, if the color 0 is used in $f$. Using the assertion above, we may consider only minimal $T$-colorings. We note that in a minimal $T$-coloring for a complete graph, each color $f(x) \neq 0$ must be an allowed distance. In other words, $f(x) \notin T$ for each vertex $x$ in $K_n$ with $f(x) \neq 0$. Furthermore, we obtain for each vertex $0 \leq f(x) \leq sp_{T,f}(G)$.

The next lemma shows that there is symmetric relation about the $T$-colorings.

**Lemma 2.** *Let $G = (V, E)$ be a graph, let $T$ be a finite set of positive integers, with $0 \in T$, and let $f$ be a minimal $T$-coloring for $G$ with span $sp_{T,f}(G)$. Then, $f'$ defined by $f'(x) = sp_{T,f}(G) - f(x)$ is also a $T$-coloring with $sp_{T,f'}(G) = sp_{T,f}(G)$.*

*Proof.* Let $\{x, y\} \in E$ with $x \neq y$. Since $f$ is a $T$-coloring, we get

$$|f'(x) - f'(y)| = |sp_{T,f}(G) - f(x) - sp_{T,f}(G) + f(y)| = |f(x) - f(y)| \notin T.$$

Since $f$ is a minimal $T$-coloring, $f'(x) = sp_{T,f}(G) - f(x) \geq 0$. Therefore, $f'$ is also a $T$-coloring and $sp_{T,f'}(G) = sp_{T,f}(G)$. □

An interesting bound for the minimum span of an arbitrary graph is given in the next Theorem. The assertion has the implication that the minimum span is bounded by a polynom in $|T|$ and $|V|$.

**Theorem 3.** *[11]*

$$sp_T(G) \leq |T| \cdot (\chi(G) - 1)$$

*where $\chi(G)$ is the chromatic number of $G$.*

# 3   Main result

In this section, we show that the $T$-coloring problem remains NP-complete for a complete graph. As decision problem, we use the following formulation.

$T$-COLORING
INSTANCE: A finite set $T$ of positive integers, with $0 \in T$, a complete graph $K_n$ with $n$ vertices and a bound $K \in \mathbb{N}$.
QUESTION: Is there a $T$-coloring $f$ with $sp_{T,f}(K_n) \leq K$?

We use a transformation from the NP-complete 3-SATISFIABILITY problem (for short 3-SAT) to the $T$-coloring problem (cf. [4] for 3-SAT). The instance of 3-SAT is given by a formula $c_1 \wedge \ldots \wedge c_m$ with $c_i = (y_{i,1} \vee y_{i,2} \vee y_{i,3})$ and $y_{i,j} \in X = \{x_0, \overline{x_0}, \ldots, x_n, \overline{x_n}\}$. We assume that $m > n$ and that for the number $n$,

$$10^{\overline{k}} \leq n < \frac{1}{2} \cdot 10^{\overline{k}+1}$$

($\overline{k} + 1$ gives the length of the decimal representation of $n$). Moreover, we assume that in the last clause $c_m$, the first literal $y_{m,1}$ must get the truth value *true*. By a simple transformation, we can show that this restricted 3-SAT problem remains NP-complete.

**Theorem 4.** *The $T$-coloring problem restricted to a complete graph is NP-complete even if max $T_{all} = O(|T_{all}|)$.*

*Proof.* First, the problem is in NP. It is enough to consider minimal $T$-colorings with possible colors $\{0, \ldots, K\}$. In the following, we construct a set $T_{all}$ of allowed distances. We use vectors $(j, \ell, k, u)$ with $0 \leq j \leq m$, $0 \leq \ell \leq 9$, $0 \leq k \leq n$ and $0 \leq u \leq 99$ to spezify the integers

$$j \cdot 10^{\overline{k}+4} + \ell \cdot 10^{\overline{k}+3} + k \cdot 10^2 + u.$$

Then, we define the set

$$T = \{x | 0 \leq x \leq (m, 3, n, 21)\} \setminus T_{all}.$$

The set $T_{all}$ consists of several distance sets $T_1, \ldots, T_6$ which are given in Table 1.

First, the integer $\overline{k}$ with $10^{\overline{k}} \leq n < \frac{1}{2} \cdot 10^{\overline{k}+1}$ can be generated in polynomial time and $\overline{k} = O(log(n))$. Clearly, the cardinality of $T$ und each integer in $T$ can be bounded by

$$(m, 3, n, 21) \leq O(m \cdot 10^{\overline{k}} + n) = O(m \cdot n).$$

Therefore, the transformation can be done in polynomial time and $max\ T_{all} = O(m \cdot n) = O(|T_{all}|)$.

Now, we show the equivalence that there is a solution of the restricted 3-SAT instance if and only if there is a $T$-coloring $f$ for a complete graph with $n + m + 3$ vertices and $sp_{T,f}(K_{n+m+3}) \leq (m, 1, n, 21)$.

**Table 1.** The allowed distance sets $T_1, \ldots, T_6$.

| | |
|---|---|
| $T_1$ | $\{(0,0,k,10),(0,0,k,11)|0 \leq k \leq n\},$ |
| $T_2$ | $\{(0,0,k-1,99),(0,0,k,0),(0,0,k,1)|1 \leq k \leq n\},$ |
| $T_3$ | $\{(0,0,n,21)\},$ |
| $T_4$ | $\{(j,\ell,n,21)|1 \leq j \leq m, 1 \leq \ell \leq 3\},$ |
| $T_5$ | $\{(j,\ell,0,0)|1 \leq j \leq m-1, 0 \leq \ell \leq 3\} \cup$ |
| | $\{(j,\ell,0,0)|0 \leq j \leq m-2, 8 \leq \ell \leq 9\} \cup \{(m,1,0,0)\},$ |
| $T_6$ | $\{(j,\ell,n-k,11)|y_{j,\ell} = x_k, 1 \leq j \leq m, 1 \leq \ell \leq 3\} \cup$ |
| | $\{(j,\ell,n-k,10)|y_{j,\ell} = \overline{x_k}, 1 \leq j \leq m, 1 \leq \ell \leq 3\} \cup$ |
| | $\{(j,\ell,n-k,10)|y_{j,\ell} \neq x_k, \overline{x_k}, 1 \leq j \leq m, 1 \leq \ell \leq 3, 0 \leq k \leq n\} \cup$ |
| | $\{(j,\ell,n-k,11)|y_{j,\ell} \neq x_k, \overline{x_k}, 1 \leq j \leq m, 1 \leq \ell \leq 3, 0 \leq k \leq n\}.$ |

## First Direction.

Suppose, we have a truth assignment that verifies the clauses. Then, we construct a feasible $T$-coloring $f$ in the following way.

$$f(1) = 0$$

$$f(i+2) = \begin{cases} (0,0,i,10) \text{ if } x_i \text{ is } true \\ (0,0,i,11) \text{ if } x_i \text{ is } false \end{cases}$$
$$\text{for } 0 \leq i \leq n$$

$$f(n+3) = (0,0,n,21)$$
$$f(n+3+j) = \begin{cases} (j,1,n,21) \text{ if } y_{i,1} \text{ is } true \\ (j,2,n,21) \text{ if } y_{j,1} \text{ is } false, y_{j,2} \text{ is } true \\ (j,3,n,21) \text{ otherwise} \end{cases}$$
$$\text{for } 1 \leq j \leq m$$

Let $A$ be the used colors in $f$. In the following, we show by case analysis that all distance pairs are allowed.

**Case 1.** $x = (0,0,k,1h) \in A$, $y = (0,0,k',1h') \in A$ and $x > y$.
Since $x, y \in A$ and $x > y$, we have $n \geq k > k' \geq 0$ and $h, h' \in \{0,1\}$. If $h \geq h'$ then $|x - y| = (0,0,k-k',0(h-h')) \in T_2 \subset T_{all}$. If $h < h'$ then $|x - y| = (0,0,k-k'-1,99) \in T_2 \subset T_{all}$.

**Case 2.** $x = (0,0,n,21) \in A$, $y = (0,0,k,1h) \in A$.
Clearly, we have $n \geq k \geq 0$ and $h \in \{0,1\}$. In this case, we obtain $|x - y| = (0,0,n-k,1(1-h)) \in T_1 \subset T_{all}$.

**Case 3.** $x = (j,\ell,n,21) \in A$, $y = (j',\ell',n,21) \in A$ and $x > y$.
In this case, $m \geq j > j' \geq 1$ and $\ell, \ell' \in \{1,2,3\}$. If $\ell \geq \ell'$, then $|x - y| = (j-j',\ell-\ell',0,0) \in T_5 \subset T_{all}$. If $\ell < \ell'$, then $|x - y| = (j-j'-1,10+\ell-\ell',0,0) \in T_5 \subset T_{all}$.

**Case 4.** $x = (j,\ell,n,21) \in A$, $y = (0,0,n,21) \in A$.

Clearly, we have $1 \leq j \leq m$ and $1 \leq \ell \leq 3$ (for $j = m$ we have $\ell = 1$). Then, we obtain $|x - y| = (j, \ell, 0, 0) \in T_5$.

**Case 5.** $x = (j, \ell, n, 21) \in A$, $y = (0, 0, k, 1h) \in A$.
In this case, $1 \leq j \leq m$, $1 \leq \ell \leq 3$, $0 \leq k \leq n$ and $h \in \{0, 1\}$. The difference $|x - y|$ is equal to $(j, \ell, n - k, 1(1 - h))$. If $y_{j,\ell} \notin \{x_k, \overline{x_k}\}$, then $|x - y| \in T_6 \subset T_{all}$. If $y_{j,\ell} = x_k$ then $x_k$ must be true. In this case, we obtain $h = 0$ and $|x - y| \in T_6 \subset T_{all}$. If $y_{j,\ell} = \overline{x_k}$ then $x_k$ must be false. Then, we get $h = 1$ and $|x - y| \in T_6 \subset T_{all}$.

Using this case analysis, we obtain that $f$ is a $T$-coloring. Since in the last clause $y_{m,1}$ is true, we get $f(n + 3 + m) = (m, 1, n, 21)$. Then, we obtain $sp_{T,f}(K_{n+m+3}) \leq (m, 1, n, 21)$.

## Second Direction.

Now, we suppose that $f$ is a $T$-coloring for a complete graph with $n + m + 3$ vertices and span bounded by $(m, 1, n, 21)$. Using Lemma 2, we assume that $f$ is a minimal $T$-coloring. We denote with $A$ the set of colors used in $f$. Since $f$ is minimal, we have $0 \in A$ and $A \setminus \{0\} \subset T_{all}$.

In the following, we analyse the structure of this color set $A$. First, we consider the form of set $A$ intersected with $T_6$, $T_5$ and $T_1 \cup T_2$. Next, we analyse which combinations of colors in different sets $T_i$ and $T_j$, $i \neq j$ are possible or not.

Given $|A| \geq n + m + 3$ we prove that $A \subset T_1 \cup T_3 \cup T_4 \cup \{0\}$ or that $A \subset T_4 \cup T_5 \cup T_6 \cup \{0\}$. Then, we transform a feasible color set $A \subset T_4 \cup T_5 \cup T_6 \cup \{0\}$ into a feasible color set $A' \subset T_1 \cup T_3 \cup T_4 \cup \{0\}$. Last, we construct for such a color set a truth assignment which verifies all clauses.

### STEP 1.
We start with the analysis of the structure of the color set $A$ intersected with $T_6$, $T_5$ and $T_1 \cup T_2$.

**Lemma 5.** *Let $A$ be a set of colors used in a $T$-coloring with $A \cap T_6 \neq \emptyset$. Then, either there is a sequence $\ell_1, \ldots, \ell_m$ with $\ell_j \in \{1, 2, 3\}$ and an integer $k$, $0 \leq k \leq n$ and an integer $h \in \{0, 1\}$ such that*

$$A \cap T_6 \subset \{(j, \ell_j, k, 1h) | 1 \leq j \leq m\}$$

*or there is a sequence $h_0, \ldots, h_n$ with $h_k \in \{0, 1\}$ and an integer $j$, $1 \leq j \leq m$ and an integer $\ell \in \{1, 2, 3\}$ such that*

$$A \cap T_6 \subset \{(j, \ell, k, 1h_k) | 0 \leq k \leq n\}.$$

*Proof.* Let $x = (j, \ell, k, 1h) \in A$ and $y = (j', \ell', k', 1h') \in A$ with $x > y$.

**Case 1.** $j > j'$.

If $h \neq h'$ then we get a non allowed distance, because the last two digits in $|x - y|$ are 01 or 99. If $h = h'$ and $k \neq k'$, the last two digits are 00 and the third vector component in $|x - y|$ is unequal 0. In this case, $|x - y| \notin T_5$ and, therefore, $|x - y|$ is not allowed. Therefore, for $j > j'$ we must have $k = k'$ and $h = h'$.

**Case 2.** $j = j'$.

If $\ell \neq \ell'$ then the second vector component in $|x - y|$ is 1 or 2. Since the first vector component is 0, we get a non allowed distance. This implies that for $j = j'$ we must have $\ell = \ell'$ and $k \neq k'$.

Using these observations, there are only two possibilities for the set $A \cap T_6$ described above.  □

**Lemma 6.** *Let $A$ be a set of colors used in a T-coloring, with $A \cap T_5 \neq \emptyset$. Then, there is a sequence $\ell_0, \ldots, \ell_{m-2}$ with $\ell_j \in \{0, \ldots, 5\}$ such that $A \cap T_5$ is a subset of the set*

$$\{(j, 8 + \ell_j, 0, 0) | 0 \leq j \leq m - 2\} \cup \{(m, 1, 0, 0)\}.$$

*Proof.* Let us assume that $x = (j, 8 + \ell, 0, 0) \in A$ and that $y = (j, 8 + \ell', 0, 0) \in A$ with $0 \leq \ell \neq \ell' \leq 5$. In this case, the difference of $|x - y|$ is given by $(0, |\ell - \ell'|, 0, 0)$. Since $0 < |\ell - \ell'| \leq 5$, the difference is not allowed. From this observation, we obtain directly the assertion above.  □

**Lemma 7.** *Let $A$ be a set of colors used in a T-coloring with $A \cap T_2 \neq \emptyset$. Then, there is an integer $k$, $1 \leq k \leq n$, and a sequence $h_0, \ldots, h_{k-1} \in \{0, 1, 2\}$ and $h_k, \ldots, h_n \in \{0, 1\}$ such that $A \cap (T_1 \cup T_2)$ is a subset of the set*

$$\{(0, 0, i, 99 + h_i) | 0 \leq i \leq k - 1\} \cup \{(0, 0, i, 1h_i) | k \leq i \leq n\}.$$

*Proof.* Since $A \cap T_2 \neq \emptyset$, it follows that $A \cap \{(0, 0, 0, 10), (0, 0, 0, 11)\} = \emptyset$. Clearly, at most one integer of each set $\{(0, 0, i, 99), (0, 0, i + 1, 0), (0, 0, i + 1, 1)\}$ for $0 \leq i < n$ and at most one integer of each set $\{(0, 0, i, 10), (0, 0, i, 11)\}$ for $1 \leq i \leq n$ can ly in $A \cap (T_1 \cup T_2)$.

Let $x = (0, 0, k, 1h) \in A \cap T_1$ and $y = (0, 0, k', 99 + h') \in A \cap T_2$. If $x < y$ then the last two digits of $|x - y|$ are 88, 89, 90 or 91. Therefore, the difference $|x - y| \notin T_{all}$ for $x < y$. In other words, we can combine a color $x \in T_1$ with a color $y \in T_2$ only if $x > y$. From this observation, the assertion follows directly.  □

## STEP 2.

Now, we give the combinations of sets $T_i$ and $T_j$, $i \neq j$, which are possible or not.

**Lemma 8.** *Let* $x, y$ *be two colors in a* $T$-*coloring, with* $x \neq y$. *Then, the following combinations are not possible:*

*(a)* $x \in T_1$ *and* $y \in T_5$.
*(b)* $x \in T_2$ *and* $y \in T_3$.
*(c)* $x \in T_2$ *and* $y \in T_4$.
*(d)* $x \in T_2$ *and* $y \in T_5$.
*(e)* $x \in T_3$ *and* $y \in T_5$.
*(f)* $x \in T_3$ *and* $y \in T_6$.

*Proof.*
**Case (a):** Let $x = (0, 0, k, 1h) \in T_1$ and $y = (j, \ell, 0, 0) \in T_5$. The last two digits in the difference $|y - x|$ are either 90 or 89; this depends whether $h = 0$ or $h = 1$. In both cases, the difference $|x - y| \notin T_{all}$.

**Case (b):** Let $x = (0, 0, k, 99 + h) \in T_2$ with $h \in \{0, 1, 2\}$ and let $y = (0, 0, n, 21) \in T_3$. Then, the last two digits in $|x - y|$ are 78, 79 or 80. In other words, the difference $|x - y|$ is not allowed for each pair $x \in T_2$ and $y \in T_3$.

**Case (c):** In this case, the same argument as in case (b) can be applied.

**Case (d):** Let $x = (0, 0, k, 99 + h) \in T_2$ with $h \in \{0, 1, 2\}$ and let $y = (j, \ell, 0, 0) \in T_5$. For $h = 1$, the difference $|x - y|$ is equal to $(j, \ell - 1, 10^{\bar{k}+1} - k - 1, 0)$ where $0 \leq k \leq n - 1$. Since $10^{\bar{k}+1} - k - 1 \geq 1$, the difference cannot be allowed. For $h = 0$ and $h = 2$, the last two digits of $|x - y|$ are either 99 or 01. Therefore, $|x - y| \notin T_{all}$ for each pair $x \in T_2$ and $y \in T_5$.

**Case (e):** In this case, the last two digits in $|x - y|$ are 79.

**Case (f):** In this case, the last two digits in $|x - y|$ are 89 or 90. $\qquad\square$

**Lemma 9.** *Let* $x = (j, \ell, k, 1h) \in T_6$ *and* $y = (0, 0, k', h') \in T_2$ *with* $h' \in \{-1, 0, 1\}$. *If* $k < k'$, *then* $\{x, y\} \not\subset A$ *where* $A$ *is a color set used in a* $T$-*coloring.*

*Proof.* For $k < k'$, we obtain $|x - y| = (j, \ell - 1, 10^{\bar{k}+1} - k' + k, u)$ where $u \in \{9, \ldots, 12\}$. Since $k' - k \leq n$, the third vector component of $|x - y|$ can be bounded as follows:

$$10^{\bar{k}+1} - k' + k \geq 10^{\bar{k}+1} - n.$$

Using that $10^{\bar{k}} \leq n < \frac{1}{2} \cdot 10^{\bar{k}+1}$, we get

$$10^{\bar{k}+1} - k' + k > \frac{1}{2} \cdot 10^{\bar{k}+1} > n.$$

In other words, the third vector component of $|x - y|$ is larger than $n$ and, therefore, the distance $|x - y|$ is not allowed. $\qquad\square$

**Lemma 10.** *Let $x = (j, \ell, k, 1h) \in T_6$ and $y = (0, 0, k', 1h') \in T_1$ and let $A$ be a color set used in a T-coloring. If $\{x, y\} \subset A$, then $k = k'$ and $h = h'$.*

*Proof.*
**Case 1.** $h' = 0$.
If $k \geq k'$, then $|x - y| = (j, \ell, k - k', 0h)$. Then, the distance $|x - y|$ is allowed only if $k = k'$ and $h = 0$. If $k < k'$ the third vector component of $|x - y|$ is larger than $n$. Therefore, for $k < k'$ the distance $|x - y|$ is not allowed.

**Case 2.** $h' = 1$.
If $h = 0$ then the last two digits of $|x - y|$ are 99 and $|x - y|$ is not allowed. For $h = 1$ and $k \geq k'$, the distance $|x - y|$ is equal to $(j, \ell, k - k', 0)$ and is allowed only if $k = k'$. For $h = 1$ and $k < k'$, using the same argument as above, the distance $|x - y| \notin T_{all}$. □

**Lemma 11.** *Let $x = (j, \ell, k, 1h) \in T_6$ and $y = (j', \ell', n, 21) \in T_4$ and let $A$ be a color set used in a T-coloring. If $\{x, y\} \subset A$ then $x < y$ and $\ell' \geq \ell$. If $\{x, y\} \subset A$ and $j' > j$ then $\ell' > \ell$.*

*Proof.* Let $\{x, y\} \subset A$. If $x > y$ then the last two digits of $|x - y|$ are 89 or 90. In this case, $|x - y|$ is not allowed. If $y > x$ and $\ell' < \ell$ then

$$|x - y| = (j' - j - 1, 10 + \ell' - \ell, n - k, 1(1 - h)) \notin T_{all},$$

because $10 + \ell' - \ell \in \{8, 9\}$. If $y > x$, $\ell' \geq \ell$ and $j' > j$ then

$$|x - y| = (j' - j, \ell' - \ell, n - k, 1(1 - h)).$$

Since the first vector component $j' - j$ is larger than or equal to 1, $|x - y| \in T_{all}$ only if $\ell' > \ell$. □

**Lemma 12.** *Let $x = (j, \ell, k, 1h) \in T_6$ and $y = (j', \ell', 0, 0) \in T_5$ and let $A$ be a color set used in a T-coloring. If $y > x$, then $\{x, y\} \not\subset A$. If $x > y$ and $|x - y| < (0, 8, 0, 0)$ then $\{x, y\} \subset A$ only if $j = j'$ and $\ell = \ell'$.*

*Proof.* In the first case, the last two digits of $|x - y|$ are 89 or 90. In the second case, we consider the first and second vector component of $|x - y|$. Since $|x - y| < (0, 8, 0, 0)$, the first component is zero. Then, we obtain an allowed distance only if $\ell = \ell'$ and $j = j'$. □

**Lemma 13.** *Let $x = (j, \ell, n, 21) \in T_4$ and $y = (j', \ell', 0, 0) \in T_5$ and let $A$ be a color set used in a T-coloring. If $y > x$ then $\{x, y\} \not\subset A$.*

*Proof.* If $y > x$ then the last two digits in $|x - y|$ are 79. Then, $|x - y| \notin T_{all}$. □

## STEP 3.
Let $A$ be a color set in a minimal T-coloring with $|A| \geq n + m + 3$. In the following, we

prove that $A \subset T_1 \cup T_3 \cup T_4 \cup \{0\}$ or that $A \subset T_4 \cup T_5 \cup T_6 \cup \{0\}$. If $A \subset T_4 \cup T_5 \cup T_6 \cup \{0\}$ then we transform $A$ into a feasible color set $A' \subset T_1 \cup T_3 \cup T_4 \cup \{0\}$.

We show the first assertion by case analysis.

**Case 1.** $T_2 \cap A \neq \emptyset$.

Using Lemma 8 we get $A \cap (T_3 \cup T_4 \cup T_5) = \emptyset$. In other words,

$$A \subset T_1 \cup T_2 \cup T_6 \cup \{0\}.$$

Now, we apply Lemma 7 and obtain that $A \cap (T_1 \cup T_2)$ is a subset of

$$\{(0,0,i,99+h_i)|0 \leq i \leq k-1\} \cup \{(0,0,i,1h_i)|k \leq i \leq n\}.$$

Using Lemma 5, $A \cap T_6$ is a subset either of $\{(j,\ell_j,k,1h)|1 \leq j \leq m\}$ or is a subset of $\{(j,\ell,k,1h_k)|0 \leq k \leq n\}$. In the first case, at most $m + (n+1) + 1$ colors ly in $A$. In the second case, we apply the inequality $n < m$ and obtain at most $(n+1) + (n+1) + 1 < n + m + 3$ colors in $A$. In both cases, the number of colors lying in $A$ is smaller than $n + m + 3$. This means that $T_2 \cap A = \emptyset$.

**Case 2.** $T_6 \cap A \neq \emptyset$.

Using Lemma 8, we get $T_3 \cap A = \emptyset$. First, we assume that $(T_1 \cup T_2) \cap A \neq \emptyset$. For $(T_1 \cup T_2) \cap A = \emptyset$, we have $A \subset T_4 \cup T_5 \cup T_6 \cup \{0\}$.

If $T_2 \cap A \neq \emptyset$ then using case (1.) we can bound $|A| < n+m+3$. Now, we assume that $T_1 \cap A \neq \emptyset$ and that $T_2 \cap A = \emptyset$. Again, using Lemma 8, we obtain

$$A \subset T_1 \cup T_4 \cup T_6 \cup \{0\}.$$

Clearly, there is a sequence $h_0, \ldots, h_n \in \{0,1\}$ with $A \cap T_1 \subset \{(0,0,k,1h_k)|0 \leq k \leq n\}$.

Using Lemma 10 and $A \cap T_6 \neq \emptyset$, we must have $|A \cap T_1| = 1$. Let $(0,0,k,1h) \in A \cap T_1$. Then, $A \cap T_6$ can be only a subset of $\{(j,\ell_j,k,1h)|1 \leq j \leq m\}$. In addition, $A \cap T_4 \subset \{j,\ell'_j,n,21)|1 \leq j \leq m\}$. If $x \in A \cap T_6$ and $y \in A \cap T_4$ then we get $x < y$ by Lemma 11. Therefore, the cardinality of $A$ can be bounded by $1 + (m+1) + 1 < n + m + 3$.

**Case 3.** $T_5 \cap A \neq \emptyset$.

In this case, we obtain using Lemma 8 that $A \cap (T_1 \cup T_2 \cup T_3) = \emptyset$. In other words,

$$A \subset T_4 \cup T_5 \cup T_6 \cup \{0\}.$$

Using these three cases, either $A \subset T_1 \cup T_3 \cup T_4 \cup \{0\}$ or $A \subset T_4 \cup T_5 \cup T_6 \cup \{0\}$. In the following, we consider a set of the second type and transform it to a set of the first type.

Let $A \subset T_4 \cup T_5 \cup T_6 \cup \{0\}$. If $A \cap T_4 = \emptyset$, $A \cap T_5 = \emptyset$ or $A \cap T_6 = \emptyset$ then the cardinality of $A$ is smaller than $m + n + 3$. Using Lemma 6, $A \cap T_5$ is a subset of

$$\{(j,8+\overline{\ell_j},0,0)|0 \leq j \leq m-2\} \cup \{(m,1,0,0)\}.$$

Let $x \in A \cap T_6$ and $y \in A \cap T_5$. Then, by Lemma 12, we get $x > y$. Given a color $z \in A \cap T_4$ and using Lemma 11, we obtain $z > x > y$.

In the first case, $A \cap T_6 \subset \{(j, \ell_j, k, 1h) | 1 \leq j \leq m\}$. But in this case, we can bound the cardinality of $A$ by $1 + (m+2) < m+n+3$. In the remaining case, there are integers $\bar{j}$ and $\bar{\ell}$, with $1 \leq \bar{j} \leq m$, $1 \leq \bar{\ell} \leq 3$, such that $A \cap T_6 \subset \{(\bar{j}, \bar{\ell}, k, 1h_k) | 0 \leq k \leq n\}$.

If $\bar{j} < m$, then $A \cap T_5 \subset \{(j, 8 + \ell_j, 0, 0) | 0 \leq j \leq \bar{j} - 1\}$. Using Lemma 11, it is not possible that $(m, 1, n, 21) \in A$. Therefore, $|A|$ can be bounded by $m + n + 2$. If $\bar{j} = m$ then $(m, 1, n, 21) \in A$ if $\bar{\ell} = 1$. Using Lemma 12, we obtain that $\ell_{m-1} = 3$ and that $(m, 1, 0, 0) \in A$. Moreover, we can restrict the integers $\ell_j \in \{0, 1, 2\}, 0 \leq j \leq m-2$. In total, there is a sequence $\ell_0, \ldots, \ell_{m-2} \in \{0, 1, 2\}$ and a sequence $h_0, \ldots, h_n \in \{0, 1\}$ such that

$$A = \{(j, 8 + \ell_j, 0, 0) | 0 \leq j \leq m - 2\} \cup$$
$$\{(m, 1, k, 1h_k) | 0 \leq k \leq n\} \cup$$
$$\{(m, 1, 0, 0), (m, 1, n, 21)\}.$$

Now, we apply Lemma 2 and generate a feasible color set

$$A' = \{(m, 1, n, 21) - a | a \in A\}.$$

This set $A'$ satisfies the condition $A' \subset T_1 \cup T_3 \cup T_4 \cup \{0\}$.

## STEP 4.

Given a color set $A \subset T_1 \cup T_3 \cup T_4 \cup \{0\}$ with $|A| \geq n + m + 3$ we construct a feasible truth assignment which satisfies all clauses.

Clearly, the set $A$ has the following form.

$$A \cap T_1 = \{(0, 0, k, 1h_k) | 0 \leq k \leq n\},$$
$$A \cap T_3 = \{(0, 0, n, 21)\},$$
$$A \cap T_4 = \{(j, \ell_j, n, 21) | 1 \leq j \leq m\}.$$

In total, $|A| = (n+1) + 1 + m + 1 = n + m + 3$. If $h_k = 0$ then we set $x_k$ *true* and if $h_k = 1$ then we set $x_k$ *false*. We consider the $j$.th clause $c_j = (y_{j,1} \vee y_{j,2} \vee y_{j,3})$. We assume that that $y_{j,\ell_j} \in \{x_i, \overline{x_i}\}$. Then,

$$\beta = (j, \ell_j, n, 21) - (0, 0, i, 1h_i) = (j, \ell_j, n - i, 1(1 - h_i)).$$

If $y_{j,\ell_j} = x_i$ then $\beta \in T_{all}$ only if $h_i = 0$. In this case, $x_i$ and the clause $c_j$ is *true*. If $y_{j,\ell_j} = \overline{x_i}$ then $\beta \in T_{all}$ only if $h_i = 1$. In this case, $\overline{x_i}$ and the clause $c_j$ is *true*. In total, we have a truth assignment which verifies all clauses. This completes the proof of the NP-completeness result. $\qquad \square$

# References

1. Bonias, I.: $T$-colorings of complete graphs. Ph.D. Thesis, Northeastern University, Boston, MA (1991)
2. Cozzens, M.B., Roberts, F.S.: $T$-colorings of graphs and the channel assignment problem. Congress Numerantium **35** (1982) 191–208
3. Cozzens, M.B., Roberts, F.S.: Greedy algorithms for $T$-colorings of complete graphs and the meaningfulness of conclusions about them. J. Comb. Inform. Syst. Sci., to appear

4. Garey, M.R., Johnson, D.S.: Computers and Intractability: A Guide to the Theory of NP-Completeness. Freeman, San Francisco (1979)
5. Gräf, A.: Complete difference sets and $T$-colorings of complete graphs. Bericht, Universität Mainz 7 (1993)
6. Hale, W.K.: Frequency assignment: theory and applications. Proc. IEEE 68 (1980) 1497–1514
7. Liu, D.D.: Graph homomorphisms and the channel assignment problem. Ph.D. Thesis, University of South Carolina, Columbia, SC (1991)
8. Raychaudhuri, A.: Further results on $T$-colorings and frequency assignment problems. SIAM J. Disc. Math. to appear
9. Roberts, F.S.: From garbage to rainbows: Generalizations of graph colorings and their applications. Y. Alavi, G. Chartrand, O.R. Oellermann and A.J. Schwenk (eds.): Graph Theory, Combinatorics, and Applications, Vol. 2, Wiley, New York (1991) 1031–1052
10. Roberts, F.S.: $T$-colorings of graphs: recent results and open problems. Disc. Math. 93 (1991) 229–245
11. Tesman, B.: $T$-colorings, list $T$-colorings, and set $T$-colorings of graphs. Ph.D. Thesis, Rutgers University, New Brunswick, NJ (1989)
12. Wang, D.-I.: The channel assignment problem and closed neighborhood containment graphs. Ph.D. Thesis, Northeastern University, Boston, MA (1985)

# Approximating the Chromatic Polynomial of a Graph*

Nai-Wei Lin

*Department of Computer Science*

*The University of Arizona*

*Tucson, AZ 85721*

*naiwei@cs.arizona.edu*

### Abstract

The problem of computing the chromatic polynomial of a graph is #P-hard. This paper presents an approximation algorithm for computing the chromatic polynomial of a graph. This algorithm has time complexity $O(n^2 \log n + nm^2)$ for a graph with $n$ vertices and $m$ edges. This paper also shows that the problem of computing the chromatic polynomial of a chordal graph can be solved in polynomial time. Knowledge about the chromatic polynomial of graphs can be employed to improve the performance of logic programs and deductive databases.

## 1 Introduction

This paper considers finite graphs without loops (i.e., edges joining a vertex to itself) and multiple edges between any pair of vertices. The *chromatic number* of a graph $G$, written as $\chi(G)$, is the minimum number of colors necessary to color $G$ such that no adjacent vertices have the same color. The *chromatic polynomial* of a graph $G$, denoted by $C(G, x)$, is a polynomial in $x$ representing the number of different ways in which $G$ can be colored by using at most $x$ colors.

The problem of $k$-colorability of a graph $G$ is the one of deciding whether the value $C(G, k) > 0$. Since the problem of deciding whether the value $C(G, k) > 0$ is NP-complete [7], the problem of computing the value $C(G, k)$ is #P-complete [12]. The more general problem of computing the chromatic polynomial of a graph is therefore #P-hard.

This paper presents an approximation algorithm for computing the chromatic polynomial of a graph. This algorithm is based on the *greedy* method. We first determine an ordering on the vertices of the graph using some *heuristics*. According to the determined ordering, we next derive an upper bound and a lower bound on how many different ways each vertex can be colored. The product of the upper bounds and the product of the lower bounds for all the vertices in the graph then give, respectively, an upper bound and a lower bound on the total number of different ways the entire graph can be colored.

---

*This work was supported in part by the National Science Foundation under grant number CCR-8901283.

If the number of available colors is given as a symbolic variable, then the two products are polynomials in this variable. Finally, we take a mean of these two polynomials as an approximation of the chromatic polynomial of the graph. This algorithm has time complexity $O(n^2 \log n + nm^2)$ for a graph with $n$ vertices and $m$ edges.

A graph is called *chordal* if every cycle of length greater than 3 has an edge joining two nonconsecutive vertices of the cycle. Chordal graphs arise in many contexts and contain the following families of graphs: interval graphs, cactus graphs, adjoint graphs of cactus graphs, and so on [6]. Gavril has shown that problems of finding a minimum coloring, a minimum covering by cliques, a maximum clique, and a maximum independent set, of a chordal graph can be solved in polynomial time [5]. In this paper we show that the problem of computing the chromatic polynomial of a chordal graph can also be solved in polynomial time.

Many problems in areas such as operations research and artificial intelligence require enumerating all the solutions that satisfy a set of binary equality or disequality constraints on variables ranging over a finite domain of values. The constraints are of the form $x = y$ or $x \neq y$. These constraint satisfaction problems can be reduced to the graph coloring problem as follows. Each vertex in the graph corresponds to a variable in the constraints, each edge in the graph corresponds to a disequality constraint, and the edges of two vertices are merged if there is an equality constraint between the corresponding two variables. Hence, each coloring of the graph corresponds to a solution to the original constraint satisfaction problem. Accordingly, the associated counting problem for each such constraint satisfaction problem reduces to the computation of the chromatic polynomial of its corresponding graph.

This class of constraint satisfaction problems is frequently realized in the settings such as logic programs and deductive databases, where it is easy to generate multiple solutions of a problem if required. The ability to estimate the chromatic polynomial of a graph allows us to estimate the number of solutions generated by the procedures realizing these constraint satisfaction problems. Information about the number of solutions generated by procedures has been used to estimate the cost of executing procedures, and the cost information has been further used to optimize programs and queries in these settings. For example, this information has been employed for query optimization in deductive databases by appropriately rearranging the evaluation order of subgoals [10, 13], and for process granularity control in parallel logic programming systems by properly preventing small-grain processes from spawning [3].

The remainder of this paper is organized as follows. Section 2 derives a number of upper and lower bounds on the chromatic polynomial of a graph. Section 3 discusses how the orderings on the vertices of a graph affect the bounds on its chromatic polynomial. Section 4 presents an approximation algorithm for computing the chromatic polynomial of a graph based on an ordering proposed in Section 3. Section 5 shows some experimental results on the performance behavior of the algorithm. Finally, Section 6 concludes this paper.

## 2 Bounds on Chromatic Polynomials

Since the value of the chromatic polynomial $C(G, x)$ is always nonnegative, for any graph $G$ and natural number $x$, we shall assume that any polynomial $P(x)$ really means the function $max(P(x), 0)$. Furthermore, for any two polynomials $P_1(x)$ and $P_2(x)$, we define $P_1(x) \leq P_2(x)$ if and only if $max(P_1(x), 0) \leq max(P_2(x), 0)$, for all natural numbers $x$. We shall derive bounds on the chromatic polynomial of a graph based on the greedy method.

Let $G = (V, E)$ be a graph. The *order* of $G$, denoted by $|G|$, is the number of vertices in $G$. Let $U$ be a subset of $V$. The subgraph of $G$ *induced* by $U$ is a graph $H = (U, F)$ such that $F$ consists of all the edges in $E$ both of whose vertices belong to $U$. The *neighbors*, $N_G(v) = \{w \in V | (v, w) \in E\}$, of a vertex $v$ in $G$ is the set of vertices adjacent to $v$. The *adjacency graph*, $Adj_G(v)$, of $v$ is the subgraph of $G$ induced by $N_G(v)$.

Let $G = (V, E)$ be a graph of order $n$ and $\omega = v_1, \ldots, v_n$ be an ordering on $V$. We define two sequences of subgraphs of $G$ according to $\omega$. The first is a sequence of subgraphs $G_1(\omega), \ldots, G_n(\omega)$, called *accumulating subgraphs*, where $G_i(\omega)$ is the subgraph induced by $V_i = \{v_1, \ldots, v_i\}$, for $1 \leq i \leq n$. The second is a sequence of subgraphs $G'_2(\omega), \ldots, G'_n(\omega)$, called *interfacing subgraphs*, where $G'_i(\omega)$ is the adjacency graph $Adj_{G_{i-1}(\omega)}(v_i)$, for $1 < i \leq n$. For example, consider the graph shown in Figure 1. The imposed ordering is denoted by the labels of vertices. The corresponding accumulating subgraphs and interfacing subgraphs are shown in Figure 2.

The following proposition gives an upper bound and a lower bound on the chromatic polynomial of a graph in terms of, respectively, the chromatic number and the order of interfacing subgraphs.

**Proposition 2.1** *Let $G = (V, E)$ be a graph of order $n$ and $\Omega$ be the set of all possible orderings on $V$. Suppose the interfacing subgraphs of $G$ corresponding to an ordering*

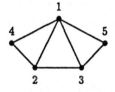

Figure 1: An example

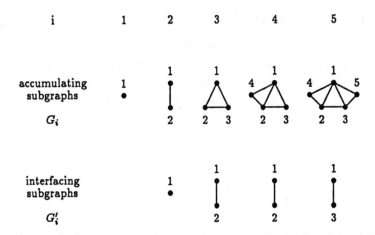

Figure 2: The accumulating and interfacing subgraphs of the graph in Figure 1

$\omega \in \Omega$ are $G'_2(\omega), \ldots, G'_n(\omega)$. Then

$$max_{\omega \in \Omega}\{x \prod_{i=2}^{n}(x - |G'_i(\omega)|)\} \leq C(G, x) \leq min_{\omega \in \Omega}\{x \prod_{i=2}^{n}(x - \chi(G'_i(\omega)))\}. \qquad (1)$$

**Proof** Suppose $G_1(\omega), \ldots, G_n(\omega)$ are the accumulating subgraphs of $G$ corresponding to an ordering $\omega$. The proof is by induction on $G_j(\omega)$, for $1 \leq j \leq n$. In base case, $G_1(\omega)$ is a graph consisting of one vertex $v_1$, so $C(G_1(\omega), x) = x$. Suppose $x \prod_{i=2}^{j}(x - |G'_i(\omega)|) \leq C(G_j, x) \leq x \prod_{i=2}^{j}(x - \chi(G'_i(\omega)))$ for some $j$, $1 \leq j < n$. Then consider adding the vertex $v_{j+1}$ and associated edges into $G_j(\omega)$ to form $G_{j+1}(\omega)$. For the lower bound, since $|G'_{j+1}(\omega)|$ is the degree of $v_{j+1}$ in $G_{j+1}(\omega)$, we have $(x - |G'_{j+1}(\omega)|) C(G_j(\omega), x) \leq C(G_{j+1}(\omega), x)$. From the hypothsis, we obtain

$$x \prod_{i=2}^{j+1}(x - |G'_i(\omega)|) \leq (x - |G'_{j+1}(\omega)|) C(G_j(\omega), x) \leq C(G_{j+1}(\omega), x).$$

For the upper bound, since $\chi(G'_{j+1}(\omega))$ is the minimum number of colors necessary for coloring $G'_{j+1}(\omega)$, we have $C(G_{j+1}(\omega), x) \leq (x - \chi(G'_{j+1}(\omega))) C(G_j(\omega), x)$. From the hypothsis, we obtain

$$C(G_{j+1}(\omega), x) \leq (x - \chi(G'_{j+1}(\omega))) C(G_j(\omega), x) \leq x \prod_{i=2}^{j+1}(x - \chi(G'_i(\omega))).$$

Since $G_n(\omega) = G$, we have the following formula

$$x \prod_{i=2}^{n}(x - |G'_i(\omega)|) \leq C(G, x) \leq x \prod_{i=2}^{n}(x - \chi(G'_i(\omega))). \qquad (2)$$

Finally, because Formula (2) holds for any ordering $\omega$, Formula (1) follows. $\square$

The upper bound in Formula (1) is in terms of the chromatic number of interfacing subgraphs. Unfortunately, the computation of the chromatic number of a graph is NP-complete. Nevertheless, it turns out that if each of the chromatic numbers in Formula (1) is replaced by any of its lower bounds, the resultant expression is still an upper bound on the chromatic polynomial. The following theorem by Bondy gives a lower bound on the chromatic number of a graph that can be computed efficiently [2].

**Theorem 2.2** *Let $G$ be a graph of order $n$, $d(1) \geq \cdots \geq d(n)$ be the degrees of nodes in $G$ and $\sigma_j$ be defined recursively by*

$$\sigma_j = n - d(\sum_{i=1}^{j-1} \sigma_i + 1).$$

*Suppose $k \leq n$ is some integer satisfying*

$$\sum_{j=1}^{k-1} \sigma_j < n. \tag{3}$$

*Then $\chi(G) \geq k$.* □

We define $\rho(G)$, called the *Bondy number* of a graph $G$, to be the largest integer $k \leq n$ satisfying Formula (3). Then an upper bound on the chromatic polynomial of a graph can be expressed in terms of the Bondy number of interfacing subgraphs.

**Proposition 2.3** *Let $G = (V, E)$ be a graph of order $n$ and $\Omega$ be the set of all possible orderings on $V$. Suppose the interfacing subgraphs of $G$ corresponding to an ordering $\omega \in \Omega$ are $G'_2(\omega), \ldots, G'_n(\omega)$. Then*

$$max_{\omega \in \Omega}\{x \prod_{i=2}^{n}(x - |G'_i(\omega)|)\} \leq C(G, x) \leq min_{\omega \in \Omega}\{x \prod_{i=2}^{n}(x - \rho(G'_i(\omega)))\}. \tag{4}$$

**Proof** By Theorem 2.2 and an argument parallel to the upper bound argument of Proposition 2.1. □

## 3  Ordering of Vertices

It is clear that carrying out the computation of the maximum and minimum among all the possible orderings in Formula (4) is impractical. As a result, we shall employ a representative ordering to compute upper and lower bounds on the chromatic polynomial.

A graph is a *clique* if every pair of vertices in the graph are adjacent. An ordering on the vertices of a graph is said to be a *perfect elimination ordering* if all the corresponding interfacing subgraphs are cliques [9]. Dirac [4] and Rose [8] have shown that a graph is chordal if and only if it has a perfect elimination ordering. The graph in Figure 1 is an example of a chordal graph, and the labels of vertices show a perfect elimination ordering. The following proposition states an appealing property of perfect elimination ordering.

**Proposition 3.1** *Let $G = (V, E)$ be a graph of order $n$ and $\omega$ be an ordering on $V$. If $\omega$ is a perfect elimination ordering, then $x \prod_{i=2}^{n}(x - |G_i'(\omega)|) = C(G, x) = x \prod_{i=2}^{n}(x - \rho(G_i'(\omega)))$.*

**Proof** By the fact, easily proved, that $\rho(K) = |K|$ for any clique $K$. $\square$

One implication of Proposition 3.1 is that if a perfect elimination ordering of a graph can be generated efficiently, then the chromatic polynomial of that graph can be computed efficiently. Unfortunately, not every graph has a perfect elimination ordering. For example, no complete bipartite graph $K_{n,m}$ with $n > 1$ and $m > 1$ is chordal.

We now describe an ordering that will be used to compute bounds on the chromatic polynomials. For obvious reasons, we shall try to generate a perfect elimination ordering whenever it is possible. The ordering generation is an iterative graph reduction process and the ordering is generated in reverse order.

At each iteration, we search for a vertex such that its adjacency graph is a clique. If such a vertex $v$ exists, it is chosen as the vertex to generate. It is clear that if the graph resulted from removing $v$ has a perfect elimination ordering, then the original graph containing $v$ also has a perfect elimination ordering. The latter ordering can be constructed by simply adding $v$ at the rear of the former ordering. The process continues by removing $v$ from the graph and proceeding to the next iteration.

On the other hand, if such a vertex $v$ does not exist, then we choose a vertex $w$ that has the smallest degree to generate. The basic idea behind this heuristic is that, to yield nontrivial lower bounds for larger number of values, we demand the maximum order of the interfacing subgraphs in Formula (2) to be as small as possible. Since the ordering generation is a graph reduction process, the degree of vertices or the order of the adjacency graph of vertices will become smaller when the process goes on. Therefore, when the generation of a perfect elimination ordering cannot continue, we greedily choose the vertex that has the smallest degree to generate. The process continues by removing the chosen vertex $w$ from the graph and proceeding to the next iteration. The whole process terminates when all the vertices are generated.

Notice also that the ordering among the vertices whose adjacency graph is a clique is not crucial. In the process, once a vertex has a clique as its adjacency graph, its later adjacency graphs will still remain as cliques. This is because the removal of vertices from a clique results in another clique.

Let $G = (V, E)$ be a graph and $U$ be a subset of $V$. Then the graph $G - U$ denotes the subgraph induced by $V - U$. We define an ordering $\omega_0$, called a *perfect-smallest-last ordering*, as follows. Let $G_1, \ldots, G_n$ be a sequence of subgraphs of a graph $G$ of order $n$, with

   1. $G_1 = (V_1, E_1) = G$ and $G_{i+1} = (V_{i+1}, E_{i+1}) = G_i - \{v_i\}$,

$$2.\ v_i = \begin{cases} v & \text{if there is a vertex } v \in V_i \text{ such that} \\ & Adj_{G_i}(v) \text{ is a clique} \\ min_{v \in V_i}\{d_{G_i}(v)\} & \text{otherwise,} \end{cases}$$

for $1 \le i \le n$, then $\omega_0 = v_n, \ldots, v_1$. Since a perfect-smallest-last ordering always chooses a vertex whose adjacency graph is a clique if such a vertex exists. It has the following property:

---

**Algorithm CP($G$)**
    **begin**
        $G_1 \equiv (V_1, E_1) := G;$
        $U(G, x) := L(G, x) := 1;$
        **for** $i := 1$ **to** n **do**
            **if** there is a vertex $v \in V_i$ such that $Adj_{G_i}(v)$ is a clique **then**
                $v_i := v$
            **else**
                $v_i := min_{v \in V_i}\{d_{G_i}(v)\}$
            **fi**
            $U(G, x) := U(G, x) \times (x - \rho(G_i(v_i)));$
            $L(G, x) := L(G, x) \times (x - |G_i(v_i)|);$
            $G_{i+1} \equiv (V_{i+1}, E_{i+1}) := G_i - \{v_i\}$
        **od**
        $\widehat{C}(G, x) :=$ the harmonic mean of $U(G, x)$ and $L(G, x)$
    **end**

Figure 3: An algorithm for estimating the chromatic polynomial of a graph

---

**Proposition 3.2** *If a graph $G$ is chordal, then a perfect-smallest-last ordering on the vertices of $G$ is a perfect elimination ordering.* □

## 4   An Approximation Algorithm

We are now ready to present an algorithm for computing an upper bound and a lower bound on the chromatic polynomial of a graph based on Formula (4) and the perfect-smallest-last ordering. The algorithm is shown in Figure 3. Apart from the bounds, this algorithm also computes a mean of the bounds. Let $L(G, x)$ denote the lower bound $x \prod_{i=2}^{n}(x - |G_i'(\omega_0)|)$ and $U(G, x)$ denote the upper bound $x \prod_{i=2}^{n}(x - \rho(G_i'(\omega_0)))$ in Formula (4), respectively, with $\omega_0$ being a perfect-smallest-last ordering on the vertices of $G$. We estimate the chromatic polynomial of a graph $G$ as

$$\widehat{C}(G, x) = \frac{2 \times U(G, x) \times L(G, x)}{U(G, x) + L(G, x)}, \tag{5}$$

the *harmonic mean* of $U(G, x)$ and $L(G, x)$. Notice that although $\widehat{C}(G, x)$ is an estimate of $C(G, x)$, $\widehat{C}(G, x)$ itself may be a rational function, but not a polynomial.

We now consider the complexity of Algorithm CP. Let $n$ and $m$ be, respectively, the number of vertices and edges in the graph. We first consider the test in the if statement. Detecting if a graph of order $k$ is a clique can be performed in time $O(k^2)$. At the worst case, when no adjacency graph is a clique and the detection has to be performed for every vertex, the time required is $O(n+\sum_{j=1}^{n} d_{G_i}^2(v_j)) = O(n+(\sum_{j=1}^{n} d_{G_i}(v_j))^2) = O(n+m^2)$. The minimum operation in the else branch can be performed in time $O(n)$. Thus, altogether, the complexity of the if statement is $O(n + m^2)$.

The symbolic multiplications of polynomials for $U(G, x)$ and $L(G, x)$ can be performed in time $O(n)$ since the order of $U(G, x)$ and $L(G, x)$ is at most $n$. The computation of $\rho(G_i(v_i))$ demands a sorting step, so it requires time $O(n \log n)$. The updating from $G_i$ to $G_{i+1}$ can be performed in time $O(n + m)$. Put together, the complexity for each iteration of the for statement is $O(n \log n + m^2)$. Taking the number of iterations into account, the time demanded for the for statement is $O(n^2 \log n + nm^2)$.

The computation of the harmonic mean needs a symbolic multiplication and a symbolic addition. It can be performed in time $O(n^2)$. Therefore, the complexity of the entire algorithm is $O(n^2 \log n + nm^2)$. This complexity analysis leads to the following theorem:

**Theorem 4.1** *The problem of computing the chromatic polynomial of a chordal graph is solvable in polynomial time.*

**Proof** By Propositions 3.1, 3.2 and the complexity analysis of Algorithm CP. □

## 5  Performance Analysis and Measurements

This section investigates the performance behavior of Algorithm CP. Since the values of $C(G, x)$ are usually very large, we shall consider the relative error $\Delta = |\widehat{C}(G, x) - C(G, x)|/C(G, x)$ in performance analysis.

The worst-case relative error of an approximation algorithm occurs at the cases where $C(G, x) = U(G, x)$ or $C(G, x) = L(G, x)$; namely, $C(G, x)$ is equal to one of the two extremes. The worst-case relative error is minimized when the relative errors for $C(G, x) = U(G, x)$ and $C(G, x) = L(G, x)$ are equal. The harmonic mean provides such an optimum because

$$\Delta \leq \frac{U(G, x) - \widehat{C}(G, x)}{U(G, x)} = \frac{\widehat{C}(G, x) - L(G, x)}{L(G, x)} = \frac{U(G, x) - L(G, x)}{U(G, x) + L(G, x)}. \tag{6}$$

Since both $U(G, x)$ and $L(G, x)$ are nonnegative, we have

$$\Delta \leq \frac{U(G, x) - L(G, x)}{U(G, x) + L(G, x)} \leq 1, \tag{7}$$

for any nonnegative integer $x$. This implies that we always have $0 \leq \widehat{C}(G, x) \leq 2C(G, x)$. This statement is always true if we can compute both a lower bound and an upper bound

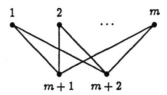

Figure 4: Graph $K_{m,2}$

on the measure we are interested in. On the other hand, we should also realize that because the problem of $k$-colorability of a graph is NP-complete, there is no polynomial time algorithm for approximating the chromatic polynomial of a graph that has a relative error less than 1 unless P = NP [11]. Therefore, if P $\neq$ NP, 1 is the best worst-case relative error we can expect. We now give an example that has a relative error of 1. Consider the complete bipartite graph $K_{m,2}$ shown in Figure 4. According to Formula (7), we have the relative error

$$\frac{x^2(x-1)^{m-1} - x^2(x-2)^{m-1}}{x^2(x-1)^{m-1} + x^2(x-2)^{m-1}},$$

which gives 1 when $x = 2$. In general, $\widehat{C}(K_{m,n}, x) = 0$ for $x \leq min(m,n)$, while $C(K_{m,n}, x) > 0$ for $x \geq 2$.

Another reason for choosing the harmonic mean is as follows. Since $U(G,x)$ becomes more significant than $L(G,x)$ as $x$ increases, we usually have $\widehat{C}(G,x) > C(G,x)$ except for some small $x$. At the mean time, the harmonic mean is always less than or equal to the arithmetic mean [1]. Therefore, in most cases the harmonic mean gives a better upper bound than the arithmetic mean.

To consider the average performance behavior, we also run some experiments on randomly generated graphs. The edges in the graph are chosen independently and with probability 0.5. Figure 5 displays the results of the average values of the relative errors over 5 graphs of order 16 and over 5 graphs of order 8. The results show that the value of the relative error $\Delta$ increases as the order of graphs increases. This is because the higher the order of a graph, the higher the degree of the estimated bounds on its chromatic polynomial, and the higher the accumulated error. The results also show that the value of the relative error decreases as the number of colors $k$ increases except for transient fluctuation for small values of $k$. This is because for each graph $G$ of order $n$, both $U(G,x)$ and $L(G,x)$ are polynomial of degree $n$ with the coefficient of $x^n$ being 1. Hence, the degree of the numerator is smaller than the degree of the denominator in Formula (7) and $\lim_{x \to \infty} \Delta = 0$. The smallest-degree heuristic employed in the perfect-smallest-last ordering aims to restrict the transient fluctuation to very small values. This allows us to have a good approximation of the chromatic polynomial of a graph in most cases.

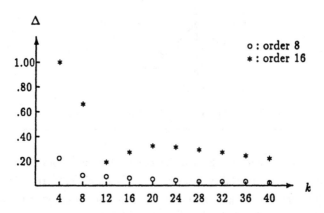

Figure 5: The average values of the relative errors of the approximate chromatic poly-nomials for random graphs of order 16 and 8

## 6   Conclusions

The problem of computing the chromatic polynomial of a graph is #P-hard. This paper has presented an approximation algorithm for computing the chromatic polynomial of a graph. This algorithm has time complexity $O(n^2 \log n + nm^2)$ for a graph with $n$ vertices and $m$ edges. This paper has also shown that the problem of computing the chromatic polynomial of a chordal graph can be solved in polynomial time. Knowledge about the chromatic polynomial of graphs can be employed to improve the performance of logic programs and deductive databases.

## Acknowledgments

Special thanks to S. Debray, P. Downey and U. Manber for many valuable comments on this work.

## References

[1] A. O. Allen, *Probability, Statistics, and Queuing Theory With Computer Science Applications*, Academic Press, Inc, 1990.

[2] J. A. Bondy, "Bounds for the Chromatic Number of a Graph," *Journal of Combinatorial Theory* 7, 1 (Jan. 1969), pp. 96–98.

[3] S. K. Debray and N.-W. Lin, "Cost Analysis of Logic Programs," to appear in *ACM Transactions on Programming Languages and Systems*.

[4] G. A. Dirac, "On Rigid Circuit Graphs," Abhandlungen aus dem Mathematischen Seminar der Universitat Hamburg, 25 (1961), pp. 71–76.

[5] F. Gavril, "Algorithms for Minimum Coloring, Maximum Clique, Minimum Covering by Cliques, and Maximum Independent Set of a Chordal Graph," *SIAM Journal of Computing* 1, 2 (June 1972), pp. 180–187.

[6] M. C. Golumbic, *Algorithmic Graph Theory and Perfect Graphs*, Academic Press, New York, 1980.

[7] R. M. Karp, "Reducibility Among Combinatorial Problems," *Complexity of Computer Computations*, edited by R. E. Miller and J. W. Thatcher, Plenum Press, New York, 1972, pp. 85–103.

[8] D. J. Rose, "Triangulated Graphs and the Elimination Process," *Journal of Mathematical Analysis and Applications* 32, (1970), pp. 597–609.

[9] D. J. Rose, R. E. Tarjan and G. S. Lueker, "Algorithmic Aspects of Vertex Elimination," *SIAM Journal of Computing* 5, 2 (June 1976), pp. 266–283.

[10] D. E. Smith and M. R. Genesereth, "Ordering Conjunctive Queries," *Artificial Intelligence* 26 (1985), pp. 171–215.

[11] L. Stockmeyer, "On Approximation Algorithms for #P," *SIAM Journal on Computing* 14, 4 (November 1985), pp. 849–861.

[12] L. G. Valiant, "The Complexity of Computing the Permanent," *Theoretical Computer Science* 8, (1979), pp. 189–201.

[13] D. H. D. Warren, "Efficient Processing of Interactive Relational Database Queries Expressed in Logic", *Proc. Seventh International Conference on Very Large Data Bases*, 1981, pp. 272–281.

# Asteroidal Triple-Free Graphs

Derek. G. Corneil[1] Stephan Olariu[2] and Lorna Stewart[3]

[1] Department of Computer Science, University of Toronto, Toronto, Ontario, Canada
[2] Department of Computer Science, Old Dominion University, Norfolk, Virginia, U.S.A.
[3] Department of Computer Science, University of Alberta, Edmonton, Alberta, Canada

**Abstract.** Many families of perfect graphs such as interval graphs, permutation graphs, trapezoid graphs and cocomparability graphs demonstrate a type of linear ordering of their vertex sets. These graphs are all subfamilies of a class of graphs called the asteroidal triple-free graphs. (An independent triple $\{x, y, z\}$ is called an asteroidal triple (AT, for short) if between any pair in the triple there exists a path that avoids the neighbourhood of the third vertex.) In this paper we argue that the property of being AT-free is what is enforcing the linear ordering of the vertex sets. To justify this claim, we present various structural properties and characterizations of AT-free graphs.

## 1 Introduction

Recently a great deal of attention has been spent on the algorithmic study of the "tree structure" of various families of graphs such as cographs [4], chordal graphs [9], and partial $k$-trees [1]. This research is an example of the general theme of understanding the underlying structure in the graphs expected for a given application. The hope is to find efficient algorithms that will exploit this structure and thereby deal adequately with very large input (for example, in the order of one million vertices). In this paper we follow this approach and examine the "linear structure" that is apparent in various families of graphs such as interval graphs, permutation graphs, trapezoid graphs and cocomparability graphs. This attention is motivated in part by their many areas of application, including molecular biology, archaeology, information retrieval, computer memory management, channel routing, and circuit layout [7, 6, 9].

In order to illustrate the common concept amongst three of these graph families, consider a *trapezoid representation* $R$ consisting of two parallel lines (denoted $L_1$ and $L_2$) and some trapezoids with two endpoints lying on $L_1$ and the other two endpoints lying on $L_2$. A graph $G$ is a *trapezoid graph* if and only if it is the intersection graph of such a representation where the vertices of $G$ are in one-to-one correspondence with the trapezoids in $R$ and two vertices in $G$ are adjacent if and only if their corresponding trapezoids in $R$ intersect. If the trapezoids degenerate with the endpoints on $L_1$ (resp. $L_2$) coinciding (i.e. the trapezoids become lines) then the intersection graph is a *permutation graph*. Similarly if the intervals on $L_1$ are the mirror image of the intervals on $L_2$ (i.e. two trapezoids intersect if and only if their intervals on $L_1$ (as well as $L_2$) intersect), then the intersection graph is an *interval graph*. It is shown in [3] that permutation graphs and interval graphs are strictly contained in trapezoid graphs.

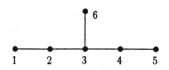

Figure 1

Note that the trapezoid representation that provides the common thread amongst the three families also indicates that, in some sense, these graphs can only "grow" linearly. For example, the graph in Figure 1 is at the same time an interval graph, a permutation graph and a trapezoid graph. For this graph, we can add new vertices adjacent to either 1 or 5 without destroying membership in any of the three families; however it seems as though we cannot add a new vertex adjacent to 6 without destroying membership in each family. Thus, these graph families seem to capture the notion of "linear" growth that is evident in the various applications mentioned previously.

A graph is a *comparability graph* if the edges may be given a transitive orientation. A *cocomparability graph* is the complement of a comparability graph. A graph is *chordal* if it does not contain an induced cycle of size $\geq 4$ and is *weakly chordal* if neither $G$ nor $\overline{G}$ contain an induced cycle of size $\geq 5$. Both cocomparability and weakly chordal graphs are perfect (for every induced subgraph the chromatic number equals the clique number); trapezoid graphs are strictly contained in the intersection of cocomparability and weakly chordal graphs.

In an attempt to identify the general property which is responsible for the linear growth of the four families of graphs discussed in the previous section, we state an early characterization of interval graphs by Lekkerkerker and Boland. They define an *asteroidal triple* (AT, for short) to be an independent set of vertices $x, y, z$ such that for each pair of vertices there exists a path joining them that avoids the neighbourhood of the third. In the early 1960s Lekkerkerker and Boland demonstrated the importance of ATs in the following theorem.

**Proposition 1.1:** [12] $G$ is an interval graph iff it is chordal and AT-free. □

Thus, we see that the condition of being AT-free prohibits a chordal graph from growing in three directions at once.

In fact, Golumbic *et al.* [10] have shown that cocomparability graphs (and thus permutation and trapezoid graphs as well) are also AT-free. Subsequently, it was shown that the perfect AT-free graphs strictly contain the cocomparability graphs [2]. Since the complements of odd holes of size greater than 3 are AT-free, the AT-free graphs are not perfect (however, they are valid for the Strong Perfect Graph Conjecture [13]).

The thesis presented in this paper is that the property of being AT-free captures the linear structure we observe in the interval, permutation, trapezoid and cocomparability graph families. To support this thesis we present various "linear" structural properties of AT-free graphs. In Section 2 we show that every connected AT-free graph contains a *dominating pair*, that is, two vertices such that *every* path between them is dominating. In Section 3 we provide a characterization of AT-free graphs with respect to dominating pairs. In Section 4 we show that an AT-free graph may be extended to another AT-free graph by adding two new degree 1 vertices each adjacent to one of the vertices in an appropriate dominating pair. This result leads to a decomposition scheme of AT-free graphs whereby an AT-free graph is reduced to a single vertex. Finally, Section 5 provides concluding remarks and poses some open problems.

## 2 Dominating paths

We now turn our attention to the first "linear" structure property of AT-free graphs. Given an AT-free graph $G = (V, E)$ and vertex $x$ we let $N'(x)$ denote the induced subgraph of $G$ defined on all vertices of $G$ except $x$ and its neighbours. We say that a vertex $u$ *intercepts a path* $P$ if $u$ is adjacent to at least one vertex on $P$ and we let $D(v, x)$ denote the set of vertices that intercept all $v - x$ paths. Finally, as mentioned previously, a pair of vertices $(u, v)$ is called a *dominating pair* (DP, for short) if every path between $u$ and $v$ is a dominating path (i.e. $D(u, v) = V$).

First we present some observations about connected AT-free graphs. Throughout we assume that $F$ is an arbitrary connected component of $N'(x)$.

**Claim 2.1:** If $u \in D(v, x)$, $w \in D(u, x)$ and $(w, u) \notin E$ then $w \in D(v, x)$.

**Proof:** Suppose to the contrary, that $w$ misses some $v - x$ path $P: v = v_0, v_1, \cdots, v_k = x$. Let $j$ be the largest subscript for which $(u, v_j) \in E$; since $u \in D(v, x)$ such a subscript must exist. But now, $w$ misses the $u - x$ path, $u, v_j, v_{j+1}, \cdots, v_k = x$ contradicting the fact that $w \in D(v, x)$. $\square$

We say that vertices $u$ and $v$ are *unrelated* if $u \notin D(v, x)$ and $v \notin D(u, x)$.

**Claim 2.2:** $F$ contains no unrelated vertices.

**Proof:** If $u$ and $v$ are unrelated vertices in $F$, then the connectedness of $F$ implies that $\{u, v, x\}$ is an AT. $\square$

**Claim 2.3:** If $u$ and $v$ are in $F$ and $v \notin D(u, x)$, then $D(u, x) \subset D(v, x)$.

**Proof:** From Claim 2.2, $u \in D(v, x)$. Let $w$ be an arbitrary vertex in $D(u, x)$. If $(w, u) \notin E$, then Claim 2.1 guarantees that $w \in D(v, x)$; if $(w, u) \in E$, then clearly $w \in F$. If $w$ misses some $v - x$ path, then in particular $(v, w) \notin E$. Thus, $P \cup \{w\}$ contains a $w - x$ path missed by $v$, where $P$ is a $u - x$ path missed by $v$. Now $v$ and $w$ are unrelated, contradicting Claim 2.2. Thus, $w \in D(v, x)$ and $D(u, x) \subseteq D(v, x)$; the inclusion is strict since $v \notin D(u, x)$. $\square$

A vertex $y$ in $F$ is called *special* if $D(u, x) \subseteq D(y, x)$ for all $u$ in $F$.

**Claim 2.4:** $y \in F$ is special iff $v \in D(y, x)$, $\forall v \in F$.

**Proof:** If $y$ is special $v \in D(y, x)$ $\forall v \in F$ since $v \in D(v, x)$. Conversely, if $v \in D(y, x)$ for all $v \in F$, then in particular, $F \subset D(y, x)$. Let $u$ be an arbitrary vertex in $F$ and let $w$ be an arbitrary vertex in $D(u, x)$. If $w \in F$, then since $F \subset D(y, x)$, $w \in D(y, x)$; if $w \in N'(x) \backslash F$ then Claim 2.1 guarantees that $w \in D(y, x)$ and so $D(u, x) \subseteq D(y, x)$. Since $u$ is arbitrary, the conclusion follows. $\square$

**Claim 2.5:** $F$ contains a special vertex.

**Proof:** Choose a vertex $y$ in $F$ with $D(y, x) \subset D(t, x)$ for no vertex $t$ in $F$. If $y$ is not special, then by Claim 2.4, we find a vertex $v \in F$ with $v \notin D(y, x)$. By Claim 2.3 $D(y, x) \subset D(v, x)$ contradicting our choice of $y$. $\square$

**Claim 2.6:** If $v \in N'(x) \backslash F$ then either:

$$v \in D(w, x) \; \forall w \in F$$
$$\text{or } v \notin D(w, x) \; \forall w \in F$$

**Proof:** Suppose not; for a suitable choice of vertices $w, w'$ in $F$ we have $v \in D(w, x)$ and $v \notin D(w', x)$. Let $P$ stand for a $w' - x$ path missed by $v$, and let $P'$ stand for a $w - w'$ path entirely within $F$. But now $P \cup P'$ contains a $w - x$ path missed by $v$ contrary to our assumption. $\square$

**Claim 2.7:** Let $v \in N'(x) \backslash F$. If $F \not\subset D(v, x)$ then for a special vertex $u^* \in F$, $u^* \notin D(v, x)$.

**Proof:** Let $U = \{u \in F \mid u \notin D(v, x)\}$. Since $F \not\subset D(v, x)$, $U \neq \emptyset$. Choose a vertex $u^* \in U$ such that $D(u^*, x) \subset D(u, x)$ for no $u \in U$. If $u^*$ is not special, then by Claim 2.4, there exists some vertex $w \in F \backslash D(u^*, x)$. By Claim 2.3, $D(u^*, x) \subset D(w, x)$; by our choice of $u^*$, $w \in F \backslash U$. This, however, implies that $w \in D(v, x)$. Since $w \notin D(u^*, x)$, Claim 2.2 implies that $u^* \in D(w, x)$. Since $(u^*, w) \notin E$, Claim 2.1 states that $u^* \in D(v, x)$ which is a contradiction, and so $u^*$ must be special. $\square$

Call a vertex $u \in N'(x)$ *strong* if $N'(x) \subset D(u, x)$. It is easy to confirm that if $u$ is a strong vertex, then $(u, x)$ is a DP in $G$. From now on we shall tacitly assume that $N'(x)$ contains no strong vertices. A pair $(y, z)$ of vertices in distinct components $Y, Z$ of $N'(x)$ is an *admissible pair* if $D(y, x) \cup D(z, x) \subset D(t, x) \cup D(t', x)$ for no vertices $t, t'$ in distinct components of $N'(x)$.

Observe that since $N'(x)$ contains no strong vertices, $N'(x)$ must be disconnected; otherwise Claim 2.5 implies that $N'(x)$ contains a special vertex, which by Claim 2.4 is strong. Consequently, the absence of strong vertices in $N'(x)$ together with the finiteness of $G$ imply that $N'(x)$ contains admissible pairs. As it turns out, admissible pairs play a crucial role in our arguments. We now study some of their properties.

**Claim 2.8:** If $(y, z)$ is an admissible pair with $y \in Y$, $z \in Z$, then $Y \not\subset D(z, x)$ and $Z \not\subset D(y, x)$.

**Proof:** Assume $Z \subset D(y, x)$; then, trivially, $z \in D(y, x)$. To see that $D(z, x) \subseteq D(y, x)$ note that for an arbitrary vertex $w$ in $D(z, x)$, $w \in D(y, x)$ whenever $w \in Z$ and that, by virtue of Claim 2.1, $w \in D(y, x)$ whenever $w \notin Z$. Since $y$ is not strong, we find a vertex $y' \in N'(x) \backslash D(y, x)$. But now either $(z, y')$ or $(y, y')$ contradicts our choice of $(y, z)$. To see this, note that if $y' \in Y$ then by Claim 2.3, $D(y, x) \subset D(y', x)$ and so $D(y, x) \cup D(z, x) = D(y, x) \subset D(y', x) \subseteq D(y', x) \cup D(z, x)$; if $y' \notin Y$, then $D(y, x) \cup D(z, x) = D(y, x) \subseteq D(y', x) \cup D(y, x)$. Since $y' \notin D(y, x)$, the inclusion is strict. $Y \not\subset D(z, x)$ follows by a similar argument. $\square$

**Claim 2.9:** If $(y, z)$ is an admissible pair, then $N'(x) \subset D(y, x) \cup D(z, x)$.

**Proof:** Assume this is false and let $w$ be an arbitrary vertex in $N'(x) \backslash (D(y, x) \cup D(z, x))$. Clearly $w \notin D(y, x)$ and $w \notin D(z, x)$. We now show that:

$$w \notin Y \cup Z. \tag{1}$$

If $w \in Y$ then by Claim 2.3, $D(y, x) \subset D(w, x)$ and since $w \notin D(z, x), D(y, x) \cup D(z, x) \subset D(z, x) \cup D(w, x)$ contradicting $(y, z)$ being an admissible pair. Similarly $w \notin Z$.

Since $G$ is AT-free, we note that:

$$\text{no distinct vertices } t, t', t'' \text{ in } N'(x) \text{ are pairwise unrelated.} \tag{2}$$

We now claim that for a suitable choice of vertices $u, v$ in $N'(x)$:

$$u \in D(y, x) \backslash (D(z, x) \cup D(w, x)) \text{ and } v \in D(z, x) \backslash (D(y, x) \cup D(w, x)). \tag{3}$$

By (2), $y$, $z$ and $w$ belong to distinct components of $N'(x)$. Since $(y, z)$ is an admissible pair, $D(y, x) \cup D(z, x) \not\subseteq D(z, x) \cup D(w, x)$; therefore the required vertex $u$ exists. A similar argument holds for the existence of $v$.

Next observe that:

$$y \in D(z, x) \cup D(w, x) \text{ and } z \in D(y, x) \cup D(w, x). \tag{4}$$

If $y \notin D(z, x) \cup D(w, x)$, then by our choice of $w$, we know that $y$ and $w$ are unrelated. Therefore, $z \in D(y, x) \cup D(w, x)$ since otherwise $y$, $z$ and $w$ would be pairwise unrelated contradicting (3). Consider the vertex $v$ specified in (4); since $z \in D(y, x) \cup D(w, x)$ and $v \in D(z, x) \backslash (D(y, x) \cup D(w, x))$, Claim 2.1 implies that $(z, v) \in E$. But now we have an AT on $\{y, v, w\}$; this follows since $y$ and $w$ are unrelated, and both $v, w$ and $v, y$ are unrelated by (4) and Claim 2.6. Similarly we see that $z \in D(y, x) \cup D(w, x)$.

We now show that:

$$u \in Y \text{ and } v \in Z. \tag{5}$$

By (5), $y \in D(z, x) \cup D(w, x)$; by (4) $u \in D(y, x) \backslash (D(z, x) \cup D(w, x))$; but $(u, y) \in E$ otherwise we contradict Claim 2.1. The fact that $v \in Z$ is proved similarly.

To complete the proof of Claim 2.9, we first observe that (6), (4) and Claim 2.6 combined, guarantee that $u \notin D(v, x)$ and $v \notin D(u, x)$ (i.e. $u$ and $v$ are unrelated). Similarly, by (6), (4) and Claim 2.6, $u$, $w$ are unrelated as are $v$ and $w$. To complete the proof, note that $u$, $v$, $w$ are pairwise unrelated, contradicting (3). $\square$

We are now in a position to prove the main result of this section.

**Theorem 2.10**: Every connected AT-free graph contains a DP.

This result is in fact a corollary of the following stronger result.

**Theorem 2.11**: Let $x$ be an arbitrary vertex in a connected AT-free graph $G$. Either $(x, x)$ is a DP or else for a suitable choice of vertices $y, z$ in $N'(x)$, $(y, x)$ or $(y, z)$ is a DP.

**Proof**: If $N'(x)$ is empty, then $(x, x)$ is a DP. If $N'(x)$ is not empty but contains a strong vertex $y$, then by definition, $(x, y)$ is a DP. Otherwise, let $(y, z)$ be an admissible pair where we assume that $y$ and $z$ belong to connected components $Y$ and $Z$ of $N'(x)$ respectively. By Claim 2.8 and Claim 2.7 we find special vertices $y^*$ in $Y$ and $z^*$ in $Z$ such that $y^* \notin D(z^*, x)$ and $z^* \notin D(y^*, x)$, i.e. $y^*$ and $z^*$ are unrelated. Furthermore, since $y^*$ and $z^*$ are special, we have $D(y, x) \cup D(z, x) \subseteq D(y^*, x) \cup D(z^*, x)$ implying that $(y^*, z^*)$ is an admissible pair.

Finally we claim that $(y^*, z^*)$ is a DP in $G$.

By Claim 2.9 any vertex $v$ not dominated by some $y^* - z^*$ path must be in $N(x)$. ($v \neq x$ since every $y^* - z^*$ path contains at least one vertex in $N(x)$.) Since $y^*$ and $z^*$ are unrelated, $y^*$ misses some $z^* - x$ path $P$ and $z^*$ misses some $y^* - x$ path $P'$. But now we have reached a contradiction; $\{y^*, z^*, v\}$ is an AT. To see this, note by assumption, $v$ misses some $y^* - z^*$ path; in addition $y^*$ misses the $z^* - v$ path $P \cup \{v\}$ and $z^*$ misses the $y^* - v$ path $P' \cup \{v\}$. $\square$

Note that Theorems 2.10 and 2.11 do not lead to a characterization of AT-free graphs. For example, antipodal vertices in $C_6$ constitute a DP. Furthermore, if we add a universal vertex to an arbitrary graph, we obtain a graph that has a DP consisting of the universal vertex and any other vertex. Clearly, any attempt to provide a characterization of AT-free graphs involving DPs must not only be based on induced subgraphs, it must also restrict the types of DPs. ($C_6$ for example has an AT, yet every induced subgraph has a DP).

## 3  Dominating pair characterization

We now provide a characterization of AT-free graphs based on dominating pairs. As indicated above, such a result must restrict the types of DPs. In particular, as the following theorem shows, we impose an adjacency condition on $G$ with DP $(x, y)$ whereby the connected component of $G \backslash \{x\}$ containing $y$ has a DP $(x', y)$ where $x'$ is adjacent to $x$. Note that $C_6$ fails this criterion.

**Theorem 3.1:** $G$ is AT-free iff every connected induced subgraph $H$ of $G$ satisfies the property:

(P)   (i) $H$ has a DP

(ii) for every non-adjacent DP $(\alpha, \beta)$, there exists $\alpha'$, a neighbour of $\alpha$, such that $(\alpha', \beta)$ is a DP of the component of $H \backslash \{\alpha\}$ containing $\beta$.

**Proof:** To settle the "only if" part, we first deal with the case where all DPs are adjacent.

**Claim 3.2:** A connected AT-free graph $G$ is a clique iff it contains no non-adjacent DP.

**Proof:** The "only if" part is trivial. To prove the "if" part, note that if $G$ is not a clique, then for some $x$, $N'(x) \neq \emptyset$. By Theorem 2.11, there exists $y, z \in N'(x)$ such that either $(x, y)$ is a DP or $(y, z)$ is a DP where $y, z$ belong to different components of $N'(x)$.   □

We now assume $(\alpha, \beta)$ is a non-adjacent DP of $H$. Let $C_\beta$ denote the connected component of $H \backslash \{\alpha\}$ that contains $\beta$. Let $A$ denote $N(\alpha) \cap C_\beta$.

Choose vertex $\tilde{\alpha} \in A$ such that $D(\tilde{\alpha}, \beta) \subset D(t, \beta)$ for no $t$ in $A$.

**Claim 3.3:** $(\tilde{\alpha}, \beta)$ is a DP in $C_\beta$.

**Proof:** Suppose not; let $t$ be a vertex in $C_\beta$ missed by some $\tilde{\alpha} - \beta$ path. $t \in A$ since otherwise this path extends to an $\alpha - \beta$ path in $H$ that misses $t$ contradicting $(\alpha, \beta)$ being a DP of $H$.

If $\tilde{\alpha}, t$ belong to the same component of $N'_H(\beta)$ then by Claim 2.3, $D_H(\tilde{\alpha}, \beta) \subset D_H(t, \beta)$, resulting in a contradiction.

Since $(t, \beta) \notin E$, $\alpha, t$ belong to the same component of $N'_H(\beta)$. The only way for $\tilde{\alpha}$ and $t$ not to be in the same component of $N'_H(\beta)$ is for $(\tilde{\alpha}, \beta) \in E$.

By our choice of $\tilde{\alpha}$ $D(\tilde{\alpha}, \beta) \not\subset D(t, \beta)$, and $t \notin D(\tilde{\alpha}, \beta)$. Thus, $\exists w \in D(\tilde{\alpha}, \beta)$ such that $w \notin D(t, \beta)$. $w \neq \tilde{\alpha}$ since $(\tilde{\alpha}, \beta) \in E$. $(w, \beta) \notin E$, thus $(w, \tilde{\alpha}) \in E$. $(\alpha, w) \in E$ since otherwise the $t - \beta$ path missed by $w$ could be extended to a $\alpha - \beta$ path missed by $w$ contradicting $(\alpha, \beta)$ being a DP. Now $w$ and $t$ are in the same component of $N'(\beta)$ and are unrelated contradicting Claim 2.2 (i.e. $\exists$ AT $\{w, t, \beta\}$).   □

We prove the "if" implication of Theorem 3.1 by contradiction. Assume the statement false, and let $G$ be a graph that satisfies (P) but has an AT. Let $H$ be an induced subgraph of $G$ such that $H$ has an AT and yet every proper induced subgraph of $H$ is AT-free.

Let $\{x, y, z\}$ be an AT of $H$ and let $(a, b)$ be a DP of $H$.

**Claim 3.4:** $\{a, b\} \cap \{x, y, z\} = \emptyset$.

**Proof:** Suppose not; without loss of generality $a = x$. Since $(a, b)$ is a DP, $b$ must be on the $y - z$ path missed by $x$ (by Lemma 2.1 this path may be assumed to be induced). Consider the $x - y$ path missed by $z$ concatenated with the $y - b$ portion of the $y - z$ path. This path misses $z$ unless $(b, z) \in E$. Similarly, $(b, y) \in E$. Since $H$ satisfies (P) we should be able to move $b$ to a neighbour $b'$ such that $(a, b')$ is a DP of $H \backslash \{b\}$. If $b'$ is on the $x - y$ path it misses $z$, if it is on the $x - z$ path it misses $y$. $\square$

Throughout the proof we will show that the existence of a DP $(a, b)$ implies the existence of another DP which has $x, y$ or $z$ as one of its vertices, thereby violating Claim 3.4.

By Lemma 2.1, we assume that all paths demonstrating the AT are chordless. Without loss of generality, let $a$ belong to the $x - y$ path and $b$ belong to the $x - z$ path. $(a, y) \in E$ (otherwise the $a - b$ path through $x$ is missed by $y$) $(b, z) \in E$ (otherwise the $a - b$ path through $x$ is missed by $z$). Either $(a, x) \in E$ or $(b, x) \in E$ (otherwise the $a - b$ path through $y$ and $z$ is missed by $x$). Without loss of generality $(a, x) \in E$.

These edges, together with Lemma 2.1, imply that the $x - y$, $y - z$ and $x - z$ paths meet only at the vertices of the AT.

**Claim 3.5:** $(x, b) \notin E$.

**Proof:** Suppose $(x, b) \in E$. Since $\{b, y\}$ cannot be a DP, some vertex $c$ on the $y - z$ path misses a path from $b$ to $y$ that does not involve $a$. Since $(a, b)$ is a DP, $(a, c) \in E$. A similar argument shows the existence of vertex $d$ on the $y - z$ path such that $(d, b) \in E$, $(d, a) \notin E$. If $(y, d) \notin E$, $\{x, y, d\}$ is an AT in $H \backslash \{z\}$ contradicting the minimality of $H$. Similarly $(c, z) \in E$. If $(c, d) \notin E$, $\{c, d, x\}$ is an AT in $H \backslash \{y, z\}$ again contradicting the minimality of $H$. But now both $\{a, z\}$ and $\{b, y\}$ are DPs contradicting Claim 3.4. $\square$

**Claim 3.6:** $(a, b) \in E$.

**Proof:** If $(a, b) \notin E$, by property (P) we must be able to move $a$ to a neighbour $a'$ such that $(a', b)$ is a DP of $H \backslash \{a\}$. Claim 3.4 prohibits $a'$ to be $x$ or $y$ and thus $a'$ is either on the $x - z$ path and $y$ is not dominated or is on the $y - z$ path in which case $x$ is not dominated. $\square$

Let $b = b_0, b_1, \cdots, b_j$ be the path from $b$ to $x$ where $j \geq 1$ and $(b_j, x) \in E$. Immediately we see that $(a, b_i) \in E$ $2 \leq i \leq j$ since $(a, b)$ is a DP.

**Claim 3.7:** $b_1$ is not adjacent to any vertex on the $y - z$ path.

**Proof:** Suppose $(b_1, c) \in E$, $c \in y - z$ path. If $(y, c) \notin E$, then $x, y, z$ is an AT in $H \backslash \{b\}$. If $(z, c) \notin E$, then $x, y, z$ is an AT in $H \backslash \{a\}$. Thus, if $b_1$ is adjacent to a vertex on the $y - z$ path we have a $P_3$: $y, c, z$. Now $(a, z)$ is a DP contradicting Claim 3.4. $\square$

**Claim 3.8:** $(a, b_1) \notin E$.

**Proof:** Suppose $(a, b_1) \in E$. Now $(a, z)$ is a DP since otherwise there is an $a - z$ path missing vertex $c$ on the $y - z$ path. Since $(a, b)$ is a DP, $(b, c) \in E$. Now $x, y, z$ is an AT in $H \backslash \{d\}$ where $d$ is any vertex on the $y - z$ path between $c$ and $z$. Such a vertex exists since $(c, z) \notin E$ (i.e. $c$ misses an $a - z$ path). $\square$

**Claim 3.9:** $j = 1$.

**Proof:** If $j > 1$ then $\{b_1, y, z\}$ is an AT in $H \backslash \{x\}$ since the path $b_1, b_2, a, y$ is missed by $z$. $\square$

Now let $c$ be the vertex on the $y - z$ path adjacent to $y$. If $(b, c) \in E$ then

$(c, z) \in E$ otherwise there is the AT $x, y, z$ in $H \backslash \{d\}$, where $d$ is adjacent to $z$ on the $y - z$ path $(d \neq c)$. If $(b, c) \notin E$ then $(a, c) \in E$. Again $(c, z) \in E$, otherwise $c, x, z$ is an AT in $H \backslash \{y\}$. But now $(x, c)$ is a DP. This completes the proof. $\square$

The above theorem could lead one to the false conjecture, that a graph is AT-free iff there is a path $x_1, x_2, \cdots, x_j$ such that $(x_i, x_j)$ is a DP of $G \backslash \{x_1, \cdots, x_{i-1}\}$. As a counter-example of this, consider the graph $G$ with vertices $\{1,2,3,4,5,6,7\}$ and edges $\{12,13,15,16,17,23,27,34,45,47,56,67\}$. This graph has an AT $\{2,4,6\}$ and path $1,7,4$ where $(1,4)$ is a DP of $G$ and $(7,4)$ is a DP of $G \backslash \{1\}$. (Furthermore, $(1,7)$ is a DP of $G \backslash \{4\}$.) Note, however, that the induced subgraph $G \backslash \{7\}$ has a DP $(1,4)$ yet $G \backslash \{1,7\}$ has no DP consisting of 4 and a neighbour of 1.

## 4  Augmenting AT-free graphs

We now address the issue of augmenting an arbitrary AT-free graph $G$ to obtain another AT-free graph. This augmentation will be accomplished by finding a particular DP $(x, y)$ and then adding new vertices $x', y'$ where $x'$ is adjacent to $x$ and $y'$ is adjacent to $y$. This "growing" of $G$ again supports our thesis of AT-free graphs possessing a linear structure since the $DP$ $(x, y)$ has been "stretched" to a new $DP$ $(x', y')$. Vertex $v$ of AT-free graph $G$ is called *pokable* if the graph $G'$ obtained from $G$ by adding a pendant vertex adjacent to $v$ is AT-free. A set is pokable if all of its vertices are pokable. We are now in a position to state the main result of this section.

**Theorem 4.1**: Every connected AT-free graph contains a pokable DP.

**Proof:** The theorem is trivial for cliques. We shall therefore assume that $G$ is not a clique. Since $G$ is not a clique, it is easy to show that $G$ has a $DP$ $(x, y_0)$ where $x$ and $y_0$ are not adjacent. Let $F$ be the connected component of $N'(x)$ containing $y_0$, and let $Y$ stand for the set of vertices $y$ in $F$ for which $(x, y)$ is a $DP$ in $G$. Note that a vertex $v$ is unpokable if there exist non-adjacent vertices $u$ and $w$ in $G$ and induced paths

$v = u_0, u_1, \cdots, u_p = u$ missed by $w$,

$v = w_0, w_1, \cdots, w_q = w$ missed by $u$, and

a $u - w$ path in $G$ that does not contain $v$.

Subsequently we will prove:

**Lemma 4.2**: $Y$ contains a pokable vertex.

To see how Theorem 4.1 follows from Lemma 4.2, let $\beta$ be a pokable vertex in $Y$ and let $X$ denote the set of vertices $x'$ in the same component of $N'(\beta)$ as $x$, for which every pair $(\beta, x')$ is a DP. Clearly $x$ is in $X$. By applying Lemma 4.2 again but now with $\beta$ as the 'anchor', we find a pokable vertex $\alpha$ in $X$. The proof of Theorem 4.1 is established by noting that $(\alpha, \beta)$ is a desired pokable DP.

**Proof of Lemma 4.2:** The proof is by induction on the order of $G$. Assume the lemma true for all connected AT-free graphs with fewer vertices than $G$. We now present various facts that are used in the proof.

**Claim 4.3**: Let $v$ be an unpokable vertex in $Y$. Then all vertices $u_i$ and $w_j$ $(1 \leq i \leq p; 1 \leq j \leq q)$ belong to $F$.

**Proof:** Without loss of generality, let $i$ be the smallest subscript for which $u_i$ lies outside $F$. Trivially, $u_i \in N(x)$. Since $w$ cannot miss the $v - x$ path, $v = u_0, u_1, \cdots, u_i, x$ and since $w$ is adjacent to none of $v = u_0, u_1, \cdots, u_i$, it follows that $w \in N(x)$.

Similarly, since $u$ cannot miss the $v - x$ path, $v = w_0, w_1, \cdots, w_q = w, x$ and since $u$ is adjacent to none of $v = w_0, w_1, \cdots, w_q$, it follows that $u \in N(x)$. But now we have an AT, $\{u, v, w\}$ contradicting $G$ being AT-free. $\square$

It is important to note that by virtue of Claim 4.3, Lemma 4.2 is established as soon as we exhibit a vertex in $Y$ that is pokable in the subgraph of $G$ induced by $F$. If $F$ and $Y$ coincide, then by the induction hypothesis, such a vertex must exist. Therefore, from now on we shall assume that

$$F \backslash Y \neq \emptyset. \tag{6}$$

A vertex $a$ in $F \backslash Y$ is called an *attractor* if $Y \subset D(a, x)$. Let $A$ denote the set of all attractors in $F \backslash Y$.

**Claim 4.4**: $A = \emptyset$.

**Proof**: Let $a^*$ be a vertex in $A$ for which $D(a^*, x) \subset D(a, x)$ for no vertex $a$ in $A$. We claim that $(a^*, x)$ is a $DP$ in $G$. Suppose not, and let $t$ miss some $a^* - x$ path. Then:

1. $t \notin A$ by our choice of $a^*$;
2. $t \notin Y$ because $Y \subset D(a^*, x)$;
3. $t \notin N'(x) \backslash F$, for otherwise $t$ would miss a $y_0 - x$ path. Such a path proceeds from $y_0$ to $a^*$ in $F$ and then to $x$ along the $a^* - x$ path missed by $t$;
4. $t \notin F \backslash (A \cup Y)$. Since $Y \subset D(a^*, x)$, $t$ must be adjacent to every vertex in $Y$, and so $t \in A$ a contradiction. $\square$

Let $Y_1, Y_2, \cdots, Y_k$ $(k \geq 1)$ be the connected components of the subgraph of $\overline{G}$ induced by $Y$.

**Claim 4.5**: Let $t$ be a vertex in $F \backslash Y$. If some vertex $z$ in $Y_i$ satisfies $z \in D(t, x)$, then $Y_i \subset D(t, x)$.

**Proof**: If the claim is false, then we find vertices $z, z'$ in $Y_i$ such that $z \in D(t, x)$ and $z' \notin D(t, x)$. Since $Y_i$ is a connected subgraph of $\overline{G}$, there exists a chordless path $z = s_1, s_2, \cdots, s_r = z'$ joining $z$ and $z'$ in $\overline{G}$, with all internal vertices in $Y_i$.

Let $j$ be the smallest subscript for which $s_j \notin D(t, x)$. Since $z' \notin D(t, x)$ such a subscript must exist. But now, in $G$, $s_{j-1}$ and $s_j$ are non-adjacent and $s_j$ misses some $t - x$ path, while $s_{j-1}$ intercepts all such paths. It follows that $s_j$ misses a $s_{j-1} - x$ path, a contradiction. $\square$

**Claim 4.6**: $Y$ induces a disconnected subgraph of $\overline{G}$.

**Proof**: First we claim that

$$| \overset{\cdot}{Y} | \geq 2. \tag{7}$$

If (8) is false, then $Y = \{y_0\}$. Let $U$ stand for the set of all vertices in $F$ adjacent to $y_0$. Note that the connectedness of $F$ guarantees that $U$ is not empty. But now, for every $u$ in $U$, $Y = \{y_0\} \subset D(u, x)$, implying that $u$ is an attractor. Therefore (8) holds. (Note that by virtue of (8) it makes sense to talk about $Y$ being disconnected in the complement.)

We now continue the proof of Claim 4.6. If $Y = Y_1$, (7) and the connectedness of $F$ imply the existence of a vertex $z$ in $Y$ adjacent to some vertex $t$ in $F \backslash Y$. Note, in particular, that $z \in D(t, x)$ and so by Claim 4.5, $Y \subset D(t, x)$ implying that $t$ is an attractor, a contradiction. $\square$

**Claim 4.7**: Let $v$ be an unpokable vertex in $Y$. Then $v \in D(u_i, x)$ $(1 \leq i \leq p)$ and $v \in D(w_j, x)$ $(1 \leq j \leq q)$.

**Proof**: Since $v$ is adjacent to $u_1$, it follows that $v \in D(u_1, x)$. Let $i$ be the smallest subscript for which $v \notin D(u_i, x)$. Let $P$ be a $u_i - x$ path missed by $v$. Note that $w$ must intercept $P$, for otherwise $w$ would miss a $v - x$ path contained in $\{v, u_1, \cdots, u_i\} \cup P$. However, now $\{u_i, v, w\}$ is an AT. The proof that $v \in D(w_j, x)$ follows by a mirror argument. $\square$

For every $i$ $(1 \leq i \leq k)$, let $T_i$ stand for the set of vertices $t$ in $F \backslash Y$ with the property that $Y_i \subset D(t, x)$. By renaming the $Y_i$'s, if necessary, we ensure that

$$| T_1 | \leq | T_2 | \leq \cdots \leq | T_k |.$$

**Claim 4.8**: Every vertex in $T_1$ is adjacent to all vertices in $Y_1$.

**Proof**: The statement is vacuosly true if $T_1$ is empty. We may, therefore, assume that $T_1$ is nonempty. Let $t$ be a vertex in $T_1$ non-adjacent to some $z$ in $Y_1$. Since $t$ cannot be an attractor, we find a subscript $j$ $(j \geq 2)$ such that for some $z'$ in $Y_j$, $z' \notin D(t, x)$. $| T_1 | \leq | T_j |$ implies that there must exist a vertex $t'$ in $F \backslash Y$ such that $z \notin D(t', x)$ and $z' \in D(t', x)$. Note that $t \notin D(t', x)$ else by Claim 2.1, $z \in D(t', x)$, a contradiction.

Since $z' \notin D(t, x)$, in particular, $z'$ is not adjacent to $t$. Since $t \notin D(t', x)$, there exists a $t' - x$ path $P'$ missed by $t$. But now $P' \cup \{z'\}$ contains a $z' - x$ path missed by $t$, contradicting $z' \in Y$. $\square$

Let $Z$ be a connected component of the subgraph of $G$ induced by $Y_1$. By the induction hypothesis, $Z$ contains a pokable vertex $v$. To complete the proof of Lemma 4.2 we only need show that $v$ is also pokable in $F$. Suppose not. By Claims 4.7 and 4.8 combined, all the vertices $u_i$ and $w_j$ $(1 \leq i \leq p; 1 \leq j \leq q)$ belong to $Y$. To clarify this last point, note that by Claim 4.7 and Claim 4.8, at most $u_1$ and $w_1$ belong to $F \backslash Y$. But now, either $u_1$ and $w$ or $w_1$ and $u$ contradict Claim 4.8.

Further, since $Y$ is disconnected in the complement, all the $u_i$'s and $w_j$'s must belong to $Y_1$. In fact, by Claim 4.3, all the $u_i$'s and $w_j$'s must belong to $Z$, a contradiction. This completes the proof of Lemma 4.2. $\square$

Note that Theorem 4.1 implies the following results:

**Corollary 4.9**: Every AT-free graph is either a clique or else contains two non-adjacent pokable vertices.

**Corollary 4.10**: (Composition Theorem) Given two AT-free graphs $G_1$ and $G_2$ and pokable DPs $(x_1, y_1)$ and $(x_2, y_2)$ in $G_1$ and $G_2$, respectively, let $G$ be the graph constructed by identifying $x_1$ and $x_2$. Then $G$ is an AT-free graph. $\square$
(Refer to Figure 2 for an illustration.)

We now show that the existence of a pokable DP $(x, y_0)$ in a connected AT-free graph $G$ leads to a decomposition scheme of $G$. If $G$ is not a clique, then the proof of Theorem 4.1 shows that $x$ and $y_0$ may be considered to be non-adjacent.

Let $G_0, G_1, \cdots, G_k$ be the sequence of graphs defined as follows:

1. $G_0 = G$.
2. For all $i$, $(0 \leq i \leq k - 1)$, let $R_i$ be the equivalence relation defined on $G_i$ by setting $u R_i v \iff D(u, x) = D(v, x)$ and let $C(y_i)$ be the equivalence class corresponding to $y_i$. The graph $G_{i+1}$ is obtained from $G_i$ by contracting $C(y_i)$

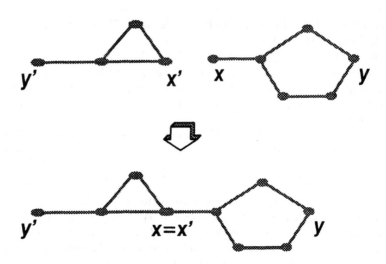

Figure 2: Illustrating the composition theorem

into a new vertex $y_{i+1}$. (i.e. $G_{i+1}$ contains all the vertices in $G_i \backslash C(y_i)$ as well as the new vertex $y_{i+1}$ where $y_{i+1}$ is adjacent to all vertices in $G_i \backslash C(y_i)$ that were adjacent to at least one vertex in $C(y_i)$.)

3. $G_k = \{y_k\}$ (i.e. $G_{k-1} \backslash C(y_{k-1}) = \emptyset$).

For obvious reasons, such a sequence $G_0, G_1, \cdots, G_k$ is called *involutive*. We are now in a position to state the following property of AT-free graphs.

**Theorem 4.11**: Every connected AT-free graph has an involutive sequence.

Before proving this theorem, we first characterize when a vertex in a $DP$ is pokable.

**Claim 4.12**: Let $G$ be a connected AT-free graph with a $DP(x, y)$. Then $x$ is pokable iff there are no unrelated vertices with respect to $x$.

**Proof**: The "if" implication is trivial. To prove the "only if" part, consider unrelated vertices $u, v$ with respect to $x$. In particular, we find a $v - x$ path missed by $u$ and a $u - x$ path missed by $v$. Since $(x, y)$ is a $DP$, $u$ and $v$ intercept every path joining $x$ and $y$. Let $P$ be such a path and let $u'$ and $v'$ be vertices on $P$ adjacent to $u$ and $v$, respectively. Trivially both $u'$ and $v'$ are distinct from $x$. But now there exists a $u - v$ path in $G$ that does not contain $x$ (this path contains $u', v'$ and a subpath of $P$) implying that $x$ is not pokable. □

**Proof of Theorem 4.11**: The proof is trivially true if $G$ is a clique. We shall, therefore, assume that $G$ is not a clique. Let $(x, y_0)$ be a non-adjacent $DP$ in $G$. We now show that for each $i$ ($0 \le i \le k - 1$), $G_{i+1}$ is AT-free and $(x, y_{i+1})$ is a pokable $DP$ in $G_{i+1}$. First we state a condition under which $G_{i+1}$ is AT-free.

**Claim 4.13**: If $C(y_i)$ is connected and $G_i$ is AT-free, then $G_{i+1}$ is AT-free.

**Proof**: Suppose $\{u, v, w\}$ forms an AT in $G_{i+1}$. We distinguish the following cases.

**Case 1:** $y_{i+1} \in \{u, v, w\}$, say $y_{i+1} = w$.

If $P$ is a $u - v$ path missed by $y_{i+1}$, then $P$ is missed, in $G_i$, by all vertices in $C(y_i)$. Let $Q$ be a $u - y_{i+1}$ path in $G'$ missed by $v$. Then there exists a vertex $c_1$ in $C(y_i)$ such that $v$ misses the path $Q - y_{i+1} + c_1$. Similarly, let $R$ be a $v - y_{i+1}$ path in $G'$ missed by $u$. Again there exists a vertex $c_2$ in $C(y_i)$ such that $u$ misses the path $R - y_{i+1} + c_2$. Since $C(y_i)$ is connected, there exists a path joining $c_1$ and $c_2$ in $C(y_i)$. Since $u$ and $v$ are not adjacent to $y_{i+1}$ in $G_{i+1}$, $u$ and $v$ both miss this path in $G_i$. Therefore, $G_i$ contains an AT consisting of $u, v$ and a suitably chosen vertex in $C(y_i)$.

**Case 2:** $y_{i+1} \notin \{u, v, w\}$.

Let $P, Q, R$ be respectively a $u - v$ path, a $u - w$ path and a $v - w$ path that demonstrate the AT. If $y_{i+1}$ does not belong to any of these paths, then $\{u, v, w\}$ forms an AT in $G_i$, a contradiction.

We may assume without loss of generality that $y_{i+1} \in P$. Since $w$ misses $P$, it is clear that $w$ is adjacent, in $G_i$, to no vertex in $C(y_i)$. We now claim that in $G_i$ there exists a $u - v$ path $P'$ that is missed by $w$. Outside of $C(y_i)$, $P'$ contains the same vertices as $P$. Inside $C(y_i)$, $P'$ contains a path between two vertices $c_1$ and $c_2$ such that:

1. $w$ misses a $u - c_1$ path consisting of a subpath of $P$.
2. $w$ misses a $c_2 - v$ path consisting of the remaining vertices in $P \backslash C(y_i)$.

If $y_{i+1}$ also belongs to $Q$ and/or $R$, a similar argument shows that in $G_i$ there exists a $v - w$ path $R'$ missed by $u$ and a $u - w$ path $Q'$ missed by $v$. Thus, $\{u, v, w\}$ is an AT in $G_i$, a contradiction. $\square$

To prove that $(x, y_{i+1})$ is a pokable $DP$ in $G_{i+1}$ and that $C(y_i)$ is connected we distinguish two cases depending on whether $x$ and $y_i$ are adjacent.

**Case 1:** $x, y_i$ are not adjacent.

The connectivity of $C(y_i)$ and the fact that $| C(y_i) | \geq 2$ follow from Claim 4.6. We now show that $(x, y_{i+1})$ is a pokable $DP$ in $G_{i+1}$. First we show $(x, y_{i+1})$ is a $DP$.

Suppose there exists a $x - y_{i+1}$ path $P$ in $G_{i+1}$ missed by some vertex $w$. Trivially, in $G$, $w$ is adjacent to no vertex in $C(y_i)$, in particular not to $y_i$. Since $C(y_i)$ is connected, $w$ misses a $x - y_i$ path in $G$ consisting of $P$ along with some internal path in $C(y_i)$. This contradicts $(x, y_i)$ being a $DP$ in $G_i$.

Now suppose that $x$ is not pokable in $G_{i+1}$. Claim 4.12 guarantees the existence of unrelated vertices $u$ and $v$ in $G_{i+1}$. Thus, in $G_{i+1}$ we have a $v - x$ path $P$ missed by $u$ and a $u - x$ path $Q$ missed by $v$. Since $(x, y_{i+1})$ is a $DP$ in $G_{i+1}$, $y_{i+1}$ belongs to neither $P$ nor $Q$. But now, these paths must have been paths in $G_i$, contradicting $(x, y_i)$ being a pokable $DP$ in $G_i$.

Finally, suppose that $y_{i+1}$ is not pokable. Again, Claim 4.12 dictates the existence of vertices $u$ and $v$ in $G_{i+1}$ and of paths $P$ (from $u$ to $y_{i+1}$ and missed by $v$) and $Q$ (from $v$ to $y_{i+1}$ and missed by $u$). In particular, neither $u$ nor $v$ is adjacent to $y_{i+1}$ and thus in $G_i$ neither $u$ nor $v$ is adjacent to any vertex in $C(y_i)$. But now in $G_i$ there is a $u - y_i$ path missed by $v$ and a $v - y_i$ path missed by $u$ contradicting that $y$ is pokable. (Recall that $(x, y_i)$ is a pokable $DP$ in $G_i$).

**Case 2:** $x, y_i$ are adjacent.

First we show that $C(y_i)$ is connected and $\mid C(y_i) \mid \geq 2$. If $x$ is adjacent to all vertices in $G_i$, then $C(y_i) = G_i$ which is connected and contains at least 2 vertices.

Now let $N_i'(x)$ be the set of vertices in $G_i$ that are not adjacent to $x$. Furthermore, let $u^*$ be a vertex in $N_i'(x)$ such that

$$D(u^*, x) \subset D(u, x) \text{ for no vertex } u \in N_i'(x).$$

We claim that $u^* \in C(y_i)$. If not, we find a vertex $u \in G_i$ and a $u^* - x$ path missed by $u$. Trivially, $u \in N_i'(x)$. Since $x$ is pokable, Claim 4.12 guarantees that $u^* \in D(u, x)$. We now show that $D(u^*, x) \subset D(u, x)$ contradicting our choice of $u^*$. Let $w$ be an arbitrary vertex in $D(u^*, x)$. If $w$ and $u^*$ are not adjacent, then $w \in D(u, x)$ by Claim 2.1. If $w$ and $u^*$ are adjacent, then if $w \notin D(u, x)$, then there is a $u - x$ path $P$ (in $G_i$) missed by $w$. Since $u \notin D(u^*, x)$ let $Q$ be a $u^* - x$ path missed by $u$. But now $w$ and $u$ are unrelated ($u$ misses some path contained in $w \cup Q$) contradicting, by Claim 4.12, that $x$ is pokable. Thus, we conclude that $u^* \in C(y_i)$.

To complete the proof that $C(y_i)$ is connected and $\mid C(y_i) \mid \geq 2$, note that every vertex in $C(y_i) \cap N_i'(x)$ must be adjacent to every vertex in $C(y_i) \cap N_i(x)$ (i.e. the neighbours of $x$ in $G_i$).

Clearly $(x, y_{i+1})$ is a $DP$ in $G_{i+1}$ since every vertex in $G_i \setminus \{x, y_i\}$ is adjacent to either $x$ or $y_i$ or both. Since $(x, y_i)$ is a pokable $DP$ in $G_i$ and any unrelated vertices with respect to $x$ (resp. $y_{i+1}$) in $G_{i+1}$ would also be unrelated with respect to $x$ (resp. $y_i$) in $G_i$, $x$ and $y_{i+1}$ are pokable in $G_{i+1}$.

Finally, we need to show that for some $k$, $G_k = \{y_k\}$. Since $\mid N_0'(x) \mid > \mid N_1'(x) \mid > \cdots$, there is a least subscript $j$ for which $C(y_j) \supset N_j'(x)$. Clearly $N_{j+1}'(x) = \emptyset$ and so $C(y_{j+1}) = G_{j+1}$ implying that $G_{j+2} = \{y_{j+2}\}$ as claimed. $\square$

The involutive sequence involves contraction of the set of vertices $v_i$ with $D(v_i, x) = V$ (i.e. each $v_i$ forms a $DP$ with $x$). Claim 4.13 shows that if $G$ is AT-free with non-adjacent $DP$ $(x, y)$ then the contraction of any connected equivalence class of $uRv \iff D(u, x) = D(v, x)$ preserves being AT-free. It is not immediately obvious that the connectivity of the equivalence class is required. We leave it to the reader to find an AT-free graphs in which the contraction of a disconnected equivalence class creates an AT. $G$ is AT-free with pokable $DP$ $(x, e)$ whereas the contraction of the set $\{a, b\}$ yields graph $G'$ that has the AT $\{a', b', w'\}$.


# 5 Concluding remarks

As mentioned previously, AT-free graphs strictly contain cocomparability graphs. Recently, Kratsch and Stewart [11] made the following observation about cocomparability graphs.

**Proposition 5.1** [11]: A graph $G = (V, E)$ is a cocomparability graph iff there is an ordering of $V$ such that, for any three distinct vertices $u < v < w$, if $(u, w) \in E$ then $(u, v) \in E$ or $(v, w) \in E$.

Thus, we see that cocomparability graphs have a linear ordering; this ordering exemplifies the linear structure we observe in interval graphs, permutation graphs and trapezoid graphs. Do AT-free graphs also possess some form of linear structure expressed as a linear ordering?

Recently Möhring [14] has added to the thesis of the linear structure of AT-free graphs by showing that the pathwidth of an AT-free graph equals its treewidth.

Just as there are many families of perfect AT-free graphs, one would expect to see a rich hierarchy of families of non-perfect AT-free graphs. So far nothing is known here. Since perfect AT-free graphs strictly contain cocomparability graphs, it would be interesting to study the perfect AT-free graphs.

Turning to algorithmic questions, we first mention that we have an $O(n^3)$ AT-free recognition algorithm. Does a faster algorithm exist? Also we would expect that the structural properties of AT-free graphs should lead to fast algorithms for various problems that are difficult for general graphs.

**Acknowledgement:**

D. G. Corneil and L. Stewart wish to thank the Natural Sciences and Engineering Research Council of Canada for financial assistance. S. Olariu was supported, in part, by the National Science Foundation under grant CCR-8909996. The authors wish to thank F. Maffray and M. Preissman of Université de Grenoble for providing us with an English translation of Gallai's paper.

# References

1. S. Arnborg and A. Proskurowski, Linear time algorithms for NP-hard problems restricted to partial $k$-trees, *Discrete Applied Mathematics* 23 (1989), 11–24.
2. F. Cheah, private communication.
3. D.G. Corneil and P.A. Kamula, Extensions of permutation and interval graphs, *Proceedings 18th Southeastern Conference on Combinatorics, Graph Theory and Computing* (1987), 267–276.
4. D.G. Corneil, H. Lerchs, L. Stewart Burlingham, Complement reducible graphs, *Discrete Applied Mathematics* 3 (1981), 163–174.
5. D. G. Corneil, S. Olariu, and L. Stewart, Asteroidal triple-free graphs, Department of Computer Science, University of Toronto, Technical Report 262/92, June 1992.
6. I. Degan, M.C. Golumbic and R.Y. Pinter, Trapezoid graphs and their coloring, *Discrete Applied Mathematics* 21 (1988), 35–46.
7. S. Even, A. Pnueli and A. Lempel, Permutation graphs and transitive graphs, *Journal of the ACM* 19 (1972), 400–410.
8. T. Gallai, Transitive orientierbare Graphen, *Acta Mathematica Academiae Scientiarum Hungaricae* 18 (1967), 25–66.
9. M.C. Golumbic, *Algorithmic Graph Theory and Perfect Graphs* Freeman, New York (1980).
10. M.C. Golumbic, C.L. Monma and W.T. Trotter Jr., Tolerance graphs, *Discrete Applied Mathematics* 9 (1984), 157–170.
11. D. Kratsch and L. Stewart, Domination on cocomparability graphs, submitted for publication, 1989.
12. C.G. Lekkerkerker and J.C. Boland, Representation of a finite graph by a set of intervals on the real line, *Fundamenta Mathematicae* 51 (1962), 45–64.
13. F. Maffray, private communication.
14. R.H. Möhring, private communication.

# The Parallel Complexity of Elimination Ordering Procedures

Elias Dahlhaus

Basser Dept. of Computer Science, University of Sydney, NSW 2006, Australia

**Abstract.** We prove that lexicographic breadth-first search is P-complete and that a variant of the parallel perfect elimination procedure of P. Klein [11] is powerful enough to compute a semi-perfect elimination ordering in sense of [10] if certain induced subgraphs are forbidden. We present an efficient parallel breadth first search algorithm for all graphs which have no cycle of length greater four and no house as an induced subgraph. A side result is that a maximal clique can be computed in polylogarithmic time using a linear number of processors.

## 1 Introduction

Various authors considered elimination orderings to characterize certain graph classes [6, 22, 7, 1]. Such elimination orderings have several applications as sparse Gauss elimination [16], efficient graph coloring [1, 22, 2], etc.. An interesting graph class are the chordal graphs. That are the graphs with the property that any cycle of length greater than three has two nonconsecutive vertices which are adjacent. Chordal graphs can also be characterized as those graphs which have a *perfect elimination ordering*, i.e. the greater neighbors of any vertex form a complete set. Sequential linear algorithms to recognize chordal graphs and to compute a perfect elimination ordering are the lexical breadth-first search method of Rose, Tarjan, and Lueker [17] and the maximum cardinality search method of Tarjan and Yannakakis [21]. An efficient parallel algorithm is due to Klein [11]. It is well known that a minimum coloring of a chordal graph can be obtained efficiently by coloring the vertices in reversal order with the color of the least number which is not the color of a greater neighbor. There is a more generalized result in behind. Chvatal [1] found out that one can proceed in the same way if the ordering is a "perfect ordering" (not perfect elimination ordering), i.e. an ordering with the property that for no path $(x_1, x_2, x_3, x_4)$, such that nonconsecutive vertices are not adjacent, $x_2 < x_1$ and $x_3 < x_4$. Note that Chvatal presented perfect orderings in the reversed order [1]. Middendorf and Pfeiffer proved that the existence of a perfect ordering is an NP-complete problem [13]. Jamison and Olariu [10] introduced the notion of *semiperfect elimination orderings*, which lie between perfect elimination orderings and perfect orderings. We can state that an ordering is a perfect elimination ordering if any vertex is not the midpoint of an induced path of length two in the given graph restricted to those vertices which are greater than or equal the given vertex. Jamison and Olariu relaxed this condition that any vertex is not the midpoint of an induced path of length three ($P_4$) in the given graph restricted to those vertices greater or equal the given vertex. Jamison and Olariu [10] stated conditions when a semi-perfect elimination ordering can be computed efficiently by the known perfect elimination algorithms in [17] and [21]. Jamison and Olariu stated some properties that an ordering ever has if one applies the algorithm in [17] or the algorithm in [21] and proved that under certain conditions, an ordering satisfying these properties is a semi-perfect elimination ordering.

Here we deal with the problem to develop parallel algorithms doing essentially the same as lexical breadth-first search or maximum cardinality search. While a variant of P. Klein's parallel elimination algorithm does essentially the same as the maximum cardinality search, lexical breadth-first search is inherently sequential. An immediate consequence is that the computation of a doubly lexicographic ordering in sense of [12] is inherently sequential. Klein [11] developed also an efficient parallel breadth-fist search tree algorithm for chordal graphs based on the existence of a perfect elimination ordering. We develop an efficient parallel breadth-first search algorithm for chordal graphs which is not based on the knowledge of a perfect elimination ordering. We shall see that this algorithm can be extended to a larger graph class. It remains open to extend Klein's depth-first search algorithm for chordal graphs to larger graph classes. In the second section, we introduce the notation necessary in the whole paper. In the third section, we describe different elimination processes and prove that lexical breadth-first search is P-complete. In the fourth section we discuss the power of a variant of P. Klein's parallel elimination procedure. Here we have to solve the problem to compute an inclusion maximal clique in parallel. As a side result, we found out that an inclusion maximal clique can be computed in polylogarithmic time with a linear processor bound in nearly the same manner as an inclusion maximal independent set [9]. In the fifth section, we discuss an efficient parallel breadth-first search algorithm which is not only applicable for chordal graphs but for graphs not containing the house or a cycle of length greater than four as an induced subgraph.

## 2  Notation

A *graph* $G = (V, E)$ consists of a *vertex set* $V$ and an *edge set* $E$. Multiple edges and loops are not allowed. The edge joining $x$ and $y$ is denoted by $xy$. We say that $x$ is a *neighbor* of $y$ iff $xy \in E$. The *neighborhood* of $x$ is the set $\{y : xy \in E\}$ consisting of all neighbors of $x$ and is denoted by $N(x)$. The neighborhood of a set of vertices $V'$ is the set $N(V') = \{y | \exists x \in V', yx \in E\}$ of all neighbors of some vertex in $V'$. A *subgraph* of $(V, E)$ is a graph $(V', E')$ such that $V' \subset V$, $E' \subset E$. An *induced subgraph* is an edge-preserving subgraph, i.e. $(V', E')$ is an induced subgraph of $(V, E)$ iff $V' \subset V$ and $E' = \{xy \in E : x, y \in V'\}$. The *clique number*, i.e. the maximum size of a clique, of a graph $G$ is denoted by $\omega(G)$. The *chromatic number* of a graph $G$ is denoted by $\chi(G)$. A graph is *perfect* if, for each induced subgraph, the chromatic number and the clique number coincide. A trivial heuristics to color a graph is to color the vertices in reversed order with respect to an enumeration $(v_i)_{i=1}^n$ of the vertex set $V$ with the available color of least index (*greedy coloring*). This heuristics gives a coloring in the bounds of the chromatic number for each induced subgraph if and only if the ordering behind the enumeration is the reversal of a *perfect ordering* [1]:

If $(v_i, v_j, v_k, v_l)$ forms an induced path then $i < j$ or $l < k$.

It is NP-complete to check, for any graph, whether it has a perfect ordering or not [13]. But in special graph classes necessarily a perfect ordering exists. Moreover in these special graph classes, a perfect ordering can be determined efficiently. Special perfect ordering are the *perfect elimination orderings*:

If $x < y, x < z, xy, xz \in E$, then $yz \in E$.

It is well known that a graph has a perfect elimination ordering iff it is *chordal*, i.e. it has no induced cycle of length greater than 3 [7]. A generalization of perfect elimination orderings are semi-perfect elimination orderings [10]. An ordering $<$ on the vertex set $V$ of a graph $G = (V, E)$ is called *semi-perfect* if each vertex $v \in V$ is not the midpoint of an induced path of length three ($P_4$) in $G$ restricted to $\{w|v \leq w\}$.

The parallel computation model is the parallel random access machine (PRAM). In most cases we use the Exclusive Read Exclusive Write PRAM (EREW-PRAM), i.e. only one processor is allowed to read from and to write into the same memory cell at the same time. Sometimes we also use the Concurrent Read Concurrent Write PRAM model (CRCW-PRAM), i.e. at the same time arbitrary many processors are allowed to read from the same memory cell and arbitrary many processors are allowed to apply to write into the same memory cell. Usually the processor with the smallest number applying to write into a certain memory cell writes into the memory cell.

## 3 Elimination Procedures

In this section we recall known elimination procedures and their power. Jamison and Olariu considered elimination orderings with the following properties [10]:

P1 If $a < b, b < c, ac \in E, bc \notin E$ then there is a vertex $b'$ such that $bb' \in E, ab' \notin E, c < b'$.

P2 If $a < b, b < c, ac \in E, bc \notin E$ then there is a vertex $b'$ such that $bb' \in E, ab' \notin E, b < b'$.

Note that property P1 implies property P2.

Tarjan and Yannakakis [21] proved that a graph is chordal iff any ordering satisfying property P2 is a perfect elimination ordering.

Note that that the maximum cardinality search procedure of Tarjan and Yannakakis [21] ever computes an ordering which satisfied property P2 and the lexicographical breadth-first search procedure of Rose, Tarjan, and Lueker ever computes an ordering satisfying property P1.

Jamison and Olariu [10] stated necessary and sufficient conditions that an ordering satisfying property P1 or P2 is semi-perfect.

We consider the following graphs:

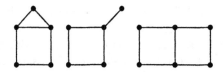

We call the first graph the *house*, the second graph the *guitar* and the third graph the *domino*.

**Lemma 1.** *[10]*

1. *For each induced subgraph $H$ of $G$, any ordering on the vertices of $H$ satisfying P1 is a semi-perfect elimination ordering iff $G$ does not contain a if it does not contain a cycle of length $> 4$ or the house or the domino as an induced subgraph.*
2. *For each induced subgraph $H$ of $G$, any ordering on the vertices of $H$ satisfying P2 is a semi-perfect elimination ordering iff $G$ does not contain a cycle of length $> 4$ or the house or the guitar as an induced subgraph.*

Lubiv [12] extended even the conditions of P1:

P0 If $a < b, ac \in E, bc \notin E$ then there is a vertex $b'$ such that $bb' \in E, ab' \notin E, c < b'$.

In the case of bipartite graphs an ordering satisfying P0 is also called a *doubly lexicographic ordering*. It can be proved that, in any strongly chordal graph [6], an ordering satisfying P0 is a strongly perfect elimination ordering [12]. Efficient sequential algorithms to compute a doubly lexicographic ordering are due to Lubiv [12], Paige and Tarjan [15], and Spinrad [19].

We shall see that even the computation of an ordering satisfying the property P1 is inherently sequential.

**Theorem 2.** *The computation of an ordering satisfying P1 is P-complete.*

**Remark:** Miyano [14] developed a general tool to prove the P-completeness of the computation of the lexicographically smallest subset of the vertex set of an ordered graph satisfying certain property. The formulation of the problem to find a lexicographic breadth-first search ordering is a little bit different. Here we would like to find an ordering on an *unordered* graph satisfying some lexicographic property.

*Proof of Theorem:* We prove that an $NC$-algorithm for the computation of an ordering satisfying the property P1 induces an $NC$-algorithm for the following problem which is P-complete [4]:

*First Maximal Clique*

*Input:* A graph $G = (V, E)$ and an ordering $<$ on the set $V$ of the vertices of $G$.

*Output:* The clique $C$ which is computed as follows:

1. The smallest element $v_0$ is in $C$,
2. put stepwise the smallest element $v \notin C$ into $C$ which is adjacent to all vertices in $C$ until no vertex not in $C$ is adjacent to all vertices in $C$.

Suppose $G = (V, E)$ with $V = \{v_1, \ldots, v_n\}$ and the ordering $<$ on $V$ with $v_i < v_j$ iff $i < j$ are given.

First we construct a (bipartite) graph $G' = (V', E')$ with a distinguished vertex $M$ such that any ordering with $M$ as maximum which satisfies P1 induces the first maximum clique in $G$.

Set $V' = \{M\} \cup \bigcup_{i=0}^{n} V_i$ with $V_0 = \{0\} \times V$, and for $i > 0$, $V_i = \{i\} \times \{v_j | j \geq i\}$.

The edge set $E'$ is defined as follows. $E' = \{Mv | v \in V_0\} \cup \{(0, v_i)(1, v_j) | i \geq j\} \cup \{(i, v_j)(i+1, v_j) | j \geq i+1\} \cup \{(i, v_i)(i+1, v_j) | v_i v_j \in E \text{ and } i \leq j\}$.

It is easily seen that $G'$ can can be computed from $G$ by an EREW-PRAM with $O(n^3)$ processors in logarithmic time.

Suppose the first maximal clique $C$ of $G$ is $\{v_{i_1}, \ldots, v_{i_k}\}$ and $1 = i_1 < \ldots < i_k$. Let $<'$ be a lexicographical breadth first search ordering on $G'$ with $M$ as its maximum. Then by induction on $i$, one can show:

**Lemma 3.** *Suppose $i_j < i \leq i_{j+1}$. Then the following statements are satisfied.*

1. *$(i, v_{i_{j+1}})$ is the largest vertex in $V_i$ with respect to $<'$,*
2. *Let $v_l \geq v_i$ be a vertex which is adjacent to all vertices $v_{i_1}, \ldots, v_{i_j}$ and $v_{l'} \geq v_i$ be a vertex which is not adjacent to all vertices $v_{i_1}, \ldots, v_{i_j}$. Then $(i, v_{l'}) <' (i, v_l)$.*
3. *If $v_l \geq v_i$ and $v_{l'} \geq v_i$ are both adjacent to all vertices $v_{i_1}, \ldots, v_{i_j}$ then $(i, v_l) <' (i, v_{l'})$ iff $l' < l$.*

*Proof of Lemma*

First note that $i < j$ and $(1, v_i) >' (1, v_j)$ are equivalent statements, because $i < j$ implies that the neighborhood of $(1, v_j)$ in $V_0$ is a proper subset of the neighborhood of $(1, v_i)$ in $V_0$. Therefore $(1, v_1) = (1, v_{j_1})$ is the $<'$-largest vertex in $V_1$. The second statement is trivially true for $i = 1$ because no $i_j$ is smaller than 1. Therefore the lemma has been proved for $i = 1$.

Suppose the lemma is true for $i = 1, \ldots, i' - 1$. To prove the lemma for $i = i'$, we have to consider two cases.

*Case 1: $i' - 1 = i_j$, for some $j$.*

By induction hypothesis, $(i' - 1, v_{i_j}) = (i' - 1, v_{i'-1})$ is the largest vertex in $V_{i'-1}$. Therefore the vertices $(i', v_j)$ with $v_{i'-1} v_j \in E$ are $<'$-larger than those vertices $(i', v_j)$ such that $v_j$ is not in the neighborhood of $v_{i'-1}$, because the neighbors of $(i' - 1, v_{i'-1})$ in $V_{i'}$ are those $(i', v_j)$ with $v_i v_j \in E$. Note that the only remaining neighbor of $(i', v_j)$ in $V_{i'-1}$ is $(i' - 1, v_j)$. Therefore for all $(i', v_j)$ and $(i', v_k)$, such that $v_j$ and $v_k$ are neighbors of $v_{i'-1}$, $(i', v_j) <' (i', v_k)$ iff $(i' - 1, v_j) <' (i' - 1, v_k)$. Therefore, since the second and the third statement is true for $i = i' - 1$, they are even true for $i = i'$. Since $v_{i_{j+1}}$ is the smallest vertex adjacent to all vertices $v_{i_1}, \ldots, v_{i_j}$, also the first statement is true for $i = i'$.

*Case 2: $i_j < i' - 1 < i_{j+1}$.*

Note that for all $v_j$ with $j \geq i'$, $(i'-1, v_j)$ is a neighbor of $(i', v_j)$ and $(i'-1, v_{i'-1})$ is the only other possible neighbor of $(i', v_j)$. Moreover $v_{i'-1}$ is not adjacent to all vertices $v_{i_1}, \ldots, v_{i_j}$. Since the second statement is true if $i = i' - 1$, $(i' - 1, v_{i'-1}) <' (i' - 1, v_k)$, and $v_k$ is adjacent to all $v_{i_1}, \ldots v_{i_j}$. Since the neighbors of $(i', v_k)$ in $V_{i'-1}$ are $(i'-1, v_k)$ and possibly $(i'-1, v_{i'-1})$, for all $v_k, v_l$ with $(i'-1, v_{i'-1}) <' (i'-1, v_k)$, and $(i' - 1, v_l)$, $(i', v_k) <' (i', v_l)$ iff $(i' - 1, v_k) < (i' - 1, v_l)$. Therefore the second and the third statement are true for $i = i'$. The first statement follows from the third statement by the definition of $v_{i_{j+1}}$.

□ (Lemma)

Therefore $C$ consists of those $v_i$ with $i \leq i_k$ such that $(i, v_i)$ is maximal in $V_i$ with respect to $<'$. Only we do not know $i_k$.

To compute $C$, we compute the set $C' = \{v_i : (i, v_i)$ is maximal with respect to $<'$ in $V_i\}$ and determine the first index $i'$ in $C'$ such that there is an index $j < i'$ in $C'$ such that $v_j v_i \notin E$. Apparently $C = \{v_j \in C' | j < i'\}$.

For the general case, we consider two copies of $G'$ and identify in both copies the vertex $M$. The maximum must be in one of the copies. Any ordering has its maximum in one of the copies of $G'$. Then $M$ must be the maximum of the other copy and there the behavior of an ordering satisfying P1 must be the same as in the case that $M$ is the maximum.

□

# 4 P. Klein's Perfect Elimination Procedure and its Application to Compute Semi-Perfect Elimination Orderings

Here present a variant of Klein's parallel perfect elimination procedure which computes, for any graph, an ordering satisfying P2 [11].

The basic idea of P. Klein's algorithm is that we begin with an initialization procedure NONE which computes levels $L_1, \ldots, L_k$ of size at most a fixed ratio of the number of vertices. Successively the levels $L_i$ are refined independently into sublevels that have again a size of at most a fixed ratio of the number of vertices in $L_i$. We stop this procedure if all levels have the cardinality one.

To compute the initial level partition, P. Klein [11] made use of the fact that the intersection of the neighborhoods of two nonadjacent vertices of a chordal graph induces a complete graph. This is not true in general. Therefore we have to develop an elimination procedure which works for any graph.

We proceed as follows:

1. We compute a maximal clique $C$ on those vertices which have a larger degree than $2/3|V|$. Suppose $C = \{x_1, \ldots, x_k\}$
2. For $i = 1, \ldots, k$, we set $L_i := \{v \in V \setminus C|$ for all $j < i$, $x_j v \in E$ and $x_i v \notin E\}$ for each $i = 1, \ldots, k$. Clearly each $L_i$ has a cardinality of at most $1/3|V|$ and $M = \{v \in V \setminus C | \forall x \in C, vx \in E\}$.
3. We compute the connected components $M_1, \ldots, M_p$ of $G$ restricted to $M$ and let $M_p$ be that connected component with the largest cardinality. Let $L_{k+i} = M_i$, for $i = 1, \ldots, p-1$
4. To refine $M_p$, we compute a spanning tree $T$ for $M_p$. We select a vertex $r$ of $M_p$ as root and compute the enumeration $(u_1, \ldots, u_q)$ of a postordering with respect to $T$ (see [20]). $v \in M_p \setminus \{r\}$ is set into $L_{k+p+i-1}$ if $i$ is the maximum index such that $vu_i \in E$ and $r$ is set into $L_{k+p+q}$.
5. We set $L_{k+p+q} + i = \{x_{k-i+1}\}$.

It is easily checked that the levels $L_1, \ldots, L_k$ have a cardinality of at most $1/3|V|$, the levels $L_{k+1}, \ldots, L_{k+p-1}$ have a cardinality of at most $1/2|V|$, and the levels

$L_{k+p}, \ldots, L_{k+p+q-1}$ have a cardinality of at most $2/3|V|$ (note that all common neighbors of $C$ have a degree of at most $2/3|V|$.

The following fact can be interpreted in such a way that the ordered partition $(L_1, \ldots, L_{2k+p+q})$ is an approximation of an ordering satisfying P2.

**Fact 1** *Suppose* $a \in L_i, b \in L_j, c \in L_l$, $i < j < l$, *and* $ac \in E, bc \notin E$. *Then there is a vertex* $b' \in L_{l'}$ *with* $ab' \notin E$, $bb' \in E$, *and* $l' > j$, *the neighborhood of* $b$ *in the levels* $L_\nu$ *with* $\nu > j$ *is not a proper subset of the neighborhood of* $a$ *in the levels* $L_\nu$ *with* $\nu > j$.

To refine the level with maximal index, we only have to apply above procedure.

To refine any other level, say $L_i$ into sublevels $L_{i,1}, \ldots, L_{i,r}$, such that the requirements of 1 are satisfied, we can proceed as in [11].

Let $G_i = (V_i, E_i)$ with $V_i = L_i \cup \{x | \exists y \in L_i, \, xy \in E \text{ and } x \in L_j \text{ for some } j > i\}$ and $E_i = \{xy \in E | x, y \in V_i, \, x \in L_i \text{ or } y \in L_i\}$. Let $S_i$ be the set of those vertices in $V_i$ with at most $2/3|L_i|$ neighbors in $L_i$ (sparse vertices) and $D_i$ be the set of those vertices in $V_i$ which have more than $2/3|L_i|$ neighbors in $L_i$ (dense vertices).

Let $B_i$ be the union of all connected components of $S_i$ which contain at least some vertex in some $L_j$ with $j > i$. Let $L_i^2$ be the set of all vertices in $L_i$ which are in $B_i$ or adjacent to some vertex in $B_i$ and $L_i^1$ be the set of vertices in $L_i$ which are not adjacent to some vertex in $B_i$.

If $L_i^2$ has a size greater than $2/3|L_i|$ then we refine the level $L_i^2$ as follows:

1. We compute a spanning forest $F_i$ for $B_i$ consisting of the trees $T_i^1, \ldots, T_i^r$.
2. For each such tree $T_i^j$, we choose root $R_i^j$ which is in some $L_{i'}$ with $i' > i$ (note that such a vertex $R_i$ ever exists).
3. We compute the concatenation $(u_1, \ldots, u_s)$ of postorders of the trees $T_i^j$.
4. For any $v \in L_i^2$, we put $v$ into $L_{i,j}$ if $u_j$ is the neighbor of $v$ in $B_i$ with the largest index $j$.
5. The new levels are $L_i^1, L_{i,j}, j = 1, \ldots, s$. $L_i^1$ is that level with the smallest index.

Now we suppose that $L_i^1$ is of size greater than $2/3|L_i|$. Note that all vertices in $L_i^2$ or in some $L_j$ with $j > i$ which have neighbors in $L_i^1$ have more than $2/3|L_i|$ neighbors in $L_i$ and are therefore in $D_i$. To get a refinement of $L_i^1$, we first fix some enumeration $(v_1, \ldots v_t)$ of all vertices in $D_i$ which are in $L_i^2$ or not in $L_i$. We set $L_{i,j} = \{v \in L_i^1 | vv_j \notin E \text{ and for all } j' < j, \, vv_{j'} \in E\}$. All these levels $L_{i,j}$ are of cardinality at most $1/3|L_i|$.

To refine the set $M_i$ of all vertices in $L_i^1$ which are adjacent to all vertices $v_j$, we apply the initial procedure to create a first refinement to $G$ restricted to $M_i$.

Note that fact 1 remains true and all levels have a cardinality of at most $2/3|L_i|$.

*Before we can make a complexity statement, we have to consider the complexity to compute an inclusion maximal clique.*

**Lemma 4.** *A maximal clique can be computed by an EREW-PRAM in $O(\log^3 n)$ time using $O(n+m)$ processors.*

*Proof:* We proceed in a similar way as in the maximal independent set algorithm of Goldberg and Spencer [9]. Note that this lemma is not an immediate consequence of the theorem of Goldberg and Spencer, because the number of edges in the complement of a graph is only bounded by $n^2$ but not by $m$.

To make the paper more self contained, we repeat this algorithm.

The main subprocedure is $FINDSET$ which computes an independent set $I$ such that $|I \cup N(I)| \geq (1/2 - o(1))(n + m)$. We apply $FINDSET$ to the original graph, remove all vertices of $I \cup N(I)$ and apply again $FINDSET$ to the rest. This is done $O(\log n)$ times.

The procedure $FINDSET$ works as follows:

At each step, the remaining vertex set is partitioned into independent sets, called color classes, and uncolored vertices.

Initially, each vertex forms a color class. Denote the degree of a vertex $x$ by $deg(x)$. The *weight* of a vertex set $X$ is defined a $\sigma(X) = \Sigma_{x \in X}(1 + deg(x))$. If there is a color class $C^*$ such that $\sigma(N(C^*)) \geq (n+m)/\log n$ then we we delete all vertices is $C^* \cup N(C^*)$ and add $C^*$ to the independent set $I$ (Phase 1). If we do not find such a $C^*$ then we enter *phase 2* defined as follows:

We suppose the color classes are $C_0, \ldots, C_{k-1}$. We partition the pairs $(i,j), i = 0, \ldots, k-1$:

If $k$ is even then $index(i,j) = (i+j) \, mod(k-1)$, for $i,j \neq k-1$ and $index(k-1,j) = index(j, k-1) = 2j$. If $k$ is odd then $index(i,j) = (i+j) \, mod \, k$.

Note that if $index(i,j) = index(i',j')$ then the unordered pairs $\{i,j\}$ and $\{i',j'\}$ are equal or disjoint.

We find a *suitable* index $l$ and mix all pairs $C_i, C_j$ with $index(i,j) = l$ as follows: We compare the weights of $C_i \cap N(C_j)$ and and of $C_j \cap N(C_i)$. Suppose $C_i \cap N(C_j)$ is of smaller weight. Then we uncolor $C_i \cap N(C_j)$ and make $(C_i \cap C_j) \setminus (C_i \cap N(C_j))$ a new color class. An index $l$ is considered as suitable if the weight of the set of uncolored vertices is minimal. The whole procedure $FINDSET$ stops if there is only one color class.

To turn the algorithm of Goldberg and Spencer into an algorithm to compute a maximal clique, we proceed as follows. Color classes are pairwise disjoint complete sets. To check some weight constraints, we have to compute the degree $deg'(x)$ of any vertex in the complement of $G$. Clearly $deg'(x) = n - deg(x) - 1$. We define the weight of a set $C$ as $\sigma'(C) = \Sigma_{x \in C}(1 + deg'(x))$. Instead of computing the neighborhood of a color class $C_i$, we have to compute the set $J_i$ of vertices that are adjacent to all vertices in $C_i$. This can be done by computing, for each $v$ and each $C_i$ with an edge $vy \in E$ with $y \in C_i$, the number of neighbors of $v$ in $C_i$. If this number coincides with $|C_i|$ then $v$ belongs to $J_i$. For all $i$ simultaneously, $J_i$ can be computed in $O(\log n)$ time using $O(n + m)$ processors. Let $N'(C)$ be the neighborhood of $C$ in the complement of $G$. To check whether phase 1 can be applied to $C_i$, we have to compute the weight $\sigma'(N'(C_i))$ of $N'(C_i)$. Note that $N'(C_i) = V \setminus (C_i \cup J_i)$. Therefore $\sigma'(N'(C_i)) = \sigma'(V) - \sigma'(C_i) - \sigma'(J_i)$. Note that the number of pairs $(i,x)$ with $x \in J_i$ is bounded by $m$. Therefore for all $i$ simultaneously, $\sigma'(J_i)$ can be computed in $O(\log n)$ time using $O(n + m/\log n)$ processors. The values $\sigma'(C_i)$ can be computed, simultaneously for all $i$, in $O(\log n)$ time with $O(n)$ processors. The step to delete $C_i$ and all vertices in $N'(C_i)$ is the deletion of all vertices which are not in $J_i$. This can be done in $O(\log n)$ time using $O(n)$ processors by checking whether $x$ is in $J_i$. Instead of finding a partition of the color classes that minimizes the weight of the uncolored vertices, we find a partition which maximizes the weight of the set of vertices which remain colored. For each $i$, we compute the weight $w_{(i,j)}$ of the nonempty sets $J_i \cap C_j$ in $O(\log n)$ time with $O(n+m)$ processors. The set of vertices of $C_i \cup C_j$ which remain colored is $w'_{(i,j)} = \sigma'(C_i) + w_{(i,j)}$. We have the choice to uncolor vertices in $C_i$ or to uncolor vertices in $C_j$. To minimize the weight of uncolored vertices or to maximize the weight of colored vertices, we consider the maximum $w''_{i,j} = \max(w'_{(i,j)}, w'_{(j,i)})$. Note that we have no direct access to the pairs $(i,j)$ such that $C_j \cap J_i = \emptyset$ and $C_i \cap J_j = \emptyset$ if we restrict the number of processors by $O(n+m)$. For such pairs, the $C_i$ or $C_j$ with the maximum weight remains colored. To find the sum of weights of such color classes, we first sort the color classes $C_i$ with respect to their weights (in $O(\log n)$ time with $O(n)$ processors [3]). We consider now the set $S_l$ of all $i$ such that for that $j$ with $index(i,j) = l$, $i > j$. If $k$ is odd then $S_l$ consists of those $i$ with $l/2 \le i < l$ and those $i$ with $(k+l)/2 \le i < k$, i.e. of the intervals $[l/2, l)$ and $[(k+l)/2, k)$. If $k$ is even then $S_l$ consists of those $i$ with $i = k - 1$ or $l/2 < i < l$ or $(k+l-1)/2 < i < k - 1$, i.e of the intervals $(l/2, l)$ and $((k+l-1)/2, k)$. Note that for $i \in S_l$ and with $index(i,j) = l$, the weight of $C_i$ is at least as large as the weight of $C_j$. We call $C_i$ also a *maximum weight class* modulo $l$ if $i \in S_l$.

To compute the weight of the set of vertices which remain colored if we choose those pairs with $index(i,j) = l$, we compute the sum of weights of the maximum weight classes modulo $l$ and for each pair $i, j$ with $(C_i \cap J_j) \cup (C_j \cap J_i) \ne \emptyset$ and $index(i,j) = l$, we add the difference $d''(i,l)$ of $w''_{i,j}$ and the maximum weight of the color classes $C_i$ and $C_j$. Note that the number of pairs $i, j$ with $(C_i \cap J_j) \cup (C_j \cap J_i)$ is bounded by $m$. Therefore it is possible, to compute, for all such pairs $i, j$ simultaneously, the maximum weight $m(i,j)$ of $C_i$ and of $C_j$ in constant time with $O(m)$ processors. We get the same time and processor bound to compute $d(i,l) = d(i, index(i,j)) = w''(i,j) - m(i,j)$. It remains to compute the sum of weights of the maximum weight classes modulo $l$, for all $l$ simultaneously. For this purpose, we compute the prefix sums $Sum_i = \Sigma_{j<i} \sigma'(C_j)$, for all $i = 0, \ldots k-1$ simultaneously, in $O(\log n)$ time with $O(n/\log n)$ processors (see for example [8]). Using the fact that $S_l$ is ever a union of two intervals, we get the sum of the weights of the maximum weight classes modulo $l$ in constant time with $O(n)$ processors, for all $l$ simultaneously.

Therefore we can execute phases 1 and 2 in $O(\log n)$ time with $O(n+m)$ processors.

Now we only have to follow the arguments of [9] to get a time bound of $O(\log^3 n)$ to compute an inclusion maximal clique.

□

Putting above elimination algorithm and last lemma together, we get the following.

**Theorem 5.** *The enumeration of an ordering satisfying property P1 can be computed by an EREW-PRAM in $O(\log^4 n)$ time using $O(n+m)$ processors.*

An immediate consequence is the following.

**Corollary 6.** *For any graph not containing a cycle of length greater than four, a house, or a guitar as an induced subgraph, a semi-perfect elimination ordering can be computed by an EREW-PRAM in $O(\log^4 n)$ time using $O(n+m)$ processors.*

# 5 An Efficient Parallel Breadth-First Search Algorithm for Graphs without Houses and without Cycles of Length Greater than Four

P. Klein's parallel breadth first search algorithm for chordal graphs computes a breadth first search tree by setting, for each vertex the largest neighbor with respect to a perfect elimination ordering as its parent. We shall see that the following algorithm computes a breadth first search tree of a chordal graph without the knowledge of a perfect elimination ordering.

**Algorithm BFC:**

1. Compute a spanning tree $T$ for the graph $G = (V, E)$.
2. Compute the enumeration $(u_1, \ldots, u_n)$ of a postorder on $T$.
3. $L_i = \{v | i$ is the maximal index such that $vu_i \in E\}$ and $L_{n+1} = \{u_n\}$.
4. Let $p(v)$ be the maximum $i$ such that $v$ has a neighbor in $L_i$.
5. The parent $P(v)$ is a vertex $w \in L_{p(v)}$ with a maximum number of neighbors in $\bigcup_{j > p(v)} L_j$.

We can prove the following extended result.

**Lemma 7.** *The tree $T_P$ constructed in algorithm BFC is a breadth-first search tree if $G = (V, E)$ has no house and no cycle of length $> 4$ as an induced subgraph.*

*Proof:* If $x \in L_i$, $y \in L_j$, and $x < j$ then we write also $x < y$. We say $x \leq y$ iff $x \in L_i$, $y \in L_j$, and $i \leq j$.

Clearly $v < P(v)$. By construction, if $xy \in E$ and $y \neq P(x)$ then $y$ cannot be an ancestor of $x$.

Let $r$ be the root of $T$, and therefore also the root the tree $T_P$ with parent function $P$.

Let $M_i$ be the set of vertices with distance $i$ from $r$ in $T_P$. To prove that $P$ is the parent function of a breadth-first search tree, we have to prove that if $xy \in E$, $x \in M_i$, and $y \in M_j$ then $|i - j| \leq 1$.

We prove this indirectly. Assume $xy \in E$, $x \in M_i$, $y \in M_j$, and $j - i \geq 2$. Moreover we assume that $j$ is a maximum index, such that there is an $i$ with $j - i \geq 2$ and there are $x \in M_i$ and $y \in M_j$ with $xy \in E$ ($j$ satisfies the *maximality condition*).

*Claim:* $P(x)P(y) \in E$.

*Proof of Claim:* Since $j - i \geq 2$, $y < P(x)$. We consider two cases:

1. $P(y) \leq P(x)$: Note that: $x < y < P(y) \leq P(x)$. That means there is an $l$ such that $P(x), P(y) \in \bigcup_{i > l} L_i$ and $x, y \notin \bigcup_{i > l} L_i$. Therefore $x$ and $y$ are not neighbors of the set $S_l = \{u_l, \ldots, u_n\}$ and $P(x)$ and $P(y)$ are neighbors of $S_l$. Note that $S_l$ induces a subtree of $T$ and therefore a connected subgraph of $G$. Therefore there is a path $p$ from $P(x)$ to $P(y)$ such that all inner vertices are in $S_l$. $P(x), x, y, P(y)$ together with $p$ forms a cycle $C$ of length greater than four. We an assume that $p$ has no chord. Therefore the only chord of $C$ is $P(x)P(y)$ and such a chord must exist, because we assume that $G$ has no cycle of length greater than four as an induced subgraph.

2. $P(x) < P(y)$: Note that $x < y < P(x) < P(y), P(P(x))$. Then we find an $l$ such that $x, y \notin \bigcup_{i>l} L_i$ and $P(x) \in L_l$. Therefore $x$ and $y$ are not neighbors of $u_l$ and $P(x)$ is a neighbor of $u_l$. Note that $P(x) < u_l$. Therefore we find a $k$ such that $x, y, P(x) \notin \bigcup_{i \geq k} L_i$ and $P(y), u_l \in \bigcup_{i \geq k} L_i$. Therefore $P(y)$ and $u_l$ are in the neighborhood of $S_k = \{u_k, \ldots, u_n\}$ but $x, y, P(x)$ are not in the neighborhood of $S_k$. Since $S_k$ is connected, there is a (chordless) path $p$ from $P(y)$ to $u_l$ with inner vertices in $S_k$. $P(y), y, x, P(x), u_l$ together with $p$ forms a cycle $C$.

Assume $P(y)P(x) \notin E$. Since even a chord $yu_l$ does not exist, we are forced to choose chords $P(y)u_l$ or $yP(x)$. But if we choose only the chord $P(y)u_l$ then we still have an induced cycle $P(y), u_l, P(x), x, y$ of length greater five, and if we have only the chord $yP(x)$ then $P(y), y, P(x), u_l$ together with $p$ forms an induced cycle of length at least five. Therefore both possible chords of $C$ must exist as edges. But then $\{x, y, P(y), u_l, P(x)\}$ induces a house. This is a contradiction.

□ (Claim)

Note that $P(x) \in M_{i+1}$ and $P(y) \in M_{j+1}$. The fact that $P(x)P(y) \in E$ leads to a contradiction to the maximality condition of for $j$.

□(Lemma)

Note that all steps of the algorithm BFC with the exception of the first step can be executed by an EREW-PRAM in $O(\log n)$ time using $O(n+m)/\log n$ processors. The first step can be done by a CRCW-PRAM in $O(\log n)$ time with $O(n+m)$ processors [18]. Therefore we get the following result.

**Theorem 8.** *Suppose $G$ is a graph not containing the house or a cycle of length greater than 4 as an induced subgraph. Then a breadth-first search tree can be computed in the same time and processor bound as a spanning tree, i.e. in $O(\log^2 n)$ time with $O(n+m)$ processors on a EREW-PRAM or with the same processor bound in $O(\log n)$ time on a CRCW-PRAM.*

The depth-first search tree algorithm of P. Klein cannot be improved in the same way. It remains an open problem to compute a depth first search tree efficiently in parallel for certain graph classes beyond chordal graphs or planar graphs.

## 6 Conclusions

We did not give a characterization of graphs having a semiperfect elimination ordering. It is also not clear how to check efficiently (with a linear processor number in polylogarithmic time) that the ordering we just computed is a semiperfect elimination ordering. It might be an interesting problem to check efficiently (in parallel) that a given ordering on the vertices of a graph is a semiperfect elimination ordering.

## References

1. V. Chvatal, *Perfectly Ordered Graphs*, Topics on Perfect Graphs, C. Berge, V. Chvatal eds., North Holland, Amsterdam(1984), pp. 63-65.
2. V. Chvatal, C. Hoang, N. Mahadev, D. de Werra, *Four Classes of Perfectly orderable Graphs*, Journal of Graphs Theory 11 (1987), pp. 481-495.

3. R. Cole, *Parallel Merge Sort*, 27. IEEE-FOCS (1986), pp. 511-516.
4. S. Cook, *A Taxonomy of Problems with Fast Parallel Algorithms*, Information and Control 64 (1985), pp. 2-22.
5. E. Dahlhaus, *Chordale Graphen im besonderen Hinblick auf parallele Algorithmen*, Habilitation thesis, 1991, University of Bonn.
6. M. Farber, *Characterizations of Strongly Chordal Graphs*, Discrete Mathematics 43 (1983), pp. 173-189.
7. D. Fulkerson, O. Gross, *Incidence Matrices and Interval Graphs*, Pacific Journal of Mathematics 15 (1965), pp.835-855.
8. A. Gibbons, W. Rytter, *Efficient Parallel Algorithms*, Cambridge University Press, Cambridge, 1989.
9. M. Goldberg, T. Spencer, *Constructing a Maximal Independent Set in Parallel*, SIAM Journal on Discrete Mathematics 2 (1989), pp. 322-328.
10. B. Jamison, S. Olariu, *On the Semi-Perfect Elimination*, Advances in Applied Mathematics 9 (1988), pp. 364-376.
11. P. Klein, *Efficient Parallel Algorithms for Chordal Graphs*, 29. IEEE-FOCS (1988), pp. 150-161.
12. A. Lubiw, *Doubly Lexical Orderings of Matrices*, SIAM Journal on Computing 16 (1987), pp. 854-879
13. M. Middendorf, F. Pfeiffer, *On the Complexity of Recognizing Perfectly Orderable Graphs*, Discrete Mathematics 80 (1990), pp. 327-333.
14. S. Miyano, *The Lexicographically First Maximal Subgraph Problems: P-Completeness and NC-Algorithms*, Mathematical Systems Theory 22 (1989), pp. 47-73.
15. R. Paige, R. Tarjan, *Three Partition Refinement Algorithms*, SIAM Journal on Computing 16 (1987), pp. 973-989.
16. D. Rose, *Triangulated Graphs and the Elimination Process*, Journal of Mathematical Analysis and Applications 32 (1970), pp. 597-609.
17. D. Rose, R. Tarjan, G. Lueker, *Algorithmic Aspects on Vertex Elimination on Graphs*, SIAM Journal on Computing 5 (1976), pp. 266-283.
18. Y. Shiloach, U. Vishkin, *An O(log n) Parallel Connectivity Algorithm*, Journal of Algorithms 3 (1982), pp. 57-67.
19. J. Spinrad, *Doubly Lexical Ordering for Dense 0-1 Matrices*, to appear.
20. R. Tarjan, U. Vishkin, *Finding Biconnected Components in Logarithmic Parallel Time*, SIAM-Journal on Computing 14 (1984), pp. 862-874.
21. R. Tarjan, M. Yannakakis, *Simple Linear Time Algorithms to Test Chordality of Graphs, Test Acyclicity of Hypergraphs, and Selectively Reduce Acyclic Hypergraphs*, SIAM Journal on Computing 13 (1984), pp. 566-579. Addendum: SIAM Journal on Computing 14 (1985), pp. 254-255.
22. D. Welsh, M. Powell, *An Upper Bound on the Chromatic Number of a Graph and its Application to Timetabling Problems*, Computer Journal 10 (1967), pp. 85-87.

# Dually chordal graphs

Andreas Brandstädt [1], Feodor F. Dragan [2], Victor D. Chepoi [2] and Vitaly I. Voloshin [2]

[1] Universität –GH– Duisburg FB Math. FG Inf. D 47048 Duisburg Germany
[2] Dept. of Math. and Cybern. Moldavian State University Chisinau A. Mateevici str. 60 Kishinev 277009 Moldova

**Abstract.** Recently in several papers ([10],[22],[42]) independently graphs with maximum neighbourhood orderings were characterized and turned out to be algorithmically useful.

This paper gives a unified framework for characterizations of those graphs in terms of neighbourhood and clique hypergraphs which have the Helly property and whose line graph is chordal. These graphs are dual (in the sense of hypergraphs) to chordal graphs.

By using the hypergraph approach in a systematical way new results are obtained, some of the old results are generalized and some of the proofs are simplified.

## 1   Introduction

The class of chordal graphs is a by now classical and well-understood graph class which is algorithmically useful and has several interesting characterizations. In the theory of relational database schemes there are close relationships between desirable properties of database schemes, acyclicity of corresponding hypergraphs and chordality of graphs which corresponds to tree and Helly properties of hypergraphs ([2],[9],[17],[26],[33]).

Chordal graphs arise also in solving large sparse systems of linear equations ([30],[40]) and in facility location theory ([15], [35]).

Recently a new class of graphs was introduced and characterized in several papers ([10],[22],[42]) which is defined by the existence of a maximum neighbourhood ordering. These graphs appeared first in [22] and [19] (in Russian) under the name *HT–graphs* but only a part of the results has been published. [38] also introduces maximum neighbourhoods but only in connection with chordal graphs (chordal graphs with maximum neighbourhood ordering are called there *doubly chordal graphs*).

It turns out that graphs with maximum neighbourhood orderings can be characterized by hypergraph properties which are dual (in the sense of hypergraphs) to those characterizing chordal graphs. Thus graphs with maximum neighbourhood orderings are in this sense dual to chordal graphs (therefore we call them *dually chordal graphs*) but have very different properties – thus they are in general not perfect and not closed under induced subgraphs.

Graphs which have a maximum neighbourhood ordering are a generalization of strongly chordal graphs (a well-known subclass of chordal graphs for which not only a maximum neighbourhood but a linear ordering of neighbourhoods of neighbours is required – this leads to the fact that strongly chordal graphs are exactly the hereditary dually chordal graphs i.e. graphs for which each induced subgraph is a dually chordal graph). Observe also that doubly chordal graphs are nothing else than chordal and dually chordal graphs.

Maximum neighbourhood orderings are also algorithmically useful, especially for domination-like problems and problems which base on distances. Many problems remaining NP–complete on chordal graphs have efficient algorithms on strongly chordal graphs. In some cases this is due to the existence of maximum neighbours (and not to chordality). Therefore many problems efficiently solvable for strongly chordal and doubly chordal graphs remain polynomial-time solvable for dually chordal graphs too. In a forthcoming paper of the authors of this paper this will be treated systematically.

Graphs with maximum neighbourhood ordering seem to represent an important supplement of the world of classical graph classes.

One of our theorems shows that a graph has a maximum neighbourhood ordering iff the neighbourhood hypergraph of G is a *hypertree*, i.e. it has the Helly property and its line graph is chordal. Due to the self–duality of neighbourhood hypergraphs this is also equivalent to the $\alpha$-acyclicity of the hypergraph which implies a linear time recognition of the graph class. This contrasts with the fact that the best known recognition algorithms for strongly chordal graphs have complexity $O(|E|log|V|)$ [39] and $O(|V|^2)$ [41].

There are several interesting generalizations of this class. Theorem 7 shows that a graph G has a maximum neighbourhood ordering iff the clique hypergraph (or the disk hypergraph) of G has the Helly property and its line graph is chordal. It is known from [8],[20] that G is a disk–Helly graph (i.e. a graph whose disk hypergraph has the Helly property) iff G is a dismantlable clique–Helly graph, and in [6] it is shown that G is an absolute reflexive retract iff G is a dismantlable clique–Helly graph. Thus dually chordal graphs are properly contained in the classes of disk–Helly and clique– Helly graphs.

One can substitute the Helly property also by weaker properties demanding nonempty intersection not for all but only for subsystems of bounded size k. This is done for k=3 and chordal graphs in [38], which leads to the chordal pseudo–modular graphs of [7].

The paper is organized as follows. In sections 2 and 3 we give standard hypergraph notions and properties. There we define some types of hypergraphs associated with graphs and show some helpful properties of these hypergraphs. Section 4 is devoted to characterizations of graphs with different types of maximum neighbourhood orderings via hypergraph properties. Section 5 deals with doubly chordal graphs and strongly chordal graphs. In section 6 we return from hypergraphs to graphs. In section 7 some results confirming the duality between chordal graphs and dually chordal graphs are established.

We conclude with two diagrams which presents relationships between classes of graphs, hypergraphs and some bipartite graphs.

## 2  Standard hypergraph notions and properties

We mainly use the hypergraph terminology of Berge [11]. A finite hypergraph $\mathcal{E}$ is a family of nonempty subsets (the *edges* of $\mathcal{E}$) from some finite underlying set V (the *vertices* of $\mathcal{E}$). The *subhypergraph* induced by a set $A \subseteq V$ is the hypergraph $\mathcal{E}_A$ defined on A by the edge set $\mathcal{E}_A = \{e \cap A : e \in \mathcal{E}\}$. The *dual hypergraph* $\mathcal{E}^*$ has $\mathcal{E}$ as its vertex set and $\{e \in \mathcal{E} : v \in e\}$ $(v \in V)$ as its edges. The *2–section graph* $2SEC(\mathcal{E})$ of the hypergraph $\mathcal{E}$ has vertex set V and two distinct vertices are adjacent iff they are contained in a common edge of $\mathcal{E}$. The *line graph* $L(\mathcal{E}) = (\mathcal{E}, E)$ of $\mathcal{E}$ is the intersection graph of $\mathcal{E}$, i.e. $ee' \in E$ iff $e \cap e' \neq \emptyset$.

A hypergraph $\mathcal{E}$ is *reduced* iff no edge $e \in \mathcal{E}$ is contained in another edge of $\mathcal{E}$.

A hypergraph $\mathcal{E}$ is *conformal* iff every clique C in $2SEC(\mathcal{E})$ is contained in an edge $e \in \mathcal{E}$. A *Helly hypergraph* is one whose edges satisfy the Helly property, i.e. any subfamily $\mathcal{E}' \subseteq \mathcal{E}$ of pairwise intersecting edges has a nonempty intersection.

First we give a list of well-known properties of hypergraphs (for these and other properties cf. [11]):

(i) Taking the dual of a hypergraph twice is isomorphic to the hypergraph itself i.e. $(\mathcal{E}^*)^* \sim \mathcal{E}$.

(ii) $L(\mathcal{E}) \sim 2SEC(\mathcal{E}^*)$

(iii) $\mathcal{E}$ is conformal iff $\mathcal{E}^*$ has the Helly property.

A hypergraph $\mathcal{E}$ is a *hypertree* iff there is a tree T with vertex set V such that every edge $e \in \mathcal{E}$ induces a subtree in T.

A hypergraph $\mathcal{E}$ is a *dual hypertree* iff there is a tree T with vertex set $\mathcal{E}$ such that for all vertices $v \in V$ $T_v = \{e \in \mathcal{E} : v \in e\}$ induces a subtree of T. Observe that $\mathcal{E}$ is a hypertree iff $\mathcal{E}^*$ is a dual hypertree.

A sequence $C = (e_1, e_2, ..., e_k, e_1)$ of edges is a *hypercycle* iff $e_i \cap e_{i+1(mod k)} \neq \emptyset$ for $1 \le i \le k$. The *length* of C is k. A *chord* of the hypercycle C is an edge e with $e_i \cap e_{i+1(mod k)} \subseteq e$ for at least three indices $i$, $1 \le i \le k$. A hypergraph $\mathcal{E}$ is $\alpha - acyclic$ iff it is conformal and contains no chordless hypercycles of length at least 3. (Note that the notion of $\alpha - acyclicity$ was introduced in [9] in a different way but the notion give above is equivalent to that given in [9] (cf. [31]).

In a similar way, a graph G is *chordal* iff it does not contain any induced (chordless) cycles of length at least 4.

## Theorem 1

(i) ([24], [28]) $\mathcal{E}$ is a hypertree iff $\mathcal{E}$ is a Helly hypergraph and its line graph $L(\mathcal{E})$ is chordal.

(ii) ([9],[26],[33],[31]) $\mathcal{E}$ is a dual hypertree iff $\mathcal{E}$ is $\alpha - acyclic$.

Because of the dualities between hypertrees and dual hypertrees, the conformality and the Helly property, the line graph of a hypergraph and the 2-section graph of the dual hypergraph, Theorem 1 can be expressed also in many other variants by switching between a property and its dual.

A particular instance of hypertrees are totally balanced hypergraphs. A hypergraph is *totally balanced* iff every cycle of length greater than two has an edge containing at least three vertices of the cycle.

**Theorem 2** *[36] A hypergraph $\mathcal{E}$ is totally balanced iff every subhypergraph of $\mathcal{E}$ is a hypertree.*

There is a close connection between totally balanced hypergraphs, strongly chordal graphs and chordal bipartite graphs ([1], [27], [13]); see [12] for a systematic treatment of these relations. Motivated by these results, we will establish similar connections between hypertrees, dually chordal graphs and some classes of bipartite graphs.

Hypergraphs can be represented in a natural way by incidence matrices:

Let $\mathcal{E} = \{e_1, ..., e_m\}$ be a hypergraph and $V = \{v_1, ..., v_n\}$ be its vertex set.

The *incidence matrix* $\mathcal{IM}(\mathcal{E})$ of the hypergraph $\mathcal{E}$ is a matrix whose (i,j) entry is 1 iff $v_i \in e_j$ and 0 otherwise. The (bipartite vertex–edge) *incidence graph* $\mathcal{IG}(\mathcal{E}) = (V, \mathcal{E}, E)$ of the hypergraph $\mathcal{E}$ is a bipartite graph with vertex set $V \cup \mathcal{E}$ where two vertices $v \in V$ and $e \in \mathcal{E}$ are adjacent iff $v \in e$.

Note that the transposed matrix $\mathcal{IM}(\mathcal{E})^T$ is the incidence matrix of the dual hypergraph $\mathcal{E}^*$, while $\mathcal{IG}(\mathcal{E}) \sim \mathcal{IG}(\mathcal{E}^*)$ if the sides of the bipartite graph are not marked.

Following [37] a matrix M is in *doubly lexical order* iff rows and columns as 0-1-vectors are in increasing order.

Two rows $r_1 < r_2$ and columns $c_1 < c_2$ form a $\Gamma$ iff the crossing points of these rows and columns define the submatrix $\begin{bmatrix} 1 & 1 \\ 1 & 0 \end{bmatrix}$.

An ordered 0-1 matrix M is *supported* $\Gamma$ iff for every pair $r_1 < r_2$ of rows and pair $c_1 < c_2$ of columns which form a $\Gamma$ there is a row $r_3 > r_2$ with $M(r_3, c_1) = M(r_3, c_2) = 1$ ($r_3$ *supports* $\Gamma$).

A *subtree matrix* is the incidence matrix of a collection of subtrees of a tree T. A *totally balanced matrix* is the incidence matrix of a totally balanced hypergraph.

**Theorem 3** *Let M be a 0-1 matrix.*

*(i) [37] M is a subtree matrix iff it has a supported $\Gamma$-ordering.*

*(ii) [1], [34], [37] M is a totally balanced matrix iff it has a $\Gamma$-free ordering.*

(Part (i) of this Theorem yields not only a matrix characterization of chordal graphs but due to the duality shown later also a matrix characterization of dually chordal graphs by transposing the incidence matrix).

# 3 From graphs to hypergraphs

Let $G=(V,E)$ be a finite connected simple (i.e. without loops and multiple edges) and undirected graph. For two vertices $x, y \in V$ the *distance* d(x,y) is the length of a shortest path connecting x and y. Let
$$I(x,y) = \{v \in V : d(x,v) + d(v,y) = d(x,y)\}$$
be the *interval* between vertices x and y.
By $N(v) = \{u : uv \in E\}$ we denote the *open neighbourhood* of v. By $N[v] = N(v) \cup \{v\}$ we denote the *closed neighbourhood* of v.
Let $\mathcal{N}^0(G) = \{N(v) : v \in V\}$ be the *open neighbourhood hypergraph* of G and
let $\mathcal{N}(G) = \{N[v] : v \in V\}$ be the *closed neighbourhood hypergraph* of G.
Let $\mathcal{C}(G) = \{C : C \text{ is a maximal clique in } G\}$ be the *clique hypergraph* of G.
It is easy to see that the following holds::

(i) $2SEC(\mathcal{C}(G))$ is isomorphic to G (and thus $\mathcal{C}(G)$ is conformal).

(ii) $(\mathcal{N}(G))^*$ is isomorphic to $\mathcal{N}(G)$ (where it is assumed that the hypergraph $\mathcal{N}(G) = \{N[v] : v \in V\}$ is a multiset) and the same holds for $\mathcal{N}^0(G)$.

Concerning clique hypergraphs of chordal graphs, from Theorem 1 we have the following well-known equivalence:

(iii) A graph G is chordal iff its clique hypergraph $\mathcal{C}(G)$ is $\alpha$-acyclic iff $\mathcal{C}(G)$ is a dual hypertree.

Let v be a vertex of G. The *disk* centered at v with radius k is the set of all vertices having distance at most k to v: $N^k[v] = \{u : u \in V \text{ and } d(u,v) \le k\}$. Denote by $\mathcal{D}(G) = \{N^k[v] : v \in V, k \text{ a positive integer}\}$ the *disk hypergraph* of G.
Following [5],[6] the sets $HD_{odd}(v) = \{u \in V : d(u,v) \le k \text{ and d(u,v) is odd}\}$ and $HD_{even}(v) = \{u \in V : d(u,v) \le k \text{ and d(u,v) is even}\}$ are called the *half-disks* centered at v with radius k. By $\mathcal{HD}(G)$ we denote the family of all half-disks of G and call it the *half-disk hypergraph* of the graph G.

For bipartite graphs $B = (X,Y,E)$ there are also standard hypergraph constructions: $\mathcal{N}^X(B) = \{N(y) : y \in Y\}$ denotes the *X-sided neighbourhood hypergraph* of B (analogously define $\mathcal{N}^Y(B)$).
Note that $(\mathcal{N}^X(B))^*$ is isomorphic to $\mathcal{N}^Y(B)$ and the same for X and Y exchanged. In addition, $\mathcal{N}^0(B) = \mathcal{N}^X(B) \cup \mathcal{N}^Y(B)$.
The half-disks of a bipartite graph B are defined as follows:
For $z \in X$ let $HD^X(z,k) = \{x : x \in X \text{ and } d(z,x) \le k \text{ and } d(z,x) \text{ even}\}$ and for $z \in Y$ let $HD^X(z,k) = \{x : x \in X \text{ and } d(z,x) \le k \text{ and } d(z,x) \text{ odd}\}$ (the half-disks in X). Analogously define the half-disks in Y. The *half-disk hypergraph* $\mathcal{HD}(B)$ of the bipartite graph B splits into two components $\mathcal{HD}^X(B) = \{HD^X(y, 2k+1) : y \in Y \text{ and k a positive integer}\}$

$\} \cup \{ HD^X(x, 2k) : x \in X$ and k a positive integer $\}$ called the *X-sided half-disk hypergraph* (consisting of subsets of X) and $\mathcal{HD}^Y(B)$ (defined analogously) called the *Y-sided half-disk hypergraph* (consisting of subsets of Y) i.e. $\mathcal{HD}(B) = \mathcal{HD}^X(B) \cup \mathcal{HD}^Y(B)$.

A bipartite graph B=(X,Y,E) is called *X-conformal* [2] iff for any set $S \subseteq Y$ with the property that all vertices of S have pairwise distance 2 there is a vertex $x \in X$ with $S \subseteq N(x)$.
B is *X-chordal* [2] iff for every cycle C in B of length at least 8 there is a vertex $x \in X$ which is adjacent to at least two vertices in C whose distance in C is at least 4 (a *bridge vertex*).
Analogously define Y-chordality and Y–conformality.
In [2] it is also shown that the following connections hold (which justifies the use of the corresponding notions):

**Proposition 1** *[2] Let B=(X,Y,E) be a bipartite graph. Then*

(i) *B is X-chordal iff $2SEC(\mathcal{N}^Y(B))$ is chordal.*

(ii) *B is X-conformal iff the hypergraph $\mathcal{N}^Y(B)$ is conformal.*

*Thus B is X-chordal and X-conformal iff $\mathcal{N}^Y(B)$ is a dual hypertree iff $\mathcal{N}^X(B)$ is a hypertree.*

An analogous proposition holds if X and Y are permuted.
An important standard bipartite graph construction which we use throughout this paper is the following:
Let $G = (V, E)$ be a graph. The *bigraph* $B(G) = (V', V'', E')$ with
$V' = \{v' : v \in V\}$, $V'' = \{v'' : v \in V\}$,
$E' = \{\{v', v''\} : v \in V\} \cup \{\{u', v''\} : \{u, v\} \in E\} \cup \{\{u'', v'\} : \{u, v\} \in E\}$
is the (vertex – closed-neighbourhood) incidence graph of G, i.e. $B(G) = \mathcal{IG}(\mathcal{N}(G))$. In a similar way we define the bipartite graph $B_C(G) = \mathcal{IG}(\mathcal{C}(G))$. From Proposition 1 and the definitions we have

**Proposition 2** *For a graph $G = (V, E)$ the following conditions are equivalent:*

(i) *$\mathcal{N}(G)$ is conformal*

(ii) *$\mathcal{N}^Y(B(G))$ is conformal*

(iii) *$B(G)$ is X-conformal*

(iv) *$B(G)$ is X-conformal and Y-conformal.*

Because of the symmetry in B(G) X and Y can again be exchanged.

**Proposition 3** *If $\mathcal{N}(G)$ is conformal then $(\mathcal{C}(G))^*$ is so. In particular, the X–conformality of the graph $B(G)$ implies the X-conformality of the graph $B_C(G)$.*

**Proposition 4** *For a graph G=(V,E) the following conditions are equivalent:*

(i) *$L(\mathcal{N}(G))$ is chordal,*

(ii) *$L(\mathcal{D}(G))$ is chordal,*

(iii) *$L(\mathcal{N}^Y(B(G)))$ is chordal.*

**Proposition 5** *If $L(\mathcal{C}(G))$ is chordal then also $L(\mathcal{N}(G))$ is chordal. Furthermore the X-chordality of $B_C(G)$ implies the X—chordality of $B(G)$.*

The proofs of the last two propositions base on the subsequent lemmas from [22]. Let a maximal induced cycle of G be an induced cycle of G with a maximum number of edges. Let $l(G)$ be the number of edges of a maximal induced cycle of G.

**Lemma 1** *For any graph G $l(L(\mathcal{D}(G))) = l(L(\mathcal{N}(G)))$ holds.*

**Lemma 2** *For any graph G $l(L(\mathcal{N}(G))) \leq l(L(\mathcal{C}(G)))$ holds.*

From Propositions 2 and 4 it follows

**Corollary 1** *$\mathcal{N}(G)$ is a hypertree iff $\mathcal{N}^Y(B(G))$ is a hypertree.*

Lemma 1 gives a connection between the closed neighbourhood and the disk hypergraphs of a given graph. The next lemma establishes a similar connection between the open neighbourhood hypergraph and the half–disk hypergraph of a graph G.

**Lemma 3** *For any graph G $l(L(\mathcal{HD}(G))) = l(L(\mathcal{N}^0(G)))$ holds.*

Thus we have

**Proposition 6** *For any graph G the following conditions are equivalent:*

(i) $L(\mathcal{N}^0(G))$ is chordal,

(ii) $L(\mathcal{HD}(G))$ is chordal.

# 4 Maximum neighbourhood orderings of graphs and bipartite graphs

Let G=(V,E) be a graph. A vertex $v \in V$ is *simplicial* in G iff N[v] is a clique in G. Let $G_i = G(\{v_i, v_{i+1}, ..., v_n\})$ be the subgraph induced by $\{v_i, v_{i+1}, ..., v_n\}$ and $N_i[v]$ be the closed neighbourhood of v in $G_i$.
A linear ordering $(v_1, ..., v_n)$ of V is a *perfect elimination ordering* of G iff for all $i \in \{1, ..., n\}$ $N_i[v_i]$ is a clique i.e. $v_i$ is simplicial in $G_i$.
It is known from [29] that a graph G is chordal iff G has a perfect elimination ordering. Moreover, every non–complete chordal graph has two nonadjacent simplicial vertices (see [30]).
A vertex $u \in N[v]$ is a *maximum neighbour of v* iff for all $w \in N[v]$ $N[w] \subseteq N[u]$ holds (note that u=v is not excluded).
A linear ordering $(v_1, v_2, ..., v_n)$ of V is a *maximum neighbourhood ordering* of G ([10]) iff for all $i \in \{1, ..., n\}$ there is a maximum neighbour $u_i \in N_i[v_i]$:
$$\text{for all } w \in N_i[v_i] \; N_i[w] \subseteq N_i[u_i] \text{ holds.}$$
Note that graphs with maximum neighbourhood orderings are in general not perfect:
Let G=(V,E) be any graph and $x \notin V$ be a new vertex.
Then for $G' = (V \cup \{x\}, E \cup \{vx : v \in V\})$ the ordering $(v_1, ...v_n, x)$ is a maximum neighbourhood ordering . Thus e.g. the $C_5$ with an additional dominating vertex (the wheel $W_5$) has a maximum neighbourhood ordering and is not perfect.
Now let B=(X,Y,E) be a bipartite graph. A vertex $y \in N(x)$ is a *maximum neighbour of x* iff for all $y' \in N(x)$ $N(y') \subseteq N(y)$ holds.
Let $B_i^Y = B(X \cup \{y_i, y_{i+1}, ..., y_n\})$ and $N_i(x)$ be the neighbourhood of $x \in X$ in $B_i^Y$.
A linear ordering $(y_1, ..., y_n)$ of Y is a *maximum X–neighbourhood ordering* of B ([10]) iff for all $i \in \{1, ..., n\}$ there is a maximum neighbour $x_i \in N(y_i)$ of $y_i$:
$$\text{for all } x \in N(y_i) \; N_i(x) \subseteq N_i(x_i) \text{ holds.}$$
Analogously define a *maximum Y–neighbourhood ordering* .
It is easy to see that the following holds:

**Proposition 7** *Let G=(V,E) be a graph. Then G has a maximum neighbourhood ordering iff B(G) has a maximum X–neighbourhood ordering (maximum Y–neighbourhood ordering ).*

There are other notions similar to maximum neighbourhood orderings called extremal ([22]) and b-extremal orderings ([23]):

Let G=(V,E) be a graph. A vertex $v \in V$ is *extremal* iff there is a vertex $u \in V$ such that $N^2[v] = N^1[u]$.

A linear ordering $(v_1, ..., v_n)$ of V is an *extremal ordering* of G iff for all $i \in \{1, ..., n\}$ $v_i$ is extremal in $G_i$.

Obviously, for any graph G a linear ordering $(v_1, ..., v_n)$ is a maximum neighbourhood ordering of G iff it is an extremal ordering of G.

In a graph G a vertex v is *dominated* by another vertex $u \neq v$ iff $N(v) \subseteq N(u)$.

A vertex v is *b-extremal* iff it is dominated by another vertex and there exists a vertex w such that $N(N(v)) = N(w)$.

The ordering $(v_1, ..., v_n)$ of $X \cup Y$ is a *b-extremal ordering* of G iff for all $i \in \{1, ..., n\}$ $v_i$ is b-extremal in $G_i$.

**Lemma 4** *If the graph G=(V,E) has an ordering $(v_1, ..., v_n)$ of V such that for all i=1,...,n $v_i$ is dominated in $G_i$ then G is bipartite.*

As follows from results of [23] and [10] a bipartite graph B=(X,Y,E) has a b-extremal ordering iff B has a maximum X–neighbourhood ordering and a maximum Y–neighbourhood ordering (cf. also Theorem 9).

Now to a characterization of bipartite graphs with maximum X–neighbourhood ordering .

**Theorem 4** *[10] Let B=(X,Y,E) be a bipartite graph. Then the following conditions are equivalent:*

(i) *B has a maximum X–neighbourhood ordering .*

(ii) *B is X-chordal and X-conformal.*

*Furthermore $(y_1, ..., y_n)$ is a maximum X–neighbourhood ordering of G iff $(y_1, ..., y_n)$ is a perfect elimination ordering of $2SEC(\mathcal{N}^Y(G))$.*

**Theorem 5** *Let B=(X,Y,E) be a bipartite graph. Then the following conditions are equivalent:*

(i) *B has a maximum X–neighbourhood ordering .*

(ii) $\mathcal{N}^X(B)$ *is a hypertree.*

(iii) *The X-sided half-disk hypergraph $\mathcal{HD}^X(B)$ is a hypertree.*

Now to graphs with maximum neighbourhood ordering .

**Theorem 6** *[22] Let G=(V,E) be a graph. Then the following conditions are equivalent:*

(i) *G has a maximum neighbourhood ordering .*

(ii) $\mathcal{N}(G)$ *is a dual hypertree.*

(iii) $\mathcal{N}(G)$ *is a hypertree.*

In [43] a linear time algorithm for recognizing $\alpha$–acyclicity of a hypergraph is given. Since dual hypertrees are exactly the $\alpha$– acyclic hypergraphs by Theorem 6 we have

**Corollary 2** *It can be recognized in linear time $O(|V| + |E|)$ whether a graph G has a maximum neighbourhood ordering .*

**Theorem 7** *[22] For a graph G the following conditions are equivalent:*

(i) *G has a maximum neighbourhood ordering*

(ii) *there is a spanning tree T of G such that any maximal clique of G induces a subtree in T*

(iii) *there is a spanning tree T of G such that any disk of G induces a subtree in T.*

From Theorem 7 it also follows that G has a maximum neighbourhood ordering iff $C(G)$ is a hypertree. Recall that the graph G is chordal iff $(C(G))^*$ is a hypertree. Thus graphs with maximum neighbourhood ordering are dual to chordal graphs in this sense. Therefore we call them *dually chordal graphs*. The further results will confirm this term and will show the deepness of this duality. Note that unlike for chordal graphs where the number of maximal cliques is linearly bounded, this is not the case for dually chordal graphs.
Furthermore from Theorem 7 it follows that G has a maximum neighbourhood ordering iff $\mathcal{D}(G)$ is a hypertree.

# 5 Doubly chordal and strongly chordal graphs

A vertex v of a graph G is *simple* [27] iff the set $\{N[u] : u \in N[v]\}$ is totally ordered by inclusion. A linear ordering $(v_1, ..., v_n)$ of V is a *simple elimination ordering of G* iff for all $i \in \{1, ..., n\}$ $v_i$ is simple in $G_i$. A graph is *strongly chordal* iff it admits a simple elimination ordering . A *k-sun* [13], [16], [27] is a graph with 2k vertices for some $k \geq 3$ whose vertex set can be partitioned into two sets $U = \{u_1, u_2, ..., u_k\}$ and $W = \{w_1, w_2, ..., w_k\}$ such that U induces a complete graph, W forms an independent set, and $u_i$ is adjacent to $w_j$ iff i=j or $i = j + 1 (mod k)$.

**Corollary 3** *For a graph G the following conditions are equivalent:*

(i) *G is a strongly chordal graph*

(ii) *G is a sun-free chordal graph*

(iii) *G is a hereditary dually chordal graph, i.e. any induced subgraph of G is dually chordal*

The equivalence of (iii) with (i) and (ii) was observed already in [19].
The *k-th power* $G^k, k \geq 1$ of G has the same vertices as G, and two distinct vertices are joined by an edge in $G^k$ iff their distance in G is at most k.

**Corollary 4** *[22] Any power of a dually chordal graph is dually chordal.*

By Proposition 3 conformality of $\mathcal{N}(G)$ implies conformality of $(C(G))^*$. Moreover in [19], [20] it has been shown that that for chordal graphs $\mathcal{N}(G)$ is a Helly hypergraph iff $C(G)$ is so. By Lemma 2 we know that $L(\mathcal{N}(G))$ is chordal if $L(C(G))$ is chordal. The following result shows that for chordal graphs the converse is also true.

**Lemma 5** *[22] For a chordal graph G the following conditions are equivalent:*

(i) $G^2 = L(\mathcal{N}(G))$ *is chordal*

(ii) $L(C(G))$ *is chordal*

A graph is *power-chordal* iff all of its powers are chordal. For the next theorem we need the following lemma:

**Lemma 6** *[22] Let G be a non-complete graph. If both graphs G and $G^2$ are chordal then there exist two non-adjacent vertices of G which are simplicial in G and $G^2$.*

**Theorem 8** *[22] For a graph G the following conditions are equivalent:*

*(i) G is power–chordal*

*(ii) G and $G^2$ are chordal*

*(iii) there exists a common perfect elimination ordering of G and $G^2$ (i.e. an ordering $(v_1, ..., v_n)$ of V such that $v_i$ is simplicial in both graphs $G_i$ and $G_i^2$)*

Let G be a graph. A vertex v of G is *doubly simplicial* [38] iff v is simplicial and has a maximum neighbour. A linear ordering $(v_1, ..., v_n)$ of V is a *doubly perfect ordering* iff for all $i \in \{1, ..., n\}$ $v_i$ is a doubly simplicial vertex of $G_i$. A graph G is *doubly chordal* [38] iff it admits a doubly perfect ordering. The following result justifies the term "doubly chordal graphs".

**Corollary 5** *[22], [38] For a graph G the following conditions are equivalent:*

*(i) G is doubly chordal*

*(ii) G is chordal and dually chordal*

*(iii) both hypergraphs $C(G)$ and $(C(G))^*$ are hypertrees*

From these results we conclude that powers of doubly chordal graphs are doubly chordal. Note also that for strongly chordal graphs a similar result was shown in [37]: powers of strongly chordal graphs are strongly chordal.

**Theorem 9** *For a graph G the following conditions are equivalent:*

*(i) $\mathcal{N}^0(G)$ is a hypertree*

*(ii) $\mathcal{HD}(G)$ is a hypertree*

*(iii) [10] G is bipartite and G has a maximum X–neighbourhood ordering and a maximum Y–neighbourhood ordering*

*(iv) [23] G has a b–extremal ordering*

Recall [32] that a graph G is *chordal bipartite* if G is bipartite and any induced cycle of G has length 4.

**Corollary 6** *[23] For a graph the following conditions are equivalent:*

*(i) Every induced subgraph of G admits a b–extremal ordering.*

*(ii) G is a chordal bipartite graph.*

# 6    From hypergraphs to graphs

Recall that in section 2 the notion of the incidence graph $\mathcal{IG}(\mathcal{E}) = (V, \mathcal{E}, E)$ of the hypergraph $\mathcal{E}$ was given. Note that for the one–sided neighbourhood hypergraphs $\mathcal{N}^V(\mathcal{IG}(\mathcal{E})) = \mathcal{E}$ and $\mathcal{N}^{\mathcal{E}}(\mathcal{IG}(\mathcal{E})) = \mathcal{E}^*$ holds. According to Proposition 1 we have

**Proposition 8** *1) $\mathcal{IG}(\mathcal{E})$ is X-chordal iff $2SEC(\mathcal{N}^Y(\mathcal{IG}(\mathcal{E})))$ is chordal. Analogously $\mathcal{IG}(\mathcal{E})$ is Y-chordal iff $2SEC(\mathcal{N}^X(\mathcal{IG}(\mathcal{E})))$ is chordal. 2) $\mathcal{IG}(\mathcal{E})$ is X-conformal iff $\mathcal{N}^Y(\mathcal{IG}(\mathcal{E}))$ is conformal. Analogously $\mathcal{IG}(\mathcal{E})$ is Y-conformal iff $\mathcal{N}^X(\mathcal{IG}(\mathcal{E}))$ is conformal.*

**Corollary 7** *1) $\mathcal{E}$ is a hypertree iff $\mathcal{IG}(\mathcal{E})$ has a maximum X–neighbourhood ordering*
*2) $\mathcal{E}^*$ is a hypertree iff $\mathcal{IG}(\mathcal{E})$ has a maximum Y–neighbourhood ordering*

There is yet another way to construct graphs from hypergraphs. Let B=(X,Y,E) be a bipartite graph. Then a graph $split_X(B)=(X, Y, E_X)$ with $E_X = E \cup \{xx' : x, x' \in X\}$ is obtained from B by completing X to a clique.
(assume that X is a maximal clique in $split_X(B)$ i.e. for no $y \in Y$ $X \subseteq N(y)$).
Note that the set of maximal cliques in $split_X(B)$ is
$C(split_X(B)) = \{\{y\} \cup N(y) : y \in Y\} \cup \{X\}$ (this is a set very similar to $\mathcal{N}^X(B)$).

**Proposition 9** *1) $\mathcal{N}^X(B)$ has the Helly property iff $C(split_X(B))$ has the Helly property (analogously for Y instead of X).*
*2) $L(\mathcal{N}^X(B))$ is chordal iff $L(C(split_X(B)))$ is chordal.*

Thus also $\mathcal{N}^X(B))$ is a hypertree iff $C(split_X(B))$ is a hypertree.

**Corollary 8** *1) Let B=(X,Y,E) be a bipartite graph. Then B is X–chordal and X–conformal iff $split_X(B)$ is doubly chordal.*
*2) $\mathcal{E}$ is a hypertree iff $split_Y(\mathcal{IG}(\mathcal{E}))$ has a maximum neighbourhood ordering .*

# 7 The duality between chordal and dually chordal graphs

In this section we give a systematic treatment of the duality between chordal and dually chordal graphs. Summarizing the previous results we have

**Theorem 10** *Let G=(V,E) be a graph.*

*(1) G is chordal iff $B_C(G)$ has a maximum Y–neighbourhood ordering .*

*(2) G is dually chordal iff $B_C(G)$ has a maximum X–neighbourhood ordering .*

*(3) G is doubly chordal iff $B_C(G)$ has a maximum X–neighbourhood ordering and a maximum Y–neighbourhood ordering iff $B_C(G)$ has a b–extremal ordering.*

It is well–known [14] that chordal graphs are exactly the intersection graphs of subtrees of a tree. The next result shows that a dual property characterizes the class of dually chordal graphs.

**Theorem 11** *Let G=(V,E) be a graph.*

*(1) [14] G is chordal iff it is the line graph of some hypertree iff it is the 2–section graph of some $\alpha$–acyclic hypergraph*

*(2) G is dually chordal iff it is the line graph of some $\alpha$–acyclic hypergraph iff it is the 2–section graph of some hypertree iff it is the 2–section graph of paths of a tree*

*(3) G is doubly chordal iff it is the line graph of some $\alpha$– acyclic hypertree iff it is the 2–section graph of some $\alpha$–acyclic hypertree.*

As was shown in [37] the matrix $M^T M$ ($M^T$ the transpose of M) is totally balanced provided that M is so. Unfortunately a similar property does not hold for subtree matrices; see Figure 1 (Appendix).
The graph $\Gamma$ is dually chordal. So the incidence matrix $M = \mathcal{IM}(C(\Gamma))$ is a subtree matrix.
The matrix $M^T M$ is the neighbourhood matrix $M = \mathcal{IM}(\mathcal{N}(L(C(\Gamma))))$ of the clique graph

$L(\mathcal{C}(\Gamma))$ of $\Gamma$. Since $L(\mathcal{C}(\Gamma))$ is not dually chordal $M^T M$ is not a subtree matrix. Nevertheless the following is true.

**Corollary 9** *If M is a subtree matrix then so is $MM^T$.*

The graph $\Gamma$ of Figure 1 shows that the clique graph of a dually chordal graph is not necessarily dually chordal. The results below characterize the clique graphs of chordal, dually chordal and doubly chordal graphs.

Subsequently we use the following notions:
A graph G is *clique–Helly* iff $\mathcal{C}(G)$ has the Helly property. G is *Helly chordal* iff G is chordal and clique–Helly. G is *clique–chordal* iff $L(\mathcal{C}(G))$ is chordal.

**Corollary 10** *G is a Helly chordal graph iff G is the clique graph of some dually chordal graph $G'$ i.e. $G = L(\mathcal{C}(G'))$.*

**Corollary 11** *[42] G is a dually chordal graph iff G is the clique graph of some chordal graph iff G is the clique graph of some intersection graph of paths in a tree.*

The proof follows from Theorem 11 by using similar arguments as in the proof of Corollary 10.
Combining Corollaries 10 and 11 we obtain

**Corollary 12** *G is a doubly chordal graph iff G is the clique graph of some doubly chordal graph.*

Our duality results are established using the clique hypergraph $\mathcal{C}(G)$ of a graph G. The following four properties of this hypergraph play a crucial role:

- conformality of $\mathcal{C}(G)$

- chordality of $G = 2SEC(\mathcal{C}(G))$

- Helly property of $\mathcal{C}(G)$

- chordality of $L(\mathcal{C}(G))$

The conformality of $\mathcal{C}(G)$ is fulfilled for all graphs. Chordal graphs are a well–investigated class – see for instance [30]. Clique–Helly graphs are characterized in [8], [19], [20].
In different combinations the four conditions above characterize the graph classes considered in this paper:

| dually chordal | = | clique–Helly | ∩ | clique–chordal | | |
|---|---|---|---|---|---|---|
| doubly chordal | = | clique–Helly | ∩ | clique–chordal | ∩ chordal | |
| Helly chordal | = | clique–Helly | ∩ | chordal | | |
| Power–chordal | = | clique–chordal ∩ | chordal | | | |

We conclude with the hint to two diagrams (Figures 2 and 3 (Appendix)) which show the relations between graph classes and hypergraphs associated with these graphs.

# 8 Concluding remarks

We have shown the close relationship of graphs with maximum neighbourhood ordering and hypergraph properties as the Helly property and tree-like representations of maximal cliques and neighbourhoods.

Thus in the sense of hypergraph duality these graphs are dual to chordal graphs but have different properties – especially they are in general not perfect.

On the other hand maximum neighbourhood orderings turn out to be very useful for domination–like problems (see [38], [21], [10]).

In a forthcoming paper of the authors of this paper the algorithmic use of maximum neighbourhood orderings will be treated systematically.

**Acknowledgement** The authors are grateful to Dr. H.-J. Bandelt for suggestions on the structure of this paper and a lot of discussions on this topic which also led to the term "dually chordal graphs". Furthermore the first author also thanks his colleague Falk Nicolai for several useful discussions.

# References

[1] R.P. ANSTEE and M. FARBER, Characterizations of totally balanced matrices, *J. Algorithms*, 5(1984), 215–230

[2] G. AUSIELLO, A. D'ATRI, and M. MOSCARINI, Chordality properties on graphs and minimal conceptual connections in semantic data models, *Journal of Computer and System Sciences* vol. 33, (1986),179–202

[3] H.-J. BANDELT, Neighbourhood–Helly Powers, *manuscript* 1992, University of Hamburg

[4] H.-J. BANDELT, Hereditary modular graphs, *Combinatorica* 8 (2) (1988), 149–157

[5] H.-J. BANDELT, A. DÄHLMANN, and H. SCHÜTTE, Absolute retracts of bipartite graphs, *Discrete Applied Mathematics* 16 (1987), 191–215

[6] H.-J. BANDELT, M. FARBER, and P. HELL, Absolute reflexive retracts and absolute bipartite retracts, *Discrete Mathematics*, to appear

[7] H.-J. BANDELT, and H. M. MULDER, Pseudo-modular graphs, *Discrete Mathematics*, 62 (1986), 245–260

[8] H.-J. BANDELT, and E. PRISNER, Clique graphs and Helly graphs, *Journal of Combinatorial Theory B*, 51,1 (1991),34–45

[9] C. BEERI, R. FAGIN, D. MAIER, and M. YANNAKAKIS, On the desirability of acyclic database schemes, *Journal of the ACM*, 30,3 (1983), 479–513

[10] H. BEHRENDT and A. BRANDSTÄDT, Domination and the use of maximum neighbourhoods, TECHNICAL REPORT SM–DU–204, University of Duisburg 1992

[11] C. BERGE, Hypergraphs, *North Holland*, 1989

[12] A. BRANDSTÄDT, Classes of bipartite graphs related to chordal graphs, *Discrete Applied Mathematics*, 32 (1991), 51–60

[13] A. E. BROUWER, P. DUCHET, and A. SCHRIJVER, Graphs whose neighbourhoods have no special cycles, *Discrete Mathematics* 47 (1983), 177–182

[14] P. BUNEMAN, A characterization of rigid circuit graphs, *Discr. Math.*, 9 (1974), 205–212

[15] R. CHANDRASEKARAN and A. TAMIR, Polynomially bounded algorihtms for locating p–centers on a tree, *Math. Programming*, 22 (1982), 304–315

[16] G.J. CHANG and G.L. NEMHAUSER, The k–domination and k–stability problems on sun–free chordal graphs, *SIAM J. Algebraic and Discrete Methods*, 5(1984), 332–345

[17] A. D'ATRI and M. MOSCARINI, On hypergraph acyclicity and and graph chordality, *Inf. Proc. Letters*, 29,5 (1988), 271–274

[18] G.A. DIRAC, On rigid circuit graphs, *Abh. Math. Sem. Univ. Hamburg*, 25(1961), 71–76

[19] F. F. DRAGAN, Centers of graphs and the Helly property, (in Russian) Ph.D. Thesis, Moldova State University 1989

[20] F.F. DRAGAN, Conditions for coincidence of local and global minimums for the eccentricity function on graphs and the Helly property, (in Russian), *Res. in Appl. Math. and Inform. (Kishinev)*, 1990, 49–56

[21] F.F. DRAGAN, HT–graphs: centers, connected r–domination and Steiner trees, manuscript 1992

[22] F. F. DRAGAN, C. F. PRISACARU, and V. D. CHEPOI, Location problems in graphs and the Helly property (in Russian), *Discrete Mathematics, Moscow*, 4(1992), 67–73 (the full version appeared as preprint: F.F. Dragan, C.F. Prisacaru, and V.D. Chepoi, r–Domination and p–center problems on graphs: special solution methods and graphs for which this method is usable (in Russian), Kishinev State University, preprint Mold-NIINTI, N. 948–M88, 1987)

[23] F. F. DRAGAN and V. I. VOLOSHIN, Hypertrees and associated graphs, manuscript Moldova State University 1992

[24] P. DUCHET, Propriete de Helly et problemes de representation, *Colloqu. Intern. CNRS 260*, Problemes Combin. et Theorie du Graphes, Orsay, France 1976, 117–118

[25] P. DUCHET, Classical perfect graphs: an introduction with emphasis on triangulated and interval graphs, *Ann. Discr. Math.*, 21(1984), 67–96

[26] R. FAGIN, Degrees of acyclicity for hypergraphs and relational database schemes, *J. ACM*, 30 (1983), 514–550

[27] M. FARBER, Characterizations of strongly chordal graphs, *Discr. Math.*, 43 (1983), 173–189

[28] C. FLAMENT, Hypergraphes arbores, *Discrete Mathematics*, 21 (1978), 223–227

[29] D.R. FULKERSON and O.R. GROSS, Incidence matrices and interval graphs, *Pacif. J. Math.* 15 (1965), 835-855

[30] M.C. GOLUMBIC, Algorithmic Graph Theory and Perfect Graphs, *Academic Press*, New York 1980

[31] M.C. GOLUMBIC, Algorithmic aspects of intersection graphs and representation hypergraphs, *Graphs and Combinatorics*, 4 (1988), 307–321

[32] M.C. GOLUMBIC and C.F. GOSS, Perfect elimination and chordal bipartite graphs, J. Graph theory, 2(1978), 155–163

[33] N. GOODMAN and O. SHMUELI, Syntactic characterization of tree database schemes, *Journal of the ACM* 30 (1983),767–786

[34] A.J. HOFFMAN, A.W.J. KOLEN, and M. SAKAROVITCH, Totally balanced and greedy matrices, *SIAM J. Alg. Discr. Methods*, 6(1985), 721–730

[35] A.W.J. KOLEN, Duality in tree location theory, *Cah. Cent. Etud. Rech. Oper.*, 25 (1983), 201–215

[36] J. LEHEL, A characterization of totally balanced hypergraphs, *Discr. Math.*, 57 (1985), 59–65

[37] A. LUBIW, Doubly lexical orderings of matrices, *SIAM J. Comput.* 16 (1987), 854–879

[38] M. MOSCARINI, Doubly chordal graphs, Steiner trees and connected domination, *Networks*, 23(1993),59–69

[39] R. PAIGE and R.E. TARJAN, Three partition refinement algorithms, *SIAM J. Comput.*, 16(1987), 973–989

[40] D.J. ROSE, R.E. TARJAN, and G.S. LUEKER, Algorithmic aspects of vertex elimination on graphs, *SIAM J. Comput.*, 5(1976), 266–283

[41] J.P. SPINRAD, Doubly lexical ordering of dense 0–1– matrices, manuscipt 1988, to appear in *SIAM J. Comput.*

[42] J.L. SZWARCFITER and C.F. BORNSTEIN, Clique graphs of chordal and path graphs, manuscript 1992

[43] R.E. TARJAN and M. YANNAKAKIS, Simple linear time algorithms to test chordality of graphs, test acyclicity of hypergraphs, and selectively reduce acyclic hypergraphs, *SIAM J. Comput.* 13,3 (1984), 566–579

[44] V.I. VOLOSHIN, Properties of triangulated graphs (in Russian), Issledovaniye operaziy i programmirovanie (Kishinev), 1982, 24–32

**Figure 1**

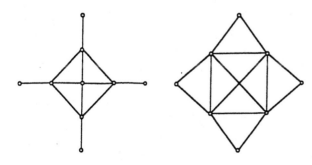

$\Gamma$            $L(C(\Gamma))$

**Figure 2**

| Bipartite graphs | Hypergraphs | Graphs |
|---|---|---|

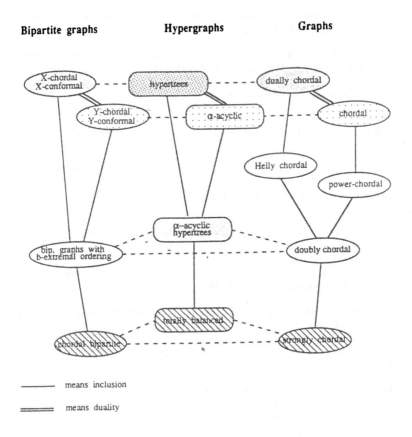

——————  means inclusion

══════  means duality

**Figure 3**

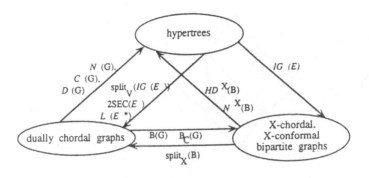

# The Size of Reduced OBDDs and Optimal Read-once Branching Programs for Almost all Boolean Functions

Ingo Wegener[*]

FB Informatik, LS II, Univ. Dortmund,
44221 Dortmund, Fed. Rep. of Germany

**Abstract.** Boolean functions are often represented by ordered binary decision diagrams (OBDDs) introduced by Bryant [2]. Liaw and Lin [7] have proved upper and lower bounds on the minimal OBDD size of almost all Boolean functions. Now tight bounds are proved for the minimal OBDD size for arbitrary or optimal variable orderings and for the minimal read-once branching program size of almost all functions. Almost all Boolean functions have a sensitivity of almost 1, i.e. the minimal OBDD size for an optimal variable ordering differs from the minimal OBDD size for a worst variable ordering by a factor of at most $1 + \varepsilon(n)$ where $\varepsilon(n)$ converges exponentially fast to 0.

## 1 Introduction

Efficient representations of Boolean functions are important e.g. for the logic design process, for test pattern generation algorithms or as part of CAD tools. Perhaps the most popular representation is the ordered binary decision diagram due to Bryant [2].

OBDDs are highly restricted BDDs or branching programs. A branching program for a Boolean function $f$ on the variables $x_1, \ldots, x_n$ can be represented by a directed acyclic graph with one source (fan-in 0) and at most two sinks (fan-out 0). Each sink is labeled by a Boolean constant, 0 or 1, and every other node (inner node) is labeled by one of the Boolean variables $x_1, \ldots, x_n$. An inner node has two outgoing edges, one labeled by 0 and the other by 1. The computation path for an input $a \in \{0,1\}^n$ starts at the source. If an inner node with label $x_i$ is reached, the outgoing edge with label $a_i$ is used. The computation path ends at a sink. The branching program represents the Boolean function $f$, if the computation path for each input $a$ ends at the sink with label $f(a)$.

In read-once branching programs computation paths containing two nodes with the same label are not allowed.

Finally, an OBDD with respect to some variable ordering is a read-once branching program where the list of the labels on a computation path is a sublist of the prescribed variable ordering. In the special case of the ordering $x_1, \ldots, x_n$, the label $x_j$ of a successor of a node with label $x_i$ has to fulfil the property $j > i$.

---

[*] Supported in part by DFG grant We 1066/7–1.

OBDDs are such a powerful representation, since reasonably many important functions have efficient OBDD representations and since the most important operations on representations of Boolean functions can be performed for OBDDs by efficient algorithms. The list of such operations contains the satisfiability test, the synthesis of two OBDDs with respect to Boolean operators, the equality test and the replacement of variables by constants or functions. The most fundamental advantage of OBDD representations is the fact that the OBDD of minimal size for a fixed Boolean function $f$ and a fixed variable ordering is unique (up to isomorphisms) and can be computed efficiently. Minimal OBDDs are also called reduced. The algorithms for the operations mentioned above can be designed in such a way (see Brace, Rudell and Bryant [1]) that they work efficiently and produce only reduced OBDDs.

Two reduction rules are sufficient for the reduction of OBDDs.

**R1** (deletion rule). If the two edges leaving some node $v$ lead to the same node $w$, the node $v$ can be deleted, i.e. all edges leading to $v$ may lead directly to $w$.

**R2** (merging rule). If two nodes $v$ and $w$ have the same label, the same 0-successor and the same 1-successor, $v$ and $w$ can be merged.

It is known (Bryant [2]) that an OBDD is reduced if and only if R1 and R2 are not applicable and all nodes are reachable from the source.

The size of the reduced OBDD for a Boolean function $f$ may obviously depend on the variable ordering. The function $x_1 x_2 \vee x_3 x_4 \vee \ldots \vee x_{2n-1} x_{2n}$ has an OBDD of size (number of inner nodes) $2n$ for the variable ordering $x_1, x_2, \ldots, x_{2n}$ while the reduced OBDD for the variable ordering $x_1, x_3, x_5, \ldots, x_{2n-1}, x_2, x_4, \ldots, x_{2n}$ has a size larger than $2^n$. It can be proved that the fraction of variable orderings which lead to a reduced OBDD of polynomial size for this function is converging fast to 0. No efficient algorithm is known which guarantees the computation of an optimal or even only a good variable ordering. Hence, it is still crucial to gain more insight in the behavior of the reduced OBDD size with respect to variable orderings. In this paper general Boolean functions or equivalently random Boolean functions or almost all Boolean functions are studied. In particular the sensitivity $s(f)$ of Boolean functions $f$ is investigated. The sensitivity of $f$ is defined as the quotient of the size of a reduced OBDD with respect to a worst variable ordering and the size of a reduced OBDD with respect to a best variable ordering.

In Section 2 the known results are presented. In Section 3 the structure of optimal OBDDs is described. This leads to a simple combinatorial description of the size of reduced OBDDs making the powerful probabilistic methods (introduced by Erdös and Spencer [4]) applicable. In Section 4 the power of the deletion rule is investigated. It turns out that the merging rule is strong enough to obtain OBDDs of almost minimal size. Probabilistic methods are used in Section 5 first to estimate the size of a single level of reduced OBDDs. A level is defined as the set of nodes with the same label. These results are generalized and tight bounds on the size of reduced OBDDs of almost all Boolean functions with respect to a fixed variable ordering are obtained. In Section 6 it is proved that the sensitivity of almost all functions is bounded by $1 + \varepsilon(n)$ where $\varepsilon(n) \to 0$ exponentially fast. The size of reduced OBDDs is almost the same for all variable orderings.

In recent time it has been shown (Sieling and Wegener [9]) that generalized OBDDs called graph driven BDDs can be used as representations for Boolean functions. Their minimal size for a Boolean function $f$ equals the minimal size of a read-once branching program for $f$. These representations of minimal size are also unique up to isomorphisms (for a given oracle which replaces the variable ordering for OBDDs) and efficient algorithms for the most important operations are known. There are functions, e.g. the hidden weighted bit function (Bryant [3]), whose minimal OBDD size is exponential even for optimal variable orderings while their minimal size with respect to graph driven BDDs is bounded by a polynomial of small degree. In Section 7 it is shown that for almost all functions the minimal size of graph driven BDDs or read-once branching programs is almost the same as for OBDDs.

## 2   Comparing the Known and the New Results

We cite the results of Liaw and Lin [7].

1. A tight upper bound for the worst case size of reduced OBDDs, $(2^n/n)(2+\varepsilon)$ is derived.
2. Starting with complete binary decision trees the merging rule contributes the reduction factor of $1/n$ in the worst case size $O(2^n/n)$ while the deletion rule contributes no more than 1% for large $n$.
3. Almost all Boolean functions require at least $2^n/(2n)$ nodes even with respect to an optimal variable ordering.
4. The sensitivity of almost all Boolean functions is bounded by $4 + \varepsilon$.

The first result is optimal and cannot be improved. The second result is improved in Section 4. It is shown that the deletion rule can reduce an OBDD where the merging rule has been applied for almost all Boolean functions only by an exponentially small amount. Also the third and fourth result are replaced by optimal results. The size of reduced OBDDs with respect to arbitrary or optimal variable orderings is calculated exactly up to terms of lower order. This leads to the conclusion that the sensitivity of almost all Boolean functions is bounded by $1 + \varepsilon(n)$ where $\varepsilon(n) \to 0$ exponentially fast, i.e. $\varepsilon(n) = O(2^{-\delta n})$ for some $\delta > 0$. Furthermore, the size of optimal read-once branching programs is calculated exactly up to terms of lower order.

## 3   On the Structure of Reduced OBDDs

In order to obtain our exact results it is necessary to discuss the structure of reduced OBDDs. Let us fix a Boolean function $f$ and the variable ordering $x_1, \ldots, x_n$. We consider a complete binary decision tree for $f$. This is a binary tree of depth $n$, the edges to left sons are labeled by 0 and the edges to right sons are labeled by 1. The nodes on level $i$ are labeled by $x_i$ and the leaves are labeled by Boolean constants in such a way that the decision tree considered as a branching program represents $f$. The node $v$ with label $x_i$ is reached for all inputs $a$ where $(a_1, \ldots, a_{i-1})$ is the labeling of the edges from the source to $v$. The decision tree with source $v$ represents the subfunction $f_{|x_1=a_1,\ldots,x_{i-1}=a_{i-1}}$ of $f$. We obtain an OBDD if we merge all 0-leaves and also all 1-leaves. By bottom-up induction on the levels of the so-constructed OBDD the following simple facts can be proved.

1. The node $v$ with label $x_i$ can be deleted if and only if the two sons $v_0$ and $v_1$ of $v$ with label $x_{i+1}$ represent the same subfunction of $f$, i.e. the subfunction represented at $v$ does not depend essentially on $x_i$.
2. The nodes $v$ and $w$ with label $x_i$ can be merged if and only if they represent the same subfunction of $f$.

This leads to the following description of the $x_i$-level of the quasi-reduced (only the merging rule is available) and the reduced OBDD.

**Theorem 1.** *The $x_i$-level of the quasi-reduced OBDD for $f$ and the variable ordering $x_1, \ldots, x_n$ contains exactly as many nodes as there are different subfunctions $f_{|x_1=a_1,\ldots,x_{i-1}=a_{i-1}}$ of $f$ for $(a_1, \ldots, a_{i-1}) \in \{0,1\}^{i-1}$. The $x_i$-level of the reduced OBDD for $f$ and the variable ordering $x_1, \ldots, x_n$ contains exactly as many nodes as there are different subfunctions $f_{|x_1=a_1,\ldots,x_{i-1}=a_{i-1}}$ of $f$ for $(a_1, \ldots, a_{i-1}) \in \{0,1\}^{i-1}$ which depend essentially on $x_i$.*

Obviously there are $2^{i-1}$ different vectors $(a_1, \ldots, a_{i-1}) \in \{0,1\}^{i-1}$. The number of different functions on $n - (i-1)$ variables equals $2^{2^{n-i+1}}$. This leads to the following corollary.

**Corollary 2.** *The $x_i$-level of a quasi-reduced OBDD for $f$ and the variable ordering $x_1, \ldots, x_n$ contains no more than*

$$W(i,n) := \min\left\{2^{i-1}, 2^{2^{n-i+1}}\right\}$$

*nodes. The total size of the quasi-reduced OBDD is bounded by $W(n) := W(1,n) + \cdots + W(n,n)$.*

For later purposes we remark that the $2^{i-1}$ subfunctions $f_{|x_1=a_1,\ldots,x_{i-1}=a_{i-1}}$ are defined on $2^{i-1}$ disjoint subsets of $\{0,1\}^n$, each of size $2^{n-i+1}$. Geometrically, the cube $\{0,1\}^n$ is partitioned into $2^{i-1}$ disjoint subcubes isomorphic to $\{0,1\}^{n-i+1}$.

It is known and easy to see that the upper bounds of Corollary 2 are best possible, i.e. there are functions whose complexity equals $W(n)$. The complexity of almost all Boolean functions has been considered for many computation models (see Wegener [10]). It happens quite often that almost all functions have the same complexity (up to terms of lower order) as the hardest function. Such an effect is called (strong) Shannon effect because of the fundamental paper [8]. The effect that almost all functions have (up to terms of lower order) the same complexity but a smaller complexity than the hardest function is called weak Shannon effect. Our results will be expressed also in terms of the strong and weak Shannon effect.

# 4 The Power of the Deletion Rule

In order to measure the power of the deletion rule we compare the size of quasi-reduced OBDDs with the size of reduced OBDDs. Quasi-reduced OBDDs are unique (up to isomorphisms), if we start with the OBDD obtained from the complete binary decision tree by merging all 0-sinks and all 1-sinks. We remember that the deletion rule can be applied to a node $v$ with label $x_i$ representing $g := f_{|x_1=a_1,\ldots,x_{i-1}=a_{i-1}}$ if and only if $g$ does not depend essentially on $x_i$, i.e. $g_{|x_i=0} = g_{|x_i=1}$.

As in the rest of this paper we argue with random Boolean functions. Each of the $2^{2^n}$ Boolean functions has the same probability of being chosen. Since $g_{|x_i=0}$ and $g_{|x_i=1}$ are defined on disjoint input sets, they are independent. The probability that these independent random functions on $n-i$ variables are the same functions equals $p_i := 2^{-2^{n-i}}$.

We like to prove results which hold for almost all Boolean functions or equivalently for a random function with overwhelming probability. First only one level of the OBDDs is investigated. The mean value and the variance of the considered random variable are computed. By Tschebyscheff's inequality (see e. g. Feller [5]) it can be shown that the random variable for the $i$-th level takes values in a small enough interval with an overwhelming probability of at least $1 - \varepsilon_i(n)$. Since the different levels of OBDDs are not independent, the probability that all random variables for all levels take values in the considered intervals is estimated below by the trivial bound $1 - \varepsilon_1(n) - \cdots - \varepsilon_n(n)$.

We have to ensure that this lower bound is large enough.

We consider a quasi-reduced OBDD with $N_i$ nodes on the $x_i$-level. The $N_i$ functions represented on this level are different, each is a random one out of the class of all Boolean functions on $n-i+1$ variables $x_i, \ldots, x_n$. We are interested in the number $X_i$ of functions which do not depend essentially on $x_i$. The corresponding experiment can be described as follows. There is an urn with $B_i := 2^{2^{n-i+1}}$ balls representing the different Boolean functions on $x_i, \ldots x_n$, among them are $W_i := p_i B_i = 2^{2^{n-i}}$ "white" balls representing the different Boolean functions not depending essentially on $x_i$. Then $N_i$ different balls are drawn. Hence, $X_i$ the random number of drawn white balls has a hypergeometrical distribution. It is known (see e. g. Feller [5]) that

$$E(X_i) = N_i p_i \qquad \text{(mean value)}$$
$$V(X_i) = N_i p_i (1 - p_i)(B_i - N_i)/(B_i - 1) \qquad \text{(variance)}.$$

It follows that $V(X_i) \leq E(X_i)$.

We recall *Tschebyscheff's inequality.* Let $X$ be a random variable with mean value $E(X)$ and variance $V(X)$. For $\varepsilon > 0$

$$\text{Prob}(|X - E(X)| \geq \varepsilon) \leq V(X)/\varepsilon^2.$$

Now we apply our approach to the random variables $X_i$. We distinguish three cases. In the top part of OBDDs the probability of white balls is so small that almost no nodes are deleted. In a small middle part the number of deleted nodes is not small but very small in comparison to the number of nodes in the quasi-reduced OBDD. In the bottom part the number of nodes in the quasi-reduced OBDD is so small that we even may assume that all nodes become deleted. We omit here the detailed description of the estimations. The proof of Theorem 7 contains more complicated calculations and is presented in detail. We obtain the following theorem.

**Theorem 3.** *Let $Q(f)$ be the quotient of the size of the quasi-reduced OBDD for $f$ and the size of the reduced OBDD for $f$ both with respect to the variable ordering $x_1, \ldots, x_n$. Then $Q(f) \leq 1 + O(2^{-n/3}n)$ for all but a fraction of $O(2^{-n/3})$ of all Boolean functions.*

Up to now we have investigated the power of the deletion rule only for a fixed variable ordering. It is still possible that for many functions there exists some ordering of the variables where the deletion rule is powerful. The number of nodes deleted on the $i$-th level depends only on the choice of the set of $i - 1$ variables tested before and the choice of the variable tested on the $i$-th level. Hence we have to distinguish only

$$\binom{n}{i-1}(n - i + 1) = O\left(\min\left\{n^i, n^{n-i+2}, n^{1/2}2^n\right\}\right)$$

choices.

For the levels of Case 1 we now choose $\varepsilon_i = 2^{2n/3}n^{-1}$. It follows that the probability that more than $n^{-1} + 2^{2n/3}n^{-1}$ nodes are deleted on the $x_i$-level for some variable ordering is bounded by $O(2^{-n/3}n^{3/2})$. For the levels of Case 2 the error probability is increased only by the maximal number of choices

$$n^{\log n + 2} = O(2^{\delta n})$$

for each $\delta > 0$. Nothing has to be changed in Case 3. Altogether we obtain the following result.

**Theorem 4.** *Let $Q^*(f)$ be the maximal quotient of the size of the quasi-reduced OBDD and the size of the reduced OBDD for $f$ and the same variable ordering where the maximum is taken over all variable orderings. Then $Q^*(f) \leq 1 + O(2^{-n/3}n)$ for all but a fraction of $O(2^{-n/3+\delta n})$ of all Boolean functions where $\delta$ is an arbitrary positive constant.*

*Remark.* Here and in the following sections it is possible to obtain even better bounds, e. g. by considering more cases. In order to keep the considerations simple we leave these improvements to the interested reader.

# 5 The Size of Reduced OBDDs for a Fixed Variable Ordering

We discuss the size of quasi-reduced OBDDs with respect to the variable ordering $x_1, \ldots, x_n$. By Theorem 3 it is easy to obtain the results on the size of reduced OBDDs.

First we consider the $x_{i+1}$-level. As already discussed the $2^i$ functions $f_{|x_1=a_1,\ldots,x_i=a_i}$ for $(a_1, \ldots, a_i) \in \{0,1\}^i$ are drawn independently from the set of all $2^{2^{n-i}}$ Boolean functions on the variables $x_{i+1}, \ldots, x_n$. Each function is chosen with respect to the uniform distribution. The number of nodes on the $x_{i+1}$-level of the quasi-reduced OBDD is equal to the number of different functions among the $2^i$ functions. We think of the $2^i$ functions $f_{|x_1=a_1,\ldots,x_i=a_i}$ as balls and of the $2^{2^{n-i}}$ functions on $x_{i+1}, \ldots, x_n$ as buckets. The balls are thrown randomly and independently into the buckets. We are interested in the random number $Y_i$ of non-empty buckets which equals the random number of nodes on the $x_{i+1}$-level of the quasi-reduced OBDD. The random number

$$W(i+1,n) - Y_i = \min\{2^i, 2^{2^{n-i}}\} - Y_i$$

describes the number of nodes saved by the merging rule compared to the maximal size of the $x_{i+1}$-level. Following our general approach we are interested in the mean value and the variance of $Y_i$.

**Lemma 5.** Let $k = 2^i$, $m = 2^{2^{n-i}}$ and $\alpha = k/m$.

i) $E(Y_i) = m \left(1 - \left(1 - \dfrac{1}{m}\right)^k\right) = m - me^{-\alpha} + \dfrac{\alpha}{2}e^{-\alpha} - O\left(\dfrac{\alpha(1-\alpha)}{m}e^{-\alpha}\right).$

ii) $V(Y_i) = me^{-\alpha}(1 - (1+\alpha)e^{-\alpha}) + O\left(\alpha(1+\alpha)e^{-\alpha}\left(e^{-\alpha} + \dfrac{1}{m}\right)\right).$

*Proof.* The proof of this lemma is due to Kolchin, Sevast'yanov and Christyakov [6].

**Theorem 6.** *Let the variable ordering $x_1, \ldots, x_n$ be fixed. The expected size of the quasi-reduced OBDD for a random Boolean function is equal to*

$$S(n) := \sum_{0 \le i \le n-1} 2^{2^{n-i}} \left(1 - \left(1 - 2^{-2^{n-i}}\right)^{2^i}\right).$$

*Proof.* The result follows directly from the Lemma by summing up the expected sizes for the different levels. □

**Theorem 7.** *The fraction of Boolean functions whose reduced OBDD size with respect to the variable ordering $x_1, \ldots, x_n$ differs more than $O(2^{2n/3})$ from $S(n)$ is bounded by $O(2^{-n/3})$. The weak Shannon effect holds for Boolean functions and reduced OBDDs with a fixed ordering of the variables.*

*Proof.* We partition the OBDD levels into five parts.

**Case 1:** The top part of the OBDD: $k^2/(2m) \to \lambda < \infty$.
By Lemma 5, the considerations above and, since $\alpha \to 0$,

$$E(Y_i) = k - \lambda + o(1)$$

$$V(Y_i) = me^{-\alpha}(1 - (1+\alpha)(1 - \alpha + \frac{\alpha^2}{2} - \ldots)) + o(1)$$

$$= m\left(1 - \alpha + \frac{\alpha^2}{2} - \ldots\right)\left(\frac{\alpha^2}{2} + O(\alpha^3)\right) + o(1)$$

$$= m\frac{\alpha^2}{2} + O(m\alpha^3) + o(1) = \frac{m\alpha^2}{2} + o(1)$$

and $V(Y_i) \to \lambda$.

The variance is bounded by a constant. By Tschebyscheff's inequality the probability that the number of nodes on such a level differs more than $2^{n/2}$ from the expected value is bounded by $O(2^{-n})$.

**Case 2:** The subcritical levels at the bottom of the top part of the OBDD: $k^2/(2m) \to \infty$ and $\alpha \to 0$.
By Lemma 5

$$E(Y_i) = m - me^{-\alpha} + o(1)$$

$$= \frac{k}{\alpha}\left(1 - 1 + \alpha - \frac{\alpha^2}{2!} + \ldots\right) + o(1)$$

$$= k\left(1 - \frac{\alpha}{2!} + \frac{\alpha^2}{3!} - \ldots\right) + o(1)$$

$$\geq k - k\alpha/2 + o(1) \ .$$

Furthermore, as in Case 1

$$V(Y_i) = m\frac{\alpha^2}{2} + O(m\alpha^3) + o(1)$$

$$= k\alpha\left(\frac{1}{2} + o(1)\right) \ .$$

Hence, the variance is smaller than $k \leq 2^n$. We apply Tschebyscheff's inequality for $\varepsilon = 2^{2n/3}$. The probability that the number of nodes on such a level differs more than $2^{2n/3}$ from its expected value is bounded by $2^n \varepsilon^{-2} = 2^{-n/3}$.

**Case 3:** The critical level of the OBDD (if existent): $\alpha = k/m \to \beta > 0$.
By Lemma 5

$$E(Y_i) = m - me^{-\alpha} + \frac{\alpha}{2}e^{-\alpha} - O\left(\frac{\alpha(1+\alpha)}{m}e^{-\alpha}\right) \ .$$

For $\alpha = \beta$

$$E(Y_i) = m - me^{-\beta} + \frac{\beta}{2}e^{-\beta} - O\left(\frac{1}{m}\right) = m(1 - e^{-\beta}) + O(1) \quad \text{and}$$

$$V(Y_i) = me^{-\beta}(1 - (1 + \beta)e^{-\beta}) + O(1) \ .$$

Since $\beta > 0$, $V(Y_i) < m = \alpha k \le 2^n$ for large $n$. Hence, Tschebyscheff's inequality can be applied as in Case 2.

**Case 4:**   The subcritical levels at the top of the bottom part of the OBDD: $\alpha \to \infty$ and $me^{-\alpha} \to \infty$.
By Lemma 5

$$E(Y_i) = m - me^{-\alpha} + o(1)$$
$$V(Y_i) = me^{-\alpha}(1 - o(1)) \ .$$

Since $\alpha \to \infty$ and $\alpha = k/m$, we have $m < k \le 2^n$ for large $n$ and $V(Y_i) \le 2^n$. Again Tschebyscheff's inequality can be applied as in Case 2.

**Case 5:**   The bottom part of the OBDD: $\alpha \to \infty$ and $me^{-\alpha} \to \lambda < \infty$.
By Lemma 5

$$E(Y_i) = m - \lambda$$
$$V(Y_i) = \lambda \ .$$

Since the variance is bounded by a constant, Tschebyscheff's inequality can by applied as in Case 1.

For the first statement of the theorem we use the results of the five cases above and the fact that only $O(1)$ levels belong to the Cases 2, 3 and 4. Furthermore we use the results of Section 4 for the step from quasi-reduced to reduced OBDDs. For the last statement we remember that $S(n) = \Theta(2^n/n)$. Hence, the probability that the reduced OBDD size is outside the interval $S(n)(1 \pm O(n2^{-n/3}))$ is bounded by $O(2^{-n/3})$. $\qquad\Box$

We finish the section with some further results and remarks. Considering the five cases we see that $S(n) \ge W(n)(1 - o(1))$ for almost all $n$. This relation does not hold only if (not necessarily if)

$$(\log m)^{-1} \le \alpha \le \log m$$

for some level. This is equivalent to

$$2^{2^{n-i}}2^{i-n} \le 2^i \le 2^{2^{n-i}}2^{n-i} \quad \text{or}$$

$$n - \log n \le i \le n - \log(2i - n)$$

This condition can be relaxed to

$$n - \log n \le i \le n - \log(n - 2\log n).$$

The length of this interval for $i$ is $\log \frac{n}{n-2\log n} = o(1)$. Hence, for a random $n \in \{2^l, \ldots, 2^{l+1} - 1\}$ the probability that $S(n) \geq W(n)(1 - o(1))$ and that the strong Shannon effect holds tends to 1 as $l \to \infty$.

By results of Kolchin, Sevast'yanov and Christyakov [6] it follows that the distribution of the random number of non-empty buckets is for the critical and subcritical levels (Cases 2, 3 and 4) normal (with respect to the mean value and the variance given in Lemma 5). For the levels of the top part the distribution of the difference between $k$ and the number of non-empty buckets is Poisson, while for the levels of the bottom part the distribution of the number of empty buckets is Poisson.

## 6  The Size of Optimal OBDDs

We have already seen in Section 4 how results on OBDDs with a fixed variable ordering can be generalized to results on OBDDs with optimal variable orderings. We change our estimations only for the top part. The probability that the number of nodes on such a level differs more than $2^{2n/3}$ from its expected value is bounded by $O(2^{-4n/3})$. The number of different situations on such a level is bounded by $O(n^{1/2}2^n)$, see Section 4. Hence, the probability that for some variable ordering the number of nodes on some level of the top part differs by more than $2^{2n/3}$ from its expected value is bounded by $O(n^{3/2}2^{-n/3})$.

For all other levels we know that $k^2 > 2m$ and therefore $i > n - \log n - 1$. The number of different situations is bounded by

$$\binom{n}{i} (n - i) \leq n^{n-i}(\log n + 1) = O(2^{\delta n})$$

for each $\delta > 0$. We obtain the following result.

**Theorem 8.** *The fraction of Boolean functions whose optimal OBDD size differs more than $O(n2^{2n/3})$ from $S(n)$ is bounded by $O(2^{-n/3+\delta n})$ for arbitrary constants $\delta > 0$. The weak Shannon effect holds for Boolean functions and OBDDs. The fraction of Boolean functions whose sensitivity is larger than $1 + O(n^2 2^{-n/3})$ is bounded by $O(2^{-n/3+\delta n})$ for $\delta > 0$.*

## 7  The Size of Optimal Read Once Branching Programs

Functions with small read-once branching program complexity can be represented by small graph driven BDDs (see Sieling and Wegener [9]). Hence, we are interested whether this new approach helps not only for many important functions ([9]) but for a positive fraction of all Boolean functions.

In order to keep the calculations simple we do not try to prove the best possible result and are satisfied with the following theorem.

**Theorem 9.** *For a random $n \in \{2^l, \ldots, 2^{l+1} - 1\}$ the probability that the following property P holds tends to 1 as $l \to \infty$.*
*P: the fraction of Boolean functions whose minimal read-once branching program size is less than $W(n)(1-o(1))$ is bounded by $O(2^{-n/3+\delta n})$ for an arbitrary $\delta > 0$ and the strong Shannon effect holds.*

*Proof.* By the application of the inverse deletion rule and the inverse merging rule each read-once branching program can be unfolded into a complete read-once decision tree of depth $n$. By first applying the inverse deletion rule we can ensure that the nodes created by the inverse merging rule lie on the same level of the decision tree. Afterwards we may obtain back the given read-once branching program using first the merging rule and then the deletion rule.

If a node $v$ becomes deleted during this process, the subfunction realized at $v$ does not depend essentially on that variable which is the label of $v$. Hence, there is some variable ordering for which a node representing the same subfunction and being labeled by the same variable as $v$ becomes deleted. In Section 4 we have shown that the number of nodes which may be deleted is bounded by $O(2^{2n/3})$ for all but a fraction of $O(2^{-n/3+\delta n})$ of all Boolean functions.

If the nodes $v$ and $w$ become merged, they represent the same subfunction and they are labeled by the same variable. Because of the read-once property and since the nodes are on the same level of the decision tree, there is a variable ordering where a node of the quasi-reduced OBDD represents the situation represented here by $v$ and $w$. Hence, there is a variable ordering where we count this merging. In Section 5 we have shown that the number of mergings does not differ for most functions much from the expected value. For the read-once decision tree considered here the variable orderings may be different for the different paths. Hence, it is possible that the number of mergings has to be summed up for the different situations. We distinguish again different parts of the read-once branching programs.

For the levels $i \leq n - \sqrt{n}$ we assume that all nodes are merged. The total number of nodes on these levels is $O(2^{n-\sqrt{n}})$ and small enough.

For the levels $i > n - \sqrt{n}$ which belong to the top part (Case 1 in Section 5) the number of different situations is bounded by $2^{O(\sqrt{n}\log n)}$. Using the results of Section 5 it follows that the probability that the number of nodes on such a level after the merging phase is smaller than

$$2^i - 2^{n/2+O(\sqrt{n}\log n)}$$

is bounded by $O(2^{-n+\delta n})$ for $\delta > 0$.

From the subcritical levels at the bottom of the top (Case 2 in Section 5) part we consider only those where

$$k \, 2^{(\log\log k)^3} = O(m).$$

Then, since $k \leq 2^n$,

$$k\alpha = k^2/m = O(k \, 2^{-(\log\log k)^3}) = O(2^n n^{-(\log n)^2}).$$

The number of different situations is, since $i > n - \log n - 1$ in this part, bounded by $n^{\log n + 2}$. Using the results of Section 5 it follows that the probability that the number of nodes on such a level after the merging phase is smaller than

$$2^i - O(2^n n^{-\log n})$$

is bounded by $O(2^{-n/3 + \delta n})$ for $\delta > 0$.

For the levels of the bottom part (Case 5 in Section 5) we assume that all nodes are merged. The total number of nodes on these levels is $O(2^n n^{-2})$ and small enough.

We have proved the property $P$ for all those $n$ such that we have considered all levels.

By a somehow tedious but nevertheless simple calculation it can be shown that the fraction of all $n \in \{2^l, \ldots, 2^{l+1} - 1\}$ for which we have not proved the assertion is bounded by

$$\log((n + O(\log^3 n))/(n - O(\log n))) = o(1).$$

$\square$

# References

1. K.S. Brace, R.L. Rudell and R.E. Bryant. Efficient implementation of a BDD package. In 27th ACM/IEEE Design Automation Conference, 1990, pp. 40–45.

2. R.E. Bryant. Graph-based algorithms for Boolean function manipulation. IEEE Trans. on Computers 35(8), 1986, pp. 677–691.

3. R.E. Bryant. On the complexity of VLSI implementations and graph representations with application to integer multiplication. IEEE Trans. on Computers 40(2), 1991, pp. 205–213.

4. P. Erdös and J. Spencer. Probabilistic methods in combinatorics. New York, Academic Press, 1974.

5. W. Feller. An introduction to probability theory and its applications. New York, Wiley, 1968.

6. V.F. Kolchin, B.A. Sevast'yanov and V.P. Christyakov. Random allocations. New York, Wiley, 1978.

7. H.-T. Liaw and C.-S. Lin. On the OBDD-representation of general Boolean functions. IEEE Trans. on Computers 41(6), 1992, pp. 661–664.

8. C.E. Shannon. The synthesis of two-terminal switching circuits. AT&T Bell Syst. Tech. J. 28, 1949, pp. 59–98.

9. D. Sieling and I. Wegener. Graph driven BDD's – a new data structure for Boolean functions. Submitted to Theoretical Computer Science.

10. I. Wegener. The complexity of Boolean functions. New York, Wiley, 1987.

# Regular Marked Petri Nets

Jörg Desel

Institut für Informatik, Technische Universität München,

D-80290 München

**Abstract.** A class of Petri nets called *regular marked nets* is introduced. Its definition refers to the linear algebraic representation of nets. It is shown that every regular marked net is live – i.e. no transition can get deadlocked – and bounded – i.e. its state space is finite. In turn, live and bounded marked extended free choice nets are a proper subclass of regular marked nets. A series of results concerning behavioural properties – i.e. properties of the corresponding state graph – which are known for live and bounded marked extended free choice nets are shown to hold for regular marked nets.

## 1 Introduction

Behavioural properties of marked Petri nets can be defined as properties of the corresponding state graphs. Most of such properties are decidable. Their enormous complexity, however, does not allow for an automatic analysis in general (see [Jan87] for an overwiew). For verification purposes it often suffices to consider semi-decision algorithms, which are based on necessary *or* sufficient structural conditions for behavioural properties. Prominent constructs for such conditions are place-invariants, deadlocks and traps (see [Rei85] for examples). Stronger results can only be obtained for restricted classes of Petri nets. One example is the class of conflict-free nets and derivatives (see e.g. [AF92]). Another example, which we shall refer to in the present paper, are extended free choice nets. The conjunction of liveness (lack of total or partial deadlocks) and boundedness (finiteness of the state graph) of extended free choice nets is characterized by the Rank Theorem [CCS91, Des92] which only refers to the linear algabraic representation of a net.

We shall prove the following generalization of the Rank Theorem: its conditions provide a sufficient condition for liveness and boundedness, applicable to arbitrary nets. Marked nets enjoying the properties of the Rank Theorem will be named regular. It is shown that most of the results known for live and bounded marked extended free choice nets can be transferred to regular marked nets. These are in particular

- the existence of home states, i.e. markings that can always be reached again,

- for cyclic marked nets (i.e. marked nets with strongly connected state graphs), a structural characterization of reachable markings,

- a structural characterization of the maximal number of tokens a place can get,

- a property of the state graph which states that every two connected reachable markings are connected by a path with polynomial length, although the total number of reachable markings can be exponential.

The significance of these results depend on the relative expressive power of regular marked nets. This class contains all live and bounded marked extended free choice nets. The converse does not hold: The running example, shown in Figure 1, is an example of a regular marked net which is not extended free choice. At the end of the paper we argue that regular marked nets are not just a technical generalization of extended free choice nets but have in fact more expressive power.

## 2  Definitions and Preliminaries

We use the standard notions for nets, as defined e.g. in [Rei85]. The same reference can be consulted for the basic results given in this chapter.

Let $S$ and $T$ be disjoint sets and let $F \subseteq (S \times T) \cup (T \times S)$. Then $N = (S, T, F)$ is called a *Petri net* (or just *net*), $S$ is the set of *places* of $N$ and $T$ is the set of *transitions* of $N$. We shall only consider nets with finite and nonempty sets of places and transitions. Moreover, we require that every net is *connected*, i.e. the least equivalence relation which includes $F$ is $(S \cup T) \times (S \cup T)$. *Pre-* and *post-sets* of elements are denoted by the dot-notation: $^\bullet x = \{y \mid (y, x) \in F\}$ and $x^\bullet = \{y \mid (x, y) \in F\}$.

A *marking* of $N$ is a mapping $M: S \to I\!N$. It *marks* a place $s$ if $M(s) > 0$. It *enables* a transition $t$ if it marks every place of $^\bullet t$. The *occurrence* of an enabled transition $t$ leads to the *successor marking* $M'$ (written $M \xrightarrow{t} M'$) which is defined for every place $s$ by

$$M'(s) = \begin{cases} M(s) - 1 & \text{if } s \in {}^\bullet t \setminus t^\bullet \\ M(s) + 1 & \text{if } s \in t^\bullet \setminus {}^\bullet t \\ M(s) & \text{if } s \notin {}^\bullet t \cup t^\bullet \text{ or } s \in {}^\bullet t \cap t^\bullet \end{cases}$$

If $M \xrightarrow{t_1} M_1 \xrightarrow{t_2} \cdots \xrightarrow{t_n} M_n$, then $\sigma = t_1 t_2 \ldots t_n$ is called *occurrence sequence* and we write $M \xrightarrow{\sigma} M_n$. This notion includes the empty sequence $\epsilon$: $M \xrightarrow{\epsilon} M$ for each marking $M$. We write $M \longrightarrow M'$ and call $M'$ *reachable* from $M$ if $M \xrightarrow{\sigma} M'$ for some occurrence sequence $\sigma$. $[M\rangle$ denotes the set of all markings reachable from $M$.

A marked net is a pair $(N, M)$ where $N$ is a net and $M$ is a marking of $N$. A marked net $(N, M)$ is called *live* if, for every $L \in [M\rangle$ and every $t \in T$, there exists a marking $K \in [L\rangle$ that enables $t$. $(N, M)$-is called *bounded* if, for each place $s \in S$, there is an integer $k$ such that $L(s) \leq k$ for every $L \in [M\rangle$. $(N, M)$ is called *safe* if $k = 1$ satisfies this condition for every place. $(N, M)$ is called *cyclic* if $M$ is reachable from every marking $K \in [M\rangle$.

The *state graph* of a marked net $(N, M)$ is a directed graph with the set of vertices $[M\rangle$ and the set of edges $\{(K, L) \mid K \xrightarrow{t} L \text{ for some transition } t\}$. Behavioural properties of a marked net are structural properties of its state graph, e.g. a marked net is bounded iff its state graph has a finite set of vertices, a bounded marked net is cyclic iff its state graph is strongly connected.

Figure 1 shows a marked net together with its state graph (the labels of the vertices used in the figure denote safe markings in the obvious way). Places are denoted by circles, transitions by squares and the interconnecting relation by arcs. The marking is represented by dots (*tokens*) in places. The marked net is live and safe and, hence, bounded. It is moreover cyclic.

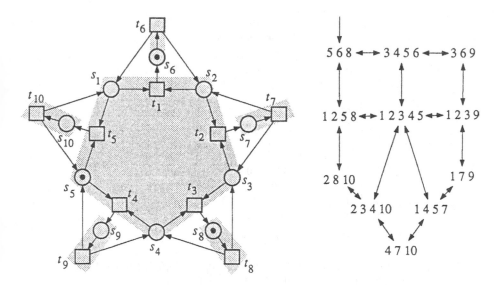

**Fig. 1**  A marked net and its state graph

Let $S$ and $T$ be arbitrarily but fixed ordered. Then every marking $M$ can be described by a vector, which we shall also denote by $M$.

The *incidence matrix* $[N]: S \times T \to \{-1, 0, 1\}$ of $N$ is defined by

$$[N](s,t) = \begin{cases} -1 & \text{if } (s,t) \in F \text{ and } (t,s) \notin F \\ 1 & \text{if } (t,s) \in F \text{ and } (s,t) \notin F \\ 0 & \text{otherwise} \end{cases}$$

Hence, the matrix entry $[N](s,t)$ denotes the change of the number of tokens on a place $s$ caused by the occurrence of a transition $t$. For every occurrence sequence $M_0 \xrightarrow{\sigma} M_1$ we obtain the following *Marking Equation* by the definition of $[N]$ :

$$M_0 + [N] \cdot x = M_1$$

where $x: T \to \mathbb{N}$ such that $x(t)$ denotes the number of occurrences of $t$ in $\sigma$.

We denote every vector $(0, 0, \ldots, 0)$ by $\mathbf{0}$ and every vector $(1, 1, \ldots, 1)$ by $\mathbf{1}$. We write $I \geq I'$ for vectors $I$ and $I'$ if the inequality holds for every component.

Every solution for $i$ of $i \cdot [N] = \mathbf{0}$ is an *S-invariant* of $N$. Salient properties of S-invariants follow immediately from the Marking Equation:

- If $M_0 \longrightarrow M_1$ then we have $I \cdot M_0 = I \cdot M_1$ for every S-invariant $I$.

- Assume that there exists an S-invariant $I \geq \mathbf{1}$ and let $s$ be a place. Then we have $M_1(s) \leq I(s) \cdot M_1(s) \leq I \cdot M_1$. Hence, for every initial marking $M$, the marked net $(N, M)$ is bounded, with the bound $k \leq I \cdot M$ for every place.

- Assume that $(N, M)$ is live and let $I > \mathbf{0}$ be an S-invariant. Then there exists a place $s$ satisfying $I(s) > 0$. Since $N$ is connected there is a transition $t \in {}^\bullet s \cup s^\bullet$. By liveness, we find markings $M_1, M_2$ such that $M \longrightarrow M_1 \xrightarrow{t} M_2$. Either $M_1(s) > 0$ or $M_2(s) > 0$. Hence $I \cdot M = I \cdot M_1 = I \cdot M_2 > 0$.

$$(0,0,0,0,1,1,0,1,0,0)$$

The marking vector

|        | $t_1$ | $t_2$ | $t_3$ | $t_4$ | $t_5$ | $t_6$ | $t_7$ | $t_8$ | $t_9$ | $t_{10}$ |
|--------|-------|-------|-------|-------|-------|-------|-------|-------|-------|----------|
| $s_1$  | −1    | 0     | 0     | 0     | −1    | 1     | 0     | 0     | 0     | 1        |
| $s_2$  | −1    | −1    | 0     | 0     | 0     | 1     | 1     | 0     | 0     | 0        |
| $s_3$  | 0     | −1    | −1    | 0     | 0     | 0     | 1     | 1     | 0     | 0        |
| $s_4$  | 0     | 0     | −1    | −1    | 0     | 0     | 0     | 1     | 1     | 0        |
| $s_5$  | 0     | 0     | 0     | −1    | −1    | 0     | 0     | 0     | 1     | 1        |
| $s_6$  | 1     | 0     | 0     | 0     | 0     | −1    | 0     | 0     | 0     | 0        |
| $s_7$  | 0     | 1     | 0     | 0     | 0     | 0     | −1    | 0     | 0     | 0        |
| $s_8$  | 0     | 0     | 1     | 0     | 0     | 0     | 0     | −1    | 0     | 0        |
| $s_9$  | 0     | 0     | 0     | 1     | 0     | 0     | 0     | 0     | −1    | 0        |
| $s_{10}$ | 0   | 0     | 0     | 0     | 1     | 0     | 0     | 0     | 0     | −1       |

The incidence matrix

$(1,1,0,0,0,1,0,0,0,0),$  $(0,1,1,0,0,0,1,0,0,0),$  $(0,0,1,1,0,0,0,1,0,0)$
$(0,0,0,1,1,0,0,0,1,0),$  $(1,0,0,0,1,0,0,0,0,1)$

A base of S-invariants

$(1,0,0,0,0,1,0,0,0,0),$  $(0,1,0,0,0,0,1,0,0,0),$  $(0,0,1,0,0,0,0,1,0,0)$
$(0,0,0,1,0,0,0,0,1,0),$  $(0,0,0,0,1,0,0,0,0,1)$

A base of T-invariants

**Fig. 2**  Marking representation, incidence matrix, and invariants of the net of Figure 1

Every solution for $j$ of $[N] \cdot j = 0$ is a *T-invariant* of $N$.

- If $M_1 \xrightarrow{\sigma} M_1$, then the vector corresponding to $\sigma$ is a T-invariant.

- If a marked net $(N, M)$ is live and bounded then there exists an occurrence sequence $M \longrightarrow M_1 \xrightarrow{\sigma} M_1$ such that $\sigma$ contains every transition at least once. Hence, then there exists a T-invariant $J$ which satisfies $J \geq 1$.

By definition, the set of S-invariants and the set of T-invariants are vector spaces.

Figure 2 shows the vector representation of the marking, the incidence matrix and bases of the S-invariants and T-invariants of the marked net of Figure 1. Note that we do not formally distinguish row vectors from column vectors. The product of two vectors is always the scalar product.

A transition $t$ is in a *conflict situation* with some other transition $t'$ if both transition are in the post-set of some place $s$; if $s$ is only marked once then both transitions are enabled but only one of it can occur. This motivates the definition of *clusters*: For every element $x \in S \cup T$, $[x]$ denotes the smallest set containing $x$ which includes $s^\bullet$ for every place $s \in [x]$ and $^\bullet t$ for every transition $t \in [x]$. $[x]$ is called a *cluster* of $N$. The set of all clusters of $N$ constitutes a partition of $S \cup T$.

The clusters of the net of Figure 1 are depicted by shaded areas.

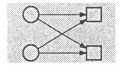 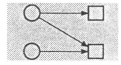

**Fig. 3** An extended free choice cluster and a violation of the extended free choice property

## 3 Extended Free Choice Nets and Regular Marked Nets

**Definition 3.1**

A net is an *extended free-choice net* if, for every cluster $c$, every place $s \in c$ and every transition $t \in \bar{c}$ there is an arc $(s, t)$.

Figure 3 illustrates this definition. The net of Figure 1 is not extended free choice; there are arcs missing in the inner cluster, namely $(s_1, t_2), (s_1, t_3), (s_1, t_4), (s_2, t_3), \ldots$

The following theorem characterizes live and bounded marked extended free choice nets by means of linear algebraic properties. This result and consequences were introduced for different net classes in [CCS91, Esp90]. A proof for extended free choice nets can be found in [Des92].

**Theorem 3.2** *The Rank Theorem*

Let $N$ be an extended free choice net and $M$ a marking of $N$. Then $(N, M)$ is live and bounded if and only if

(1) $N$ has an S-invariant $I \geq 1$,

(2) $N$ has a T-invariant $J \geq 1$,

(3) the number of clusters exceeds the rank of the incidence matrix of $N$ by 1,

(4) every S-invariant $I > 0$ of $N$ satisfies $I \cdot M > 0$. ∎

The Rank Theorem does not hold for arbitrary nets. However, [CCS90] gives a necessary condition for liveness which generalizes one direction of the Rank Theorem. In [Des92], it is shown that the conditions of the Rank Theorem provide a sufficient condition for liveness and boundedness for the class of *feedback-free nets*. In the present paper we consider arbitrary nets which satisfy the conditions of the Rank Theorem.

**Definition 3.3**

A marked net $(N, M)$ (not necessarily extended free choice) is called *regular* if it satisfies the conditions (1) to (4) of the Rank Theorem.

Consider again the marked net of Figure 1. $(1,1,1,1,1,2,2,2,2,2) \geq 1$ is an S-invariant. $(1,1,1,1,1,1,1,1,1,1) \geq 1$ is a T-invariant. The number of clusters is 6. The number of places is 10 and the dimension of the space of S-invariants is 5 (Figure 2 shows a base of S-invariants). Hence, the rank of the incidence matrix is 5. So conditions (1) to (3) hold. All S-invariants $I > 0$ are nonnegative linear combinations of the S-invariants shown in Figure 2. All these S-invariants satisfy $I \cdot M > 0$ for the depicted marking $M$, which proves condition (4). So the marked net is regular.

As shown in [ES91], regularity is decidable in polynomial time. The proof uses the fact that conditions (1), (2), and (4) can be verified using Linear Programming techniques.

Fig. 4   A non-proper and a proper feedback

## 4   The Liveness Theorem

Every regular marked net is bounded since it has an S-invariant $I \geq 1$ by definition. The aim of this chapter is to prove that every regular marked net is also live.

The proof is divided in two parts: First we relate regular marked nets to extended free choice nets. This construction is only possible for nets without *proper feedbacks*. In the second part we show that regular marked nets have no proper feedbacks.

### Definition 4.1

A *feedback* of a net is an arc $(t, s)$ from a transition $t$ to a place $s \in [t]$. A feedback $(t, s)$ is called *proper* if there is no arc $(s, t)$.

Figure 4 illustrates this definition. Note that extended free choice nets may have feedbacks but have no proper feedbacks by definition.

### Definition 4.2

Let $N = (S, T, F)$ be a net without proper feedback. Its *efc-representation* $N'$ is defined by $N' = (S, T, F \cup \widehat{F} \cup \widehat{F}^{-1})$ where

$$\widehat{F} = \{(s,t) \in S \times T \mid [s] = [t] \wedge (s,t) \notin F\} \text{ and } \widehat{F}^{-1} = \{(t,s) \mid (s,t) \in \widehat{F}\}$$

The net of Figure 1 has no proper feedbacks. Figure 5 shows the efc-representation (with the marking of the original net) and the corresponding marking graph.

### Proposition 4.3

Let $N$ be a net without proper feedback and $N'$ the efc-representation of $N$.

(1) $N'$ is an efc-net.

(2) $[N] = [N']$, i.e. both nets have identical incidence matrices.

(3) If $(N, M)$ is regular for some marking $M$ then so is $(N', M)$.

(4) Each occurrence sequence $M \xrightarrow{\sigma} M'$ of $N'$ is an occurrence sequence of $N$.

(5) If $(N', M)$ is live and bounded, then so is $(N, M)$.

### Proof:

(1) Follows immediately from the definition of $N'$.

(2) By definition, $F \cap \widehat{F} = \emptyset$. By the lack of proper feedbacks, $F \cap \widehat{F}^{-1} = \emptyset$. The result follows since $(x,y) \in \widehat{F}$ iff $(y,x) \in \widehat{F}^{-1}$.

(3) Follows from (2) and the definition of regularity.

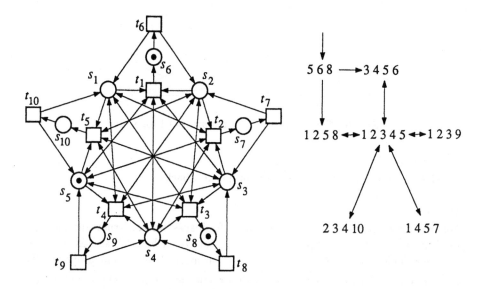

**Fig. 5**  The efc-representation of the marked net of Figure 1

(4) We show the proposition for $\sigma = t$ where $t$ is a transition. The general case
follows then inductively. By construction, the pre-set of $t$ in $N$ is a subset of
the pre-set of $t$ in $N'$. Hence, since $M$ enables $t$ in $N'$, $M$ enables $t$ in $N$. The
successor markings coincide because $[N] = [N']$ and the successor markings
are determined by the Marking Equation.

(5) $N'$ is extended free choice by (1). Assume that $(N', M)$ is live and bounded.
By the Rank Theorem, $N'$ has an S-invariant $I \geq 1$. Since $[N] = [N']$, $I$ is
also an S-invariant of $N$. So $(N, M)$ is bounded.
Let $M \longrightarrow L$ in $N$ and let $t$ be a transition. For proving liveness we have to
show that there exists a marking $K$, reachable from $L$, which enables $t$.
Since $(N', M)$ is live and bounded, it satisfies the conditions of the Rank Theorem. Hence $(N', L)$ satisfies conditions (1) to (3) of the Rank Theorem, because
these conditions do not refer to the marking. Since $M \longrightarrow L$ in $N$ and since
S-invariants of $N'$ are S-invariants of $N$, $(N', L)$ also satisfies (4). Again by
the Rank Theorem, $(N', L)$ is live and bounded. Hence there is an occurrence
sequence $L \xrightarrow{\sigma} K$ of $N'$ such that $K$ enables $t$. By (4), $L \xrightarrow{\sigma} K$ is also an
occurrence sequence of $N$ and $K$ enables $t$ in $N$.  ∎

**Proposition 4.4**

Let $N$ be a net without proper feedback and $M$ a marking of $N$. If $(N, M)$ is
regular then it is live and bounded.

**Proof:**

We apply Proposition 4.3. Let $(N, M)$ be regular. By (3), $(N', M)$ is regular. By
(1) and the Rank Theorem, $(N', M)$ is live and bounded. By (5) $(N, M)$ is live and
bounded.  ∎

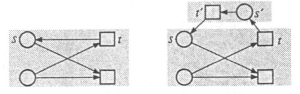

**Fig. 6**  Construction used in the proof of Proposition 4.5

## Proposition 4.5
Every regular marked net is free of proper feedbacks.

**Proof:**
For every regular marked net with proper feedbacks we construct another regular marked net which has less proper feedbacks, but which has still at least one proper feedback. This proves inductively that no regular marked net has any proper feedback.

Let $(N, M)$ be a regular marked net with a proper feedback $(t, s)$, i.e. an arc from a transition $t$ to a place $s$ where both nodes belong to the same cluster and there is no arc $(s, t)$. We modify $(N, M)$ as follows: we add a new place $s'$, a new transition $t'$ and replace the arc $(t, s)$ by arcs $(t, s')$, $(s', t')$ and $(t', s)$. The new place $s'$ remains unmarked (see Figure 6). The modified marked net is denoted by $(\overline{N}, \overline{M})$.

We show next that $(\overline{N}, \overline{M})$ is regular.
The incidence matrix $[\overline{N}]$ of $\overline{N}$ can be represented as

|       | $t'$ | $t$ |   |   |   |
|-------|------|-----|---|---|---|
| $s'$  | $-1$ | 1   | 0 | $\cdots$ | 0 |
| $s$   | 1    | 0   |   |   |   |
|       | 0    |     |   |   |   |
| $\vdots$ |   |     | $[N]$ |   |   |
|       | 0    |     |   |   |   |

$([\overline{N}](s,t) = 0$ and $[\overline{N}](x,y) = [N](x,y)$ for all other pairs $(x,y)$ of nodes of $N)$.

(1) Let $I \geq 1$ be an S-invariant of $N$, $I = (I(s) \ldots)$. Then $\overline{I} = (I(s)\ I\ )$ is an S-invariant of $\overline{N}$. $\overline{I} \geq 1$ since $I \geq 1$ and $\overline{I}(s') = I(s) \geq 1$.

(2) Let $J \geq 1$ be a T-invariant of $N$, $J = (J(t) \ldots)$. Then $\overline{J} = (J(t)\ J\ )$ is a T-invariant of $\overline{N}$. $\overline{J} \geq 1$ since $J \geq 1$ and $\overline{J}(t') = J(t) \geq 1$.

(3) The rank of $[\overline{N}]$ exceeds the rank of $[N]$ by one. To see this, add the first row of $[\overline{N}]$ to the second row and obtain

|         | $t'$ | $t$ |   |   |   |
|---------|------|-----|---|---|---|
| $s'$    | $-1$ | 1   | 0 | $\cdots$ | 0 |
| $s + s'$| 0    |     |   |   |   |
|         | $\vdots$ |  | $[N]$ |   |   |
|         | 0    |     |   |   |   |

$\overline{N}$ has one more cluster than $N$, namely $\{s', t'\}$. So $\overline{N}$ satisfies condition (3).

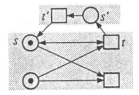 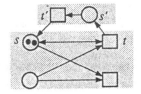 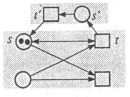

**Fig. 7** The marking $K$, the marking after the occurrence of $t$ and the marking $K'$

(4) Let $\overline{I} > 0$ be an S-invariant of $\overline{N}$. Then $\overline{I}(s) = \overline{I}(s')$ (consider the first column of $(\overline{N})$). It is easily seen that the vector $I$ composed by all entries of $\overline{I}$ except the first (which corresponds to $s'$) is an S-invariant of $N$. $I > 0$ since $\overline{I} > 0$ and $\overline{I}(s) = \overline{I}(s')$. Since $\overline{I} \cdot \overline{M} > 0$ and $\overline{M}(s') = 0$ we get $I \cdot M > 0$.

Every proper feedback of $\overline{N}$ is a proper feedback of $N$. Since $(t, s)$ is a proper feedback of $N$ but no feedback of $\overline{N}$, $\overline{N}$ has less proper feedbacks than $N$.

Finally we show that $\overline{N}$ has at least one proper feedback. Assume the contrary. Let $\overline{N}'$ be the efc-representation of $\overline{N}$. By Proposition 4.3(3), $(\overline{N}', \overline{M})$ is regular. Hence it has an S-invariant $I \geq 1$. We prove that for every marking $K$ of $\overline{N}'$ such that $(\overline{N}', K)$ is regular there is a marking $L$ satisfying $I \cdot L \leq I \cdot K - 1$ such that $(\overline{N}', L)$ is also regular. But, since $I \geq 1$ and no marking has negative entries, the scalar product always remains nonnegative – a contradiction.

So let $(\overline{N}', K)$ be regular. By the Rank Theorem, $(\overline{N}', K)$ is live. Let $K \longrightarrow K'$ such that $K'$ enables the transition $t$. Let $K' \xrightarrow{t\,t'} K''$. Then $K''(s) \geq 2$ (see Figure 7). Define the marking $L$ which coincides with $K''$ on all places except $s$ and satisfies $L(s) = 1$. We get $I \cdot L \leq I \cdot K - 1$ because $I \cdot K = I \cdot K''$, $I(s) \geq 1$ and $L(s) \leq K''(s) - 1$. For showing that $(\overline{N}', L)$ is again regular we only have to prove condition (4), because the other conditions do not refer to the marking. For every S-invariant $I' > 0$ with $I'(s) = 0$ we have $I' \cdot L = I' \cdot K''$ by the construction of $L$, $I' \cdot K'' = I' \cdot K$ since $K \longrightarrow K''$ and $I' \cdot K > 0$ since $(\overline{N}', K)$ is regular. For every S-invariant $I' > 0$ with $I'(s) > 0$ we have $I' \cdot L \geq I'(s) \cdot L(s) > 0$ since $L(s) = 1$. ∎

Finally we obtain the main result of this chapter:

**Theorem 4.6**
Every regular marked net is live and bounded.

**Proof:**
Follows from Propositions 4.4 and 4.5. ∎

## 5 Behavioural Properties

In this chapter, properties of regular marked nets are directly derived from respective results on extended free choice nets. The proofs base on the fact that occurrence sequences of the efc-representation of a regular marked net are occurrence sequences of the regular marked net itself (Propositions 4.3(4) and 4.5).

The following two theorems are shown in [DE91] for extended free choice nets:

**Theorem 5.1**

Let $(N, M)$ be a regular marked net. Let $K$ and $L$ be two markings satisfying $I \cdot K = I \cdot L = I \cdot M$ for every S-invariant $I$ of $N$. Then $[K\rangle \cap [L\rangle \neq \emptyset$.  ∎

A consequence of this result is the existence of *home states*:

**Theorem 5.2**

Every regular marked net $(N, M)$ has a marking $K$ that is reachable from every marking $L \in [M\rangle$ (a *home state*).  ∎

This theorem also follows easily from the identical property of live and bounded extended free choice nets (see [BV84] for a proof for the restricted class of live and safe marked *free choice nets*) since every home state of the efc-represenation is also a home state of the regular marked net. The converse does not hold. The marking of the net given in Figure 1 is a home state. The same marking is not a home state of the efc-representation shown in Figure 5. This example thus also shows that the proof of the characterization of home states given in [BDE92] does not carry over to regular marked nets. However, at least every home state of the efc-representation is a home state of the regular marked net.

A marked net with the initial marking being a home state is cyclic. For cyclic regular marked nets we get the following characterization of reachable markings as a consequence of Theorem 5.1.

**Theorem 5.3**

Let $(N, M)$ be a cyclic regular marked net. Then $M \longrightarrow K$ iff $I \cdot M = I \cdot K$ for every S-invariant $I$ of $N$.  ∎

By this characterization of reachable markings, a marking $K$ is reachable from $M$ iff $M + [N] \cdot x = K$ has some rational-valued solution for $x$. So this property is decidable in polynomial time.

In [BD90] it is shown that the maximal number of tokens on a place – the bound of the place – can be calculated by means of S-invariants. Again, the same result holds for regular marked nets:

**Theorem 5.4**

Let $(N, M)$ be a regular marked net where $N = (S, T, F)$. Then the bound of a place $s$ equals $\min \{I \cdot M \mid I$ is an S-invariant satisfying $I > 0$ and $I(s) = 1\}$  ∎

As a consequence, we can characterize safety in terms of S-invariants.

**Theorem 5.5**

A regular marked net $(N, M)$ is safe if and only if for every place $s$ there exists an S-invariant $I > 0$ satisfying $I(s) = 1$ and $I \cdot M = 1$.  ∎

Even for safe marked nets, the number of reachable markings can grow exponentially in the size of the net. However, a recent result [DE93] proves for live and safe extended free choice nets that between every two markings of the state space there is either no path at all or a path with polynomial length. This result carries over to regular marked nets:

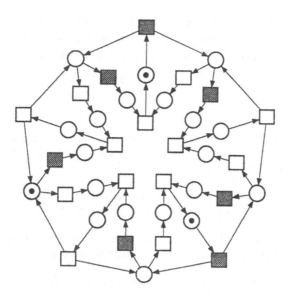

**Fig. 8**   The released version of the net of figure 1

**Theorem 5.6**

Let $(N, M)$ be a regular marked net and let $M_1$, $M_2$ be markings reachable from $M$ such that $M_1 \longrightarrow M_2$. Then there exists an occurrence sequence $M_1 \overset{\sigma}{\longrightarrow} M_2$ such that the number of transition occurrences in $\sigma$ is at most

$$\frac{|T| \cdot (|T| + 1) \cdot (|T| + 2)}{6}$$

where $T$ is the set of transitions of $N$.   ∎

## 6   Conclusion

We have shown that regularity is a sufficient condition for liveness and boundedness of marked Petri nets. Moreover, we have shown that a series of results for live and bounded extended free choice nets can be reformulated as results of regular marked nets.

The class of regular marked nets contains all extended free choice nets. It is however properly greater, as is proved by the example in Figure 1. The shown marked net is not a "hidden free choice net" but exhibits a behaviour which differs significantly from the behaviour of free choice nets. For a more formal argumentation we conclude with showing that the standard free choice translation of a net does not provide a proof for the liveness of regular marked nets.

The idea to relate a free choice net to an arbitrary net is due to Hack [Hac76] who defined the *released version* of a marked net. Figure 8 shows the released version of the marked net of Figure 1. A marked net is live whenever its released version is live.

However, the marked net of Figure 8 is not live: Consider any occurrence sequence that starts with the initial marking and uses every shaded transition exactly once. The marking reached by this sequence enables no transition at all. So, in this example, the released version can not be used for proving liveness of the marked net.

# References

[AF92]  ALIMONTI,P.; FEUERSTEIN,E.: *Petri Nets, Hypergraphs and Conflicts.* Proc. Graph-Theoretical Concepts in Computer Science WG'92, LNCS 657 (1992) 293-309

[BD90]  BEST,E.; DESEL,J.: *Partial Order Behaviour and Structure of Petri Nets.* Formal Aspects of Computing 2 (2) (1990) 123-138

[BDE92] BEST,E.; DESEL,J.; ESPARZA,J.: *Traps Characterise Home States in Free Choice Systems.* Theoretical Computer Science 101 (1992) 161-176

[BV84]  BEST,E.; VOSS,K.: *Free Choice Systems have Home States.* Acta Informatica 21 (1984) 89-100

[CCS91] CAMPOS,J.; CHIOLA,G.; SILVA,M.: *Properties and Performance Bounds for Closed Free Choice Synchronized Monoclass Queueing Networks.* IEEE Transactions on Automatic Control AC-36 (12) (1991) 1368-1382

[CCS90] COLOM,J.M.; CAMPOS,J.; SILVA,M.: *On Liveness Analysis Through Linear Algebraic Techniques.* Departamento de Ingenieria Electrica e Informatica, Universidad de Zaragoza, Research Report GISI-RR-90-11 (1990)

[Des92] DESEL,J.: *A Proof of the Rank Theorem for Extended Free Choice Nets.* Proc. Application and Theory of Petri Nets 1992, LNCS 616 (1992) 134-153

[DE91]  DESEL,J.; ESPARZA,J: *Reachability in Cyclic Extended Free-Choice Systems.* Theoretical Computer Science 114 (1993) 93-118

[DE93]  DESEL,J.; ESPARZA,J: *Shortest Paths in Reachability Graphs.* Proc. Application and Theory of Petri Nets 1993, LNCS 619 (1993) 224-241

[Esp90] ESPARZA,J.: *Synthesis Rules for Petri Nets, and How they Lead to New Results.* Proc. CONCUR'90, LNCS 458 (1990) 182-198

[ES91]  ESPARZA,J.; SILVA,M.: *On the Analysis and Synthesis of Free Choice Systems.* Advances in Petri Nets 1990, LNCS 483 (1991) 243-286

[Hac76] HACK,M.: *Petri Net Languages.* MIT Laboratory Computer Science TR-159 (1976)

[Jan87] JANTZEN,M.: *Complexity of Place/Transition Nets.* Petri Nets: Central Models and their Properties, LNCS 254 (1987) 413-434

[Rei85] REISIG,W.: *Petri Nets.* EATCS Monographs on Theoretical Computer Science 4, Springer-Verlag (1985)

# The Asynchronous Committee Meeting Problem

Javier Esparza[1] and Bernhard von Stengel[2]

[1] Institut für Informatik, Universität Hildesheim, Samelsonplatz 1, 31141 Hildesheim, Germany. email: esparza@informatik.uni-hildesheim.de

[2] Informatik 5, Universität der Bundeswehr München, 85577 Neubiberg, Germany. email: i51bbvs@rz.unibw-muenchen.de

**Abstract.** The committee meeting problem consists in finding the earliest meeting time acceptable to every member of a committee. We consider an asynchronous version of the problem that does not presuppose the existence of a global clock, where meeting times are maximal antichains in a poset of 'local times', and propose an efficient algorithm to solve it. A generalization, the private meeting problem, where the earliest time for a meeting *without* some committee members is sought, turns out to be NP-complete. However, it can be solved in polynomial time if the poset is N-free, that is, representing a precedence of the arcs (and not the nodes) of an acyclic directed graph. This special case is relevant, because it allows to improve the key algorithm of the model checking technique developed in [4] for Petri nets.

**Key words.** Committee meeting problem, maximal antichains, N-free posets, Petri net unfolding

## 1 The asynchronous committee meeting problem

The committee meeting problem is an elementary, well-known problem of program design. It is for instance the first problem used by Chandy and Misra [3] to illustrate their UNITY programming methodology. The problem is to find the earliest meeting time acceptable to every member of a committee. Time instants are linearly ordered, and there exists a zero time without predecessor. Every committee member labels a subset of the time instants as 'free'. The desired output is the earliest time which is labeled as 'free' by all members.

A simple algorithm (with several variants, as shown in [3]) can be used to solve the problem: time zero is taken as the first candidate to a solution; if some person cannot meet at that time, she suggests her next free time as new candidate, and the procedure is iterated until nobody makes a new suggestion. It is not difficult to prove that, if some meeting time exists, then the algorithm yields the earliest one.

This problem presupposes the existence of a global clock, which can be viewed as a totally ordered set of times that are common to every committee member. We study an asynchronous version of the problem in which no global clock exists. In this version, every person has her own totally ordered set of local times, or, as we shall say, *states*. The states can also be seen as pauses between the execution of two tasks. The time taken by a task is not known in advance, and therefore it is not possible to ensure that, for instance, the first pause will simultaneously take place for all persons. However, in addition to the (fixed) sequential order of the tasks of one person, there may be dependencies between certain tasks of different persons, which impose additional precedences between the states. Thereby, the union of the (pairwise disjoint) sets of states is partially ordered.

The total order of the special synchronous case is thereby represented by introducing for each time instant separate states, one for each person, that are mutually incomparable and otherwise precede all states corresponding to later time instants. Furthermore, *any* partially ordered set (poset) may arise from the asynchronous meeting problem by partitioning the poset into chains (totally ordered subsets) which are assigned to committee members; here, only finite posets will be considered.

Figure 1.1 shows a case in which the committee consists of three people, $A$, $B$ and $C$, whose sets of states contain 4, 5 and 3 elements, respectively. As in the synchronous case, every person labels a subset of her own states as 'free'. The free states are shown in boldface.

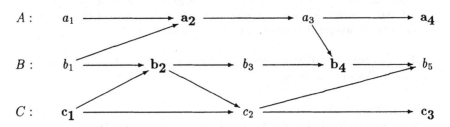

**Fig. 1.1.** An instance of the asynchronous version of the meeting problem.

Two states where one precedes the other are called *comparable*. An *antichain* is a set of pairwise incomparable states. Here, a maximal antichain with respect to set inclusion is called a *cut*. A cut can be interpreted as a snapshot of the committee, which tells which members have finished which tasks at a certain moment. For instance, $\{a_3, b_3, c_2\}$ is a cut of the partial order of Fig. 1.1, but $\{a_3, b_1, c_1\}$ is not, because $a_3$ and $b_1$ are comparable (and so $A$ and $B$ cannot simultaneously be in $a_3$ and $b_1$).

Cuts are the asynchronous counterpart of the global time instants in the asynchronous case. There is a natural notion of temporal precedence between cuts, which corresponds to the precedence between global time instants of the synchronous case: We say that a cut is *earlier* than another if every state of the first precedes (not necessarily strictly) a suitable state of the second. Independently of the relative speeds of the local clocks, if a cut $Y$ is earlier than a cut $Z$, then $Z$ can only be observed after $Y$. In Fig. 1.1, $\{a_4, b_3, c_2\}$ is earlier than $\{a_4, b_5, c_3\}$, but not earlier than $\{a_3, b_3, c_3\}$. It is easy to see that the 'earlier than' relation on cuts is a partial order.

The asynchronous committee meeting problem is to find an earliest cut containing one 'free' state for each member of the committee. In the example of Fig. 1.1, the reader can check that the unique earliest cut satisfying this condition is $\{a_4, b_4, c_3\}$.

## 2  Formalization and algorithm

Before proceeding to give an algorithm for this problem, we introduce further notations and conventions.

The set of states is denoted by $X$, which is a finite set to guarantee termination of the algorithm, possibly with a negative result if no common meeting time exists. The precedence relation between states is a partial order and denoted by $\leq$, so the poset under consideration is $(X, \leq)$. A state $x$ is *smaller* than, or *precedes*, a state $y$ if $x \leq y$. For a subset $Y$ and an element $x$ of $X$, say that $x$ precedes $Y$ if $x$ precedes some element of $Y$. Given two cuts $A$, $B$, we denote by $A \sqsubseteq B$ that $A$ is earlier than $B$, that is, each element of $A$ precedes $B$. A *smallest* state is a $\leq$-minimal state (with no predecessor), but not necessarily smaller than all states. Similarly, an *earliest* cut is $\sqsubseteq$-minimal.

The problem is stated as follows. Each state is labeled with a label from a finite set $\mathcal{S}$, or unlabeled. Any two states that carry the same label (from $\mathcal{S}$) are comparable. Then, find an earliest cut containing one $s$-labeled element for every label $s$ in $\mathcal{S}$.

The interpretation is that the set $X$ is the disjoint union of the sets of states of the given persons. Each person has a separate label in $\mathcal{S}$ to mark certain of her states as 'free', while the others are unlabeled and thus marked as 'blocked' (these are not distinguished per person). The states of any one person are totally ordered.

We shall consider a slightly more general problem: Given a subset $\mathcal{G}$ of $\mathcal{S}$ of 'prescribed' labels, find an earliest cut containing one $s$-labeled element for every label $s$ in $\mathcal{G}$. The asynchronous committee meeting problem corresponds to the case $\mathcal{G} = \mathcal{S}$.

A cut containing an (obviously at most one) $s$-labeled element for every label $s$ in $\mathcal{G}$ shall be called a $\mathcal{G}$-*cut*. Notice that the set of $\mathcal{G}$-cuts may be empty, so that it has no $\sqsubseteq$-minimal element, and then the problem has no solution. However, if such a cut exists, then it is unique, so it is *the* earliest $\mathcal{G}$-cut:

**Proposition 2.1.** *Let the set of $\mathcal{G}$-cuts be nonempty. Then it has exactly one $\sqsubseteq$-minimal element.*

*Proof.* Let $A$, $B$ be two $\mathcal{G}$-cuts, and $C$ be the set of largest states preceding both $A$ and $B$. It is easy to see that the elements of $A$ preceding $B$ and the elements of $B$ preceding $A$ belong to $C$, which comprise at least one of the $s$-labeled states in $A$ or $B$ for each label $s$ in $\mathcal{G}$. A further consequence is that $C$ is a cut and in fact the lattice-theoretic *meet*, the greatest lower bound with respect to $\sqsubseteq$, of $A$ and $B$ (it is well known [1] that the cuts of any poset form a lattice). So $C$ is also a $\mathcal{G}$-cut. The meet of all $\mathcal{G}$-cuts is the unique earliest $\mathcal{G}$-cut. $\qquad\square$

To simplify the algorithm, we assume that the set of $\mathcal{G}$-cuts is nonempty, and denote by $C$ its earliest element. This can be achieved for any poset $(X, \leq)$ by 'stacking' an extra $\mathcal{G}$-cut $D$ on top of $X$, that is, $D$ is a set of additional states (not in $X$), one for each label in $\mathcal{G}$, which are mutually incomparable and larger than any element of $X$. The algorithm is then applied to $X \cup D$ instead of $X$, and if it produces an earliest $\mathcal{G}$-cut containing elements of $D$, then $X$ itself has no $\mathcal{G}$-cut, so the original problem has no solution.

The problem is then to design an algorithm that yields $C$ as output. We use a variable $A$, which is initialized to the earliest cut of $(X, \leq)$, namely the set of smallest elements of $X$. A loop increases $A$ until it becomes a $\mathcal{G}$-cut, while preserving $A \sqsubseteq C$. When the loop has ended, $A$ is the output.

The algorithm uses the following notation: given an element $x$ of $X$, the set $\downarrow x$ (read 'proper predecessors of $x$') consists of the states smaller than $x$ and different from it. For a subset $A$ of $X$, the set $\downarrow A$ is defined as the union of the sets $\downarrow x$ for every element $x$ of $A$. The set of labels of $A$ is denoted by $l(A)$.

$A :=$ set of smallest elements of $X$;
**do**   { *invariant*: $A$ is a cut and $A \sqsubseteq C$ }
      there exists $s$ in $\mathcal{G} \setminus l(A) \longrightarrow$
           $x :=$ smallest $s$-labeled state not preceding $A$;
           $A :=$ set of smallest elements of $X \setminus \downarrow(A \cup \{x\})$
**od**

Let us prove the correctness of the algorithm. The loop guard simply states that $A$ is not a $\mathcal{G}$-cut. The invariant then guarantees that, upon termination of the algorithm, $A$ is a $\mathcal{G}$-cut and $A \sqsubseteq C$. By the $\sqsubseteq$-minimality of $C$, $A = C$. The invariant holds before the first execution of the loop because the set of minimal elements of $X$ is the unique $\sqsubseteq$-minimal cut. The following lemma and proposition show that the invariant is preserved by the body of the loop.

**Lemma 2.2.** *Let $Y$ be a subset of $X$, and let $Z$ be set of smallest elements of $X \setminus \downarrow Y$. Then $Z$ is a cut.*

*Proof.* Since the smallest elements of any set are pairwise incomparable, $Z$ is an antichain. To prove that $Z$ is a cut, let $x$ be an arbitrary element of $X$. If $x \in X \setminus \downarrow Y$, then $x$ is larger than some element of $Z$ by the definition of $Z$. If $x \in \downarrow Y$, then $x$ is smaller than some largest element $y$ of $Y$. So $y \in Z$ by the definition of $Z$. In both cases, $x$ is comparable with some element of $Z$, so $Z$ is a cut. $\qquad\square$

**Proposition 2.3.** *Let $A$ be a cut so that $A \sqsubseteq C$ and so that for some label $s$, no element of $A$ is $s$-labeled. Let $x$ be the smallest $s$-labeled state not preceding $A$. Let $B$ be the set of smallest states of $X \setminus {\downarrow}(A \cup \{x\})$. Then $B$ is a cut and $B \sqsubseteq C$.*

*Proof.* $B$ is a cut by Lemma 2.2. It remains to show $B \sqsubseteq C$, i.e., every element of $B$ is smaller than some element of $C$. By the definition of $B$, it suffices to prove that $C$ is a subset of $X \setminus {\downarrow}(A \cup \{x\})$, or, equivalently, $C \cap ({\downarrow}A \cup {\downarrow}x) = \emptyset$.

$C \cap {\downarrow}A = \emptyset$ because $A \sqsubseteq C$ and $C$ is a cut. To prove $C \cap {\downarrow}x = \emptyset$, observe that $C$ contains one $s$-labeled element $y$, which is comparable with $x$ and not smaller than an element of $A$, so $x \leq y$ by the definition of $x$; since $C$ is an antichain, it contains no proper predecessor of $x$. $\qquad\square$

To prove termination, it suffices to show that the body of the loop strictly increases $A$ with respect to the partial order $\sqsubseteq$. In other words, that $A$ is earlier than, and different from, the set of smallest elements of $X \setminus {\downarrow}(A \cup \{x\})$. The first part is trivial. For the second part, observe that $x \notin A$ (because $A$ contains no $s$-labeled element), but $x$ is one of the smallest elements of $X \setminus {\downarrow}(A \cup \{x\})$.

To estimate the running time of the algorithm, its input has to be made precise. We choose as input the *diagram* $(X, \prec)$ of the poset, where $x \prec y$ if $y$ is an *immediate successor* of $x$, that is, $x < y$ and there exists no state $z$ with $x < z < y$. Since $(X, \leq)$ is finite, $\leq$ can be recovered as the reflexive and transitive closure of $\prec$.

The cut $A$ can be increased with respect to $\sqsubseteq$ at most $|X|$ times. So the number of iterations of the loop is at most $|X|$. An iteration requires to compute the element $x$ and the set of smallest elements of $X \setminus {\downarrow}(A \cup \{x\})$, which can be done in $O(|X| + |\prec|)$ time. So the algorithm requires at most $O\big(|X| \cdot (|X| + |\prec|)\big)$ time.

## 3   The asynchronous private meeting problem

A subset of the committee members wish to meet, for instance those belonging to a certain political party. However, they also want to make sure that, when they meet, the committee members of another party are busy, i.e., that none of them is in a 'free' state. To this end, a new set $\mathcal{H}$ of 'forbidden' labels is introduced besides the set $\mathcal{G}$. The asynchronous private meeting problem is to find an earliest cut $C$ of $(X, \leq)$ so that:

(1) for every label $s \in \mathcal{G}$, some element of $C$ is $s$-labeled, and

(2) for every label $s \in \mathcal{H}$, no element of $C$ is $s$-labeled.

The set of cuts satisfying these two conditions may be empty. Furthermore, even if the set is nonempty it may happen that it has no unique $\sqsubseteq$-minimal element. Figure 3.1 shows such a poset (in the conventional way of drawing poset diagrams, with smaller elements further down). There, let the states $a$ and $b$ carry one label and $a'$ and $b'$ another, both labels belonging to $\mathcal{G}$, and the state $x$, indicated by a white dot, have a forbidden label in the set $\mathcal{H}$. Then both $\{a, a'\}$ and $\{b, b'\}$ are cuts satisfying (1) and (2), but their meet $\{a, x, b'\}$ is not.

**Fig. 3.1.** A poset with no earliest cut fulfilling (1) and (2).

Thus, for a general poset, the asynchronous private meeting problem may not have a unique answer. Even finding one possible solution is much harder than in the previous case. This question turns out to be NP-complete already for $\mathcal{G} = \emptyset$. In this case, the condition on $\mathcal{G}$ is vacuously fulfilled, and the problem reduces to finding an earliest cut of $(X, \leq)$ containing no element with a label in $\mathcal{H}$. An even weaker question is whether there exists such a cut at all: if we color the elements with labels in $\mathcal{H}$ white and the rest black, the problem is to find a cut containing only black elements.

**Proposition 3.1.** *The following problem is NP-complete: Given a finite poset $(X, \leq)$ and each element of $X$ being either black or white, find a cut in $X$ containing only black elements.*

*Proof.* As before, call the elements of $X$ states. A cut $C$ in $X$ is an antichain so that every state in $X$ compares with an element of $C$. Since this is verified fast, the problem belongs to the complexity class NP.

The NP-complete problem 3SAT will be reduced to this problem, which will show that it is NP-complete [5]. An instance of 3SAT is a Boolean expression in conjunctive normal form, given by a conjunction of clauses, each of which is a disjunction of exactly three literals, which are either positive, given by a Boolean variable, or negative, given by a negated variable. The problem is to decide if this Boolean expression is satisfiable, that is, to find an assignment of truth values to the variables so that each clause evaluates to true.

Given an instance of 3SAT, construct the poset $(X, \leq)$ as follows. For each variable $x$ of the expression, introduce two black states $p(x)$ and $n(x)$ and a white state $w(x)$ with $n(x) < w(x) < p(x)$; between these states, considered for all variables $x$, these are the only comparabilities, and the white states $w(x)$ will compare with no other states. Therefore, a cut $C$ that is a solution to the problem contains for each variable $x$ either $p(x)$ or $n(x)$ (which is interpreted as $x$ assuming the value true or false, respectively), since each white node must compare with some element of $C$. Additional white states will be used to encode the clauses. If a clause $c$ contains only positive literals $x$, $y$ and $z$, say, introduce a white state $w(c)$ with $w(c) < p(x)$, $w(c) < p(y)$ and $w(c) < p(z)$ and no further comparabilities, so $C$ is a solution only if it contains at least one of the states $p(x)$, $p(y)$ or $p(z)$, that is, if the clause $c$ is true. Similarly, a clause $c$ containing only negative literals $\bar{x}$ ($x$ negated), $\bar{y}$ and $\bar{z}$ is represented by a white node $w(z)$ larger than $n(x)$, $n(y)$ and $n(z)$.

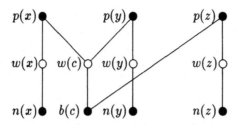

**Fig. 3.2.** Encoding of the clause $c$ given by $x \vee y \vee \overline{z}$.

A 'mixed' clause $c$ contains the literals $x$, $y$ and $\overline{z}$ for variables $x, y, z$ (or literals $x$, $\overline{y}$ and $\overline{z}$, in which case $c$ will be encoded analogously). As shown in Fig. 3.2, a white node $w(c)$ is introduced with $w(c) < p(x)$ and $w(c) < p(y)$, and an additional black node $b(c)$ with $b(c) < w(c)$ and $b(c) < p(z)$ (note that $n(z) < w(c)$ would not work since then $n(z) < p(x)$). Then, a solution $C$ has to contain at least $p(x)$, $p(y)$ or $b(c)$ to contain a state comparing with $w(c)$. The clause $c$ is true if $x$ or $y$ is true, represented by $p(x) \in C$ respectively $p(y) \in C$, or if $z$ is false, represented by $n(z) \in C$; only in this case, $b(c)$ can belong to $C$, since otherwise $p(z) \in C$ and $p(z)$ compares with $b(c)$. This encodes mixed clauses.

The poset is constructed in polynomial time from the 3SAT expression, which is satisfiable if and only if there is a cut containing only black nodes. □

## 4 The private meeting problem for N-free posets

In this section, we consider the special class of N-free posets, where the asynchronous private meeting problem always has a unique answer, and can be solved in polynomial time. These are the (always finite) posets $(X, \leq)$ satisfying the following condition:

(*) If $x, y, z$ are states with $x \prec y$ and $x \prec z$, then $\downarrow y = \downarrow z$.

Notice that the poset of Fig. 3.1 does not satisfy this condition: $b$ and $a'$ are immediate successors of $x$, and $a \in \downarrow b$ but $a \notin \downarrow a'$.

The following equivalent conditions to (*) are better known in the literature: $X$ contains no four distinct elements $a, b, c, d$ with $a < b$, $c \prec b$, $c < d$ and otherwise being incomparable [6], or, more specially, with $a \prec b$, $c \prec b$, $c \prec d$, otherwise being incomparable [7]. In the latter case, the states $a, b, c, d$ form an N-shaped part of the poset diagram, which is why such posets are called N-free. Another equivalent condition 'CAC' is chain-antichain-completeness: every maximal chain intersects every maximal antichain [6]. Finally, ordering the arcs of an acyclic directed graph (with $x \prec y$ if $y$ is an arc starting at the endpoint of the arc $x$) yields an N-free poset, and conversely, for each N-free poset such a directed graph is easily constructed.

In the following, consider an N-free poset. First, we establish that the private meeting problem has a unique solution. That is, there is a unique earliest cut satisfying the conditions (1) and (2) in Sect. 3 above, where the labels of the elements of the cut include all labels from $\mathcal{G}$ and none from $\mathcal{H}$, if there is such a cut at all.

**Proposition 4.1.** *For an N-free poset, let the set of cuts satisfying conditions (1) and (2) be nonempty. Then it has exactly one $\sqsubseteq$-minimal element.*

*Proof.* Let $A$, $B$ be two cuts satisfying (1) and (2). As in Proposition 2.1, the meet $C$ of the cuts $A$ and $B$ consisting of the maximal states preceding both $A$ and $B$ is a $\mathcal{G}$-cut. We claim that $C$ is also a subset of $A \cup B$ (in fact, it equals the set of smallest elements of $A \cup B$). Assume that some $x$ in $C$ belongs neither to $A$ nor to $B$, and consider $a \in A$, $b \in B$ and states $y$, $z$ such that $x \prec y \leq a$ and $x \prec z \leq b$. Since $y$ precedes $A$ but not $B$ by the maximality of $x$, and $B$ is a cut, $b' \in \downarrow y$ for some $b'$ in $B$, but $b' \notin \downarrow z$, contradicting (*). Thus, since neither $A$ nor $B$ has elements with a label in $\mathcal{H}$, neither does $C$. So $C$ satisfies (2) as well. $\qquad\qquad\square$

We assume that the set of cuts satisfying conditions (1) and (2) is nonempty, and denote by $C$ its earliest element. The algorithm to compute $C$ is an extension of that described in Sect. 2. We use a variable $A$, which is initialized to the set of smallest elements of $X$. A loop increases $A$ with respect to $\sqsubseteq$ until it satisfies (1) and (2), while preserving $A \sqsubseteq C$. As before, $l(A)$ denotes the set of labels of $A$.

$A :=$ set of smallest elements of $X$;
**do** { *invariant*: $A$ is a cut and $A \sqsubseteq C$ }
$\qquad$ there exists $s$ in $\mathcal{G} \setminus l(A)$ $\longrightarrow$
$\qquad\qquad x :=$ smallest $s$-labeled state not preceding $A$;
$\qquad\qquad A :=$ set of smallest elements of $X \setminus \downarrow(A \cup \{x\})$
$[]\qquad$ there exists $s$ in $\mathcal{H} \cap l(A)$ $\longrightarrow$
$\qquad\qquad x :=$ an immediate successor of the unique $s$-labeled element of $A$;
$\qquad\qquad A :=$ set of smallest elements of $X \setminus \downarrow(A \cup \{x\})$
**od**

The algorithm is proved correct along the lines of the first algorithm. The loop guards (both of which have to be false to terminate the loop) now state that (1) or (2) does not hold for $A$. So the invariant guarantees that, upon termination, $A$ is a cut satisfying (1) and (2), and $A \sqsubseteq C$. By the $\sqsubseteq$-minimality of $C$, $A = C$.

The invariant holds before the first execution of the loop because the set of smallest elements of $X$ is the unique earliest cut. The first alternative (i.e., the sequence of statements after the first guard) preserves the invariant by Proposition 2.3. The following proposition shows that the second alternative also preserves the invariant.

**Proposition 4.2.** *Let $A$ be a cut with $A \sqsubseteq C$ that contains an element $y$ carrying a label $s$ from $\mathcal{H}$, and let $x$ be an immediate successor of $y$. Let $B$ be the set of smallest elements of $X \setminus \downarrow(A \cup \{x\})$. Then $B$ is a cut and $B \sqsubseteq C$.*

*Proof.* $B$ is a cut by Lemma 2.2. It remains to show $B \sqsubseteq C$. By the definition of $B$, it suffices to prove $C \subseteq X \setminus \downarrow(A \cup \{x\})$, that is, $C \cap (\downarrow A \cup \downarrow x) = \emptyset$.

$C \cap \downarrow A = \emptyset$ because $A \sqsubseteq C$. To prove $C \cap \downarrow x = \emptyset$, note that since $C$ does not contain any $s$-labeled element and $A \sqsubseteq C$, some immediate successor $z$ of $y$ is smaller than some element of $C$. Since $C$ is an antichain, $C \cap \downarrow z = \emptyset$. By assumption (*), $\downarrow x = \downarrow z$, so $C \cap \downarrow x = \emptyset$. □

To prove termination, it suffices to show that the body of the loop strictly increases $A$ with respect to the partial order $\sqsubseteq$. This was already shown for the first alternative in Sect. 2. The proof for the second alternative is analogous.

The running time does not change compared to the first algorithm. Again, the number of iterations of the loop is at most $|X|$. The element $x$ can be computed, for both alternatives, in $O(|X| + |\prec|)$ time, and so can the set of smallest elements of $X \setminus \downarrow(A \cup \{x\})$.

# 5 The private meeting problem for nonsequential processes

In this section we consider Petri nets, a well-known formal model of concurrent systems, thoroughly studied in the literature [8]. Petri nets can be given a partial order semantics in terms of certain labeled posets called nonsequential processes [2]. We shall observe that the private meeting problem for nonsequential processes corresponds to a problem stated in [4], and that nonsequential processes satisfy the condition of Sect. 4. As will be indicated, the new algorithm is more efficient than the algorithm described in [4].

A *net* is a triple $N = (S, T, F)$, where $S \cap T = \emptyset$ and $F \subseteq ((S \times T) \cup (T \times S))$. The elements of $S$ are called *places*, the elements of $T$ *transitions*, and the elements of $F$ *arcs*. The *preset* (*postset*) of an element $x$ of $S \cup T$ is the set of nodes $y$ such that $(y, x) \in F$ $((x, y) \in F)$.

A *marking* of $N$ is a mapping $M$ from $S$ to the nonnegative integers. A Petri net is a pair $(N, M_0)$, where $N$ is a net and $M_0$ an initial marking. Graphically, the places of a Petri net are represented by circles, and its transitions by boxes. If $(x, y)$ is a arc, an arrow is drawn from $x$ to $y$. Finally, if $M_0(s) = n$ for a place $s$, then $n$ black dots or tokens are drawn inside the circle corresponding to $s$. Figure 5.1 shows a Petri net. A transition $t$ is *enabled* at a marking $M$ if $M(s) > 0$ for every place $s$ in the preset of $t$. If a transition $t$ is enabled at a marking $M$, then it can *occur*, yielding a marking obtained by removing one token from each place in the preset of $t$ and adding one token to each place in the postset of $t$. Given a Petri net $(N, M_0)$, a marking $M$ of $N$ is *reachable* if there exists a finite sequence of successively occurring transitions initially enabled at $M_0$ and eventually yielding $M$.

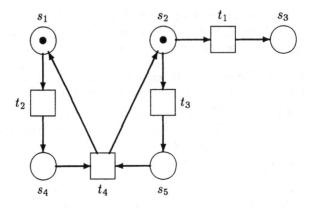

**Fig. 5.1.** A Petri net.

A *nonsequential process* (or just a process) of a Petri net $(N, M_0)$ with $N = (S, T, F)$ is a pair $(N', p)$, where $N'$ is an acyclic net $N' = (B, E, F')$, and $p$ is a labeling function $B \to S$ and $E \to T$. To avoid confusions, the elements of $B$ are called *conditions*, instead of places, and the elements of $E$ are called *events*, instead of transitions. The labeling $p$ satisfies certain properties that allow to interpret $N'$ as an 'unfolding' of the Petri net $(N, M_0)$, corresponding to one of its possible concurrent runs. Most of these properties are not relevant here; the interested reader is referred to [2].

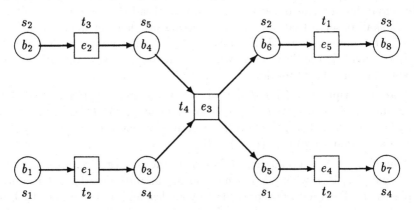

**Fig. 5.2.** A nonsequential process of the Petri net of Fig. 5.1.

Figure 5.2 shows a nonsequential process of the Petri net of Fig. 5.1. The names of the conditions and events are written inside the circles and boxes; the names of the places and transitions with which they are labeled are written close to them. If we put one token in the conditions $b_1$ and $b_2$, then we can simulate in the process a concurrent run of the Petri net; the occurrence of an event corresponds to the occurrence in the Petri net of the transition it is labeled with (for example, the occurrence of the event $e_1$ in the process corresponds to the occurrence of $t_2$ in the net); the set of conditions marked after the occurrence of the events corresponds via the labeling to a reachable marking of the Petri net. The process of Fig. 5.2 corresponds to a run in which the transitions $t_2$ and $t_3$ can occur independently, then transition $t_4$ can occur, and finally transitions $t_1$ and $t_2$ can occur independently.

Since the net $N'$ of a process is acyclic, the reflexive and transitive closure of $F'$ is a partial order, in the following denoted by $\leq$. This partial order will be of interest for the conditions of the process, considering only the poset $(B, \leq)$. This is a special poset, because $N'$ has the property that the preset and postset of a condition contain at most one event. If we extend the process with an 'initial' event preceding all smallest conditions and with a 'terminal' event succeeding all largest conditions, then every condition has a unique event in its preset and its postset, so $B$ can be seen as the set of *arcs* of a directed graph whose nodes are the events.

In this paper we restrict our attention to *safe* Petri nets, in which every reachable marking puts at most one token in a place (if this holds for the initial marking); then, a marking can be identified with the set of places with a token. The following proposition (see [2]) shows that the markings of the Petri net reached along the run correspond to the cuts of the poset $(B, \leq)$.

**Proposition 5.1.** Let $(N', p)$ be a nonsequential process of a safe Petri net $(N, M_0)$, where $N' = (B, E, F')$. Then
*(a)* two conditions carrying the same label are comparable with respect to $\leq$,
*(b)* if $A$ is a cut of $(B, \leq)$, then $p(A)$ defines a reachable marking of $(N, M_0)$.

In [4], a new model checking technique for safe Petri nets is presented, based on nonsequential process semantics. The paper shows that the verification of an arbitrary property expressible in a certain temporal logic can be reduced to the following problem: given a finite process of the safe Petri net $(N, M_0)$, and two sets $\mathcal{G}$ and $\mathcal{H}$ of places of $N$, find an earliest cut such that its associated marking puts a token in each place of $\mathcal{G}$ and no token in the places of $\mathcal{H}$. (Actually, the problem discussed in [4] is that of finding a *latest* cut. It is easily reduced to the problem of finding an earliest cut.) For instance, if in the example of Fig. 5.2 we take $\mathcal{G} = \{s_1\}$ and $\mathcal{H} = \{s_2\}$, then the solution is the cut $\{b_1, b_4\}$, which corresponds to the reachable marking which puts tokens on $s_1$ and $s_5$.

As shown in [4], if this problem concerning processes can be solved in polynomial time, then some verification problems which required exponential time with the existing algorithms could be solved in polynomial time.

To reduce this problem to the asynchronous private meeting problem, construct the poset $(B, \leq)$. Choose as the set of labels $\mathcal{S}$ the set of places of the net $N$, and label a condition $b \in B$ with $p(b)$. Any two conditions carrying the same label are comparable by Proposition 5.1(a). The following proposition shows that the poset $(B, \leq)$ satisfies the condition (*) of Sect. 4, which allows to apply the polynomial algorithm given there. The proposition is also implicit in the above remarks on directed graphs.

**Proposition 5.2.** Let $(B, \le)$ be a poset obtained from a nonsequential process. Let $b, c, d \in B$ be conditions such that $c$ and $d$ are immediate successors of $b$. Then $\downarrow c = \downarrow d$.

*Proof.* Let $N' = (B, E, F')$ be the net from which the poset $(B, \le)$ is obtained. Every condition of $B$ has a unique event in its poset. Let $e$ be the unique event of the postset of $b$, which is the unique event in the preset of $c$ and $d$. So both $\downarrow c$ and $\downarrow d$ are equal to the set of conditions smaller than $e$.  □

In [4], the problem of finding the earliest marking was solved by means of a reduction to Linear Programming. This reduction shows that the problem is solvable in polynomial time, although still with large exponents for the known polynomial algorithms for Linear Programming. The algorithm proposed here is more efficient, and does not require to construct a system of inequalities, because it works directly on the poset.

## Acknowledgements

An early version of the algorithm was developed in discussions with Stefan Hougardy and Christoph Hundack, Bonn. Eike Best gave very helpful suggestions on exposition. A referee indicated the connection to N-free posets. Partial support was provided by the ESPRIT basic research working group CALIBAN and by the Volkswagen foundation.

## References

[1]  G. Behrendt: Maximal antichains in partially ordered sets. Ars Combinatoria 25 C (1988), 149–157.

[2]  E. Best and C. Fernández: Nonsequential Processes – A Petri Net View. EATCS Monographs on Theoretical Computer Science Vol. 13, Springer Verlag (1988).

[3]  K. M. Chandy and J. Misra: Parallel Program Design – A Foundation. Addison-Wesley (1988). (The meeting problem is considered on pages 14–18.)

[4]  J. Esparza: Model Checking Using Net Unfoldings. TAPSOFT'93: Theory and Practice of Software Development, M. C. Gaudel and J. P. Jouannaud (eds.), Lecture Notes in Computer Science 668 (1993), 613–628.

[5]  M. R. Garey and D. S. Johnson: Computers and Intractability – A Guide to the Theory of NP-Completeness. Freeman (1979).

[6]  P. A. Grillet: Maximal chains and antichains. Fundamenta Mathematicae 65 (1969), 157–167.

[7]  B. Leclerc and B. Monjardet: Ordres "C.A.C.". Fundamenta Mathematicae 79 (1973), 11–22.

[8]  W. Reisig: Petri Nets – An Introduction. EATCS Monographs on Theoretical Computer Science Vol. 4, Springer Verlag (1985).

# Gossiping in Vertex-Disjoint Paths Mode in Interconnection Networks *

## Extended Abstract

Juraj Hromkovič†, Ralf Klasing, Elena A. Stöhr
Department of Mathematics and Computer Science
University of Paderborn
33095 Paderborn, Germany

### Abstract

The communication modes (one-way and two-way mode) used for disseminating information among processors of interconnection networks via vertex-disjoint paths in one communication step are investigated. The complexity of communication algorithms is measured by the number of communication steps (rounds). Since optimal broadcast and accumulation algorithms for these modes can be achieved in a straightforward way for almost all interconnection networks used, the paper concentrates on the gossip problem. The main results are listed below:

1. Optimal gossip algorithms for paths, complete graphs and flakes in both modes.

2. For hypercubes, cube-connected cycles, butterfly networks, etc., gossip algorithms which are only about $O(\log_2 \log_2 n)$ rounds slower than the optimal gossip algorithms on complete graphs are designed. Furthermore, it is shown that at least $\Omega(\log_2 \log_2 \log_2 n)$ rounds more than needed by the optimal gossip algorithm on the complete graph are necessary for every "well-structured" gossip algorithm on networks with "small" degree.

Note that the results achieved have also practical application, because the vertex-disjoint paths mode can be implemented in several hardware realisations of computing networks.

---

*This work was partially supported by the German Research Association (DFG) and by the ESPRIT Basic Research Action No. 7141 (ALCOM II).

†This author was partially supported by SAV Grant No. 88 and by EC Cooperation Action IC 1000 Algorithms for Future Technologies.

# 1 Introduction and Definitions

This paper is devoted to the problem of information dissemination in prominent interconnection networks. We investigate the three basic communication tasks, **broadcast**, **accumulation**, and **gossip** which can be described as follows. Assume that each vertex (processor) in a graph (network) has some piece of information. The **cumulative message** of $G$ is the set of all pieces of information originally distributed in all vertices of $G$. To solve the **broadcast** [**accumulation**] problem for a given graph $G$ and a vertex $u$ of $G$, we have to find a communication strategy (using the edges of $G$ as communication links) such that all vertices in $G$ learn the piece of information residing in $u$ [that $u$ learns the cumulative message of $G$]. To solve the **gossip** problem for a given graph $G$, a communication strategy such that all vertices in $G$ learn the cumulative message of $G$ must be found. Since the above stated communication problems are solvable only in connected graphs, we note that from now on we use the notion "graph" for connected undirected graphs.

The meaning of a "communication strategy" depends on the communication mode. A communication strategy is realized by a **communication algorithm** consisting of a number of **communication steps (rounds)**. The rules describing what can happen in one communication step (round) are defined exactly by the communication mode. Here, we consider the following two modes:

1. **One-way [Two-way] vertex-disjoint paths mode (1VDP** mode [**2VDP** mode])

    One round can be described as a set $\{P_1, \ldots, P_k\}$ for some $k \in \mathbb{N}$, where $P_i = x_{i,1}, \ldots, x_{i,\ell_i}$ is a simple path of length $\ell_i - 1$, $i = 1, \ldots, k$, and the paths are vertex-disjoint. The executed communication of this round in one-way mode consists of the submission of the whole actual knowledge of $x_{i,1}$ to $x_{i,\ell_i}$ via path $P_i$ for any $i = 1, \ldots, k$. [The executed communication of this round in two-way mode consists of the complete exchange of the actual knowledge between $x_{i,1}$ and $x_{i,\ell_i}$ for any $i = 1, \ldots, k$]. The inner nodes of path $P_i$ (nodes different from the end points $x_{i,1}$ and $x_{i,\ell_i}$) do not learn the message submitted from $x_{i,1}$ to $x_{i,\ell_i}$ [exchanged between $x_{i,1}$ and $x_{i,\ell_i}$] they are only used to realize the connection from $x_{i,1}$ to $x_{i,\ell_i}$.

2. **Listen-in vertex-disjoint paths mode (LVDP** mode)

    As in the previous mode, a round can be described as a set of vertex-disjoint paths. The difference is that here, for each active path $P$ (realizing some connection between two accumulation points), one end-vertex of $P$ broadcasts (submits) its whole knowledge to all other vertices in $P$. Thus, after the execution of the round all vertices of each path $P$ know the message submitted from the sending endpoint.

The one-way vertex-disjoint paths mode was introduced (but not thoroughly investigated) by Farley [Fa80]. The listen-in vertex-disjoint paths mode was introduced in [FHMMM92], where optimal (or almost optimal) broadcast, accumulation, and gossip algorithms for almost all basic interconnection networks were found. The reason to consider the listen-in vertex-disjoint paths (LVDP) mode here is that there is some relation between the LVDP mode and the 1VDP mode, which will enable us to get some optimal communication algorithms in the 1VDP mode. We note that the 1VDP/2VDP mode has not been investigated until now (apart from some small results in [Fa80, FHMMM92]) despite of the fact that this mode is very practical (some of the algorithms were implemented on a Transputer system).

Now, let us fix the notations used in this paper. Let for any graph $G = (V, E)$, $V(G) = V$ denote the set of vertices of $G$, and $E(G) = E$ denote the set of edges of $G$. In what follows we will denote broadcast, accumulation and gossip as problems $B$, $A$, and $R$. For any given graph $G$ and a vertex $u$ of $G$, let $B_u(G)$ [$\bar{B}_u(G)$] denote the number of rounds (complexity) of the optimal broadcast algorithm from $u$ in $G$ in the 1VDP [LVDP] mode. This means that $B_u(G)$ [ $\bar{B}_u(G)$] for a graph $G$ and a vertex $u$ in $G$, is the necessary and sufficient number of rounds of the 1VDP [LVDP] mode to broadcast the piece of information originally residing in vertex $u$ to all other vertices in $G$. Furthermore, we define for any graph $G$

$$
\begin{aligned}
\mathbf{B(G)} &= \max\{B_u(G) \mid u \in V(G)\} \\
\mathbf{B}_{\min}(\mathbf{G}) &= \min\{B_u(G) \mid u \in V(G)\} \\
\bar{\mathbf{B}}(\mathbf{G}) &= \max\{\bar{B}_u(G) \mid u \in V(G)\} \\
\bar{\mathbf{B}}_{\min}(\mathbf{G}) &= \min\{\bar{B}_u(G) \mid u \in V(G)\}.
\end{aligned}
$$

Similarly, $A_u(G)$ [$\bar{A}_u(G)$] denotes the number of rounds (complexity) of the optimal accumulation algorithm for $G$ and $u$ in the 1VDP [LVDP] mode. For any graph $G$ we define

$$
\begin{aligned}
\mathbf{A(G)} &= \max\{A_u(G), \mid u \in V(G)\} \\
\mathbf{A}_{\min}(\mathbf{G}) &= \min\{A_u(G) \mid u \in V(G)\} \\
\bar{\mathbf{A}}(\mathbf{G}) &= \max\{\bar{A}_u(G) \mid u \in V(G)\} \\
\bar{\mathbf{A}}_{\min}(\mathbf{G}) &= \min\{\bar{A}_u(G) \mid u \in V(G)\}.
\end{aligned}
$$

Note that we do not need an extra definition of the complexity of broadcasting and accumulation in 2VDP mode, because there is no difference between one-way mode and two-way mode for the broadcast and accumulation problem (for a proof of this simple fact, see e.g. [HKMP92]). Finally, for any graph $G$, let $R(G), R^2(G), \bar{R}(G)$ be the number of rounds (complexity) of the optimal gossip algorithm for $G$ in the 1VDP, 2VDP, LVDP mode respectively. Moreover, for any communication algorithm $A$, let $round(A)$ be the number of rounds of $A$. Let $b = (1 + \sqrt{5})/2$ and $\log_b 2 \sim 1.44...$ throughout the paper.

The paper is organized as follows. Section 2 presents some simple observations which together with the fact $A_u(G) = \bar{A}_u(G)$, for every graph $G$ and each node $u$ of $G$, proved in [FHMMM92], provide optimal broadcast (accumulation) 1VDP algorithms for almost all fundamental interconnection networks. Thus, the rest of the paper (Section 3) is devoted to the gossip problem.

Section 3 is organized as follows. Subsection 3.1 informally presents the main algorithmic ideas and the lower bound proof techniques used in the subsequent subsections. Subsection 3.2 starts with the simplest case – gossiping in a path $P_n$ of $n$ nodes and in the complete graph $K_n$ of $n$ nodes. It is proved:

(1.a)  $2 \cdot \lfloor \log_2 n \rfloor \leq R(P_n) \leq 2 \cdot \lceil \log_2 n \rceil$  for any $n \in \mathbb{N}$,

(1.b)  $2 \cdot \lfloor \log_2 n \rfloor - 1 \leq R^2(P_n) \leq 2 \cdot \lceil \log_2 n \rceil - 1$  for any $n \in \mathbb{N}$,

(2.a)  $\log_b(\lfloor n/2 \rfloor) + 2 \leq R(K_n) \leq \log_b(\lfloor n/2 \rfloor) + 4$  for any $n \in \mathbb{N}$,

$\quad\quad R(K_n) = \lceil \log_b(n/2) \rceil + 2$  for infinitely many even $n \in \mathbb{N}$,

(2.b)  $R^2(K_n) = \lceil \log_2 n \rceil$  for any even $n \in \mathbb{N}$,

$\quad\quad R^2(K_n) = \lceil \log_2 n \rceil + 1$  for any odd $n \in \mathbb{N}$.

Giving a gossip scheme working on any graph $G_n$ of $n$ nodes, we get

(3.a)  $\log_b(\lfloor n/2 \rfloor) + 2 \leq R(G) \leq 2n - 2$,

(3.b)  $\lceil \log_2 n \rceil \leq R^2(G) \leq 2n - 4$.

Note that also the upper bounds of (3) are tight, because for the flake $F_n$ of $n$ nodes (the tree with one node of degree $n - 1$) we have $R(F_n) = 2n - 2$ and $R^2(F_n) = 2n - 4$.

Following (1) and (2), we get that for every graph $G$ of $n$ nodes with a Hamiltonian path

(4.a)  $1.44... \log_2 n + 2 = \lceil \log_b(\lfloor n/2 \rfloor) \rceil + 2 \leq R(G) \leq 2 \cdot \lceil \log_2 n \rceil$,

(4.b)  $\lceil \log_2 n \rceil \leq R^2(G) \leq 2 \cdot \lceil \log_2 n \rceil - 1$.

Since most of the interconnection networks used have a Hamiltonian path, the main aim of this paper is to search for the gossip complexity of these structures in the above stated ranges (4.a) and (4.b).

The results we get in Subsection 3.3 may be a little surprising, because we show that for the networks $G_n \in$ {hypercube, permutation network, cube-connected cycles, butterfly network, etc.}

(5.a)  $R(G_n) \leq R(K_n) + O(\log_2 \log_2 n) \leq \log_b n + O(\log_2 \log_2 n)$,

(5.b)  $R^2(G_n) \leq R^2(K_n) + O(\log_2 \log_2 n) \leq \log_2 n + O(\log_2 \log_2 n)$.

Thus, we have constant-degree networks in which one can gossip in time very close to $R(K_n)$ $[R^2(K_n)]$. The designed gossip algorithms are "well-structured" (called three-phase algorithms – defined later) and easy to implement. We will continue to deal with the task of proving that these algorithms are really effective, i.e. that there exist no essentially faster ones. Note that this task is far from being trivial, because even for the standard one-way mode we only know a few non-trivial lower bounds on gossiping [HJM90, KCV92, EM89]. For the hypercube-like networks there are no non-trivial lower bounds and so we were not able to give non-trivial lower bounds for our more powerful communication mode. But if we restrict the class of all gossip algorithms to a class of all "well-structured" gossip algorithms, we are able to show that no "well-structured" gossip algorithm working in time smaller than $R(K_n) + \Omega(\log_2 \log_2 \log_2 n)$ $[R^2(K_n) + \Omega(\log_2 \log_2 \log_2 n)]$ can be designed for networks with bounded degree.

We conclude by discussing the results obtained and formulate the main problems left open.

Note that most of the proofs are omitted in this extended abstract.

# 2 Broadcasting and Accumulation

First, let us observe that $\bar{A}(G) - \bar{B}(G)$ may be large for some graphs $G$ (see [FHMMM92]). For instance, for any path $P_n$ of $n$ nodes it is obvious that $\bar{A}(G) = \lceil \log_2 n \rceil$ and $\bar{B}(G) = 1$. On the other hand we have:

**Observation 2.1** *For every graph $G$, and every $u \in V(G)$:*

$$A_u(G) = B_u(G) .$$

**Proof.**    Each accumulation algorithm for $u$ and $G$ can be "reversed" (the sequence of rounds is reversed and also the direction of information flow in each round) to obtain a broadcast algorithm for $u$ and $G$, and vice versa.    □

Furthermore, the following result has been established in Theorem 7.1 of [FHMMM92]:

**Theorem 2.2** *For every graph $G$ and every $u \in V(G)$:*

$$\bar{A}_u(G) = A_u(G).$$    □

Using optimal accumulation algorithms designed in [FHMMM92] we get:

**Theorem 2.3**

(i) *For every graph $G$ of $n$ vertices with a Hamiltonian path, $A_{min}(G) = A(G) = B(G) = B_{min}(G) = \lceil \log_2 n \rceil$ (Theorem 2.1 in [FHMMM92] claims $\bar{A}_{min}(G) = \bar{A}(G) = \lceil \log_2 n \rceil$).*

*(ii)* For any complete binary tree, $CT_h^2$, of depth $h$, $A_{min}(CT_h^2) = A(CT_h^2) = B(CT_h^2) = B_{min}(CT_h^2) = h + 2$ *(Theorem 3.3 in [FHMMM92] claims $\bar{A}_{min}(CT_h^2) = \bar{A}(CT_h^2) = h + 2$).*

*(iii)* For any complete $k$-ary tree, $CT_h^k$, of the depth $h$, $A_{min}(CT_h^k) = B_{min}(CT_h^k) = (k-1) \cdot h + 1$ and $A(CT_h^k) = B(CT_h^k) = (k-1) \cdot h + 2$ *(Theorem 3.4 in [FHMMM92] claims $\bar{A}_{min}(CT_h^k) = (k-1) \cdot h + 1$ and $\bar{A}(CT_h^k) = (k-1) \cdot h + 2$).*

□

Since most of the basic interconnection networks (hypercube $Q_n$, permutation network $PN_n$, cube-connected cycles $CCC_n$, butterfly $BF_n$, DeBruijn $DB_n$, shuffle-exchange $SE_n$, grid $Gr_n^d$; for the definition and the notation of these networks, see e.g. [MS90, FL91, Le91]) are known to possess a Hamiltonian path [Le91, FM92], we have optimal broadcast and accumulation algorithms for these networks in 1VDP mode due to Theorem 2.3. Obviously, these algorithms are also optimal for 2VDP mode because two-way communication cannot help to decrease the complexity of broadcasting and accumulation. Thus, we will only deal with the gossip problem in what follows.

# 3 Gossiping

## 3.1 Introduction

In this section we investigate the gossip problem for several prominent interconnection networks. As already mentioned above, we start by estimating $R(P_n)$ and $R(K_n)$, and then continue to try designing gossip algorithms for some fundamental networks working in time closer to $R(K_n)$ than to $R(P_n)$. To do this, we develop a new method for designing so-called "three-phase algorithms". Let us informally define what a three-phase algorithm is. Let $G$ be any graph. Let $a(G)$ be any subset of nodes of $G$. We call $a(G)$ the **set of accumulation nodes/points**, and every node in $a(G)$ is called an **accumulation node/point**. A **three-phase gossip algorithm** for $G$ according to $a(G)$ works in the following three phases:

1. **Accumulation Phase**

   Divide $G$ into $|a(G)|$ connected components, each component containing exactly one accumulation node of $a(G)$. These components are called **accumulation components**. Each $v \in a(G)$ accumulates the information from the nodes lying in its component.

   {After the first phase, the nodes in $a(G)$ together know the cumulative message of $G$.}

2. **Gossip Phase**

Perform a gossip algorithm among the nodes in $a(G)$ in 1VDP (2VDP) mode (i.e. all nodes in $V(G) - a(G)$ are considered to have no information, and they are only used to build disjoint paths between receivers and senders from $a(G)$).

{After the second phase, every node in $a(G)$ knows the cumulative message of $G$.}

3. **Broadcast Phase**

Every node in $a(G)$ broadcasts the cumulative message in its component.

{After this, all nodes of $G$ know the cumulative message of $G$.}

In order to construct a really effective gossip algorithm, we shall search for an $a(G)$ in $G$ such that

a) Phase 2 can be performed (almost) as quickly as gossiping in a complete graph of $a(G)$ nodes.

b) The maximal size of a component is as small as possible, which minimizes the time for the first and third phase.

Obviously, every second phase of a three-phase algorithm $A$ corresponds unambiguously to a gossip algorithm $C$ in a complete graph of $|a(G)|$ nodes. We say that $C$ **is implemented in the second phase of** $A$. All algorithms designed here are three-phase algorithms with second phases implementing an optimal (or almost optimal) gossip algorithm on graphs of $|a(G)|$ nodes. But we will not directly design three-phase algorithms for all networks, we also use some standard techniques based on embeddings to transport gossip algorithms from one network to another.

The lower bound techniques used here are of two distinct kinds. Roughly speaking, the first one, used for paths, is based on the estimations how much pieces of information must and can flow through the edge dividing the path into two equal-sized parts. The second technique only works for three-phase algorithms and is based on some measurement of some quantitative parameters of the whole information flow in every gossip algorithm. This technique is used to show that gossiping in graphs of bounded degree is essentially harder than gossiping in complete graphs (this is not straightforward in vertex-disjoint modes, for LVDP mode gossiping can be done in $\lceil \log_2 n \rceil + 1$ rounds in some graphs of degree three [FHMMM92]).

## 3.2 Gossiping in Complete Graphs and in Paths

In this section we will first derive general upper and lower bounds for gossiping in any connected graph. We will see that the obtained bounds are tight by looking at the easiest case, the complete graph, and the hardest case, the $n$-flake, a graph we will define further down. Afterwards, we establish upper and lower bounds for gossiping in paths. This result implies an upper bound for any graph with a Hamiltonian path.

Let us start with the complete graph. The following lemma can be derived very easily by applying results for gossiping in the standard mode from [EM89, Kn75]:

**Lemma 3.1** *Let $K_n$ be the complete graph of $n$ nodes.*

*(1)* $R^2(K_n) = \begin{cases} \lceil \log_2 n \rceil & \text{for any even } n \in I\!N, \\ \lceil \log_2 n \rceil + 1 & \text{for any odd } n \in I\!N. \end{cases}$

*(2) Let $b = \frac{1+\sqrt{5}}{2}$, $m = \lfloor \frac{n}{2} \rfloor$. Then*

    *(a)* $\log_b m + 2 \leq R(K_n) \leq \log_b m + 3$   *for even $n \in I\!N$,*

    *(b)* $R(K_n) = \lceil \log_b m \rceil + 2$   *for infinitely many even $n \in I\!N$,*

    *(c)* $\log_b m + 2 \leq R(K_n) \leq \log_b m + 4$   *for any odd $n \in I\!N$.*

**Proof.** The upper bounds were described in [ES79, EM89] (for the one-way mode) and in [Kn75] (for the two-way mode).

For deriving the lower bounds, we transform any gossip algorithm communicating via node-disjoint paths into a "regular" gossip algorithm by substituting each communication path $x_{i,1}, \ldots, x_{i,\ell_i}$ by a single edge $x_{i,1}, x_{i,\ell_i}$. Thus, we see that the well-known lower bounds from [EM89, LW90, KCV92, SW93] (for the one-way mode) and [Kn75] (for the two-way mode) also hold for communication via vertex-disjoint paths.    □

Using Lemma 3.1, we immediately obtain general lower bounds for gossiping in any connected graph. In the following theorem, we also establish general upper bounds.

**Theorem 3.2** *Let $G$ be a connected graph of $n$ nodes, $n \geq 4$. Then*

*(1)* $\log_b(\lfloor n/2 \rfloor) + 2 \leq R(G) \leq 2n - 2,$

*(2)* $\lceil \log_2 n \rceil \leq R^2(G) \leq 2n - 4.$

**Proof.** The lower bounds follow directly from Lemma 3.1. To prove the upper bounds, number the nodes of $G$ by 1 to $n$.

For the one-way mode, we can gossip in $2n-2$ rounds in $G$ by first accumulating the information of the nodes $2, 3, \ldots, n$ in node 1 in the first $n-1$ rounds and then by broadcasting the cumulative message again from node 1 to the other nodes in the next $n-1$ rounds.

For the two-way mode, we can gossip in $2n - 4$ rounds by first accumulating the information of the nodes $5, 6, \ldots, n$ in the set of nodes $\{1, 2, 3, 4\}$ in the first $n - 4$ rounds, then by exchanging the information between nodes 1,2,3,4 in 4 rounds, and finally by broadcasting the cumulative message from nodes 1,2,3,4 to the other nodes in $n - 4$ rounds. □

To see that the upper bounds are also tight, we introduce the $n$-flake $F_n$, which is a graph of $n$ nodes with one node in the center and all other nodes connected to it. Formally, $F_n = (V_n, E_n)$ where $V_n = \{1, \ldots, n\}$ and $E_n = \{(1, 2), (1, 3), \ldots, (1, n)\}$.

**Lemma 3.3** *For every $n \geq 4$:*

(1) $R(F_n) = 2n - 2$,

(2) $R^2(F_n) = 2n - 4$.

Let us now take a look at a path $P_n$ of $n$ nodes. Simple upper bounds for gossiping in $P_n$ in the VDP mode can be derived from the broadcasting schemes in [Fa80]. For the lower bounds, we use a cut argument bounding the number of pieces of information which must and can flow through the center edge of the path in a given number of rounds in order to complete the gossip task.

**Lemma 3.4** *Let $P_n$ be a path of $n$ nodes. Then*

(1) $2 \cdot \lfloor \log_2 n \rfloor \leq R(P_n) \leq 2 \cdot \lceil \log_2 n \rceil$,

(2) $2 \cdot \lfloor \log_2 n \rfloor - 1 \leq R^2(P_n) \leq 2 \cdot \lceil \log_2 n \rceil - 1$.

Lemma 3.1 and 3.4 imply the following theorem for any graph $G$ with a Hamiltonian path (e.g. hypercube, permutation network, cube-connected cycles, butterfly, DeBruijn, shuffle-exchange, grid, etc.):

**Theorem 3.5** *Let $G$ be a graph of $n$ nodes with a Hamiltonian path. Then*

(1) $\log_b(\lfloor n/2 \rfloor) + 2 \leq R(G) \leq 2 \cdot \lceil \log_2 n \rceil$,

(2) $\lceil \log_2 n \rceil \leq R^2(G) \leq 2 \cdot \lceil \log_2 n \rceil - 1$.

Now, the main question is where exactly in the range given in Theorem 3.5 does the complexity for gossiping in fundamental networks lie. In the next section we show that in several cases the complexity of gossiping is much closer to $R(K_n)$ [$R^2(K_n)$] than to $R(P_n)$.

## 3.3 Gossiping in Hypercube-Like Networks

In this section, we construct effective gossip algorithms for the hypercube $Q_k$ and hypercube-like networks, namely the cube-connected cycles $CCC_k$, butterfly network $BF_k$, etc. (for the definition and the notation of these networks, see e.g. [MS90, FL91, Le91]). The algorithms for the cube-connected cycles and the butterfly network show that there are constant-degree networks in which one can gossip in time very close to $R(K_n)$ $[R^2(K_n)]$. All the designed algorithms are three-phase algorithms. Our strategy will be to design such algorithms for some suitably chosen structures, and then transfer them to further known networks by using embedding techniques.

We start with the two-way mode. First, we design effective algorithms for cube-connected cycles and butterfly network. For these purposes, we will apply the well-known gossip algorithm for the hypercube.

**Lemma 3.6 (Kn75)** *Let $Q_k$ be the hypercube of dimension $k$ (and $n = 2^k$ nodes) for a positive integer $k$. Then*

$$R^2(Q_k) = k = \log_2 n = R^2(K_n).$$ $\square$

For convenience, we state the algorithm from [Kn75]:

### Knödel (Hypercube) Algorithm

> begin
> > for $i = 1$ to $k$ do
> > > for all $\alpha \in \{0,1\}^{i-1}$, $\beta \in \{0,1\}^{k-i}$ do in parallel
> > > > exchange information between $\alpha 0 \beta$ and $\alpha 1 \beta$ ;
> end;

We use the hypercube algorithm for the second phase of the cube-connected-cycles algorithm:

**Theorem 3.7** *Let $CCC_k$ be the cube-connected-cycles network of dimension $k \geq 2$ (and $n = k \cdot 2^k$ nodes). Then*

$$R^2(CCC_k) \leq \log_2 n + \log_2 \log_2 n = R^2(K_n) + O(\log_2 \log_2 n).$$

**Proof.** Construct a three-phase algorithm by choosing as accumulation nodes the vertices $(0, \alpha), \alpha \in \{0,1\}^k$, at level 0 of the cube-connected-cycles network. As accumulation components we take the cycles $C_\alpha = \{(0, \alpha), (1, \alpha), \ldots, (k - 1, \alpha)\}, \alpha \in \{0,1\}^k$. Accumulating the messages of $C_\alpha$ in $(0, \alpha)$ in Phase 1 takes $\log_2 k$ rounds, the same holds for broadcasting the cumulative message from $(0, \alpha)$ to $C_\alpha$ in Phase 3. In Phase 2 of the three-phase algorithm, we simulate the hypercube algorithm on all accumulation nodes as follows: Instead of having a direct communication between $\alpha 0\beta$ and $\alpha 1\beta$ in round $i$ of the hypercube algorithm, we communicate via the path $(0, \alpha 0\beta) - (1, \alpha 0\beta) - \ldots - (i-1, \alpha 0\beta) - (i, \alpha 0\beta) - (i, \alpha 1\beta) - (i-1, \alpha 1\beta) - \ldots - (1, \alpha 1\beta) - (0, \alpha 1\beta)$. It is obvious that all paths in round $i$ are vertex-disjoint. Hence, Phase 2 takes $k$ rounds. Therefore, the whole 3-phase algorithm takes

$$k + 2\log_2 k \leq \log_2 n + \log_2 \log_2 n$$

rounds. □

As the cube-connected cycles is a subgraph of the butterfly network [FU92], we immediately obtain:

**Theorem 3.8** *Let $BF_k$ be the butterfly network of dimension $k \geq 2$ (and $n = k \cdot 2^k$ nodes). Then*

$$R^2(BF_k) \leq \log_2 n + \log_2 \log_2 n = R^2(K_n) + O(\log_2 \log_2 n).$$ □

Now, let us turn to the one-way mode, which turns out to be a little more complicated, because the above ideas cannot be directly transferred. Similar to Theorem 3.7, it is easy to show that there exists a graph of degree 3 with a very efficient gossip scheme. Using this we can establish the following results:

**Theorem 3.9** *For each network $X_k \in \{BF_k, CCC_k, Q_k\}$ of $n$ nodes and dimension $k$:*

$$R(X_k) \leq R(K_n) + O(\log_2 \log_2 n).$$

Above, we have designed effective gossip algorithms for some fundamental networks. As usual, the main technical difficulty lies in proving that the proposed algorithms are really effective, i.e. to prove lower bounds on gossiping in these networks. How complex this task is can be realized by the fact that nobody was able to prove a non-trivial (higher than the diameter or the broadcast complexity of the given interconnection network) lower bound on gossiping in hypercube-like networks in standard one-way mode. Optimal lower bounds for standard one-way mode are known only for complete graphs, cycles and some trees [EM89, KCV92, HJM90]. Obviously, to prove a lower bound on gossiping in VDP modes is even harder. Additionally, the lower bounds provided by the diameter do not work.

In what follows, we tackle this problem by developing a lower bound technique allowing at least to show that three-phase algorithms with a "quick" second phase must use essentially more rounds than $R(K_n)$ $[R^2(K_n)]$ for interconnection networks with small degree.

**Theorem 3.10** *Let A (B) be a 2VDP (1VDP) three-phase gossip algorithm for a graph $G_n$ of n nodes and degree bounded by a constant independent of n. Let the second phase of A (B) be the implementation of the Knödel (Fibonacci) gossip algorithm on complete graphs. Then*

*(1)* round$(A) \geq R^2(K_n) + \Omega(\log_2 \log_2 \log_2 n)$, *and*

*(2)* round$(B) \geq R(K_n) + \Omega(\log_2 \log_2 \log_2 n)$.

# 4  Conclusion

In this paper we have studied the power of VDP communication modes. Due to the previous results on the LVDP mode in [FHMMM92], we have shown that there are no important problems for broadcasting and accumulation in VDP modes left open. For some fundamental interconnection networks we have designed "well-structured" three-phase algorithms for gossiping. The gossip algorithms presented seem to be effective, and we conjecture that they are even optimal despite of the fact that we were not able to prove this for many cases. Thus, the main open problems left are concerned with lower bounds. Note again that this is not surprising, because also for the much weaker, standard one-way mode there are only a few non-trivial lower bounds on gossiping (and consequently only a few optimal gossip algorithms are known).

# References

[EM89]    S. Even, B. Monien, "On the number of rounds necessary to disseminate information", *Proc. 1st ACM Symp. on Parallel Algorithms and Architectures*, Santa Fe, June 1989, pp. 318-327.

[ES79]    R.C. Entringer, P.J. Slater, "Gossips and telegraphs", *J. Franklin Institute* 307 (1979), pp. 353-360.

[Fa80]    A.M. Farley, "Minimum-Time Line Broadcast Networks", *Networks*, Vol. 10 (1980), pp. 59-70.

[FHMMM92] R. Feldmann, J. Hromkovič, S. Madhavapeddy, B. Monien, P. Mysliwietz, "Optimal algorithms for dissemination of information in generalized communication modes", *Proc. PARLE'92*, Lecture Notes in Computer Science 605, Springer Verlag 1992, pp. 115-130.

[FL91]    P. Fraigniaud, E. Lazard, "Methods and problems of communication in usual networks", Technical Report, Université de Paris-Sud, 1991, submitted to *Discrete Applied Mathematics*, special issue on "Broadcasting and Gossiping".

300

[FM92] R. Feldmann, P. Mysliwietz, "The Shuffle Exchange Network has a Hamiltonian Path", *Proc. of 17th Math. Foundations of Computer Science* (MFCS'92), Springer LNCS 629, pp. 246-254.

[FU92] R. Feldmann, W. Unger, "The Cube-Connected Cycles Network is a Subgraph of the Butterfly Network", *Parallel Processing Letters*, Vol. 2, No. 1 (1992), pp. 13-19.

[HHL88] S.M. Hedetniemi, S.T. Hedetniemi, A.L. Liestman, "A survey of gossiping and broadcasting in communication networks", *Networks*, Vol. 18, pp. 319-349, 1988.

[HJM90] J. Hromkovič, C. D. Jeschke, B. Monien, "Optimal algorithms for dissemination of information in some interconnection networks (extended abstract)", *Proc. MFCS'90*, Lecture Notes in Computer Science 452, Springer Verlag 1990, pp. 337-346.

[HKMP92] J. Hromkovič, R. Klasing, B. Monien, R. Peine, "Dissemination of Information in Interconnection Networks (Broadcasting and Gossiping)", manuscript, University of Paderborn, Germany, Feb. 1993, to appear as a book chapter in: F. Hsu, D.-Z. Du (Eds.), *Combinatorial Network Theory*, Science Press & AMS, 1994.

[HS74] F. Harary, A.J. Schwenk, "The communication problem on graphs and digraphs", *J. Franklin Institute* 297 (1974), pp. 491-495.

[KCV92] D.W. Krumme, G. Cybenko, K.N. Venkataraman, "Gossiping in minimal time", *SIAM J. Comput.* 21 (1992), pp. 111-139.

[KMPS92] R. Klasing, B. Monien, R. Peine, E. Stöhr, "Broadcasting in Butterfly and DeBruijn networks", *Proc. STACS'92*, Lecture Notes in Computer Science 577, Springer Verlag 1992, pp. 351-362.

[Kn75] W. Knödel, "New gossips and telephones", *Discrete Math.* 13 (1975), p. 95.

[Le91] F.T. Leighton, "Introduction to Parallel Algorithms and Architectures: Array, Trees, Hypercubes", *Morgan Kaufmann Publishers* (1991).

[LW90] R. Labahn, I. Warnke, "Quick gossiping by multi-telegraphs", In R. Bodendiek, R. Henn (Eds.), *Topics in Combinatorics and Graph Theory*, pp. 451-458, Physica-Verlag Heidelberg, 1990.

[MS90] B. Monien, I.H. Sudborough, "Embedding one Interconnection Network in Another", *Computing Suppl.* 7, pp. 257-282, 1990.

[SW93] V.S. Sunderam, P. Winkler, "Fast information sharing in a complete network", *Discrete Applied Mathematics* 42 (1993), pp. 75-86.

# The Folded Petersen Network : A New Versatile Multiprocessor Interconnection Topology

Sabine R. Öhring  and  Sajal K. Das *

Department of Computer Science
University of Würzburg
D 97074 Würzburg, Germany

Department of Computer Science
University of North Texas
Denton, TX 76203-3886, USA

**Abstract.** We introduce and analyze a new interconnection topology, called the $n$–folded Petersen network ($FP_n$), which is constructed by iteratively applying the cartesian product operation on the well–known Petersen graph itself. The $FP_n$ topology provides regularity, node– and edge–symmetry, optimal connectivity (and therefore maximal fault–tolerance), logarithmic diameter, modularity, and simple routing and broadcasting algorithms even in the presence of faults. With the same node–degree and connectivity, $FP_n$ has smaller diameter and accommodates more nodes than the $3n$–dimensional binary hypercube.

This paper also emphasizes the versatility of $FP_n$ as a multiprocessor interconnection topology by providing embeddings of many computationally important structures such as rings, multi–dimensional meshes, hypercubes, complete binary trees, X-trees, tree machines, pyramids and dynamically evolving binary trees.

**Keywords:** broadcasting, tree, embedding, fault–tolerance, hypercube, interconnection network, mesh, Petersen graph, routing.

## 1 Introduction

An interconnection network of a multiprocessor system can be modelled as an undirected graph $G = (V, E)$, where the set $V$ of nodes represents the processors and the set $E$ of edges represents the bidirectional communication links among the processors. Existing static interconnection networks include complete binary trees, X-trees, meshes, hypercubes, butterflies, cube–connected cycles, shuffle-exchange and de Bruijn networks, Stirling networks, and so on [BP89, DGD92, Lei92, MS90].

In this paper, we propose a new interconnection topology, called the $n$–folded Petersen network ($FP_n$). It is constructed by an iterative cartesian product on the Petersen graph [CW85]. It turns out that $FP_n$ is better than the widely used hypercubes with respect to the usual performance metrics of a multiprocessor architecture. The proposed network is extremely symmetric, regular, optimally fault–tolerant with logarithmic diameter, and permits self–routing and broadcasting algorithms. We also show that rings, multi–dimensional meshes, hypercubes, complete binary trees, dynamically evolving trees, X-trees, pyramids and tree machines can be embedded efficiently in $FP_n$.

* This work is partially supported by Texas Advanced Research Program Grant under Award No. 003594003. The authors can be reached via E-mail at oehring@informatik.uni-wuerzburg.de and das@cs.unt.edu

The paper is organized as follows. In Section 2, various graph-theoretic terminology and notations are defined. Section 3 gives a definition of the $n$–folded Petersen network and analyzes its topological properties. In Section 4 we give routing and broadcasting algorithms amenable to faulty processors. Section 5 is devoted to embeddings into $FP_n$.

## 2  Notations and Definitions

The *distance*, $dist_G(u,v)$, between two nodes $u$ and $v$ in the interconnection topology, $G = (V, E)$, is the length of a shortest path between them. Two paths are said to be *node-disjoint* if they have no common nodes, except for the source and destination nodes. The *diameter*, $d$, of a network is the maximum distance among all node–pairs. It is a measure of the network performance in terms of worst-case communication delay. The *node-connectivity*, $\kappa$, is the number of nodes whose removal results in a disconnected network. It is a measure of fault-tolerance of the network. The *f–fault diameter* is defined as the maximum of diameters over all possible graphs obtained by removing at most $f$ nodes. The *degree*, $deg(u)$, of a node $u$ is the number of communication links incident on it. If $deg(u) = \delta$ for all nodes $u \in V$, then $G$ is called $\delta$-*regular*. The *cost* of a network can be defined as $C = d * \delta$.

The *Petersen graph*, $P$, with ten nodes has an outer 5-cycle, an inner 5-cycle and five spokes joining them (see Figure 1a)). It is known that $P$ is the smallest 3-regular graph having girth 5. It provides node– and edge–symmetry, with diameter $d = 2$, and node-connectivity $\kappa = 3$. For details, see [CW85].

An $n$-dimensional *binary hypercube*, $Q_n$, has node-set $V_n = \mathbf{Z}_2^n$ (the set of binary strings of length $n$) and there is an edge between two nodes $x$ and $y$, iff their binary labels differ in exactly one bit.

A *generalized hypercube* [BA84], of base $b$ and dimension $n$ is a graph $GQ_b^n = (V, E)$ where, $V = \mathbf{Z}_b^n = \{x_{n-1}\ldots x_0 \mid x_i \in \mathbf{Z}_b, 0 \leq i \leq n-1\}$ and $E = \{(x_{n-1}\ldots x_0, y_{n-1}\ldots y_0) \mid \exists j, 1 \leq j \leq n, \ x_j \neq y_j \text{ and } x_i = y_i \text{ for } i \neq j\}$.

The *undirected de Bruijn graph* [dB46] of base $d$ and order $n$, $DG(d,n)$, is defined by the node-set $V_{d,n} = \mathbf{Z}_d^n = \{x_{n-1}\ldots x_0 \mid x_i \in \mathbf{Z}_d \text{ for } 0 \leq i \leq n-1\}$ and the edge-set $E_{d,n} = \{\{x_{n-1}\ldots x_0, x_{n-2}\ldots x_0 p\} \mid p, x_i \in \mathbf{Z}_d, \text{ and } 0 \leq i \leq n-1\}$.

For the definition of the $n$–folded Petersen network, we need the *cartesian product* of $n$ graphs $G_1, G_2, \ldots, G_n$ with $G_i = (V_i, E_i)$ for $1 \leq i \leq n$. The resulting network $G = G_1 \times G_2 \ldots \times G_n$ has the node-set $V = V_1 \times V_2 \ldots \times V_n$. Let $(u_1, \ldots, u_n)$ and $(v_1, \ldots, v_n)$ be two nodes in $V$. Then $\{(u_1, \ldots, u_n), (v_1, \ldots, v_n)\}$ is an edge in $G$ if and only if there exists $k \in \{1, \ldots, n\}$ such that $(u_k, v_k)$ is an edge in $E_k$, and $u_i = v_i$ for $1 \leq i \leq n, i \neq k$.

By definition, $Q_n = Q_{n-1} \times K_2$ is a product graph but $DG(2, n)$ is not. Similarly, $GQ_b^n = GQ_b^{n-1} \times K_b$, where $K_b$ is a complete graph of order $b$. The cartesian product operation has recently been used to define *hyper-de Bruijn* networks $HD(m,n) = Q_m \times DG(2, n)$ [GP91] and *hyper Petersen* networks $HP_n = Q_{n-3} \times P$ [DB92].

## 3  Folded Petersen Network: Topological Properties

The proposed $n$-folded Petersen graph, $FP_n = P \times \ldots \times P = F_{n-1} \times P$, is the iterative cartesian product on the Petersen graph ($P$) itself. That is, $FP_n$ is recursively constructed from ten $FP_{n-1}$'s. Figure 1b) gives a schematic representation of $FP_2$.

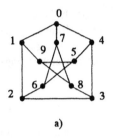
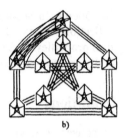

a)                        b)

**Fig. 1.** a) Petersen graph $P$      b) A schematic representation of $FP_2$

More formally, $FP_n = (V, E)$, where $V = \{x = (x_1, x_2, \ldots, x_n) \mid x_i \in \{0, 1, \ldots, 9\}$ for $1 \leq i \leq n\}$, and $E = \{\{(x_1, x_2, \ldots, x_n), (x_1, \ldots, x_{i-1}, y, x_{i+1}, \ldots, x_n)\} \mid y \in \{0, 1, \ldots, 9\}$ and $dist_P(y, x_i) = 1$ for $1 \leq i \leq n\}$.

One advantage of constructing a large network with the help of the cartesian product operation is that the orthogonality of the component networks is preserved. This implies that $FP_n$ is node– and edge–symmetric because of the node– and edge–symmetry of the Petersen graph $P$ [CW85]. Furthermore, $FP_n$ is $3n$–regular and its diameter is $2n$, since the Petersen graph is $3$–regular and has a diameter of two.

Multiple disjoint paths between two nodes provide a measure of fault–tolerance of a network. In a non–faulty Petersen graph, there exists one path of length 2 as well as two node–disjoint paths of length 3 for any two non–adjacent nodes $u$ and $v$. If the nodes are adjacent, there exists one path of length 1 and two paths of length 4 between them.

**Theorem 1.** *Let $x$ and $y$ be two arbitrary nodes in $FP_n$. There exist $3n$ node–disjoint paths between $x$ and $y$ of length at most $dist_{FP_n}(x, y) + 3$. The node–connectivity of $FP_n$ is $3n$ and thus $FP_n$ is maximally fault–tolerant (since the fault–tolerance is one less than the node–degree). The $(3n - 1)$–fault diameter of $FP_n$ is $2n + 3$.*

**Proof :** Let $x = (x_1, \ldots, x_n)$ and $y = (y_1, \ldots, y_n)$ be two nodes in $FP_n$. There exist $\lambda_\mu \in \{1, \ldots, n\}$ such that $x_{\lambda_\mu} = y_{\lambda_\mu}$ for $1 \leq \mu \leq k$. Hence, $n - k$ components of $x$ and $y$ are different, say $x_{i_1}, \ldots, x_{i_{n-k}}$. Let $(x_{i_j}, x'_{i_j}, y_{i_j})$ denote the unique shortest path in the Petersen graph $P$ of length at most 2 between the nodes $x_{i_j}$ and $y_{i_j}$, and let $x'_{\lambda_\mu}$ denote the $l$th, $1 \leq l \leq 3$, neighbor of $x_{\lambda_\mu}$ in $P$. The paths whose node–disjointness is proven in [ÖD93d] can be divided into four subsets :

1. $n - k$ shortest paths $\mathcal{P}^1_m, 1 \leq m \leq n - k$, between $x$ and $y$ (of length $dist_{FP_n}(x, y)$) by equalizing the differing components in the order $i_m, i_{m+1}, \ldots, i_{n-k}, i_1, \ldots i_{m-1}$.
2. $3k$ paths $\mathcal{P}^2_i = (x, (\ldots, x'_{\lambda_\mu}, \ldots), \ldots$), components are switched as in $\mathcal{P}^1_i$ but with $\lambda_\mu$th component $x'_{\lambda_\mu}, (\ldots, x'_{\lambda_\mu}, \ldots, y_{i_1}, \ldots, y_{i_{n-k}}, \ldots), (\ldots, x_{\lambda_\mu}, \ldots) = y)$, for $1 \leq \mu \leq k, l = 1, 2, 3$, and $i = 1, \ldots, 3k$. Each of these paths has length $dist_{FP_n}(x, y) + 2$.
3. $n - k$ paths $\mathcal{P}^3_j = (x, (\ldots, x^2_{i_j}, \ldots), \ldots$), components are switched as in $\mathcal{P}^1_j$ except for the $i_j$th component, $(\ldots, x^2_{i_j}, \ldots, y_{i_s}, \ldots)$ for $s \neq j$, switching $i_j$th component as in path $p^2_{i_j}, y)$. Path $\mathcal{P}^3_j$ has length $dist_{FP_n}(x, y) + \delta_{i_j}$, where $\delta_{i_j} = 3$ if $dist_P(x_{i_j}, y_{i_j}) = 1$, $\delta_{i_j} = 1$ otherwise. Also $p^2_{i_j} = (x_{i_j}, x^2_{i_j}, \ldots, y_{i_j})$ denotes the non–shortest path in $P$ between $x_{i_j}$ and $y_{i_j}$ that routes over the neighbor $x^2_{i_j}$ of $x_{i_j}$.
4. $n - k$ paths $\mathcal{P}^4_j$, $1 \leq j \leq n - k$ of length $dist_{FP_n}(x, y) + \delta_{i_j}$, where $(x_{i_j}, x^3_{i_j}, \ldots, y_{i_j})$ denotes the non–shortest path $p^3_{i_j}$ in $P$ between $x_{i_j}$ and $y_{i_j}$ that routes over the neigh-

bor $x_{i_j}^3$ of $x_{i_j}$. $\mathcal{P}_j^4 = (x, (\ldots, x_{i_j}^3, \ldots))$, components are switched as in $\mathcal{P}_j^1$ except for the $i_j$th components, $(\ldots, x_{i_j}^3, \ldots y_{i_\bullet}, \ldots)$ for $s \neq j$, switching $i_j$th component as in path $p_{i_j}^3, y)$.

All paths $\mathcal{P}_j^1$ have length $dist_{FP_n}(x, y)$, paths $\mathcal{P}_j^2$ have length $dist_{FP_n}(x, y) + 2$, while paths $\mathcal{P}_j^3$ and $\mathcal{P}_j^4$ have length at most $dist_{FP_n}(x, y) + 3$. $\square$

Figure 2 shows the 9 node–disjoint paths between nodes $(0, 0, 0)$ and $(0, 1, 2)$ in $FP_3$.

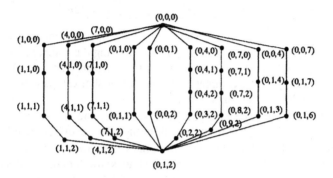

**Fig. 2.** The 9 node–disjoint paths in $FP_3$

Table 1 compares various topological properties of the $n$–folded Petersen network with those of the binary hypercube, generalized hypercube of base $b$, hyper Petersen network, de Bruijn graph, and hyper–de Bruijn network of the closest sizes. Clearly, the $FP_n$ network with $10^n$ nodes, diameter of $2n$ and cost $C = 6n^2$ has the same degree and connectivity as the binary hypercube $Q_{3n}$ which consists of only $8^n$ nodes, has diameter $3n$ and cost $9n^2$.

Covering the same number of nodes as $FP_n$, the generalized hypercube $GQ_{10}^n$ has degree $9n$ and cost $9n^2$. On the other hand, considering the diameter $(d)$ as the criterion for comparison, $FP_n$ accommodates $10^n$ nodes with degree $3n$ compared to $GQ_3^{2n}$ with $9^n$ nodes and higher degree $4n$. Again, maintaining the same degree $3n$, $GQ_4^n$ covers only $4^n$ nodes with reduced diameter of $n$.

Let us compare $FP_n$ network with two other product graphs. With the same degree $3n$, the hyper Petersen network $HP_n$ connects only $(\frac{5}{4})8^n$ nodes with a larger diameter $3n - 1$. For the same diameter $2n$, the hyper–de Bruijn network $HD(n, n)$ connects only $4^n$ nodes with a lower degree of $n + 4$ and inferior connectivity. Thus the ratio of the number of nodes to degree is larger in $FP_n$.

The problem with the de Bruijn network compared to $FP_n$ or $Q_{3n}$ is that the connectivity does not grow with the size of the network and that it is neither symmetric nor modular.

Further performance properties like container quality, reliability and packing density have been presented in [ÖD93c].

## 4   Fault-Tolerant Routing and Broadcasting

An interconnection network should provide a simple and fast routing algorithm that tolerates faulty nodes and edges. Let $x = (x_1, \ldots, x_n)$ and $y = (y_1, \ldots, y_n)$ denote the source and destination nodes, respectively. Furthermore, let $i_1, \ldots, i_{n-k}$ be the indices

**Table 1.** Comparison of topological properties

| Networks | # Nodes | degree ($\delta$) | diameter ($d$) | connectivity ($\kappa$) | Cost= $d * \delta$ |
|---|---|---|---|---|---|
| $FP_n$ | $10^n$ | $3n$ | $2n$ | $3n$ | $6n^2$ |
| $Q_{3n}$ | $8^n$ | $3n$ | $3n$ | $3n$ | $9n^2$ |
| $GQ_{10}^n$ | $10^n$ | $9n$ | $n$ | $9n$ | $9n^2$ |
| $GQ_3^{2n}$ | $9^n$ | $4n$ | $2n$ | $4n$ | $8n^2$ |
| $HP_{3n}$ | $(\frac{5}{4})8^n$ | $3n$ | $3n-1$ | $3n$ | $9n^2 - 3n$ |
| $DG(10,n)$ | $10^n$ | $20$ | $n$ | $18$ | $20n$ |
| $HD(2n,n)$ | $8^n$ | $2n+4$ | $3n$ | $2n+2$ | $6n^2 + 12n$ |

such that $x_{i_j} \neq y_{i_j}$ for $1 \leq j \leq n - k$ and $i_j \in \{1, \ldots, n\}$. A (shortest path) self–routing algorithm in a non–faulty $FP_n$ is given by equalizing the components $x_{i_j}$ and $y_{i_j}$ from $j = 1$ to $j = n - k$ using self–routing in the Petersen graph, $P$. Due to the symmetry, the self–routing in $P$ is completely defined by the shortest–path routing from node 0 to any other node in $P$ (cf. Figure 4a)).

By supplying each node with the status of all $10^n - 1$ other nodes, a source node can choose a fault–free path of length at most $2n + 3$ through the use of Theorem 1.

Next, we describe a routing algorithm which tolerates upto $3n - 1$ node failures in $FP_n$, where each node $x = x_1 \ldots x_n$ is required to know the status of only $9n$ other nodes in $\{x[i, z_i] | 1 \leq i \leq n, z_i \in \{0, 1, \ldots, 9\} \setminus \{x_i\}\}$, where $x[i, z_i] = x_1 \ldots x_{i-1} z_i x_{i+1} \ldots x_n$.

**Routing–Algorithm** $\{$in $FP_n$ with at most $3n - 1$ faulty nodes$\}$

1   $I := \{i_1, \ldots, i_{n-k}\}$ /* indices of elements different in $x$ and $y$ */
2   **While** $x \neq y$ **do**
3       Search the first index $i_l \in I$ such that there exists a path between $x$ and $x[i_l, y_{i_l}]$;
4       **If** such an $i_l$ exists **then**
5           $x := x[i_l, y_{i_l}]$ ; $I := I \setminus \{i_l\}$ ;
6       **else**
7           search an index $i_l \in I$ such that a step from $x$ on a path routing in the $i_l$th element
8           to $x[i_l, y_{i_l}]$ is possible;
9           **If** such an $i_l$ exists **then**
10        Go from $x$ on a path routing in the $i_l$th element to $x[i_l, y_{i_l}]$ as far as possible (last
11        node is $x'$); $x := x'$;
12        **else**
13           From all neighbors $\tilde{x}^\lambda$ of $x$ in an element $\lambda \notin I$, choose those from which
14           one can go at least one step on a path routing in an $i_l$th element to $\tilde{x}^\lambda[i_l, y_{i_l}]$,
15           $i_l \in I$; among those neighbors choose an $\tilde{x}^{\lambda'}$ whose index $\lambda'$ has not been
16           used for the longest time; $I := I \cup \{\lambda'\}$;
17 **end.**

We prove in [ÖD93f] that this algorithm terminates and tolerates up to $3n - 1$ faults. Figure 3 gives an example for routing in $FP_3$ from node $(0, 0, 0)$ to $(3, 0, 1)$ with faulty nodes $(0, 0, 1), (0, 0, 4), (0, 0, 7), (2, 0, 0), (3, 0, 0,), (2, 0, 1), (9, 0, 1)$ and $(2, 1, 1)$. The thick drawn path is the path routed by the above algorithm.

Next, we develop a broadcasting algorithm in $FP_n$, used for disseminating a message starting from one source node to all others. Because of the node–symmetry, let us restrict to node $(0, \ldots, 0)$ in $FP_n$ as the source node. The broadcasting is done by combining the schemes for binary hypercube and Petersen graph as in Figures 4a) and b). In this

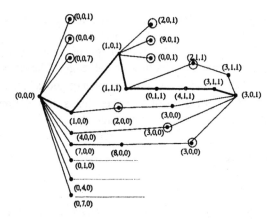

**Fig. 3.** Routing in $FP_3$ (faulty nodes are drawn circled)

figure, the labels on the edges $\{x, y\}$ in the broadcast spanning tree of $Q_n$ give the bit positions in which nodes $x$ and $y$ differ.

Starting at node $(0, \ldots, 0)$, we go in each position to its three neighbors 1, 4 and 7 in $P$. Subsequently, from each node $x$ at the current level of the broadcasting tree, we go to all of its neighbors in $P$ in the current dimension (due to the broadcasting in $P$) and to all the neighbors in the next dimension (due to the broadcasting in $Q_n$). An example is shown in Figure 4c), in which $P^j$ represents the broadcasting–tree of Figure 4a) with labels $(x, j), 0 \le x \le 9$. The height of this spanning tree is $2n$ and therefore optimal.

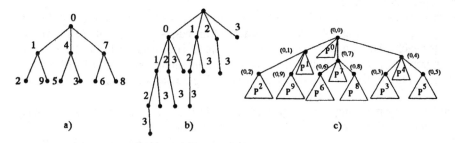

**Fig. 4.** Broadcasting a) in $P$, b) in $Q_4$, and c) in $FP_2$

In [ÖD93c] we compute also the average distance of $FP_n$. Additional optimal algorithms for communication primitives like gossiping, scattering, accumulation and others are presented by us in [ÖD93f].

## 5 Embeddings in Folded Petersen Network

This section is devoted to mappings of important structures like linear arrays, rings, meshes, hypercubes and trees in the $n$–folded Petersen network. One motivation for studying graph embeddings is the simulation of one interconnection network on another to efficiently implement an algorithm developed on one interconnection network onto another. Also, the computation may have an interdependence structure that has to be mapped on the network.

An *embedding* of a *guest* graph $G = (V(G), E(G))$ into a *host* graph $H$ is a mapping of their nodes along with a mapping $\psi : E(G) \to \{paths\ in\ H\}$, from edges in $G$ to paths in $H$. The *dilation* of an edge, $e \in E(G)$, is the length of the path $\psi(e)$ in $H$. The dilation of an embedding is the maximum dilation over all edges. The *expansion* of an embedding is $\frac{|V(H)|}{|V(G)|}$, while the *load* is the maximum number of nodes in $G$ mapped to a single processor in $H$. The *edge–congestion* of an embedding is the maximum number of edges in $G$ routed by the mapping $\psi$ over a single edge of $H$.

## 5.1 Linear Arrays and Rings

**Lemma 2.** *$FP_n$ contains a Hamiltonian path.*

**Proof** : $FP_1$ contains the linear array $L_{10} = (0, 1, 2, \ldots, 9)$. Assume that the linear array $L_{10^k}$ of length $10^k$ is a subgraph of $FP_k$ as an induction hypothesis. Then successively pass through the nodes $((0, L_{10^k}), (1, r(L_{10^k})), (2, L_{10^k}), (3, r(L_{10^k})), \ldots, (9, r(L_{10^k})))$. Here the notation $r(L_{10^k})$ means the mirror–image of the linear array $L_{10^k}$. □
Contrary to the hypercube, rings of odd lengths are also subgraphs of $FP_n$.

**Theorem 3.** *For $n > 1$, $FP_n$ has a ring $R(k)$ of length $k$ as a subgraph for $4 \leq k \leq 10^n$. In $FP_1$ there exist subgraphs $R(k)$, for only $k = 5, 6, 8, 9$.*

**Proof** : In [CW85] it is shown that the Petersen graph (i.e. $FP_1$) has only cycles of length 5, 6, 8 and 9 as subgraphs, but no cycles of length 3, 4, 7 and 10.
The network $FP_2$ contains subgraphs $\{i, i+1\} \times P \cong Q_1 \times P = FPQ_{1,1} = HP_4$, the hyper Petersen graph, for $i = 0, 2, 4, 6, 8$.
We have shown by construction [DÖ92] that any cycle of length $l$, where $4 \leq l \leq 20$, is isomorphic to a subgraph of $HP_4$. Because of the edge–symmetry of $HP_4$ it follows that for each edge $e$ in $HP_4$, there exists a cycle $R(l)$ of length $l$ for $4 \leq l \leq 20$, as a subgraph of $HP_4$ containing this edge $e$. By combining rings $R(l_1)$ in $\{0, 1\} \times P$ and $R(l_2)$ in $\{2, 3\} \times P$ in the way depicted in Figure 5a), we obtain a ring of length $4 \leq l_1 + l_2 \leq 40$ in $\{0, 1, 2, 3\} \times P$.
Similarly, by connecting cycles $R(l_3)$ and $R(l_4)$ in $\{6, 7\} \times P$ and $\{8, 9\} \times P$, respectively, a cycle of length $4 \leq l_3 + l_4 \leq 40$ is obtained in $\{6, 7, 8, 9\} \times P$.
Thus, due to the edge–symmetry of $FP_k$, one can find for the edge $e = \{(3, 0), (3, 1)\}$ for example, a cycle of length $4 \leq l \leq 40$, in $\{0, 1, 2, 3\} \times P$, containing edge $e$. By connecting $R(L_1)$ in $\{0, 1, 2, 3\} \times P$, for $4 \leq L_1 \leq 40$, with $R(L_2)$ in $\{4, 5\} \times P$, for $4 \leq L_2 \leq 20$, we get another cycle of length $L_1 + L_2$, where $4 \leq L_1 + L_2 \leq 60$. This cycle can then be connected with another in $\{6, 7, 8, 9\} \times P$ to form an $R(l)$, for $4 \leq l \leq 100$. Figure 5b) illustrates a cycle of length 27 in $FP_2$.
Recursively applying the same procedure to various subgraphs of $FP_n$, we can construct any ring $R(l)$ in $FP_n$, where $4 \leq l \leq 10^n$ and $n \geq 2$. □

## 5.2 Meshes

**Theorem 4.** *There exists an embedding of a $q$–dimensional $(2^{i_1} \cdot 5^{j_1} \times 2^{i_2} \cdot 5^{j_2} \times \ldots \times 2^{i_q} \cdot 5^{j_q})$ mesh in $FP_n$, such that $i_k, j_k \in N_0, 1 \leq k \leq q$, and $\Pi_{k=1}^q 2^{i_k} \cdot 5^{j_k} = 10^n$. Such embedding has load and expansion 1, and dilation and edge–congestion 2. An $n$–dimensional $(10 \times \ldots \times 10)$ mesh is a subgraph of $FP_n$.*

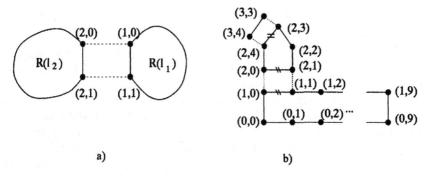

a)                                    b)

**Fig. 5.** a) Combining rings $R(l_1)$ and $R(l_2)$        b) Cycle $R(27)$ in $FP_2$

**Proof :** In $P$, one can easily construct an embedding of a $1 \times 10$ mesh as subgraph and of a $2 \times 5$ mesh with load 1, dilation and edge–congestion 2 (cf. Figure 6a)). Assume that the theorem is true for $k \leq m$. All possible meshes $M_{k+1}$ embedded in $FP_{k+1}$ result from taking ten copies $(f_k(M_k), j)$ for $0 \leq j \leq 9$, of a mesh $M_k$ which is embeddable in $FP_k$ by the node–mapping $f_k$ with expansion 1. These are then put together either one after another in one dimension (Case 1 of Figure 6b)), or two of these ten copies in one dimension and five in another dimension (Case 2 of Figure 6b)).

In Figure 6b), $m_i$ (or $m_{ij}$) denotes the mirror image by reflection in the dimension $x_i$ (or in both $x_i$ and $x_j$), where $x_i$ and $x_j$ are two different dimensions. Also, $M_k$ denotes a mesh with $10^k$ nodes and is used as the building block for the embedding of $M_{k+1}$. Since the linear array $L_{10}$ is a subgraph of $P$, an $n$–dimensional $(10 \times \ldots \times 10)$ mesh can be embedded as a subgraph in $FP_n$. □

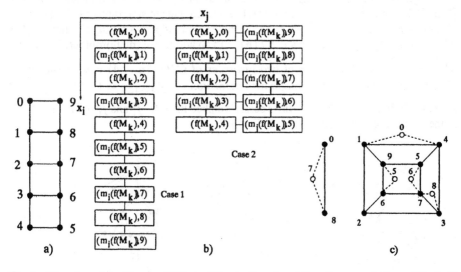

a)                    b)                    c)

**Fig. 6.** a) $2 \times 5$ mesh in $P$        b) Embedding in Case 1 and Case 2        c) $Q_1$ and $Q_3$ in $P$

## 5.3  Hypercube

**Theorem 5.** In $FP_n$, $\binom{n}{i}$ instances of the hypercube $Q_{3(n-i)+i}$, for $0 \leq i \leq n$, can be embedded with load 1, dilation and edge–congestion 2, and expansion 1. Also $Q_m$, *where $m \leq n$, can be embedded in $FP_n$ as a subgraph, but $Q_{n+i}$, $i \geq 1$, cannot be embedded as a subgraph in $FP_n$.*

**Proof** : $Q_1$ is a subgraph of $P$. Therefore, $Q_1 \times \ldots \times Q_1 \cong Q_n$ is a subgraph of $FP_n$. Assume, $f$ is a subgraph–embedding of $Q_{n+i}$, $i \geq 1$, in $FP_n$. Each node $v = v_1 \ldots v_{n+i}$ in $Q_{n+i}$ has $n+i$ neighbors given as $v(k) = v_1 \ldots v_{k-1}\overline{v_k}v_{k+1} \ldots v_{n+i}$, where $1 \leq k \leq n + i$. Because of the definition of the hypercube, two neighbors $v(k)$ and $v(l)$ have the common neighbor $v(k,l)$, where the $k$–th and $l$–th bits are complemented. Let $f(v_1, \ldots, v_{n+i}) = (x_1, \ldots x_n)$. Since $v$ has $n + i$ neighbors, but $(x_1, \ldots, x_n)$ has only $n$ components, there exists $j$, $1 \leq j \leq n$, with $f(v(k)) = (x_1, \ldots, x_{j-1}, y_j, x_{j+1}, \ldots, x_n)$ and $f(v(l)) = (x_1, \ldots, x_{j-1}, z_j, x_{j+1}, \ldots, x_n)$. Since $v(l)$ and $v(k)$ have the common neighbor $v(k,l)$, both $f(v(l))$ and $f(v(k))$ must also have a common neighbor, which needs to have the form $f(v(k,l)) = (x_1, \ldots, x_{j-1}, u_j, x_{j+1}, \ldots, x_n)$. This implies that we have a 4–cycle $(x_j, y_j, u_j, z_j, x_j)$ in $P$, which is a contradiction.

As Figure 6c) shows, $Q_1$ and $Q_3$ can be embedded with load 1, dilation and edge–congestion 2 in $P$. Thus, we can inductively show, by using cartesian products of hypercubes embedded in $FP_{n-1}$ and in $P$, that $\binom{n}{i}$ instances of $Q_{3(n-i)+i}$, $0 \leq i \leq n$, are embeddable in $FP_n$. Since $\sum_{i=0}^{n} \binom{n}{i} \cdot (2^3)^{n-i} \cdot 2^i = 10^n$, the expansion is 1. $\square$

## 5.4 Binary trees

**Complete binary trees** A *complete binary tree*, $CBT(n)$, of height $n$ is the graph whose nodes are all binary strings of length at most $n$ with root $e$ (empty string). The edges connect each string $x$ of length $m$ to strings $xa$, $a \in Z_2$, of length $m + 1$ [MS90]. An *arbitrary binary tree* of height $n$ is a connected subgraph of $CBT(n)$ with at least one node at level $n$. We define $level(x) = k$, if $x$ is a string of length $k$. Let $MetaCBT(h)$, $h \geq 0$, denote a $CBT(h)$ with an additional (meta)root (cf. Figure 7).

**Theorem 6.** $CBT(3n - 1)$ *and two instances of* $CBT(3n - 4)$ *can be embedded simultaneously in* $FP_n$ *as subgraph with expansion* $\frac{10^n}{2^{3n}+2^{3n-2}-3} \doteq \left(\frac{10}{8}\right)^{n-1}$.

**Proof** (by induction) : For $k = 1$, three possible embeddings $f_1^1$, $f_1^2$ and $f_1^3$ of $MetaCBT(2)$ in the Petersen graph ($P$) are shown in Figure 7.

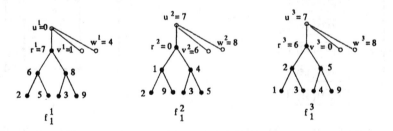

**Fig. 7.** $CBT(2)$ in $P$

As an induction hypothesis, suppose that $MetaCBT(3k - 1)$ and two instances of $MetaCBT(3k-4)$ can be embedded as subgraphs in $FP_k$ using one of the embeddings $f_k^i$, $i = 1, 2, 3$, and $f_k'$, respectively. Then the mappings $f_{k+1}^1$ and $f_{k+1}'$ for one instance of $CBT(3k + 2)$ and two $CBT(3k - 1)$'s, respectively, in $FP_{k+1}$ can be obtained by using the scheme shown in Figure 8. Here, we use $r, u, v$ instead of $r^1, u^1, v^1$ for brevity of notation. Also $f_{k+1}^2$ and $f_{k+1}^3$ can be constructed from $f_{k+1}^1$ by automorphisms of the edges in $FP_{k+1}$ that map the edge $(u^1, r^1)$ to the edges $(u^2, r^2)$ with $r^2 = u^1$ and $u^2 = r^1$ and $(u^3, r^3)$ with $u^3 = r^1$, respectively. □

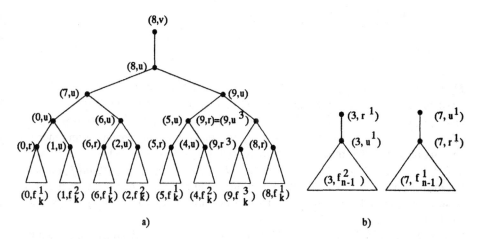

**Fig. 8.** a) $MetaCBT(3k+2)$ in $FP_{k+1}$      b) Two $CBT(3k-4)$'s in $FP_{k+1}$

**Dynamic Trees** We present in [ÖD93d] a dynamic task–allocation algorithm to dynamically embed an $M$–node tree, $T(M)$, arising for example in divide–and–conquer and branch–and–bound algorithms in $FP_n$. By a randomized flip–bit algorithm similar the one for embedding dynamic trees in butterflies, hypercubes and de Bruijn networks [BC89, Lei92, Öhr92], we prove the following

**Theorem 7.** *A dynamically evolving arbitrary binary tree with $M$ nodes can be dynamically embedded in $FP_n$ with dilation 1 and load (with high probability) of at most* $\mathcal{O}(\frac{M}{10^n} + n)$.

The term $x$ is less than $\mathcal{O}(f)$ with high probability means $\forall c \exists \vartheta$ ($\vartheta = $ const, independent of $N$) : $P(x \geq \vartheta \cdot f) < N^{-c}$, where $P(E)$ denotes the probability of the event $E$ (cf. [LNRS89]).

**X-tree** An *X-tree*, $X(n)$, is a tree $CBT(n)$ of height $n$ with additional edges connecting consecutive nodes at each level. $MetaX(n)$ denotes an X-tree with an additional *(meta)*root.

**Theorem 8.** *There exists an optimal embedding of $\binom{n}{k}$ $MetaX(3n - 2k - 1), 0 \leq k \leq n$, simultanously in $FP_n$ with load 1, dilation and edge–congestion 2, and expansion 1.*

**Proof** (by induction): An embedding of $X(k)$, $1 \leq k \leq 3n - 1$, with dilation 1 in $FP_n$ is not possible, since $FP_n$ contains no cycle of length 3. $MetaX(2)$ and $MetaX(0)$ can be embedded in $FP_1 = P$ with load 1, dilation and edge–congestion 2, and expansion 1 (cf. Figure 9a)).

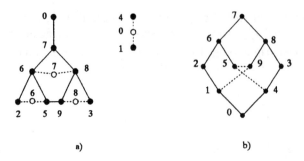

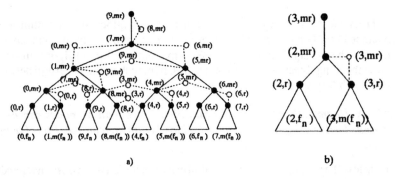

**Fig. 9.** a) $MetaX(2), MetaX(0)$ in $P$        b) Representation of $P$

Assume in $FP_i$, $i \leq n$, one can embed $\binom{i}{k}$ $MetaX(3i - 2k - 1)$, $0 \leq k \leq i$, with claimed cost–measurements. Construction scheme I depicted in Figure 10a) embeds $\binom{n}{k}$ $MetaX(3(n + 1) - 2k - 1)$, $0 \leq k \leq n$, in $FP_{n+1}$ while the construction scheme II shown in Figure 10b) embeds $\binom{n}{k}$ $MetaX(3(n + 1) - 2(k + 1) - 1)$, $0 \leq k \leq n$, that is $\binom{n}{k-1}$ $MetaX(3(n + 1) - 2k - 1)$, $0 \leq k - 1 \leq n$. In Figure 10, $m(f_n)$ denotes the mirror image of $f_n$ by vertical reflection, $r$ and $mr$ the image of the root and the metaroot in $f_n$. In total, construction schemes I and II embed $\binom{n}{k} + \binom{n}{k-1} = \binom{n+1}{k}$ $MetaX(3(n + 1) - 2k - 1)$ in $FP_{n+1}$.

Since $FP_{n+1}$ has $(2^3 + 2^1)^{n+1} = \sum_{k=0}^{n+1} \binom{n+1}{k}(2^3)^{n+1-k} \cdot (2^1)^k$ nodes, the embedding has expansion 1. $\square$

**Fig. 10.** a) Construction Scheme I        b) Construction Scheme II

## 5.5 Tree Machine

A *tree machine* of dimension $n$, denoted as $TM(n)$, is the graph consisting of two complete binary trees $CBT(n)$ and $CBT(n - 1)$ connected back to back along the leaves.

**Theorem 9.** *There exists an embedding of $TM(3n - 1)$ in $FP_n$ with load 1, dilation and edge–congestion 2, and expansion $\frac{2}{3} \cdot (\frac{10}{8})^n$ for $n \geq 2$ (expansion is 1 for $n = 1$).*

**Proof :** As Figure 9b) shows, Theorem 9 holds for $n = 1$. A dilation 1–embedding of $TM(2)$ in $P$ is not possible, since $P$ contains no cycle of length 4 as subgraph. Assume that $TM(3n - 1)$ can be embedded in $FP_n$ with the claimed cost–measurements using the embedding $f_{7,0}^n$. The subscript $(7, 0)$ indicates that the upper root of $TM(3n - 1)$

is mapped to a node $(7, *^{n-1})$ and the lower root to a node $(0, *^{n-1})$. The string $*^k$ denotes an arbitrary string of length $k$ with symbols in $\{0, 1 \ldots, 9\}$. Because of the symmetry in $P$, for any pair of adjacent nodes $x$ and $y$ in $P$ we have an embedding $f_{x,y}^n$ of $TM(3n-1)$ in $FP_n$ that maps the upper root to a vertex $(x, *^{n-1})$ and the lower root to a vertex $(y, *^{n-1})$. The generic scheme in Figure 11 describes the embedding of $TM(3n+2)$ in $FP_{n+1}$. $\square$

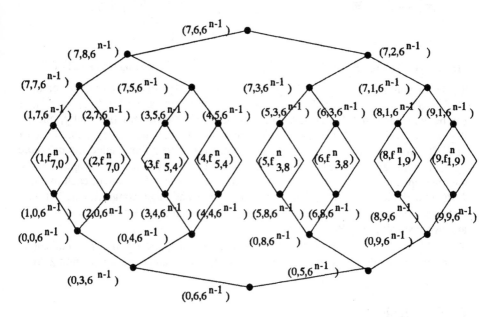

**Fig. 11.** $TM(3n+2)$ in $FP_{n+1}$

## 5.6 Pyramids

Informally, the *pyramid* network $PR(n)$ is a 4-ary complete tree of height $n$ and the nodes at each level form a square mesh. The circled nodes and the thick drawn edges in Figure 12a) form a $PR(1)$.

**Theorem 10.** *There exists an embedding of* $\binom{n-1}{i} \cdot 2^i + \binom{n-1}{i-1} \cdot 2^{i-1}$ *pyramids* $PR(i)$, $0 \le i \le n$, *in* $FP_n$ *with load 1, dilation 2 and edge–congestion 2.*

**Proof :** Since a cycle of length three is not a subgraph of $FP_n$ for all $n \ge 0$, $PR(i)$ cannot be embedded as subgraph into the folded Petersen network.

Figure 12a) shows that $PR(1)$ with an alternate apex and two $PR(0)$'s with an alternate apex can be embedded with load and expansion 1, and edge–congestion and dilation 2 in the Petersen graph, $P$. There exists an automorphism of $P$, interchanging in this mapping the apex and alternate apex, having the same cost-measurements.

For $n \ge 2$, we use the construction as in Figure 12 to embed pyramids in $FP_n$, where $f_i$ denotes the mapping of a pyramid $PR(i)$, $0 \le i \le n-1$, into $FP_{n-1}$ and

$$f_i'(x) := \begin{cases} f_i(alternate\ apex) & \text{for } x = \text{apex of } PR(i) \\ f_i(apex) & \text{for } x = \text{alternate apex of } PR(i) \\ f_i(x) & \text{for } x \ne \text{apex, alternate apex of } PR(i). \end{cases}$$

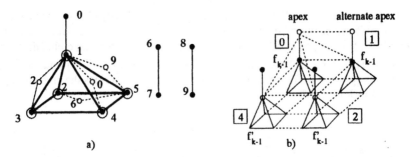

**Fig. 12.** a) Embedding $PR(1)$ and two $PR(0)$ in $P$       b) Embedding pyramids in $FP_n$

The construction in Figure 12b) for prefixes in $\{0,1,2,4\}$ can also be done for prefixes $\{3,5,8,9\}$. For prefixes $\{6,7\}$ two further pyramids embedded by $f_{n-1}$ in $FP_{n-1}$ can be embedded in $FP_n$. For each $n$, it holds due to the construction that $f_i$ and $f_i'$, $0 \le i \le n$, have load 1, dilation 2 and edge–congestion 2, and that $f_i$ maps the apex of $PR(i)$ into a node of $FP_n$, that has an unassigned neighbor, called the **alternate apex**, such that at most one edge is routed through the edge connecting the apex and the alternate apex. Due to the construction scheme, we map in $FP_{i+1}$ out of each pyramid $PR(m)$ embedded in $FP_i$ (with load 1, dilation 2 and edge–congestion 2, but with distance 1 between apex and alternate apex) two instances of a pyramid $PR(m+1)$ and two instances of $PR(m)$.

Thus, $\binom{n-1}{i} \cdot 2^i + \binom{n-1}{i-1} \cdot 2^{i-1}$ $PR(i)$'s with alternate apex, $0 \le i \le n$, can be embedded node– and edge–disjointly in $FP_n$ with load 1, dilation 2 and edge–congestion 2. □

# 6 Conclusion

In this paper, we have presented a new interconnection topology, called the $n$–folded Petersen network. It is defined by the iterative cartesian product of the Petersen graph, and possesses many important properties such as regularity, symmetry, modularity, small diameter and degree, simple routing and broadcasting and maximal fault–tolerance. A topological comparison with several popular networks indicates that the proposed topology is an attractive candidate for multiprocessor systems. Furthermore, it is versatile since it can efficiently simulate rings, multi–dimensional meshes, hypercubes, complete binary and dynamically evolving binary trees, X-trees, pyramids and tree machines.

We introduce in [ÖD93b] a generalization of the folded Petersen network, the *folded Petersen cube network* $FPQ_{n,k} = Q_n \times FP_k$, having the hypercube $Q_n$ ($k = 0$), the folded Petersen graph $FP_k$ ($n = 0$), and the hyper Petersen network $HP_{n+3} = Q_n \times P$ ($k = 1$) as special cases. The results in [ÖD93b] are generalizations of those obtained for the folded Petersen and the hyper Petersen networks [DÖ92].

Our current research [ÖD93e] involves dynamic embedding of several types of dynamic grids that arise in dynamic programming or multigrid methods, in this new network following our earlier approach considering hyper-de Bruijn network as host [ÖD93a]. Another direction of our research includes designing incomplete folded Petersen networks consisting of an arbitrary number of nodes. It will also be worthwhile to design multiprocessor topologies using generalized Petersen graph (with an inner and outer ring of length $k \ge 7$) as the basic building block.

# References

[BA84]   L. Bhuyan and D.P. Agrawal. Generalized hypercubes and hyperbus structures for a computer network. *IEEE Transactions on Computers*, C-33, pp. 323–333, 1984.

[BC89]   S.N. Bhatt and Jin-Yi Cai. Take a walk, grow a tree. In *Proc. of the 29th Annual Symposium on Foundations of Computer Science*, pp. 469–478, Los Alamitos, USA, 1989.

[BP89]   J.C. Bermond and C. Peyrat. de bruijn and kautz networks: a competitor for the hypercube? In *Hypercubes and Distributed Computers*, pp. 279–293, North–Holland, 1989.

[CW85]   G. Chartrand and R.J. Wilson. The petersen graph. *Graphs and Applications* (Eds. F. Harary and J.S. Maybee), pp. 69 –100, 1985.

[dB46]   N.G. de Bruijn. A combinatorial problem. In *Proceedings*, volume 49, part 20, pp. 758–764. Akademe Van Wetenschappen, 1946.

[DB92]   S.K. Das and A.K. Banerjee. Hyper petersen network : yet another hypercube–like topology. In *Proc. of the 4th Symposium on the Frontiers of Massively Parallel Computation (Frontiers' 92)*, pp. 270 – 277, McLean, Virginia, Oct. 1992.

[DGD92] S.K. Das, J. Ghosh, and N. Deo. Stirling networks : a versatile combinatorial topology for multiprocessor systems. *Discrete Applied Mathematics*, 37/38, special double volume on interconnection networks, pp. 119–146, Jul. 1992.

[DÖ92]   S.K. Das and S. Öhring. *Embeddings of tree–related topologies in hyper Petersen networks*. Tech. Rep. CRPDC-92-16, Univ. North Texas, Univ. Würzburg, Sep. 1992.

[GP91]   E. Ganesan and D.K. Pradhan. The hyper-de bruijn multiprocessor networks. In *Proc. Conf. on Distributed Computing Systems*, pp. 492–499, Arlington, TX, May 1991.

[Lei92]  F.T. Leighton. *Introduction to Parallel Algorithms and Architectures : Arrays – Trees – Hypercubes*. Morgan Kaufmann Publishers, San Mateo, CA, 1992.

[LNRS89] F.T. Leighton, M. Newmann, A.G. Ranade, and E. Schwabe. Dynamic tree embeddings in butterflies and hypercubes. In *Proc. of the 1989 ACM Symposium on Parallel Algorithms and Architectures*, pp. 224–234, Santa Fe, NM, Jun. 1989.

[MS90]   B. Monien and H. Sudborough. Embedding one interconnection network in another. In *Computational Graph Theory*, pp. 257–282, Wien, 1990. Springer Verlag.

[ÖD93a]  S. Öhring and S.K. Das. Dynamic embeddings of trees and quasi–grids into hyper–de bruijn networks,. In *Proc. of the 7th International Parallel Processing Symposium*, Newport Beach, CA, pp. 519–523, Apr. 1993.

[ÖD93b]  S. Öhring and S.K. Das. The folded Petersen cube networks: new competitors for the hypercube. to appear in *Proc. of the Fifth IEEE Symposium on Parallel and Distributed Processing*, Dallas, TX, Dec. 1993.

[ÖD93c]  S. Öhring and S.K. Das. The folded Petersen network: a new communication–efficient multiprocessor topology. In *Proc. of the 1993 International Conference on Parallel Processing, Volume I*, pp. 311 – 314, St. Charles, IL, Aug. 1993.

[ÖD93d]  S. Öhring and S.K. Das. *Folded Petersen Cube Network: New Competitors for the Hypercubes*. Tech. Rep. CRPDC-93-8, Univ. Würzburg, Univ. North Texas, Jun. 1993.

[ÖD93e]  S. Öhring and S.K. Das. *Mapping dynamic data structures on product networks*. Tech. Rep. CRPDC-93-5, Univ. Würzburg, Univ. North Texas, Mar. 1993.

[ÖD93f]  S. Öhring and S.K. Das. *Optimal communication primitives on the folded Petersen networks*. Tech. Rep. CRPDC-93-13, Univ. Würzburg, Univ. North Texas, Sep. 1993.

[Öhr92]  S. Öhring. Dynamic embeddings of trees into de Bruijn graphs. In *Proc. Second Joint International Conference on Vector and Parallel Processing, CONPAR 92–VAPPV*, Lyon, France, Sep. 1992. LNCS 634, pp. 783–784, 1992.

# Fast Load Balancing in Cayley Graphs and in Circuits

Jacques E. Boillat

Institute of Engineering
Department of Computer Sciences
Morgartenstrasse 2c
CH-3014 Berne

**Abstract.** We compare two load balancing techniques for Cayley graphs based on information and load exchange between neighbouring vertices. In the first scheme, called natural diffusion, each vertex gives (or receives) a fixed part of the load difference to (from) its direct neighbours. In the second scheme, called Cayley diffusion, each vertex successively gives (or receives) a part of the load difference to (or from) direct neighbours incident to the edges labelled by the elements of the generator set of the Cayley graph. We prove that the convergence of the Cayley diffusion is faster than the natural diffusion, at least for some particular graphs (cube, circuit with an even number of vertices, graphs from the symmetric group). Furthermore we compute the fastest possible way to distribute load in a circuit using local load balancing strategies.

## 1 Introduction

To achieve high performance with a parallel computer it is imperative to balance the work-load of the processors. Various strategies have been proposed to solve this problem [Hać89]. We will focus here on parallel load balancing algorithms for interconnected multiprocessor networks based on local load exchange. Furthermore, we will assume that at any moment, each processor has only information about the load of its direct neighbours. ⁻

There are many possibilities for local load exchange [ELZ86b], [ELZ86a], [MTS90], [Cyb89], etc. The *natural* concept is to give (or receive) in parallel a fixed part of the load difference to (from) all direct neighbours at the same time [Cyb89], [Boi90b]. This strategy should be applied if the structure of the underlying network (graph) is unknown. In [Boi90b] we have shown that this method is equivalent to a time discrete Poisson equation in a finite undirected graph. We have also shown how to choose the load amount that has to be exchanged for ensuring the convergence of the algorithm in any connected undirected graph.

Each load exchange step with one particular neighbour needs 2 communications. The first for getting the neighbour's load and the second for exchanging the load. In a multiprocessor system, communication may be cost intensive and excessive synchronization steps may result in poor efficiency.

In this paper we will show that it is possible, at least for some Cayley graphs, to achieve faster convergence by exchanging load in a round robin fashion with the neighbours without increasing the number of communications.

## 2  Natural load balancing

In this section, we define the natural load balancing technique. Furthermore, we show how to choose the fixed amount of local load exchange to achieve the fastest possible convergence of the natural load balancing strategy in circuit graphs.

In this paper, a graph $G = (V, E)$ is a non-directed, connected, and regular finite graph without loops. We will denote the adjacency matrix of G by A. Furthermore, we will denote the degree of G by k.

**Definition 1 (Natural diffusion in regular graphs).** Let G be a regular graph of degree k, and let $\alpha \in [0, 1]$ be a real number. Consider the symmetric stochastic matrix

$$P_\alpha = \alpha I + \frac{1 - \alpha}{k} A \tag{1}$$

Let $x_t$ denote the load vector at iteration t. The *natural diffusion* is defined by the matrix $P_\alpha$, or more precisely by solving the equation

$$x_{t+1} = P_\alpha x_t = P_\alpha^t x_0 \quad (t \in \mathbb{N}) \tag{2}$$

where the series of powers of $P_\alpha$ denote the successive load balancing steps.

Let $i \in V$ be a vertex of G. We denote by $N_i$ the set of neighbouring vertices of i. The natural diffusion corresponds to the following parallel algorithm:

*Algorithm 1 (Natural load balancing algorithm).*

> WHILE TRUE
> > PAR i = 1 FOR processors
> > > PAR j ∈ $N_i$
> > > > give $\frac{1-\alpha}{k}$ times the load difference to direct neighbour j

Algorithm 1 always converges towards the uniform distribution, provided $\alpha \in ]0, 1[$. If $\alpha = 0$, it converges if and only if G is not bipartite. Given a graph G it is always possible to find $\alpha \in [0, 1]$ such that the convergence of algorithm 1 is the fastest possible.

In practice, the load of a processor consists in a finite sum of time complexities of (atomic) processes. If the number of atomic processes is much larger than the number of processors, we may consider the load of a processor as a real number. Tests have shown that algorithm 1 behaves well, if the processes have discrete loads, provided there are many more processes than processors [BBK91]. Moreover, if all atomic processes have the same load, algorithm 1 may be interpreted as a random walk of independent particles in graph G with transition probabilities $P_{ij}$. Provided there are more atomic processes than processors, it is possible to prove that the diffusion converges towards the uniform distribution (see [Boi90a] for details).

As an example, we show how to choose $\alpha$ such that the convergence of the natural diffusion scheme is optimal in a circuit. Note that the method used here is the same for any regular graph.

Let A denote the adjacency matrix of the undirected circuit with $n$ vertices ($n \geq 3$), and let $\alpha \in [0, 1]$ be a real number. Recall that the associated stochastic matrix is

$$P_\alpha = \alpha I + \frac{1 - \alpha}{2} A \tag{3}$$

The convergence speed of this Markov chain is a function of the stochastic eigenvalue with the greatest modulus. The eigenvalues of $P_\alpha$ are easy to compute once the eigenvalues of the circuit graph (see e.g. [CDS79]) are known.

$$\lambda(P_\alpha)_j = \alpha + (1 - \alpha) \cos \frac{2\pi}{n} j \qquad j = 0, ..., n - 1 \tag{4}$$

If $\alpha = 1$ and $P_\alpha = I$ then the diffusion process does not converge. If $n$ is even and $\alpha = 0$, the smallest eigenvalue is equal to $-1$ and the diffusion process does not converge. For all other values of $\alpha$, the diffusion process will converge to the uniform load distribution.

**Theorem 2.** Let $\lambda_\alpha = \max_{j=1,...,n-1} |\lambda(P_\alpha)_j|$ and set $\lambda_{nat} = \min_{\alpha \in [0,1]} \lambda_\alpha$. Moreover let $\alpha_{nat}$ be such that $\lambda_{nat} = \lambda_{\alpha_{nat}}$, then

$$\lambda_{nat} = \begin{cases} \frac{1 + \cos \frac{2\pi}{n}}{3 - \cos \frac{2\pi}{n}} & n \text{ even} \\ \frac{\cos \frac{\pi}{n} + \cos \frac{2\pi}{n}}{2 + \cos \frac{\pi}{n} - \cos \frac{2\pi}{n}} & n \text{ odd} \end{cases} \qquad \alpha_{nat} = \begin{cases} \frac{1 - \cos \frac{2\pi}{n}}{3 - \cos \frac{2\pi}{n}} & n \text{ even} \\ \frac{\cos \frac{\pi}{n} - \cos \frac{2\pi}{n}}{2 + \cos \frac{\pi}{n} - \cos \frac{2\pi}{n}} & n \text{ odd} \end{cases} \tag{5}$$

Note that if $n = 3$, $\alpha_{nat} = \frac{1}{3}$ and $\lambda_{opt} = 0$

*Proof.* It is easy to see, that $\lambda_\alpha = \max\{|2\alpha - 1|, \alpha + (1 - \alpha) \cos \frac{2\pi}{n}$ if $n$ is even, and $\lambda_\alpha = \max\{|\alpha - (1 - \alpha) \cos \frac{\pi}{n}|, \alpha + (1 - \alpha) \cos \frac{2\pi}{n}\}$ if $n$ is odd. Let $n$ be even. If $\alpha \leq \frac{1}{2}$ then it is easy to see that $1 - 2\alpha \geq \alpha + (1 - \alpha) \cos \frac{2\pi}{n}$ if and only if $\alpha \leq \frac{1 - \cos \frac{2\pi}{n}}{3 - \cos \frac{2\pi}{n}}$ If $\alpha \geq \frac{1}{2}$ then $2\alpha - 1 \geq \alpha + (1 - \alpha) \cos \frac{2\pi}{n}$ if and only if $\alpha = 1$. If $n$ is odd, the proof is similar  $\square$

# 3 Load diffusion in Cayley Graphs

**Definition 3 (Cayley Graph).** Let $\Gamma$ be a finite group and and let $S$ be a symmetric set of generators, i.e. $s \in S \Leftrightarrow s^{-1} \in S$. The Cayley Graph $G(\Gamma, S)$ is the undirected graph with vertex set $\Gamma$ and where $(g_1, g_2) \in E(G)$ if and only if $g_1^{-1} g_2 \in S$

$G(\Gamma, S)$ is a connected regular graph of order $k = |S|$. Thus at each vertex, the edges may be labelled with the elements of $S$.

1. If $\Gamma = (\mathbb{Z}_2)^d$, and $S = \{(1, 0, ..., 0), (0, 1, ..., 0), ..., (0, 0, ..., 1)\}$, then $G(\Gamma, S) = H_d$ the d-dimensional Cube Graph.
2. Let $\Gamma = \sigma_3$ the symmetric group of the 3 element set $\{1, 2, 3\}$ and set $S = \{(1, 2), (1, 2, 3), (3, 2, 1)\}$. Then $G(\Gamma, S)$ is a prism with triangular bottom.
3. Consider $n \geq 2$ be an integer, the dihedral group $\Gamma = \; < s, t | s^2 = t^2 = (st)^n = 1 >$ of order $2n$ and set $S = \{s, t\}$. Then $G(\Gamma, S)$ is a circuit with $2n$ vertices.

We define in $S$ an equivalence relation by identifying $s$ with $s^{-1}$ ($s \in S$). Let $\hat{S}$ be the set of equivalence classes. We call $\hat{S} = \{\hat{s}_1, \hat{s}_2, ..., \hat{s}_k\}$ the set of **dimensions** of G. Note that if $s \in S$ is an element of order 2 then $\hat{s} = \{s\}$, if not, then $\hat{s} = \{s, s^{-1}\}$. Using $\hat{S}$, we may build a partition $\hat{E} = \{\hat{E}_1, \hat{E}_2, ..., \hat{E}_k\}$ of the edge set $E(G)$ by defining

$$(g_1, g_2) \in \hat{E}_i \text{ if and only if } g_1^{-1} g_2 \in \hat{s}_i \quad i = 1, ..., k \tag{6}$$

If $\hat{s}_i$ corresponds to an element of $S$ of order 2, then $\hat{E}_i$ is a disjoint set of edges. If $\hat{s}_i$ corresponds to an element of $S$ of order $d > 2$, then $\hat{E}_i$ is the union of disjoint circuits (of length d) of G.

1. For the d-dimensional Cube Graph, $\hat{E}_i$ is the set of all edges which are parallel to the $i^{th}$ axis, i.e. $(g_1, g_2) \in \hat{E}_i$ if and only if $g_1$ and $g_2$ differ only in the $i^{th}$ component.
2. For the symmetric group of the 3 element set $\{1, 2, 3\}$ with generator set $S = \{(1, 2), (1, 2, 3), (3, 2, 1)\}$, $G(\Gamma, S)$ is a prism with triangular bottom. $\hat{E}_{(1, 2)}$ corresponds to the vertical edges of G and $\hat{E}_{(1, 2, 3)}$ corresponds to the bottom and the top of G.
3. Consider the dihedral group with generator set $S = \{s, t\}$. If the edges of G are numbered from 1 to $2n$, then $\hat{E}_s$ corresponds to the even numbered edges and $\hat{E}_t$ corresponds to the odd numbered edges.

Let $\hat{G}_j$ denote the graph with the same vertex set as G and with edges $\hat{E}_j$. The natural decomposition $\{\hat{E}_1, \hat{E}_2, ..., \hat{E}_k\}$ of a Cayley graph $G(\Gamma, S)$ leads to following load balancing strategy:

319

**Algorithm 2 (Cayley diffusion).**

```
WHILE TRUE
    PAR i = 1 FOR processors
        SEQ j = 1 FOR dimensions
            exchange load using the fastest natural strategy
            with all direct neighbours in the graph Ĝⱼ
```

Instead of distributing the load to all neighbours at the same time, the load is distributed successively in all dimensions. Thus a diffusion step consists in balancing the load (using algorithm 1) along all edges of the elements of the partition $\hat{E}$ successively. A single step of the WHILE loop in the Cayley diffusion strategy requires exactly the same number of communications and synchronizations than the corresponding step of the natural diffusion strategy.

Two cases must be considered. If the direction $s_j$ corresponds to a generator $s$ of order 2, then each vertex gives (or receives) half of the load difference to (from) its single direct neighbour in direction $\hat{s}_j$. If the direction $s_j$ corresponds to a generator $s$ of order $d \geq 2$, then each vertex gives (or receives) $\frac{1-\alpha_0}{2}$ times the load difference to (from) both neighbours in direction $\hat{s}_j$, where $\alpha_0$ is defined as in theorem 2. Indeed, in this case $\hat{E}_j$ consists of disjoint circuits of length d.

Let $\hat{A}_j$ denote the adjacency matrix of the graph $\hat{G}_j$. $\hat{A}_j$ is the adjacency matrix of a disjoint union of circuits of length d (if $\hat{s}_j$ corresponds to a generator of order $d > 2$) or the adjacency of a disjoint union of edges (matching) if $\hat{s}_j$ corresponds to a generator of order $d = 2$. The stochastic matrix corresponding to the diffusion in dimension $j$ is the matrix

$$\hat{P}_j = \begin{cases} \alpha I + (1-\alpha)\hat{A}_j & \text{(if)} d = 2 \\ \alpha I + \frac{1-\alpha}{2}\hat{A}_j & \text{(if)} d > 2 \end{cases} \quad \text{where} \quad \alpha = \begin{cases} \frac{1-\cos\frac{2\pi}{d}}{3-\cos\frac{2\pi}{d}} & d \text{ even} \\ \frac{\cos\frac{\pi}{d}-\cos\frac{2\pi}{d}}{2+\cos\frac{\pi}{d}-\cos\frac{2\pi}{d}} & d \text{ odd} \end{cases} \quad (7)$$

The stochastic matrix $\hat{P}$ corresponding to one step of the Cayley diffusion is the product of all partial diffusion matrices in all directions:

$$\hat{P} = \prod_{j=1}^{k} \hat{P}_j \quad (8)$$

$\hat{P}$ is a stochastic matrix, however $\hat{P}$ is normally not symmetric.

**Theorem 4.** *The Cayley diffusion scheme always converges towards the uniform load distribution.*

*Proof.* $\hat{P}$ is a product of doubly stochastic matrices, thus $\hat{P}$ itself is doubly stochastic. Since the diagonal elements of the $\hat{P}_j$ are strictly positive, the diagonal elements of $\hat{P}$. are positive too. Since G is connected, it follows that $\hat{P}$ is primitive, i.e. $\hat{P}$ has only one eigenvalue of modulus 1. $\qquad\square$

Note that the eigenvalues of $\hat{P}$ are not independent of the numbering of the subgraphs $\hat{G}_j$. However, it is well known that given any two square matrices A and B, AB and BA have the same eigenvalues. Thus cyclic permutations of the product $\hat{P}$ will have the same eigenvalues.

Let now compare both diffusion schemes for some particular Cayley graphs, e.g. the d-dimensional cube, the circuit with an even number of vertices and some Cayley graphs from the symmetric group. We show that the Cayley diffusion is faster than the natural diffusion in all cases.

1. Let G be square graph, i.e. the Cayley graph of $(\mathbb{Z}_2)^2$ with generators $s_1 = (1,0)$ and $s_2 = (0,1)$, then

$$A = \begin{pmatrix} 0 & 1 & 1 & 0 \\ 1 & 0 & 0 & 1 \\ 1 & 0 & 0 & 1 \\ 0 & 1 & 1 & 0 \end{pmatrix} \quad \hat{A}_1 = \begin{pmatrix} 0 & 0 & 1 & 0 \\ 0 & 0 & 0 & 1 \\ 1 & 0 & 0 & 0 \\ 0 & 1 & 0 & 0 \end{pmatrix} \quad \hat{A}_2 = \begin{pmatrix} 0 & 1 & 0 & 0 \\ 1 & 0 & 0 & 0 \\ 0 & 0 & 0 & 1 \\ 0 & 0 & 1 & 0 \end{pmatrix} \tag{9}$$

This results in following stochastic matrices

$$P = \begin{pmatrix} \frac{1}{3} & \frac{1}{3} & \frac{1}{3} & 0 \\ \frac{1}{3} & \frac{1}{3} & 0 & \frac{1}{3} \\ \frac{1}{3} & 0 & \frac{1}{3} & \frac{1}{3} \\ 0 & \frac{1}{3} & \frac{1}{3} & \frac{1}{3} \end{pmatrix} \quad \hat{P} = \begin{pmatrix} \frac{1}{4} & \frac{1}{4} & \frac{1}{4} & \frac{1}{4} \\ \frac{1}{4} & \frac{1}{4} & \frac{1}{4} & \frac{1}{4} \\ \frac{1}{4} & \frac{1}{4} & \frac{1}{4} & \frac{1}{4} \\ \frac{1}{4} & \frac{1}{4} & \frac{1}{4} & \frac{1}{4} \end{pmatrix} \tag{10}$$

It follows that the Cayley diffusion process converges after a single step whereas the stochastic eigenvalue of the greatest modulus of the natural diffusion process is $\lambda = \frac{1}{3}$ resulting in slow convergence.

This result extends to the Cube graph of dimension $d \geq 3$. The stochastic eigenvalue of the greatest modulus of the best possible natural diffusion process is $\lambda = 1 - \frac{2}{d+1}$ [Cyb89]. The Cayley diffusion process converges after a single step. We give here an outline of the proof. The Cayley diffusion consists of d diffusion steps (successively along the edges of all dimensions) it follows that after j steps, all vertex pairs incident to edges in dimension $1, ..., j$ have the same load. In the next step, the load of both ends of these pairs may change, but by the same amount. The proof follows by induction.

2. In the prism, the Cayley diffusion converges in one single step. Indeed, after the diffusion step in the direction $(1,\hat{2},3)$ all the vertices of the top and all vertices of the bottom have the same load. The diffusion step in the direction of $(1,\hat{2})$ equalizes the load of the top and of the bottom. It is easy to verify that the natural diffusion scheme does not converge in finite time.

3. Let $\Gamma = \sigma_4$ the symmetric group of the 4 element set $\{1,2,3,4\}$ and set $S = \{(2,4),(1,2,3),(1,3,2)\}$. Then $G(\Gamma,S)$ is a cube with truncated vertices. The stochastic eigenvalue with the greatest modulus of the Cayley diffusion has value $\frac{2}{3}$ whereas that of the natural diffusion process has value $0.854$[1].

4. If $\Gamma = \sigma_4$ and $S = \{(1,2,3),(1,3,2),(1,2,4,3),(1,3,4,2)\}$. Then $G(\Gamma,S)$ is a cube with truncated edges. The stochastic eigenvalue with the greatest modulus of the Cayley diffusion has value $0.556$ whereas that of the natural diffusion process has value $0.75$[1].

5. The circuit graph with $2n$ ($n \geq 3$) vertices may be considered as the Cayley graph of the dihedral group with generator set $S = \{s,t\}$. The stochastic eigenvalue of the greatest modulus of the Cayley diffusion process is $\lambda = \frac{1}{2}(1+\cos\frac{2\pi}{n})$. Its is easy to verify that $\lambda$ is smaller than the corresponding eigenvalue of the natural diffusion. We will see in the next section that faster diffusion is possible in circuits with an even number of vertices.

## 4 Optimal load balancing in circuits with an even number of vertices

The Cayley diffusion strategy leads to fast convergence provided the order of the generators of the graph is small. Indeed, if $s$ is a generator of order $d$, $\hat{A}_s$ is the adjacency matrix of a disjoint union of circuits of length $d$. Unfortunately, the convergence of the natural diffusion in the circuits may be very slow if $d$ is big.

In this section, we show that it is possible to achieve faster convergence, at least in circuits with an even number of vertices. The method proposed corresponds to a subsequent labelling (colouring) of the edges in the circuits with an even number of vertices.

Let $C_{2n}$ be a circuit graph with $2n$ ($n \geq 2$) vertices, considered as the Cayley graph of the dihedral group with generator set $S = \{s,t\}$. In the examples of section 3, we have seen that the Cayley diffusion strategy may achieve faster convergence than the natural diffusion scheme. However it is not the fastest strategy.

Indeed, let $\hat{A}_s$ and $\hat{A}_t$ be the the adjacency matrices corresponding to the subgraphs in the both dimensions. For reason of simplicity, we choose the following numbering of the vertices of the circuit: $1,3,...,2n-1,2,4,...,2n$.

$$\hat{A}_s = \begin{pmatrix} & & & & 1 \\ & & & 1 & \\ & & 1 & & \\ 1 & & & & \\ & 1 & & & \\ & & 1 & & \end{pmatrix} \qquad \hat{A}_t = \begin{pmatrix} & & & & 1 \\ & & & 1 & \\ & & 1 & & 1 \\ & 1 & & & \\ & 1 & & & \\ 1 & & & & \end{pmatrix} \qquad (11)$$

Both $\hat{A}_s$ and $\hat{A}_t$ are involutions, i.e. $\hat{A}_s^2 = \hat{A}_t^2 = I$.

Let $C_n$ be the adjacency matrix of the directed circuit with $n$ vertices and let $I_n$ denote the identity matrix. Using the Cayley diffusion strategy, each edge in $\hat{A}_s$ and

---

[1] This example was computed using MAPLE

$\hat{A}_t$ would get the weight $\frac{1}{2}$ since $s$ and $t$ are generators of order 2. We solve here the more complex problem of computing the optimal diffusion speed when the edges of $\hat{A}_s$ have weight $1 - \alpha$, and those of $\hat{A}_t$ weight $1 - \beta$.

Let let $\hat{P}_s(\alpha)$ resp. $\hat{P}_t(\beta)$ be the associated diffusion matrices. It follows that

$$\hat{P}_s(\alpha) = \begin{pmatrix} \alpha I_n & (1-\alpha)I_n \\ (1-\alpha)I_n & \alpha I_n \end{pmatrix} \quad \text{and} \quad \hat{P}_t(\beta) = \begin{pmatrix} \beta I_n & (1-\beta)C_n^{-1} \\ (1-\beta)C_n & \beta I_n \end{pmatrix}$$

Let $\hat{P}(\alpha, \beta) = \hat{P}_s(\alpha)\hat{P}_t(\beta)$. Then

$$\hat{P}(\alpha, \beta) = \begin{pmatrix} \alpha\beta I_n + (1-\alpha)(1-\beta)C_n & \alpha(1-\beta)C_n^{-1} + \beta(1-\alpha)I_n \\ \alpha(1-\beta)C_n + \beta(1-\alpha)I_n & \alpha\beta I_n + (1-\alpha)(1-\beta)C_n^{-1} \end{pmatrix}$$

**Theorem 5.** *Let $c \in [-1, 1]$, $h_c = \alpha\beta + (1-\alpha)(1-\beta)c$ and $g = (1-2\alpha)(1-2\beta)$. The eigenvalues of $\hat{P}(\alpha, \beta)$ are given by*

$$\lambda_j^{\mp} = h_c \mp \sqrt{h_c^2 - g} \tag{12}$$

*where $c = \cos\frac{2\pi j}{n}$, $j = 1, ..., n$*

*Proof.* We have to solve the characteristic equation $|P(\alpha, \beta) - \lambda I_{2n}| = 0$. Before solving this equation, we recall the following well known result

**Lemma 6.** *Let A, B, C and D be commuting $n \times n$ matrices. Then*

$$\left| \begin{pmatrix} A & B \\ C & D \end{pmatrix} - \lambda I_{2n} \right| = 0 \Leftrightarrow |(A - \lambda I_n)(D - \lambda I_n) - CB| = 0 \tag{13}$$

$\square$

Since all blocks of $P(\alpha, \beta)$ commute, lemma 6 applies and we get the following equation

$$\left| (1 - 2\alpha)(1 - 2\beta)I_n + \lambda((\lambda - 2\alpha\beta)I_n - (1 - \alpha)(1 - \beta)(C_n + C_n^{-1})) \right| = 0$$

Let $C_n = C_n + C_n^{-1}$ and assume that $\lambda \neq 0$ i.e. $\alpha \neq \frac{1}{2} \wedge \beta \neq \frac{1}{2}$. Then

$$|P(\alpha, \beta) - \lambda I_{2n}| = 0 \Leftrightarrow \left| (\lambda - 2\alpha\beta + \frac{1}{\lambda}(1 - 2\alpha)(1 - 2\beta))I_n - (1 - \alpha)(1 - \beta)C_n \right| = 0$$

Let $\mu = \lambda - 2\alpha\beta + \frac{1}{\lambda}(1 - 2\alpha)(1 - 2\beta)$. It follows that $|P(\alpha, \beta) - \lambda I_{2n}| = 0 \Longleftrightarrow$ $|(1 - \alpha)(1 - \beta)C_n - \mu I_n| = 0$ Since the eigenvalues of the circuit graph $C_n$ are $2\cos\frac{2\pi j}{n}$ $(j = 1, .., n)$, it follows that $\mu = 2(1 - \alpha)(1 - \beta)\cos\frac{2\pi j}{n}$ $(j = 1, ..., n)$ i.e. $\lambda - 2\alpha\beta + \frac{1}{\lambda}(1 - 2\alpha)(1 - 2\beta) = 2(1 - \alpha)(1 - \beta)\cos\frac{2\pi j}{n}$ $(j = 1, .., n)$ hence $\lambda_j^{\mp} = h_c \mp \sqrt{h_c^2 - g}$ where $c = \cos\frac{2\pi j}{n}$, $j = 1, ..., n$ $\square$

Note that if $\alpha = \frac{1}{2}$ then it is easy to see that $\lambda_j^+ = \beta + (1-\beta)\cos\frac{2\pi j}{n}$ $(j = 1, ..., n)$ and $\lambda_j^- = 0$ $(j = 1, ..., n)$ i.e. the non zero eigenvalues are the eigenvalues of the natural diffusion in a circuit with $n$ vertices.

**Corollary 7.** *This diffusion scheme in $C_{2n}$ is at least as fast as the natural diffusion in $C_n$* □

**Corollary 8.** *If $n = 2$ and $\alpha = \beta = \frac{1}{2}$ then all stochastic eigenvalues are zero, thus the diffusion converges towards the uniform distribution in a finite number of steps. If $n = 3$ and $\alpha = \frac{1}{2}$, $\beta = \frac{1}{3}$ or $\alpha = \frac{1}{3}$, $\beta = \frac{1}{2}$ then all stochastic eigenvalues are zero, thus the diffusion converges towards the uniform distribution in a finite number of steps.* □

## 4.1 Main Results

Let $\lambda$ be the function from the unit square onto $[-1, 1]$ defined by

$$\lambda : (\alpha, \beta) \longrightarrow \max_{j=1,..,n-1}\{|\lambda_j^+|, |\lambda_j^-|\} \tag{14}$$

**Theorem 9.** *Let $n \geq 3$. The optimal diffusion is given by a pair $(\alpha_{opt}, \beta_{opt}) \in [0, 1]^2$ such that*

$$(\alpha_{opt}, \beta_{opt}) = \lambda^{-1}\left(\min_{(\alpha,\beta)\in[0,1]^2} \max\{|\lambda_1^+|, |\lambda_m^-|\}\right) \tag{15}$$

*Proof.* See [Boi93] □

**Theorem 10.** *If $n = 2m + 1$ is odd, $m \geq 3$, then the optimal diffusion is given by the two pairs $(\alpha_{opt}, \beta_{opt})$ and $(\beta_{opt}, \alpha_{opt})$ such that*

$$\alpha_{opt} = \frac{(\sqrt{1+c} + \sqrt{1+d})\left(\sqrt{1-c^2} - \sqrt{1-d^2}\right)}{(\sqrt{1-c} + \sqrt{1-d})(c-d)} \tag{16}$$

$$\beta_{opt} = \frac{(\sqrt{1+c} - \sqrt{1+d})\left(\sqrt{1-c^2} - \sqrt{1-d^2}\right)}{(\sqrt{1-c} + \sqrt{1-d})(c-d)} \tag{17}$$

*where $c = \cos\frac{2\pi}{n}$ and $d = \cos\frac{2\pi m}{n} = -\cos\frac{\pi}{n}$. Moreover*

$$\lambda_{opt} = \frac{1 - 2\sin\left(\frac{\pi}{2n}\right)}{1 + 2\sin\left(\frac{\pi}{2n}\right)} \tag{18}$$

If $n = 2m$ is even, $m \geq 1$, then the optimal diffusion is given by the pair $(\alpha_{opt}, \alpha_{opt})$ such that

$$\alpha_{opt} = \frac{\sin\left(\frac{\pi}{n}\right)}{1 - \sin\left(\frac{\pi}{n}\right)} \tag{19}$$

Moreover

$$\lambda_{opt} = \frac{1 - \sin\left(\frac{\pi}{n}\right)}{1 + \sin\left(\frac{\pi}{n}\right)} \tag{20}$$

*Proof.* (**Overview**) Let's first consider the equation $h_1(\alpha, \beta) = -h_m(\alpha, \beta)$. The solutions may be considered as a function

$$\phi : \beta \longrightarrow \frac{(1-\beta)(c+d)}{(1-\beta)(c+d) - 2\beta}$$

Moreover the solutions of the equation $\Delta_1(\alpha, \beta) = 0$ in the unit square $[0, \frac{1}{2}]^2$ may also be considered as a function

$$\psi : \beta \longrightarrow \frac{(c(1-\beta) - \beta)^2 + (1-\beta)\left(c\beta - (1-\beta) + \sqrt{1 - 2\beta}\sqrt{1 - c^2}\right)}{(c(1-\beta) - \beta)^2}$$

Let $\lambda(\alpha, \beta)$ denote the maximal modulus of the eigenvalues $\lambda_j^{\mp}$, $j = 1..n-1$ at the point $(\alpha, \beta)$ of the unit square. If $n$ is even then for any (fixed) $\beta \in [0, 1]$, then $\lambda_1^+(\phi(\beta), \beta) \leq \lambda(\alpha, \beta)$    $(\forall \alpha \in [0, 1])$ i.e. the optimal solution lies on the graph of $\phi$, moreover $\lambda_1^+(\beta, \phi(\beta)) \in \mathbb{R}$   $(\forall \beta \in [0, 1])$. If $n$ is odd then for any(fixed) $\beta \in [0, 1]$ then $\lambda_1^+(\phi(\beta), \beta) \leq \lambda(\alpha, \beta)$    $(\forall \alpha \in [0, 1])$ provided $\lambda_1^+(\phi(\beta), \beta) \in \mathbb{R}$ and $\sqrt{g(\psi(\beta), \beta)} \leq \lambda(\alpha, \beta)$    $(\forall \alpha \in [0, 1])$ provided $\lambda_1^+(\phi(\beta), \beta) \in \mathbb{C} \setminus \mathbb{R}$.
The function $h_c(\alpha, \phi(\alpha))$, $c = \cos\frac{2\pi}{n}$ takes its minimum at $\alpha_0 = \phi(\alpha_0)$, i.e. on the line $\alpha = \beta$. Moreover $\frac{\partial h_c(\alpha, \phi(\alpha))}{\partial \alpha} \leq 0$   if $\alpha \in [0, \alpha_0]$ and $\frac{\partial h_c(\alpha, \phi(\alpha))}{\partial \alpha} \geq 0$    if $\alpha \in [\alpha_0, \phi(0)]$
Since $\lambda_1^+(\alpha, \phi(\alpha)) \in \mathbb{R}$ if $n$ is even, it takes its optimal value at $\alpha_0$. If $n$ is odd, then $\lambda_1^+(\alpha, \phi(\alpha)) \in \mathbb{C} \setminus \mathbb{R}$ near the line $\alpha = \beta$. The real valued optima are thus taken at the intersection $(\alpha_0, \beta_0)$ and $(\beta_0, \alpha_0)$ of the graphs of $\phi$ and $\psi$. Assume that $\alpha_0 \leq \beta_0$.
In the complex domain $\lambda_1^+$ takes its minimal modulus on the graph of $\psi$ because $\frac{\partial g}{\partial \alpha} \leq 0$ ($\beta$ fixed) in the domain $[0, \frac{1}{2}]^2$.
In the interval $[\alpha_0, \beta_0]$, $\lambda_1^+(\alpha, \psi(\alpha))$ takes its minima at $\alpha_0$ and $\beta_0$.
See [Boi93] for detailed proofs.                                                       □

It is surprising that the edges labelled with $\hat{s}$ do not have the same weight than those labelled with $\hat{t}$. We propose the following explanation: The first weight is responsible for fast local balancing, the second for fast diffusion in the whole circuit. e.g. in $C_6$, the first weight is $\frac{1}{2}$ and leads to optimal local load balance, the second weight is $\frac{2}{3}$ and leads to fast diffusion in the rest of the circuit. Both goals are obviously contradictory.

**Theorem 11.** *Asymptotically, the Cayley diffusion in $C_{(2n)^2}$ achieves the same convergence speed than the natural diffusion in $C_{2n}$.*

*Proof.* Let $\lambda_{nat}(n)$ be the natural diffusion speed in $C_{2n}$ and $\lambda_{cal}(m)$ be the optimal (Cayley) diffusion speed in $C_{2m}$. It is easy to see that $\lambda_{nat}(n) = \frac{1+\cos\frac{\pi}{n}}{3-\cos\frac{\pi}{n}} = \frac{1-\sin^2\frac{\pi}{2n}}{1+\sin^2\frac{\pi}{2n}}$.

Since $\lambda_{nat}(n) \geq \lambda_{cal}(m) \Leftrightarrow \frac{1-\sin^2\frac{\pi}{2n}}{1+\sin^2\frac{\pi}{2n}} \geq \frac{1-\sin\frac{\pi}{2m}}{1+\sin\frac{\pi}{2m}} \Leftrightarrow \sin\frac{\pi}{2m} \geq \sin^2\frac{\pi}{2n}$, $x - \frac{x^3}{3!} \leq$ $\sin x \leq x$ ($0 \leq x$ small), and because $\sin x \geq \sin^2 y$ follows from $x \geq y^2$. It results that $2m \leq O\left(\frac{(2n)^2}{\pi}\right)$ □

# 5 Conclusion

Cayley graphs make good choices for multiprocessor networks, they are vertex symmetric, and most standard networks can be formulated in terms of Cayley graphs. Moreover, we have proposed a new load diffusion scheme for Cayley graphs and have seen that its convergence is better than the natural scheme, at least for some particular graphs. However, this strategy is not always the best one. For example, the natural diffusion in the complete graph is always better than or equal to the Cayley diffusion strategy[2]. The reason is probably that the generator set of the Cayley graph is not minimal:

**Conjecture 12.** *The convergence of the Cayley diffusion scheme in the Cayley graph $G(\Gamma, S)$ is faster than that of the natural diffusion scheme provided $S$ is a minimal symmetric set of generators.*

We have done a lot of numerical computations with Cayley graphs from the symmetric groups. The Cayley diffusion scheme was still better than the natural one.
Using the results of section 4 it is possible to improve the Cayley diffusion strategy. Indeed if the edges belonging to generators of even order $d > 2$ are subsequently labelled like in theorem 10, the convergence speed may be faster. Unfortunately, there is no canonical way to label such edges.
Sums[3] of graphs [CDS79] are also an interesting field of study. We can prove that if there is a fast diffusion scheme for each component of the sum, then there exists a fast diffusion scheme in the sum. To achieve this goal it is sufficient to use a strategy similar to the Cayley diffusion strategy (the dimensions correspond here to the components of the graph sum). An interesting result is that in any sum of hexagonal graphs, square graphs and paths with 2 vertices, uniform load distribution

---

[2] If $\Gamma$ is a group of order $n$ with unit $e$ then $G = (\Gamma, \Gamma \setminus \{e\})$ is a complete graph
[3] If $G_1 = (E_1, V_1)$ and $G_2 = (E_2, V_2)$ are graphs, the sum $G_1 + G_2$ is the graph with vertex set $V_1 \times V_2$. The vertices $(x_1, x_2)$ and $(y_1, y_2)$ are adjacent if and only if either $x_1 = y_1$ and $(x_2, y_2) \in E_2$ or $x_2 = y_2$ and $(x_1, y_1) \in E_1$. The d dimensional cube graph is the sum of d paths with 2 vertices.

is reached after a finite number of local diffusion steps. This will be discussed in detail in a subsequent paper.

A further interesting area to study would be the graphs for which there exist a partition of its edge set into perfect matchings. This would lead to similar load balancing schemes. Indeed, while writing this article, the author became aware of an article of Xu and Lau [XL92]. In this article, a similar load balancing method for coloured graphs is discussed extensively.

# References

[Boi90a]  J.E. Boillat. Distributed Load Balancing and Random Walks in Graphs. Technical Report IAM-90-013, Institute for Informatics and Applied Mathematics, University of Berne, 1990.

[Boi90b]  J.E. Boillat. Load Balancing and Poisson Equation in a Graph. *Concurrency: Practice and Experience*, 2(4), 1990.

[BBK91]  J.E. Boillat, F. Brugé, and P.G. Kropf. A Dynamic Load Balancing Algorithm for Molecular Dynamics Simulation on Multi–Processor Systems. *Journal of Computational Physics*, 96:1 – 14, 1991.

[Boi93]  J.E. Boillat. Fast Load Balancing in Cayley Graphs and in Circuits. Technical Report IAM-93-011, Institute for Informatics and Applied Mathematics, University of Berne, March 1993.

[CDS79]  D.M. Cvetkovič, M. Doob, and H. Sachs. *Spectra of graphs*. Academic Press, New York, 1979.

[Cyb89]  G. Cybenko. Dynamic Load Balancing for Distributed Memory Multiprocessors. *J. of Parallel and Distributed Computing*, 7:279-301, 1989.

[ELZ86a]  D.E. Eager, E.D. Lazowska, and J. Zahorjan. A Comparison of Receiver-Initiated and Sender-Initiated Adaptive Load Sharing. *Performance Evaluation*, 6:53-68, 1986.

[ELZ86b]  D.E. Eager, E.D. Lazowska, and J. Zahorjan. Adaptive Load Sharing in Homogenous Distributed Systems. *IEEE Transactions on Software Engeneering*, SE-12(5):662-675, 1986.

[Hać89]  A. Hać. Load balancing in distributed systems: A summary. *Performance Evaluation Review*, 16(2-4):17-19, February 1989.

[MTS90]  R. Mirchandaney, D. Towsley, and J.A. Stankovic. Adaptive Load Sharing in Heterogenous Distributed Systems. *Journal of Parallel and Distributed Computing*, 9:331-346, 1990.

[XL92]  Z. Xu and C.M. Lau. Analysis of the Generalized Dimension Exchange Method for Dynamic Load Balancing. *Journal of Parallel and Distributed Computing*, 16:385-393, 1992.

# Concurrent Flows and Packet Routing in Cayley Graphs
# (Preliminary Version)

Farhad Shahrokhi

Department of Computer Science, University of North Texas

Laszló A. Székely *

Department of Computer Science, Eötvös University

### Abstract

Let $G$ be a Cayley graph on $n$ vertices with degree $D$. We effectively compute a vertex optimal integral uniform concurrent flow in $G$ and develop an offline packet routing algorithm for routing $n(n-1)$ packets. The number of communication steps to route all packets to their destination is shown to be within a multiplicative factor of $D$ from the optimum. Our algorithm can be implemented online in many existing parallel networks, so that the number of communication steps is within a multiplicative factor $D$ from the optimum. Slight variations of our algorithm give rise to effective packet routing algorithms in perfect shuffles and deBruijn graphs. The model of computation required for our routing strategy may be assumed to be either MIMD or SIMD, as our online solutions only require the execution of the same instruction at the processors while routing.

## 1 Introduction

The uniform concurrent flow problem (UCFP) [Ma85], [SM90] is a multicommodity flow problem [Hu69], in which we supply one unit of flow between every ordered vertex pair in the underlying graph. The objective is to minimize the largest flow through any edge (congestion), or the largest flow through any vertex (vertex congestion). In the integral version of the UCFP (IUCFP), we wish to assign one path to any ordered vertex pair, so that the maximum overlapping of paths in any vertex or any edge is minimized. Unlike UCFP, which is solvable in polynomial time, IUCFP is NP-hard [Sa91].

In the current paper we investigate IUCFP in Cayley graphs and show that a vertex optimal solution, which is invariant under the action of the automorphism group, can be effectively computed in polynomial time. It is shown that for a Cayley graph of degree $D$, our solution has a congestion which is within a multiplicative factor of $D$ from the optimal congestion. (Section 3, Theorems 3.2 and 3.3.)

One application of our results is solving the all pairs packet routing problem (APPRP). In APPRP, we are given a collection of $n$ processors, which are connected together using some interconnection network with a fixed topology. Every processor must send exactly one packet to any other processor. The goal is to minimize the number of (parallel) communication steps required to send all packets to their destinations, using small queue sizes at the nodes. The APPRP appears in problems such as transposing a matrix or conversion between different data structures [JH89] on a parallel machine. Saad and Schultz [SS89], and Johnson and Ho [JH89] investigated this problem when the underlying graph is a $k$-dimensional cube and obtained solutions which are within a small (constant) multiplicative factor from the optimum. In Section 4 we specifically consider the APPRP in interconnection networks, which are Cayley graphs of degree $D$. Using the set of source-sink paths for our solution to IUCFP, we obtain an offline

*Research of the second author was supported by the A. v. Humboldt-Stiftung while he was visiting at the Institut für diskrete Mathematik, Bonn; and by the U.S. Office of Naval Research under the contract N-0014-91-J-1385.

algorithm for the movements of the packets (Theorem 4.2), so that the number of communication steps is within a multiplicative factor of $D$ from optimum. Moreover, we show that under reasonable conditions our offline algorithm can be implemented online (Corollary 4.3). In Section 5 we apply our online routing algorithm to a hypercube, a toroidal mesh, a cube connected cycles, and a butterfly. The main result in this section (Theorem 5.2) is to show that for a cube connected cycles and a butterfly, the APPRP can be solved online so that the number of communication steps is within a small constant multiplicative factor from the optimum. Finally, in Section 6 we investigate the APPRP in a perfect shuffle and a deBruijn graph (Theorems 6.2 and 6.3).

By viewing the APPRP for any graph $G =< V, E >$ as $|V| - 1$ many permutation routing problems [Ul84] and applying the best existing permutation routing algorithm, one can achieve the same order of communication steps (with large multiplicative constants) for solving APPRP as we have obtained, for the graphs discussed in this work. However, the significance of our work is in providing a general routing algorithm that applies to all graphs we have discussed in this paper, in contrast to the previous ad-hoc methods for the permutation routing. Moreover, this unified approach has the virtue of being deterministic and nevertheless permitting efficient online implementation so that the multiplicative constants are small.

## 2   Notations

We assume that the reader is familiar with the basic graph theory terminology as in [CL86] and [Wh84]. Throughout this paper but Section 6 we assume that the graph $G =< V, E >$ is undirected with no parallel edges.

Let $p$ be a path with endvertices $a$ and $b$ in graph $G =< V, E >$. Then $p$ will give rise to an oriented path from $a$ to $b$; similarly $p$ will give rise to an oriented path from $b$ to $a$. For any ordered pair of vertices $(i, j) \in V \times V$, we denote by $P_{ij}$ the set of all oriented paths from vertex $i$ to vertex $j$; any $p \in P_{ij}$ is termed an $i, j$ path (note that $P_{ii} = \emptyset$). Let $P = \cup_{(i,j) \in V \times V} P_{ij}$, thus any $p \in P$ is an oriented path. Throughout this paper the term path applies to an oriented path, unless stated otherwise. Let $P_e$ denote the collection of all paths in $P$ containing the edge $e = (x, y)$, in either $x \to y$ or $y \to x$ directions. Similarly, for any $i \in V$, let $P_i$ denote the collection of all paths in $P$ which contain $i$. For any $p \in P$ and any $(i, j) \in V \times V$, let $L(p)$ denote the number of edges in $p$, $L(i, j)$ denote $\min_{p \in P_{ij}} L(p)$ and $L$ denote $\sum_{(i,j) \in V \times V} L(i, j)$. Let $R^+$ denote the set of non-negative real numbers. A uniform concurrent multicommodity flow, or shortly a *uniform flow*, is a function $f : P \to R^+$, so that

$$\sum_{p \in P_{ij}} f(p) = 1, \text{ for any } (i, j) \in V \times V.$$

We call $f(p)$ the flow on path $p$ and call $p$ an *active* path if $f(p) > 0$. We define $f(e) = \sum_{p \in P_e} f(p)$ for any edge $e$, and call $f(e)$ the flow on edge $e$. Similarly, for any vertex $i$, we define $f(i) = \sum_{p \in P_i} f(p)$ and call $f(i)$ the flow in vertex $i$. We define the *congestion* of $f$ to be $\mu_f = \max_{e \in E} f(e)$ and the *vertex congestion* of $f$ to be $m_f = \max_{v \in V} f(v)$. A uniform flow is edge optimal (vertex optimal) if it has the smallest congestion (vertex congestion) among all uniform flows. The uniform concurrent multicommodity flow problem (UCFP) is to compute a uniform flow which is either edge optimal or vertex optimal; the preferred type of optimality is usually dictated by the application and/or context. The UCFP can be expressed as a node-arc linear program [FF62], and hence can be solved in polynomial time using the polynomial time linear programming algorithms [Ka84].

An *integral* uniform flow $f$ is a uniform flow so that $f(p) \in \{0, 1\}$, for any $p \in P$. The *integral uniform concurrent flow problem* (IUCFP) is to compute an integral flow which is either edge or vertex optimal. The integral versions of multicommodity flow problems have been well known to be NP-hard [EIS76]. In particular, computing a vertex optimal integral uniform flow is NP-hard [Sa91].

For $G = < V, E >$, let $Aut(G)$ denote the automorphism group of $G$. Let $\Gamma$ be a subgroup of $Aut(G)$, then we write $\Gamma < Aut(G)$. We say that a uniform flow $f$ is $\Gamma$-invariant [SS93], if for any $g \in \Gamma$ and any $p \in P$, we have, $f(p) = f(g(p))$, where $g(p)$ is the image of $p$ under $g$. The following result was shown in [SS92].

**Theorem 2.1** *For $G = < V, E >$, let $\Gamma < Aut(G)$ and assume that $\bar{f}$ is any $\Gamma$-invariant uniform flow so that any $ij$ path $p$ has $L(p) = L(i, j)$. Then*

$$\mu_{\bar{f}} \leq \frac{L}{|E_1|} \quad and \quad m_{\bar{f}} \leq \frac{L + |V|(|V| - 1)}{|V_1|},$$

*where $E_1$ and $V_1$ are the smallest edge and vertex orbits of $G$, respectively, under the action of $\Gamma$.* $\square$

## 3 Integral Uniform Flows in Cayley Graphs

In this section we investigate IUCFP in Cayley graphs and construct near optimal integral uniform flows. Let $\Gamma$ be a finite group with the generator set $H = \{h_1, h_2, ..., h_k\}$. The *Cayley color graph* $C_H(\Gamma)$ has the vertex set $\Gamma$; for two vertices $g, g' \in \Gamma$, there is an arc from $g$ to $g'$ colored $h$, if and only if $g' = gh$ for some $h \in H$. The undirected graph which is obtained from $C_H(\Gamma)$ by suppressing all edge directions (but not edge colors), and suppressing all multiple edges is called a *Cayley graph* and is denoted by $G_H(\Gamma)$.

Let $\Gamma = \{g_1, g_2, ..., g_n\}$ be a finite group with a generating set $H$. Throughout this paper we assume that $g_1$ is the identity element of $\Gamma$, unless stated otherwise. For each $i, 1 \leq i \leq n$, define a permutation $\alpha_i : \Gamma \to \Gamma$, by $\alpha_i(g) = g_i g$ and let $\Delta$ denote the set of all such permutations $\Delta = \{\alpha_i | g_i \in \Gamma\}$. The following result is well known [Wh84].

**Theorem 3.1** *One has $\Delta < Aut(G_H(\Gamma))$. Moreover, $\Delta$ is isomorphic to $\Gamma$ and $G_H(\Gamma)$ is vertex transitive for $\Delta$, and hence also for $\Gamma$.* $\square$

Throughout this paper we will denote the vertex degree of $G_H(\Gamma)$ (which is vertex transitive by Theorem 3.1) by $D$. For a group $\Gamma = \{g_1, g_2, ..., g_n\}$ and a generator set $H$, let $q$ be a $g_m, g_l$ walk in $G_H(\Gamma)$. We denote by $g_i(q)$ the walk from the vertex $g_i g_m$ to the vertex $g_i g_l$ in $G_H(\Gamma)$ which is the image of $q$ under $\alpha_i \in \Delta$, $1 \leq i \leq n$. Next, we construct a near optimal integral flow in $G_H(\Gamma)$.

**Theorem 3.2** *For a finite group $\Gamma = \{g_1, g_2, ..., g_n\}$ with identity element $g_1$, let $H$ be a generator set. Let $q_{i1}$ $(2 \leq i \leq n)$ be a path from vertex $g_i$ to vertex $g_1$ in the Cayley graph $G_H(\Gamma)$. Let $Q = \{g_j(q_{i1}) | 1 \leq j \leq n, 2 \leq i \leq n\}$. Then,*
*(i) $Q$ is a collection of paths in $G_H(\Gamma)$, such that for every $(a, b) \in \Gamma \times \Gamma$, $a \neq b$, there is exactly one $a, b$ path in $Q$.*
*(ii) Let $q = g_j(q_{i1})$, and $q' = g_m(q_{i1})$, $1 \leq j < m \leq n$, $2 \leq i \leq n$. Consider any vertex $g$, which is contained by both $q$ and $q'$, then $g_j^{-1} g \neq g_m^{-1} g$.*
*(iii) Let $q = g_j(q_{i1})$, and $q' = g_m(q_{i1})$, $1 \leq j < m \leq n$, $2 \leq i \leq n$. Consider any vertex $g$ which is contained by both $q$ and $q'$. Let $g_k = g_j g_i$ and $g_t = g_m g_i$; and denote by $q_1$ the $g_k, g$ subpath of $q$. Similarly, denote by $q'_1$ the $g_t, g$ subpath of $q'$. Then, $L(q_1) \neq L(q'_1)$.*
*(iv) Set $f(q) = 1$ for every $q \in Q$ and 0 elsewhere, then $f$ is an integral uniform flow which is $\Gamma$-invariant in $G_H(\Gamma)$.*

**Proof.** To verify (i), it suffices to show that for every $a, b \in \Gamma$, there exists $g \in \Gamma$ and $g^* \in \Gamma$, such that $gg_1 = a$ and $gg^* = b$. Now select $g = a$ and $g^* = a^{-1}b$, which verifies (i).
To prove (ii), assume to the contrary that $g^* = g_j^{-1} g = g_m^{-1} g$. Then, $g_j = g_m$. But this a contradiction, since we assumed $j \neq m$. To prove (iii) assume to the contrary that $L(q_1) = L(q'_1)$. Then, it is easy to show that $g$ must be the image of the same vertex of $q_{i1}$ under $g_j$ and $g_m$. That is, there exists $g^*$ a vertex of $q_{i1}$, such that $g_j g^* = g_m g^* = g$. Thus, $g_m^{-1} g = g_j^{-1} g$ which is a contradiction by Part (ii). Finally, to prove (iv), observe that the construction of $f$ implies that $f$ is $\Gamma$-invariant and integral. $\square$

For the purpose of our next result, we assume that the multiplication table for $\Gamma$ is available, so that given any two elements of $\Gamma$, we can find their product by a single table look up.

**Theorem 3.3** *The integral flow $f$ in Part (iv) of Theorem 3.2 can be computed in $\Theta(\sum_{(i,j)\in\Gamma\times\Gamma} L(i,j))$ time such that any $i,j$ active path $p$ has $L(p) = L(i,j)$. The congestion $\mu_f$ of $f$ is within a multiplicative factor of $D$ from the optimal congestion; moreover, $f$ is edge optimal, provided that $G_H(\Gamma)$ is edge transitive for $\Gamma$. Furthermore, $f$ is vertex optimal.*

**Proof.** For every vertex in $G_H(\Gamma)$ construct a shortest path from this vertex to the identity vertex in $\Gamma$ and extend this set to a set of paths $Q$ by the method of Lemma 3.2 so that $Q$ contains exactly one $i.j$ path for every $(i,j) \in \Gamma \times \Gamma$. Let $f$ be the flow that assigns exactly one unit of flow to each path in $Q$. By Part (iv) of Theorem 3.2, $f$ is a $\Gamma$-invariant integral flow in $G$. Furthermore, it is easy to see that the construction of $f$ can be done in $\Theta(\sum_{(i,j)\in\Gamma\times\Gamma} L(i,j))$ time, which is essentially the length of the output.

A simple averaging argument show that for any graph $G = <V, E>$, a lower bound of $L/|E|$ holds for the congestion of any uniform flow in $G$. For our Cayley graph $G_H(\Gamma)$, the application of Theorem 2.1 yields an upper bound of $DL/|E|$ for $\mu_f$, which is within a factor of $D$ from the lower bound, and hence, in general, $\mu_f$ is within a factor of $D$ from the optimum. Moreover, if $G_H(\Gamma)$ is edge transitive for $\Gamma$, then the upper bound of Theorem 2.1 becomes $L/|E|$ which is the lower bound, and hence, in this case $f$ is edge optimal. This verifies the claims regarding $\mu_f$. To prove the vertex optimality of $f$, we observe that for any $G = <V, E>$, a lower bound of $(L + |V|(|V| - 1))/|V|$ holds on the vertex congestion of any uniform flow in $G$, and that for $G_H(\Gamma)$ and $f$, the application of Theorem 2.1 yields an upper bound on $m_f$ which is equal to this lower bound. □

# 4 All Pairs Packet Routing Problem in Cayley Graphs

In this section we consider a packet routing problem for a network on $n$ processors, in which processor $i$ ($1 \leq i \leq n$) is sending one packet $M_{ij}$ to processor $j$ ($j \neq i, 1 \leq j \leq n$); thus each processor is the source of $n - 1$ packets and the destination for $n - 1$ packets. Furthermore, we assume that the data in $M_{ij}$ is different from the data in $M_{ik}$ for any $i, j, k$ ($j \neq k, 1 \leq i \leq n$). We call this problem the *all pairs packet routing problem* (APPRP). The goal in solving APPRP is to minimize the number of communication steps needed to send all packets to their destinations. A communication step is the time that it takes for any packet to pass through a channel (or a link). We will assume that the order of magnitude for the time associated with one communication step is much larger than the order of magnitude for the time associated with performing any arithmetic operations by CPU. In our model we assume that the channels can be used in either directions, however, at any given time only one packet can use the channel. The following obvious lemma will be used later.

**Lemma 4.0** *Let $G = <V, E>$ and $f$ be any edge optimal integral uniform flow, and $A_G$ be the minimum number of communication steps for solving APPRP in $G$. Then,*

$$A_G \geq \mu_f \geq \frac{L}{|E|}.$$

**Proof.** Observe that routing a packet form a particular source to a particular destination specifies a unique path from the source to the destination. Hence, the collection of all source-destination paths involved in routing is a feasible solution to IUCFP and has a congestion at most equal to $A_G$, since no two packets can use the same edge at the same time. Moreover, this feasible solution has a congestion at least equal to $\mu_f$, since $f$ is optimal. It follows that $A_G \geq \mu_f$. The lower bound on $\mu_f$ follows from a simple averaging argument. □

In this section we assume that the interconnection of the network is isomorphic to a Cayley graph; each vertex corresponds to a processor, whereas each edge corresponds to a channel. For the Cayley graph $G_H(\Gamma)$ with $\Gamma = \{g_1, g_2, ..., g_n\}$, let $M$ denote the set of all packets for APPRP. Consider a partition $\{M_1, M_2, ..., M_n\}$ of $M$ such that for $2 \leq i \leq n$

$$M_i = \{\text{packet with source } g_j g_i \text{ and destination } g_j | 1 \leq j \leq n\}.$$

We will show using Part (iii) of Theorem 3.2, that for any fixed $i$, all packets in $M_i$ can be routed to their destination so that no collision occurs. We assume that all packets with source $g_i$ are originally stored in an array structure at $g_i$; likewise all the packets which are destined for $g_i$ will be stored in an array upon reaching $g_i$ ($1 \leq i \leq n$). $\square$

**Lemma 4.1** Let $q_{i1}$ be a path from $g_i$ to $g_1$ in $G_H(\Gamma)$ with $\Gamma = \{g_1, g_2, ..., g_n\}$, and consider the set of paths $Q_i = \{g_j(q_{i1}) | j = 1, 2, ..., n\}$ and the packet set $M_i$, for a fixed $i$, $2 \leq i \leq n$. Then all packets in $M_i$ can be sent to their destination using the paths in $Q_i$ in

(i) $L(q_{i1})$ communication steps, if $H$ does not have an element of order 2.

(ii) $2L(q_{i1})$ communication steps, in general.

**Proof.** For all $j$ ($1 \leq j \leq n$) simultaneously, consider the path $g_j(q_{i1}) \in Q_i$ and send the packet with source $g_j g_i$ and the destination $g_j$ on this path. We claim that no two packets in $M_i$ attempt to use the same edge from the same endvertex. Assume to the contrary, that a packet with source $g_k g_i$ and destination $g_k$, and a second packet with source $g_r g_i$ and destination $g_r$ attempt to enter the edge $(g_l, g_m)$ from vertex $g_l$ at the same time. Then both packets must have arrived at $g_l$ at the same time. Thus, $L(q_1) = L(q_2)$, where $q_1$ is the $g_k, g_l$ subpath of $g_k(q_{i1})$ which describes the movements of the first packet and $q_2$ is the $g_r, g_l$ subpath of $g_r(q_{i1})$ which describes the movement of the second packet, prior to the collision. However, this is a contradiction by Part (iii) in Lemma 3.2. As a result, any edge $(g_l, g_m)$ is used by only one packet, in the direction $g_l \rightarrow g_m$, at any given time. It is possible, however, that two packets attempt to enter edge $(g_l, g_m)$ from different endpoints of the edge, and therefore, one packet must yield for the other packet. It is easy to show that this happens if and only if the edge $e$ corresponds to a generator of order 2. $\square$

Our main result is the following:

**Theorem 4.2** For $\Gamma = \{g_1, g_2, ..., g_n\}$, consider the APPRP for $G_H(\Gamma) = < V, E >$. Then a set of routes can be computed offline in $O(|V| + |E|)$ time, and can be loaded to $G_H(\Gamma)$ using $O(|V|)$ storage per each node of $G_H(\Gamma)$, such that the total number of communication steps to route all packets is,

(i) $\dfrac{D}{2|E|} \displaystyle\sum_{(g_i, g_j) \in \Gamma \times \Gamma} L(g_i, g_j)$, if $H$ does not have an element of order 2, and,

(ii) $\dfrac{D}{|E|} \displaystyle\sum_{(g_i, g_j) \in \Gamma \times \Gamma} L(g_i, g_j)$ in general.

The number of communication steps is within a factor of $D$ ($D/2$, if $H$ does not have an element of order 2) from the optimum. Furthermore, during any communication step no two packets enter the same vertex, consequently, no queues for storing the moving packets while routing is needed.

**Proof.** Compute a path $q_{i1}$ from any vertex $g_i$ ($i = 2, 3, ..., n$) to the identity vertex $g_1$ using a breadth first search from $g_1$ such that $L(g_i, g_1) = L(q_{i1})$. Observe that the set of paths $\{q_{i1} | 2 \leq i \leq n\}$ can be stored in a one dimensional array $NEXT$ of size $n - 1$ such that if any vertex $g_k$ ($1 \leq k \leq n$) is located on any path $q_{i1}$, $2 \leq i \leq n$, then $NEXT[g_k]$ is the vertex adjacent from $g_k$ on $q_{i1}$. We store array $NEXT$ at each vertex of $G_H(\Gamma)$. Furthermore, we store at any vertex $g_m \in \Gamma$, the portion of the multiplication table for $\Gamma$ which contains the values of $g_k^{-1} g_m$ and $g_m g_k^{-1}$, for all $k$, $1 \leq k \leq n$. Thus we need a total of $O(|V|)$ additional storage per each vertex to store our routing table. Our routing strategy has $n - 1$ stages; at stage $i$, $2 \leq i \leq n$, the packets in $M_i$ are sent to their destinations using the set of routes $Q_i = \{g_j(q_{i1}) | j = 1, 2, ..., n\}$. Notice that for any fixed $i$ ($2 \leq i \leq n$) the array $NEXT$ can only provide explicit information regarding the path $q_{i1}$, that is, we have only stored the information regarding one path in $Q_i$. However, it is straightforward to verify that this representation is sufficient to route all packets in $M_i$ to their destinations. Note that since we are essentially using

Lemma 4.1 for routing $M_i$, no two packets will enter the same vertex at the same communication step at any stage $i$ (recall Part (iii) of Theorem 3.2). Since by Lemma 4.1 routing the packets in $M_i$ $(2 \leq i \leq n)$ takes $2L(g_1, g_i)$ communication steps $(L(g_1, g_i)$ steps, if $H$ does not have a generator of order 2), it follows that the total number of communication steps for sending all the packets to their destinations is

$$\sum_{i=2}^{n} L(g_1, g_i) = \frac{D}{2|E|} \sum_{(g_i,g_j) \in V \times V} L(g_i, g_j),$$

provided that $H$ does not have an element of order 2, and at most

$$2\sum_{i=2}^{n} L(g_1, g_i) = \frac{2}{\Gamma} \sum_{(g_i,g_j) \in \Gamma \times \Gamma} L(g_i, g_j) = \frac{D}{|E|} \sum_{(g_i,g_j) \in V \times V} L(g_i, g_j),$$

in general. Finally, observe that by Lemma 4.0, to route all the packets takes at least $\frac{1}{|E|} \sum_{(g_i,g_j) \in V \times V} L(g_i, g_j)$ communication steps. This verifies the claim regarding the multiplicative factor. □

Next we describe how to eliminate the offline computations in Theorem 4.2, provided that certain reasonable conditions are imposed on the underlying network $G_H(\Gamma)$.

We say that the Cayley graph $G_H(\Gamma)$ with $\Gamma = \{g_1, g_2, ..., g_n\}$ has a *reasonable representation* if the following hold.
(i) The multiplication table for $\Gamma$ has an implicit representation such that for any $g_i, g_j \in \Gamma$, the representations for $g_i^{-1}$ and $g_j g_i$ can be computed using the representations of $g_i$ and $g_j$.
(ii) $g_i^{-1}$ can be computed by applying $O(1)$ arithmetic operations to $g_i$. Similarly, $g_j g_i$ can be computed by applying $O(1)$ arithmetic operations to $g_i$ and $g_j$. (In our model an arithmetic operation involves a whole record.)
(iii) There is a routing algorithm for sending a packet from the source $g_i$ $(2 \leq i \leq n)$ to the destination $g_1$ (identity vertex), using only local information at each intermediate node. More specifically, there is a function $NEXT$ such that for any packet with the source $g_i$ $(2 \leq i \leq n)$ and the destination $g_1$ which is currently at the location $g_m, 1 \leq m \leq n$, $NEXT(g_m)$ is a vertex adjacent to $g_m$ which the packet must be sent to, so that the packet eventually reaches $g_1$ by the composition of $NEXT$'s. The following results can be easily shown.

**Corollary 4.3** *Assume that $G_H(\Gamma)$ has a reasonable representation. Then the APPRP for $G_H(\Gamma)$ is solved with no offline computations and with no extra storage allocations to vertices in*
(i) *$\sum_{i=2}^{n} L(q_{i1})$ communication steps, if $H$ does not have an element of order 2,*
(ii) *$2\sum_{i=2}^{n} L(q_{i1})$ communication steps, if $H$ has an element of order 2,*
*where $q_{i1}$ $(2 \leq i \leq n)$ is the $g_i, g_1$ path which describes the movements of packet with source $g_i$ and the destination $g_1$ using the function $NEXT$.* □
The quality of packet routing in Corollary 4.3 depends on the lengths of source to destination paths which are identified by function $NEXT$. Fortunately, in many known networks, the function $NEXT$ provides for source-destination paths which are either shortest paths or very close to shortest paths. We give examples of these networks in the next section.

# 5 APPRP in Cube Family

Let $Q_k$ denote the $k$-dimensional cube and $M(n_1, n_2, ..., n_k)$ denote the toroidal mesh, that is the cartesian product of cycles on $n_i$ vertices, $i = 1, 2, ..., k$. In both of these networks the function $NEXT$ which utilizes the well known left to right coordinate matching strategy [Ul84], computes the shortest paths. An immediate consequence of Corollary 4.3 is the following.

**Theorem 5.1**

*(i) APPRP can be solved online in $k2^k$ communication steps for $Q_k$. This number of communication steps is within a multiplicative factor of $k$ from the optimum.*

*(ii) APPRP can be solved online in $(\sum_{i=1}^{k} \frac{\mu_i}{n_i})\Pi_{j=1}^{k} n_j$ communication steps for $\mathcal{M}(n_1, n_2, ..., n_k)$, where $\mu_i = n_i^2/4$, if $n_i$ is even, and $\mu_i = (n_i^2 - 1)/4$, if $n_i$ is odd. This number of communication steps is within a multiplicative factor of $k$ form the optimum.* $\square$

It is well known that the cube connected cycles and the butterfly are Cayley graphs with the same underlying group $\Gamma$ which is the *wreath product* of $Z_n$ and $Z_2$ but with different generating sets. We exploit this structure in the following. Let $N = \{0, 1, 2, ..., n-1\}$ and $\Theta_n = \{g_{W,i} | W \subseteq N, i \in N\}$. For $i, j \in N$, let $i \oplus j$ denote $i + j$ modulo $n$. For $U \subseteq N$ and $i \in N$, let $U \oplus i = \{j \oplus i | j \in U\}$. Set $V \triangle U = (V \cup U) \setminus (V \cap U)$. Now $\Theta_n$ is a group with identity $g_{\emptyset,0}$ and operations

$$g_{W,t} g_{U,i} = g_{W \triangle(U \oplus t), i \oplus t} \quad \text{and} \quad g_{W,t}^{-1} = g_{W \oplus(n-t), n-t}.$$

The cube-connected cycles $CC_n$ is a Cayley graph over $\Theta_n$ with the generating set

$$H = \{g_{\emptyset,1}, g_{\{0\},0}\}.$$

We term the edges produced by the first generator *cyclic edges* and and the edges produced by the second generator *cubic edges*. It is easy to see that $CC_n = <V, E>$, where

$$V = \{(W, i) : W \subseteq N, i \in N\}$$

and $(W, i)(U, j) \in E$ if $i = j = W \triangle U$ (cubic edges in dimension $i$) or if $|i - j| \equiv 1 \bmod n$ and $U = V$ (cyclic edges).

The butterfly $BB_n$ is a Cayley graph over $\Theta_n$ with the generating set $H = \{g_{\emptyset,1}, g_{\{0\},n-1}\}$. We term the edges arising from the first generator *cyclic edges* and the edges arising from the second generator *cubic edges*. It is easy to see that $BB_n = <V, E>$, where

$$V = \{(X, i) : X \subseteq N, i \in N\},$$

and $(X, i)(Y, j) \in E$ if $X = Y$ and $|i - j| \equiv 1 \bmod n$ (cyclic edges) or $X \triangle Y = i$ and $j \equiv i - 1 \bmod n$ (cubic edges in dimension $i$).

Next, we will show that APPRP can be solved online with a near optimal number of communication steps in $CC_n$ and $BB_n$. It is clear that Condition (i) of a reasonable representation is satisfied for $BB_n$ and $CC_n$ with the underlying group $\Theta_n$. To verify that Condition (ii) holds, we will provide a slightly different description of $\Theta_n$. Let $I$ denote the set of $n$ digit binary numbers and $\hat{\Theta}_n = \{(W, i) | W \in I, i \in N\}$. For $U, V \in I$, let $U \otimes V$ denote the *Exclusive Or* of $U$ and $V$. For $i, j \in N$, let $i \oplus j$ be $i + j$ modulo $n$. Finally, for $U \in I$, $i \in N$, let $U \oplus i$ be the $n$ digit binary number obtained by shifting circularly every bit of $U$ by $i$ positions to the right. Then the operations for $\hat{\Theta}_n$ is defined as

$$g_{W,t} g_{U,i} = g_{W \otimes(U \oplus t), i \oplus t} \quad \text{and} \quad g_{W,t}^{-1} = g_{W \oplus(n-t), n-t}.$$

The identity element for $\hat{\Theta}_n$ is $g_{I_0,0}$, where $I_0$ is the $n$ digit binary representation of 0. Clearly, $\hat{\Theta}_n$ is just an alternative representation of $\Theta_n$, we compute with the characteristic vectors of sets rather than with the sets. Using $\hat{\Theta}_n$ as the underlying group for $CC_n$ and $BB_n$, it is easy to verify that conditions (i) and (ii) of a reasonable representation are satisfied. We emphasize that $\hat{\Theta}_n$ is more appropriate for the implementation purposes, at the hardware or software level. However, for the purpose of analysis presented here, $\Theta_n$ is more convenient to use. In fact we derive our next result using $\Theta_n$, bearing in mind that the real implementation is done using $\hat{\Theta}_n$.

**Theorem 5.2**
*(i) APPRP is solved online in $\frac{14}{5}n^2 2^n(1+o(1))$ communication steps for $CC_n$; this is within a multiplicative factor of $\frac{14}{5}$ from the optimum.*
*(ii) APPRP is solved online in $\frac{5}{4}n^2 2^n(1+o(1))$ communication steps for $BB_n$; this is within a multiplicative factor of 2 from the optimum.*

**Proof.** Since by the preceding remarks to this Theorem the Condition (i) and (ii) of a reasonable representation hold for $BB_n$ and $CC_n$, we will concern ourselves with the description of the function $NEXT$. Let $C_n$ denote the cycle on $N$ given by the edge set $\{(i, i \oplus 1)| i \in N\}$. For any vertex $i$ of $C_n$, let $mate(i)$ be the vertex adjacent to $i$ on a shortest $i, 0$ graph theoretical path on $C_n$. If there are two such paths, we select $mate(i)$ to be $i \oplus 1$. Let $X = (U, i)$ be any vertex of $CC_n$, and define

$$NEXT((U, i)) = \begin{cases} (\emptyset, mate(i)), & \text{if } U = \emptyset, \\ (U - \{i\}, i), & \text{if } U \neq \emptyset \text{ and } i \in U, \\ (U, i \oplus 1), & \text{if } U \neq \emptyset \text{ and } i \notin U. \end{cases}$$

Similarly, for any vertex $(U, i)$ of $BB_n$, define

$$NEXT((U, i)) = \begin{cases} (\emptyset, mate(i)), & \text{if } U = \emptyset, \\ (U - \{i \oplus 1\}, i \oplus 1), & \text{if } U \neq \emptyset \text{ and } i \oplus 1 \in U, \\ (U, i \oplus 1), & \text{if } U \neq \emptyset \text{ and } i \oplus 1 \notin U. \end{cases}$$

Then, it is easy to verify, that by successively applying the function $NEXT$ we obtain a path between vertex $X = (U, i)$ and the identity vertex $X_0 = (\emptyset, 0)$ in $CC_n$ or $BB_n$. Thus, Condition (iii) of a reasonable description is also satisfied.

Let $V$ be the set of vertices of $CC_n$. For any $X = (U, i) \in V$, it is not hard to verify, that

$$|Cyclic(q_{XX_0})| \leq n + L(0, i) \quad \text{and} \quad |Cubic(q_{XX_0})| = |U|,$$

where $q_{XX_0}$ is the $XX_0$ path obtained by applying the function $NEXT$ and the set of cyclic (cubic) edges in $q_{XX_0}$ is denoted by $Cyclic(q_{XX_0})$ $(Cubic(q_{XX_0}))$. It follows, that for $CC_n$

$$\sum_{X \in V} L(q_{XX_0}) = \sum_{X \in V} |Cyclic(q_{XX_0})| + |Cubic(q_{XX_0})| \leq \frac{7}{4}n^2 2^n(1 + o(1)). \tag{5.1}$$

Similarly, let $V'$ be the set of vertices of $BB_n$ and $X_0 = (\emptyset, 0)$. It is not difficult to verify that,

$$\sum_{X = (U, i) \in V'} |L(q'_{XX_0})| \leq \frac{5}{4}n^2 2^n(1 + o(1)), \tag{5.2}$$

where, $q'_{XX_0}$ is the $XX_0$ path obtained by applying function $NEXT$. We conclude from (5.1) and Corollary 4.3 that for $CC_n$, the number of communication steps to route all the packets is at most $\frac{14}{4}n^2 2^n(1 - o(1))$, since one of the generators used to construct $CC_n$ has order 2. By (5.2) the number of communication steps for $BB_n$ by is $\frac{5}{4}n^2 2^n(1 + o(1))$, since it has no generator of order 2. To verify the claims regarding multiplicative factors in (i) and (ii), we note that $\mu_f = \frac{5}{4}n^2 2^n(1 - o(1))$, for an edge optimal integral flow in $CC_n$, and that $\mu_{f'} = \frac{5}{4}n^2 2^{n-1}(1 + o(1))$, for an edge optimal integral uniform flow in $BB_n$, as we have shown in [SS93]. Now use Lemma 4.0 to finish the proof. $\square$

# 6 APPRP in Perfect Shuffle and DeBruijn Graphs

Let $I$ denote the set of $n$-digit binary numbers; for $a = 0$ or 1, we denote $(1 - a)$ by $\bar{a}$. Define the permutations $Shuf: I \to I$ and $Exch: I \to I$ by

$$Shuf(a_n a_{n-1} \ldots a_1) = a_{n-1} \ldots a_1 a_n \quad \text{and} \quad Exch(a_n a_{n-1} \ldots a_1) = a_n a_{n-1} \ldots \bar{a}_1.$$

For the purposes of the deBruijn graph, define

$$Exch(a_n a_{n-1}...a_1) = a_{n-1}...a_1 \bar{a}_n.$$

The $n$-dimensional *perfect shuffle* $PS_n$ has the vertex set $I$, and the edge set $E$, where $(XY) \in E$ if $Y = Exch(X)$ or $Y = Shuf(X)$. Edges of the first type are called *exchange edges*, edges of the second type are called a *shuffle edges*. The $n$-dimensional deBruijn graph $D_n$ and its exchange and shuffle edges are defined in the same way, using the altered definition of $Exch$. We allow multiple shuffle edges, if multiple definition occurs, and do alike for exchange edges in $D_n$.

In this section we investigate APPRP in $PS_n$ and $D_n$. We will route our packets using the well known greedy paths in these networks obtained by the following algorithm [Ul84].

THE GREEDY PATH (WALK) ALGORITHM
**Input:** $A = a_n a_{n-1}...a_1$, $B = b_n b_{n-1}...b_1$, two vertices of $PS_n$.
**Output:** a walk $W_{AB}$ and a path $q_{AB}$ from $A$ to $B$.

$X \leftarrow A$, $W_{AB} \leftarrow X$
**For** $i = 1$ to $n$ **Do**
**If** $a_i \neq b_i$;
**Then**
$X \leftarrow Exch^{-1}(X)$, $W_{AB} \leftarrow W_{AB} \cup X$
$X \leftarrow Shuf^{-1}(X)$, $W_{AB} \leftarrow W_{AB} \cup X$
**Else**
$X \leftarrow Shuf^{-1}(X)$, $W_{AB} \leftarrow W_{AB} \cup X$
**EndIf**
$q_{AB} \leftarrow W_{AB}$ with loops cut out as they arise
**EndFor**
**End.**

For $D_n$ we use the same algorithm but delete the second line after **Then**.

Let

$$W : X = X_1, X_2, ..., X_m = Y$$

be an $XY$ walk in $PS_n$. We will view $W$ as a sequence of edges, each being either an exchange edge or a shuffle edge. Next, we will treat the shuffle and the exchange as an operation on each vertex, or more precisely a permutation of $I$. Thus $W$ will give rise to a sequence of permutations for $I$. The $i$th element of this sequence ($1 \leq i \leq m - 1$) is a $Shuf$, if $X_{i+1} = Shuf(X_i)$, and is an $Exch$, if $X_{i+1} = Exch(X_i)$. Let $F$ be the composition of these permutations. Then $F$ is a permutation of $I$ such that $F(X) = Y$. We term $F$ the permutation (of $I$) associated with $W$. As an example take $n = 3$ and

$$W : 000, 001, 010$$

then, $F = Shuf \circ Exch$, where $\circ$ denotes the function composition. We now state a lemma which will be used in our packet routing algorithm for $PS_n$.

**Lemma 6.1** *Let $W_1$ be the greedy $X_1, Y_1$ walk and $W_2$ be the greedy $X_2, Y_2$ walk in $PS_n$ such that $X_1 \otimes X_2 = Y_1 \otimes Y_2$. Let $W_1'$ be an $X_1, Y$ subwalk of $W_1$ and $W_2'$ be an $X_2, Y'$ subwalk of $W_2$ such that $L(W_1') = L(W_2')$. Then $F_1 = F_2$, where $F_1$ is the permutation associated with $W_1'$ and $F_2$ is the permutation associated with $W_2'$.* $\square$

**Theorem 6.2** *The APPRP for $PS_n$ is solved online in $3n2^n(1 - o(1))$ communication steps; this is within a multiplicative factor of 3 from the optimum.*
**Proof.** For any $k$, $0 \leq k \leq 2^n - 1$, let $I_k$ denote the $n$ digit binary number whose value is $k$. Consider the packet set

$$M_i = \{\text{packet with source } I_j \otimes I_i \text{ and destination } I_j | 0 \leq j \leq 2^n - 1\},$$

$1 \leq i \leq 2^n - 1$. (Recall that $\otimes$ denotes the *Exclusive Or*.) Then it is easy to verify that

$$\{M_1, M_2, ..., M_{2^n-1}\}$$

is a partition of the set of all packets. The details of routing here are identical to Theorem 4.2, except that the group multiplication in this case is replaced by *Exclusive Or*, and taking the inverse is replaced by taking the complement of a binary number. More precisely, we route the packets to their destinations in $2^n - 1$ stages. At any stage $i$ ($1 \leq i \leq 2^n - 1$) all packets in $M_i$ are sent to their destination using the set of greedy walks

$$W_i = \{W_{I_j \otimes I_i \, I_i} | 0 \leq j \leq 2^n - 1\}.$$

It can be shown that no two packets will try to enter the same edge at the same time from the same endpoint at any stage. The claim regarding the number communication steps and quality of routing can be verified using standard enumerative arguments. $\square$

Our results for APPRP in $PS_n$ can be extended to $D_n$.

**Theorem 6.3** *The APPRP can be solved online for $D_n$ in $n2^n(1 - o(1))$ communication steps; this is within a multiplicative factor of 2 from the optimum.* $\square$

# References

[CL86] Chartrand, G., and Lesniak, L., *Graphs and Digraphs*, Wadsworth and Books/Cole Mathematics Series, 1986.

[EIS76] Even, S., Itai, A., and Shamir, A., On the complexity of timetable and multicommodity flow problems, *SIAM J. Computing* 5(1976), 691–703.

[FF62] Ford, L. R., and Fulkerson, D. R., *Flows in Networks*, Princeton University Press, Princeton, 1962.

[Hu63] Hu, T. C., Multi-commodity network flows, *J. ORSA* 11(1963), 344–360.

[JH89] Johnson, L., and Ho, C. T., Optimal broadcasting and personalized communication in hypercubes, *IEEE Trans. Comput.* 38(1989), 1249–1268.

[Ka84] Karmarkar, N., A new polynomial time algorithm for linear programming, *Combinatorica* 4(1984), 373–395.

[Ma85] Matula, D. W., Concurrent flow and concurrent connectivity in graphs, in: *Graph Theory and its Applications to Algorithms and Computer Science*, eds. Alavi, Y., et al., Wiley, New York, 1985, 543–559.

[PS82] Papadimitriou, H., and Stieglitz, K, *Combinatorial Optimization: Algorithms and Complexity*, Prentice-Hall, Englewood-Cliffs, N. J., 1982.

[Ra86] Raghavan, P., Probabilistic construction of deterministic algorithms: approximating packing integer programs, *Proc. 27th IEEE Symp. on the Foundations of Computer Sci.*, 1986, 10–18.

[Sa91] Saad, R., Complexity of forwarding index problem, Universite de Paris-Sud, Lab. de Recherche en Informatique, Rapport No. 648, 1991.

[SS89] Saad, Y., and Schultz, M. H., Data communication in hypercubes, *J. Parallel and Distributed Computing* 6(1989), 115–135.

[SM90] Shahrokhi, F., and Matula, D. W., The maximum concurrent flow problem, *J. Assoc. for Computing Machinery* 37(1990), 318–334.

[SS92] Shahrokhi, F., and Székely, L. A., Effective lower bounds for crossing number, bisection width and balanced vertex separators in terms of symmetry, in: *Integer Programming and*

*Combinatorial Optimization, Proceedings of a Conference held at Carnegie Mellon University, May 25-27, 1992, by the Mathematical Programming Society*, eds. E. Balas, G. Cournejols, R. Kannan, 102–113, CMU Press, 1992.

[SS93] Shahrokhi, F., and Székely, L. A., Formulae for the optimal congestion of the uniform concurrent multicommodity flow, *submitted*.

[Ul84] Ullman, J. D., *Computational Aspects of VLSI*, Computer Science Press, Rockville, Maryland, 1984.

[Wh84] White, A. T., *Graphs, Groups and Surfaces, North-Holland, Amsterdam, 1984.*

# On Multi-Label Linear Interval Routing Schemes[1]

## (Extended Abstract)

*Evangelos Kranakis*[2]   *Danny Krizanc*[2]   *S. S. Ravi*[3]

### Abstract

We consider linear interval routing schemes studied in [3, 5] from a graph theoretic perspective. We examine how the number of linear intervals needed to obtain shortest path routings in networks is affected by the product, join and composition operations on graphs. This approach allows us to generalize some of the results in [3, 5] concerning the minimum number of intervals needed to achieve shortest path routings in certain special classes of networks. We also establish the precise value of the minimum number of intervals needed to achieve shortest path routings in the network considered in [11].

# 1   Introduction

The problem of routing messages in communication networks so that the messages travel along shortest paths has been studied by a number of researchers (see, for example, the list of references at the end of this paper). When the number of nodes in the network is large, it becomes impractical to store large amounts of routing information at a node so as to achieve a shortest path route for each message. Hence methods that use compact routing tables are of considerable practical importance.

One such method called **interval routing** was proposed in [12]. It forms the basis of routing strategy used by the Inmos T9000 transputer [10]. The basic idea of this method is the following. Each node of the network is assigned a distinct integer from the set $\{1, 2, \ldots, n\}$, where $n$ is the number of nodes in the network. This assignment is considered as producing a circular ordering of the nodes. Each link is then labeled with a subinterval of the circular interval $[1, n]$. Because of the circular ordering of the nodes, the subinterval assigned to a link may involve a wraparound. The assignment of intervals to links must satisfy the following two conditions:

1. For any node $v$ of the network, the subintervals associated with the links emanating from $v$ are pairwise disjoint.

2. For any node $v$ of the network, the union of the subintervals associated with the links emanating from $v$ covers the interval $[1, n]$.

---

[1] Research supported in part by NSERC grants OPG0137991 and OPG0122227.

[2] School of Computer Science, Carleton University, Ottawa, Ontario, CANADA K1S 5B6. Email: {kranakis, krizanc}@scs.carleton.ca

[3] Department of Computer Science, University at Albany – SUNY, Albany, NY 12222, USA. Email: ravi@cs.albany.edu.

Given an interval labeling satisfying the above two conditions, routing is carried out as follows. Suppose a message with destination $d$ arrives at a node $v$. If $d = v$, then the message has reached its destination. Otherwise, the above conditions guarantee that there is exactly one link emanating from $v$ whose associated interval contains the destination node $d$. Therefore, the node $v$ forwards the message along that link.

An interval labeling $I$ for a network $G$ is **valid** if the above routing method does not cause any message to cycle (i.e., every message is guaranteed to reach its destination). Reference [12] presents a method of obtaining a valid interval labeling for any network. For acyclic networks, they show that the resulting labeling is **optimal**; that is, for any source-destination pair, the labeling routes messages along a shortest path. For other networks, they show that the scheme routes a message along a path whose length is at most twice the diameter of the network.

Subsequent to [12], a number of other researchers have addressed the problem of obtaining optimal and near-optimal labeling schemes for many classes of networks [6, 7, 8, 13, 14]. Reference [11] presents an example of a network which does not have an optimal interval labeling scheme. More precisely, that network has the property that for every assignment of intervals to the links, there is a pair of nodes for which the scheme produces a path whose length is at least 3/2 times the diameter of the network. References [1, 2] present routing schemes which provide a trade-off between the amount of routing information stored at a node and the quality of the resulting routing (i.e., the maximum ratio of the length of a route produced by the scheme to the length of a shortest path between the endpoints of the route).

An interesting special form of interval routing, called **linear interval routing**, was proposed and studied in [3, 5]. In this scheme, the nodes of the network are arranged in a *linear* order and the interval labeling each link is also restricted to be linear; that is, wraparound is not permitted. In [3, 5] it was shown that certain classes of networks such as hypercubes and $k$-dimensional grids for any $k \geq 2$ can be labeled optimally using appropriate linear interval routing schemes. These references provide a complete characterization of the edge weighted networks which admit optimal linear interval routing schemes. They also studied the linear interval variant of the **multi-label** interval routing schemes introduced in [14]. In multi-label linear interval routing schemes, a link may be labeled with a collection of linear intervals satisfying the two conditions listed earlier. Reference [5] presents examples of networks with the following property: for every linear ordering of the nodes, there is at least one link requiring two intervals to achieve a shortest path routing. References [3, 4] study a routing scheme called *prefix routing*, which is a more general form of the linear interval routing scheme.

In this paper, we study multi-label linear interval routing schemes from a graph theoretic perspective. We examine the effect of graph theoretic operations, namely product, join and composition, on the number of interval labels needed for an optimal routing. In addition to being of independent interest, our results also generalize some of the results in [3, 5]. For example, optimal linear interval routing schemes with only one interval per link for hypercubes and $k$-dimensional grids for any $k \geq 2$ follow as corollaries of our results for graph product. We also show that complete $r$-partite graphs for any $r \geq 2$ admit optimal linear interval routing schemes with only one interval per link. Further, we establish the exact value of the minimum number of linear intervals needed to achieve an optimal routing in the network considered in [11].

The remainder of this paper is organized as follows. Section 2 presents some definitions and terminology. Section 3 considers graph theoretic operations and examines how the minimum number of linear intervals needed for optimal routing is affected by these operations. Section 4 establishes the minimum number of intervals needed for the network considered in [11]. Section 5 points out some problems for further research. In this extended abstract, proofs of many theorems are omitted. Detailed proofs will appear in a complete version of this paper.

# 2  Definitions and Terminology

Throughout this paper, a communication network is represented by a connected undirected graph. The nodes of the graph represent the sites of the network and the edges of the graph represent bidirectional communication links. We assume that the edges are *unweighted*; that is, each edge is of unit cost. Thus the length of a path between a pair of nodes is simply the number of edges in the path. All graphs considered in this paper are simple (i.e., they do not have multi-edges), connected and do not have self-loops. We assume standard graph theoretic terminology as given in [9].

Since an edge $\{u, v\}$ of the graph represents a bidirectional link, an interval routing scheme will assign one or more intervals to the edge at $u$ and one or more intervals at $v$. The reader may find it convenient to think of an undirected edge $\{u, v\}$ as a pair of oppositely directed edges, namely $(u, v)$ and $(v, u)$. Then the interval(s) assigned to the edge $\{u, v\}$ at $u$ can be thought of as being assigned to the directed edge $(u, v)$ and the interval(s) assigned at $v$ can be thought of as being assigned to the directed edge $(v, u)$. Given a graph $G$ and an interval labeling $I$ for $G$, the number of intervals used in $I$ is defined to be the maximum number of intervals assigned to an edge at a node. The **LIRS number** of $G$, denoted by $\mathrm{LIRS}(G)$, is the minimum integer $\ell$ such that there is an optimal linear interval labeling of $G$ using at most $\ell$ intervals per edge. For example, it is known [5] that for all $k \geq 5$, $\mathrm{LIRS}(C_k) = 2$, where $C_k$ is the simple cycle on $k$ nodes. Also, it follows from the results in [5, 12] that for every tree $T$, $\mathrm{LIRS}(T)$ is at most 2 and that there are trees which require two intervals. As will be seen in Section 4, for any integer $\ell$, there is a graph $G$ such that $\mathrm{LIRS}(G) \geq \ell$.

We now mention some notation for special classes of graphs which will be used in the subsequent sections of this paper. For any positive integer $n$, $K_n$ denotes the *complete graph* on $n$ nodes, $P_n$ denotes the *path graph* on $n$ nodes (i.e., $P_n$ is a simple path on $n$ nodes), and $Q_n$ denotes the $n$-dimensional hypercube. For positive integers $m$ and $n$, $K_{m,n}$ denotes the *complete bipartite graph* with $m$

and $n$ nodes respectively on the two sides of the bipartition. A generalization of the class of complete bipartite graphs is the class of *complete r-partite graphs*. For any integer $r \geq 2$, the complete $r$-partite graph with $n_i \geq 1$ nodes in group $i$ of the $r$-partition $(1 \leq i \leq r)$ is denoted by $K_{n_1, n_2, \ldots, n_r}$. Finally, for any positive integer $n$, the *n-dimensional grid* with dimensions $d_1, d_2, \ldots, d_n$, is denoted by $R_{d_1, d_2, \ldots, d_n}$.

# 3 Bounds on LIRS Numbers for Special Classes of Graphs

In this section, we first present some bounds on the LIRS numbers for certain special classes of graphs. We then examine the effect of graph theoretic operations on the LIRS number.

## 3.1 A Simple Bound

THEOREM 1 *Let $G$ be a graph with $n$ nodes. Let $\delta$ denote the minimum node degree in $G$. Then* LIRS $(G) \leq \min\{\lceil n/2 \rceil, n - \delta\}$.

PROOF Label the nodes of $G$ in a one-to-one fashion using integers $1, 2, \ldots, n$. Assign intervals to the edges emanating from each node as follows. Consider any node $v$. Let $w_1, w_2, \ldots, w_p$ be the labels of the nodes adjacent to $v$, where $p = $ degree $(v)$. Let $x_1, x_2, \ldots, x_r$ be the labels of the nodes which are *not* adjacent to $v$. Clearly, $r = n - p - 1$. For each edge $\{v, w_i\}$ emanating from $v$ $(1 \leq i \leq p)$, assign the interval $[w_i, w_i]$. For each $x_j$, $1 \leq j \leq r$, fix a shortest path from $v$ to $x_j$. Let $w_q$ be the node (adjacent to $v$) which appears immediately after $v$ in that shortest path. Add the interval $[x_j, x_j]$ to the edge $\{v, w_q\}$.

It is easy to verify that for each node $v$, the intervals assigned to the edges emanating from $v$ are pairwise disjoint and that the union of these intervals covers the interval $[1, n]$. It is also easy to verify that the resulting labeling achieves shortest path routings between every pair of vertices.

We will now bound the maximum number of intervals assigned to any edge. First notice that no edge can ever be labeled with more than $\lceil n/2 \rceil$ intervals without two intervals combining into a single larger interval. Thus LIRS $(G) \leq \lceil n/2 \rceil$. For the other part of the bound stated above, consider the node $v$ and the edge $\{v, w_i\}$. The above method assigns at most $r + 1 = n - p = n - $ degree $(v)$ intervals to the edge $\{v, w_i\}$ (one interval for the node $w_i$ and at most $r$ intervals for the nodes $x_1, x_2, \ldots, x_r$). Thus the maximum number $N_v$ of intervals assigned to any edge emanating from $v$ is $n - $ degree $(v)$. Since $\delta$ is the minimum node degree, it follows that LIRS $(G) \leq n - \delta$. ∎

There are graphs for which the above upper bound is achievable. For example, consider the graph $K_n$. Here $\delta = n - 1$ and so by the above theorem, LIRS $(K_n) = 1$.

In general, the bound provided by the above theorem is weak. For example, for the path graph $P_n$, $\delta = 1$ and so the bound provided by the theorem is $\lceil n/2 \rceil$. It is easy to see that LIRS $(P_n) = 1$.

## 3.2   Complete $r$-Partite Graphs

It was shown in [14] that for every complete bipartite graph there is an optimal interval labeling scheme using only one circular interval per edge. We first show that for every complete bipartite graph, one *linear* interval per edge suffices to obtain optimal routings. We then generalize this result to show that for any $r \geq 2$, one linear interval per edge suffices even for complete $r$-partite graphs.

THEOREM 2   *For any positive integers $m$ and $n$, LIRS $(K_{m,n}) = 1$.*

PROOF (OUTLINE)   Let $V_1$ and $V_2$ denote the bipartition of node set of $K_{m,n}$, with $|V_1| = m$ and $|V_2| = n$. Label the nodes of $V_1$ in a one-to-one fashion using the integers 1, 2, ..., $m$ and the nodes of $V_2$ also in a one-to-one fashion using the integers $m + 1$, $m + 2$, ..., $m + n$. Since the numbering of the nodes is one-to-one, in the following discussion we will not distinguish between the name of a node and the integer labeling that node.

The edges emanating from the nodes in $V_1$ are labeled as follows. For $1 \leq i \leq m$, consider the node $i$ in $V_1$. For each $j$, $1 \leq j \leq n$, $K_{m,n}$ contains the edge $\{i, m+j\}$. Label the edge $\{i, m+1\}$ with the interval $[1, m+1]$. For $2 \leq j \leq n$, label the edge $\{i, m+j\}$ with the interval $[m+j, m+j]$.

The edges emanating from the nodes in $V_2$ are labeled as follows. For $1 \leq j \leq n$, Consider the node $m + j$ in $V_2$. For $1 \leq i \leq m - 1$, label the edge $\{m + j, i\}$ with the interval $[i, i]$. Finally, label the edge $\{m + j, m\}$ with the interval $[m, m + n]$.

Note that the above labeling procedure assigns exactly one linear interval per edge. It is not difficult to verify that the labeling achieves shortest path routings between every pair of vertices. ∎

By generalizing the labeling method used in the proof of the above theorem, we can obtain the following result.

THEOREM 3   *For any integer $r \geq 2$ and positive integers $n_1$, $n_2$, ..., $n_r$, LIRS $(K_{n_1, n_2, ..., n_r}) = 1$.* ∎

## 3.3   Product of Graphs

The following definition is from [9]. Given two graphs $G_1(V_1, E_1)$ and $G_2(V_2, E_2)$, the **product** $G_1 \times G_2$, has the node set $V'$ and edge set $E'$ where

$$
\begin{aligned}
V' &= V_1 \times V_2 \quad \text{and} \\
E' &= \{\{u, v\} : u = (u_1, u_2), v = (v_1, v_2), \text{ and} \\
&\quad [(u_1 = v_1 \text{ and } \{u_2, v_2\} \in E_2) \text{ or } (u_2 = v_2 \text{ and } \{u_1, v_1\} \in E_1)]\}.
\end{aligned}
$$

In understanding the proofs of the theorems in this section, the reader will find it helpful to think of $G_1 \times G_2$ as being obtained in the following manner. Let $V_1 = \{x_1, x_2, \ldots, x_{n_1}\}$ and let $V_2 = \{y_1, y_2, \ldots, y_{n_2}\}$, where $n_1 = |V_1|$ and $n_2 = |V_2|$. Consider the node set $V_1 \times V_2$. For each $i$, $1 \leq i \leq n_1$, construct an isomorphic copy of $G_2$ on the set of nodes $\{(x_i, y_1), (x_i, y_2), \ldots, (x_i, y_{n_2})\}$ (using the obvious bijection between this set of nodes and $V_2$). Further, for each $j$, $1 \le j \le n_2$, construct an isomorphic copy of $G_1$ on the set of nodes $\{(x_1, y_j), (x_2, y_j), \ldots, (x_{n_1}, y_j)\}$. The resulting graph is $G_1 \times G_2$.

We start by considering the product of a graph $G$ with the path graph $P_r$. For expository purposes, we first present the result for the product of a graph $G$ with $P_2$.

THEOREM 4 *For any graph $G$,* LIRS $(P_2 \times G)$ = LIRS $(G)$.

PROOF (OUTLINE) We first show that LIRS $(P_2 \times G) \le$ LIRS $(G)$. That is, given a linear interval scheme for $G$, we will show how to construct a linear interval scheme for $P_2 \times G$ using at most LIRS $(G)$ intervals per edge.

Let $n$ be the number of nodes in $G$. Visualize $P_2 \times G$ as being formed by taking two copies of $G$ and adding an edge between each of the $n$ pairs of corresponding nodes in the two copies of $G$. Number the nodes in the first copy of $G$ with integers 1 through $n$ using the given linear ordering for the nodes of $G$. Number the nodes in the second copy of $G$ with integers $n+1$ through $2n$, again using the given linear ordering for the nodes of $G$. After this numbering step, note that corresponding pairs of nodes have numbers $p$ and $n + p$, $1 \le p \le n$.

For each edge $\{i, j\}$ in $P_2 \times G$, carry out the assignment of intervals as follows.

1. If both $i$ and $j$ are from the first copy of $G$ (i.e., $i \le n$ and $j \le n$), then assign all the intervals assigned to the edge $\{i, j\}$ in $G$.

2. If both $i$ and $j$ are both from the second copy of $G$, (i.e., $n + 1 \le i \le 2n$ and $n + 1 \le j \le 2n$), then consider the intervals assigned to the edge $\{i - n, j - n\}$ in $G$. For each such interval $[x, y]$, assign the interval $[x + n, y + n]$.

3. If $i$ is from the first copy of $G$ and $j$ is from the second copy of $G$, then assign the interval $[n + 1, 2n]$.

4. If $i$ is from the second copy of $G$ and $j$ is from the first copy of $G$, assign the interval $[1, n]$.

Clearly, the above scheme uses at most LIRS $(G)$ intervals per edge of $P_2 \times G$. It is easy to verify that the scheme provides shortest path routes for each pair of nodes.

Next, we now show that LIRS $(P_2 \times G) \ge$ LIRS $(G)$. To do this, we show that given a linear interval scheme for $P_2 \times G$ with at most $\ell$ intervals per edge, we can obtain a linear interval scheme for $G$ which also uses at most $\ell$ intervals per edge. ∎

The following is an interesting corollary of this theorem.

COROLLARY 5 *For any integer $n \ge 2$,* LIRS $(Q_n) = 1$.

PROOF It is well known that $Q_2 = P_2$ and $Q_i = Q_{i-1} \times P_2$ for $i \geq 3$. Since LIRS $(P_2) = 1$, the corollary follows immediately. ∎

Theorem 4 can be generalized to any path graph $P_r$ as indicated by the following result.

THEOREM 6 *For any integer $r \geq 2$ and graph $G$, LIRS $(P_r \times G) = $ LIRS $(G)$.* ∎

The above theorem has the following corollary.

COROLLARY 7 *For every n-dimensional grid $R_{d_1,d_2,\ldots,d_n}$,*

$$\text{LIRS}\left(R_{d_1,d_2,\ldots,d_n}\right) = 1.$$

PROOF It is well known that the $n$-dimensional grid can be generated by the product of an appropriate sequence of path graphs. ∎

By extending the ideas used in the proofs of Theorems 4 and 6, we can prove a general result for the LIRS number of the product $G_1 \times G_2$ of two arbitrary graphs $G_1$ and $G_2$. We begin with a lemma which points out how the lengths of shortest paths in $G_1 \times G_2$ are related to the lengths of the shortest paths in $G_1$ and $G_2$. The proof of the lemma is straightforward.

LEMMA 8 *Let $G_1(V_1, E_1)$ and $G_2(V_2, E_2)$ be two graphs, where $V_1 = \{x_1, x_2, \ldots, x_{n_1}\}$ and $V_2 = \{y_1, y_2, \ldots, y_{n_2}\}$. Let $d_1(x_i, x_j)$ denote the length of a shortest path between $x_i$ and $x_j$ in $G_1$ and let $d_2(y_i, y_j)$ denote the length of a shortest path between $y_i$ and $y_j$ in $G_2$. Then for any pair of nodes $u = \langle x_{i_1}, y_{j_1} \rangle$ and $v = \langle x_{i_2}, y_{j_2} \rangle$ in $G_1 \times G_2$, the length $d(u, v)$ of a shortest path between $u$ and $v$ in $G_1 \times G_2$ is given by $d(u, v) = d_1(x_{i_1}, x_{i_2}) + d_2(y_{j_1}, y_{j_2})$.* ∎

THEOREM 9 *Let $G_1$ and $G_2$ be graphs such that LIRS $(G_i) = \ell_i$, $i = 1, 2$. Then*

$$\max\{\ell_1, \ell_2\} \leq \text{LIRS}\,(G_1 \times G_2) \leq 1 + \max\{\ell_1, \ell_2\}.$$

PROOF (OUTLINE) Given linear interval schemes for $G_1$ and $G_2$, we first show how to construct a linear interval scheme for $G_1 \times G_2$ using at most $1 + \max\{\ell_1, \ell_2\}$ intervals per edge.

Let $f_1$ and $f_2$ be functions specifying the linear orderings of the nodes of $G_1$ and $G_2$. Visualize $G_1 \times G_2$ as being obtained by having $n_1$ copies of $G_2$ and joining the corresponding vertices in these copies to create a copy of $G_1$. The copies are considered to be ordered according to $f_1$. In each copy of $G_2$, nodes are numbered using the ordering given by $f_2$ except that the nodes in copy $i$ are numbered using the integers $(i - 1)n_2 + 1$ through $in_2$, $1 \leq i \leq n_1$.

We can think of the above linear ordering of the nodes of $G_1 \times G_2$ as assigning an ordered pair $\langle p, q \rangle$ of integers to each node $v$ of $G_1 \times G_2$, where $p$ (which satisfies the condition $1 \leq p \leq n_1$) is the number of the copy of $G_2$ that $v$ belongs to and $q$ (which satisfies the condition $1 \leq q \leq n_2$) is the position of $v$ among the nodes in that copy. The following two-part observation provides an easy way to translate an ordered pair into a position in the linear order of $G_1 \times G_2$ and vice versa.

**Observation:**

(a) Suppose a node $v$ in $G_1 \times G_2$ is assigned the ordered pair $\langle p, q \rangle$, where $1 \leq p \leq n_1$ and $1 \leq q \leq n_2$. The position of $v$ in the linear order of $G_1 \times G_2$ is $(p-1)n_2 + q$.

(b) Suppose $t$ is the position of a node $v$ in the linear ordering for $G_1 \times G_2$. The ordered pair for $v$ is $\langle p, q \rangle$, where $p = \lceil t/n_2 \rceil$ and $q = t - (p-1)n_2$.

Let us now consider how intervals can be assigned to the edges of $G_1 \times G_2$. Consider the edge $\{a, b\}$ where $a$ and $b$ are the numbers of the two nodes in the linear ordering for $G_1 \times G_2$. Let $\langle p_a, q_a \rangle$ and $\langle p_b, q_b \rangle$ denote the ordered pairs corresponding to $a$ and $b$ respectively. (Given $a$ and $b$, the above observation can be used to find the values of $p_a$, $p_b$, $q_a$ and $q_b$.) There are two cases.

**Case 1: $p_a = p_b$.**

In this case, nodes $a$ and $b$ are in the same copy, namely copy $p_a$, of $G_2$. We find the collection of intervals assigned to the edge $\{q_a, q_b\}$ in the linear interval scheme for $G_2$. For each interval $[x, y]$ in the collection, we add the interval $[(p_a - 1)n_2 + x, (p_a - 1)n_2 + y]$ to the edge $\{a, b\}$. Thus the number of intervals assigned to the edge $\{a, b\}$ is equal to the number of intervals assigned to the edge $\{q_a, q_b\}$ in the given scheme for $G_2$.

**Case 2: $p_a \neq p_b$.**

Here, nodes $a$ and $b$ are in different copies of $G_2$. (By the definition of product, $q_a = q_b$). For this case, we assign intervals to the edge $\{a, b\}$ by considering the intervals assigned to the edge $\{p_a, p_b\}$ in the given linear interval scheme for $G_1$.

Suppose $[x, y]$ is an interval assigned to the edge $\{p_a, p_b\}$ in the given scheme for $G_1$. If the integer $p_a$ does *not* appear in the interval $[x, y]$, we add the interval $[(x-1)n_2 + 1, yn_2]$ to the edge $\{a, b\}$ of $G_1 \times G_2$. If the integer $p_a$ appears in the interval $[x, y]$, there are three possibilities:

(a) If $x = p_a$, then we add the interval $[p_a n_2 + 1, yn_2]$ to the edge $\{a, b\}$ of $G_1 \times G_2$.

(b) If $y = p_a$, then we add the interval $[(x-1)n_2 + 1, (y-1)n_2]$ to the edge $\{a, b\}$ of $G_1 \times G_2$.

(c) If $x < p_a < y$, then we add two intervals, namely $[(x-1)n_2 + 1, (p_a - 1)n_2]$ and $[p_a n_2 + 1, yn_2]$ to the edge $\{a, b\}$ of $G_1 \times G_2$.

We note that in the given linear interval scheme for $G_1$, among all the intervals assigned to the edges emanating from the node labeled $p_a$, the integer $p_a$ appears in *exactly one* of the intervals. If $p_a$ appears in the middle of that interval, then step (c) given above applies and so the number of intervals assigned to the edge $\{a, b\}$ in $G_1 \times G_2$ is one more than the number of intervals assigned to the edge $\{p_a, p_b\}$ in $G_1$. Thus $\text{LIRS}(G_1 \times G_2) \leq 1 + \max\{\ell_1, \ell_2\}$.

The above scheme for $G_1 \times G_2$ routes a message from a node $a$ to $b$ in the following manner. From $a$, it first routes the message to the copy of $G_2$ which contains the node $b$. (This route uses the edges of $G_1$). Then, the message is routed to the node $b$ in that copy. (This route uses the edges of $G_2$). Using Lemma 8, it is easy to verify that the routing is optimal.

The proof that $\text{LIRS}(G) \geq \max\{\ell_1, \ell_2\}$ uses ideas similar to those used in the proof of the result for the product $P_2 \times G$. ∎

## 3.4 Composition of Graphs

Given two graphs $G_1(V_1, E_1)$ and $G_2(V_2, E_2)$, the **composition** [9] of $G_1$ with $G_2$, denoted by $G_1[G_2]$, has the node set $V'$ and edge set $E'$ where

$$\begin{aligned} V' &= V_1 \times V_2 \quad \text{and} \\ E' &= \{\{u, v\} : u = (u_1, u_2), v = (v_1, v_2), \text{ and} \\ &\quad \text{either } \{u_1, v_1\} \in E_1 \text{ or } (u_1 = v_1 \text{ and } \{u_2, v_2\} \in E_2)\}. \end{aligned}$$

A convenient way of visualizing the composition $G_1[G_2]$ is the following. Replace each node of $G_1$ by a copy of $G_2$ and then replace each edge of $G_1$ by a complete bipartite subgraph that joins every vertex in the copy corresponding to one endpoint to every vertex in the copy corresponding to the other endpoint. Using ideas similar to those used in proof of Theorem 9, we can prove the following result.

THEOREM 10 *For graphs $G_1$ and $G_2$, $\text{LIRS}(G_1[G_2]) \leq \text{LIRS}(G_1) + 4$.* ∎

## 3.5 Join of Graphs

The following definition is also from [9]. Given two graphs $G_1(V_1, E_1)$ and $G_2(V_2, E_2)$, the **join** of $G_1$ and $G_2$, denoted by $G_1 + G_2$, has the node set $V'$ and edge set $E'$ where

$$\begin{aligned} V' &= V_1 \cup V_2 \quad \text{and} \\ E' &= E_1 \cup E_2 \cup \{\{v, w\} : v \in V_1, \ w \in V_2\}. \end{aligned}$$

Informally, $G_1 + G_2$ is obtained by taking an isomorphic copy of $G_1$ along with an isomorphic copy of $G_2$, and adding an edge between every node of $G_1$ and every node of $G_2$. We note that, in general, the structure of shortest paths in $G_1 + G_2$ is very different from those of $G_1$ and $G_2$. This is because in $G_1 + G_2$, there is a path of length at most 2 between every pair of nodes. Therefore, the LIRS number of $G_1 + G_2$ may not be related to the LIRS numbers of $G_1$ and $G_2$. However, it is possible to bound the LIRS number of $G_1 + G_2$ using some parameters of $G_1$ and $G_2$ as shown by the following theorem. The proof uses ideas similar to those used in the proof of Theorem 1.

THEOREM 11 *Suppose $G_1$ and $G_2$ are graphs with $n_1$ and $n_2$ nodes, minimum degrees $\delta_1$ and $\delta_2$, and maximum degrees $\Delta_1$ and $\Delta_2$ respectively. Then*

$$\text{LIRS}(G_1 + G_2) \leq 1 + \max\{\lceil f_1/n_2 \rceil, \lceil f_2/n_1 \rceil\}$$

*where $f_i = \min\{n_i - \delta_i - 1, \Delta_i + 1\}$, $i = 1, 2$.* ∎

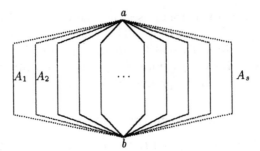

Figure 1: The globe graph $G_s^n$.

We remark that since $G_1 + G_2$ is a graph with $n_1 + n_2$ nodes, the above bound can be refined slightly using Theorem 1. In other words, an upper bound on LIRS $(G_1 + G_2)$ is the minimum of $\lceil (n_1 + n_2)/2 \rceil$ and the quantity specified in the statement of the above theorem.

# 4 Graphs Requiring Many Intervals

In this section we present a family of graphs for which any linear interval labeling scheme requires a growing number of intervals as a function of the size of the graph.

The globe graphs $G_s^n$ are constructed by joining the endpoints of $s > 0$ line segment graphs each consisting of $n > 0$ nodes. More formally, we have the following definition. Let $A_1, A_2, \ldots, A_s$ be $s$ line segment graphs, each consisting of $n$ nodes. Let the nodes of the segment $A_i$ be $a_{i1}, \ldots, a_{in}$, $i = 1, \ldots, s$ and let $a$ and $b$ be two isolated vertices. The globe graph $G_s^n$ has the vertex set

$$V = \{a_{ij} : i = 1, \ldots, s, j = 1, \ldots, n\} \cup \{a, b\},$$

and the edge set

$$E = \{\{a_{ij}, a_{i(j+1)}\} : i = 1, \ldots, s, j = 1, \ldots, n-1\}$$

$$\cup \{\{a, a_{i1}\}, \{b, a_{in}\} : i = 1, \ldots, s\}$$

(see Figure 1). Clearly, $G_s^n$ has $N := |V| = sn + 2$ nodes and $|E| = s(n + 1)$ edges. The globe graphs were first considered by Ružička [11] who proved that LIRS $(G_s^n) \geq 3$, for $n \geq 3, s \geq 14$. Here we give more precise bounds on the value of LIRS $(G_s^n)$. The main theorem of this section is the following.

**THEOREM 12** *If $s$ is odd and $n \geq s^2$, then* LIRS $(G_s^n) = \lfloor s/2 \rfloor + 1$.

**PROOF (OUTLINE)** First we prove LIRS $(G_s^n) \leq \lfloor s/2 \rfloor + 1$. Label $a$ by 1 and $b$ by $N$. The nodes of the arcs $A_1, \ldots, A_s$ are labeled as follows. Label $A_1$ from top to bottom, next $A_2$ from bottom to top, next $A_3$ from top to bottom, etc., using the labels 2 through $N - 1$ in order.

Next we prove, by contradiction, that $\text{LIRS}(G_s^n) \geq \lfloor s/2 \rfloor + 1$. Assume to the contrary $\text{LIRS}(G_s^n) \leq \lfloor s/2 \rfloor$, i.e., there is node labeling of the globe graph $G_s^n$ such that the interval labeling of each edge has at most $\lfloor s/2 \rfloor$ intervals.

Consider the edges $\{a, a_{i1}\}$, $i = 1, \ldots, s$. By assumption, each of these edges is labeled by at most $\lfloor s/2 \rfloor$ intervals. Since the shortest path from $a$ to all nodes in $A_i$ uses the edge $\{a, a_{i1}\}$ it follows that labels assigned to the nodes of each $A_i$ consist of the union of at most $\lfloor s/2 \rfloor$ disjoint intervals, for $i = 1, \ldots, s$. Let $I_i^j$, $j = 1, \ldots, k_i$, be the $k_i \leq \lfloor s/2 \rfloor$ intervals associated with interval $i$, $i = 1, \ldots, s$.

Define $c_i^j = \min_k \{a_{ik} :$ the label of $a_{ik}$ is in $I_i^j\}$ and $d_i^j = \max_k \{a_{ik} :$ the label of $a_{ik}$ is in $I_i^j\}$, for all $i$ and $j$. Since $k_i \leq \lfloor s/2 \rfloor$ for all $i$, by taking $n \geq s^2 > 2s\lfloor s/2 \rfloor$, we have by the pigeonhole principle there exists an $l$, $2 \leq l \leq n - 1$, such that $a_{il} \neq c_i^j$ and $a_{il} \neq d_i^j$ for all $i$ and $j$. Renumber the intervals $I_i^j$ so that the interval $I_i^1$ of arc $A_i$ contains the label of vertex $a_{il}$ for $i = 2, \ldots, s$. Now by considering the interval labels of the two edges adjacent to $a_{1(n+1-l)}$, it can be shown that at least $\lfloor s/2 \rfloor + 1$ intervals are required to label the edge $\{a_{1(n+1-l)}, a_{1(n-l)}\}$ or the edge $\{a_{1(n+1-l)}, a_{1(n+2-l)}\}$, contradicting our assumption. ∎

# 5 Problems for Future Research

There are several directions for further research on interval routing schemes. For example, given a graph $G$ and an integer $k$, the complexity of deciding whether $\text{LIRS}(G)$ is at most $k$ is open. A characterization for $k = 1$ for weighted graphs has been obtained in [3, 5], but to the best of our knowledge, the problem is open for graphs with unit cost links, even when we allow the nodes of $G$ to be circularly ordered and allow intervals to have wraparound. A related problem is that of obtaining bounds on the LIRS numbers of other special classes of graphs. References [6, 7, 8] address this question for planar graphs, graphs of small genus and graphs with constant separators.

The globe graphs discussed above show that for any $n$, there exists an $n$ node graph, $G$, with $\text{LIRS}(G) \geq \lfloor \sqrt[3]{n} \rfloor$. It would be interesting to know whether there are graphs with even higher LIRS number as a function of their size.

Another open question is the following. Reference [12] presents an interval routing scheme which guarantees that every message route is of length at most twice the diameter of the given graph, and reference [11] presents an example of a graph for which every interval routing scheme has at least one pair of nodes requiring a message route of length at least 1.5 times the diameter of the graph. It will be interesting to close the gap between these two bounds.

**Acknowledgement:** S. S. Ravi would like to express his sincere thanks to the faculty and staff of the School of Computer Science at Carleton University and to David and Nancy Fraser of Ottawa for their hospitality during his visits to Ottawa.

# References

[1] B. Awerbuch, A. Bar-Noy, N. Linial and D. Peleg, "Improved Routing Strategies with Succinct Tables", *J. Algorithms*, vol. 11, no. 3, Sept. 1990, pp. 307-341.

[2] B. Awerbuch and D. Peleg, "Routing with Polynomial Communication-Space Trade-off", *SIAM J. Disc. Math.*, vol. 5, no. 2, May 1992, pp. 151-162.

[3] E. M. Bakker, "Combinatorial Problems in Information Networks and Distributed Datastructuring", *Ph.D. Thesis*, Dept. of Computer Science, Utrecht University, The Netherlands, 1991.

[4] E. M. Bakker, J. van Leeuwen and R. B. Tan, "Prefix Routing Schemes in Dynamic Networks", Tech. Report RUU-CS-90-10, Dept. of Computer Science, Utrecht University, The Netherlands, Mar. 1990.

[5] E. M. Bakker, J. van Leeuwen and R. B. Tan, "Linear Interval Routing Schemes", Tech. Report RUU-CS-91-7, Dept. of Computer Science, Utrecht University, The Netherlands, Feb. 1991.

[6] G. N. Frederickson and R. Janardan, "Designing Networks with Compact Routing Tables", *Algorithmica*, vol. 3, 1988, pp. 171-190.

[7] G. N. Frederickson and R. Janardan, "Efficient Message Routing in Planar Networks", *SIAM J. Comput.*, vol. 18, no. 4, Aug. 1989, pp. 843-857.

[8] G. N. Frederickson and R. Janardan, "Space-Efficient Message Routing in c-Decomposable Networks", *SIAM J. Comput.*, vol. 19, no. 1, Feb. 1990, pp. 164-181.

[9] F. Harary, *Graph Theory*, Addison-Wesley Publishing Co., Reading, MA, 1969.

[10] *The T9000 Transputer Products Overview Manual*, Inmos, 1991.

[11] P. Ružička, "On the Efficiency of Interval Routing Algorithms", in *Proceedings of MFCS*, Carlsbad, Chechoslovakia, Aug. - Sep. 1988, Lecture Notes in CS vol. 324 (Edited by M. P. Chytil, L. Janiga and V. Koubek), pp. 492 - 500, 1988.

[12] N. Santoro and R. Khatib, "Labelling and Implicit Routing in Networks", *The Computer Journal*, vol. 28, no. 1, 1985, pp. 5-8.

[13] J. van Leeuwen and R. B. Tan, "Computer Networks with Compact Routing Tables", in *The Book of L*, Edited by G. Rozenberg and A. Salomaa, Springer-verlag, Berlin 1986, pp 259-273.

[14] J. van Leeuwen and R. B. Tan, "Interval Routing", *The Computer Journal*, vol. 30, no. 4, 1987, pp. 298-307.

# An 'All Pairs Shortest Paths' Distributed Algorithm Using $2n^2$ Messages *

S. Haldar

Department of Computer Science
Utrecht University, PO Box 80.089
3508 TB Utrecht, The Netherlands

**Abstract.** In a distributed program execution, processes communicate among themselves by exchanging messages. The execution speed of the program could be expedited by a faster message delivery system, transmitting messages to their destinations through their respective shortest paths. Some distributed algorithms have been proposed in recent years for determining all pairs shortest paths for an arbitrary computer network. The best known algorithm uses $O(n^2 \log n)$ messages, where $n$ is the network size. This paper presents a new distributed algorithm for the same problem using $2n^2$ messages in the worst case. This algorithm uses a strategy quite different from those of the other algorithms for the same problem.

## 1 Introduction

A computer network is an interconnected collection of autonomous computers, referred to as *nodes*. An application program, also called distributed program, accesses resources (hardware, data, functions) from many nodes. The network system assigns many processes at different nodes to execute a distributed program. We define a *distributed algorithm* for a collection of processes to be a collection of *local algorithms*, one for each process. Each process executes its local algorithm and cooperates with the other processes to achieve the objective of the distributed program. To communicate among themselves, the processes exchange information through *messages*. An underlying message passing system (MPS, in short) handles the exchanging of messages between processes.

In wide area networks the nodes are connected by point-to-point communication links (also referred to as *edges*) for exchanging messages between two nodes, called *neighbors*. Providing edges between all pairs of nodes is not cost effective.

---

*This research is partially supported by NWO through NFI Project ALADDIN under Contract Number NF 62-376, and partially by ESPRIT through Project ALCOM II under Basic Research Action Number 7141.

Consequently, communication between some pairs is done through intermediate node(s). We assume that both nodes and edges are reliable. Transmitting a message through an edge incurs some cost, that is, each edge has some *weight* assigned to it. Naturally, it is desirable that MPS routes messages through shortest paths. (A shortest path between a pair of nodes is any path between them with the minimum weight.) Hence, the MPS should know all pairs shortest paths in advance. It uses a *routing table* that stores the all pairs shortest paths information in some coded form.

In this paper we investigate how to build a routing table for an arbitrary network efficiently. There are two broad approaches to building a routing table, namely, centralized and distributed. In the centralized approach, inputs from all nodes are collected in a single node, a sequential algorithm is executed on the inputs, and finally, outputs are dispersed to all the nodes. There are two lucrative strategies in developing sequential algorithms, namely the greedy method [5] and the dynamic programming [2]. The single node all shortest paths algorithm of Dijkstra [4] uses a greedy method, and requires $O(n^2)$ time units, where $n$ is the number of nodes in the network. This algorithm can be executed separately for each node to obtain all pairs shortest paths in $O(n^3)$ time units. The all pairs shortest paths algorithm of Floyd and Warshall [6, 10] uses a dynamic programming strategy, taking $O(n^3)$ time units.

In this paper, we are interested in distributed approach in which each node builds its part of the routing table, and helps other nodes in building their parts. One such algorithm is presented by Toueg [9], based on the Floyd-Warshall sequential algorithm. It assumes edge weights to be positive, and each node knows identities of all the nodes, identities of its neighbors and weights of its incident edges. This algorithm exchanges $O(n)$ messages per edge, $O(mn)$ in total, where $m$ is the number of edges in the network. (There are various complexity measures for a distributed algorithm: message complexity, bit complexity, time complexity, space complexity. We restrict our attention to message complexity, namely the number of messages exchanged in the algorithm in the worst case.) Recently, Afek and Ricklin [1] have proposed a transformation scheme for distributed algorithms. It takes a distributed algorithm whose message complexity is $O(f \cdot m)$, where $f$ is an arbitrary function of $m$ and $n$, and produces a new *semi-distributed* algorithm for the same problem with $O(f \cdot n \log n + m \log n)$ message complexity. Applying this scheme (for example, to Segall's algorithm [7]) an all pairs shortest paths algorithm with upper bound of $O(n^2 \log n)$ message complexity can be obtained.

In this paper we present a new fully distributed algorithm that uses $2n^2$ messages in the worst case. It is based on the single node all shortest paths sequential algorithm of Dijkstra. We first modify his algorithm and make it more efficient by a constant factor, and then transform the modified algorithm into a distributed algorithm. In this distributed algorithm we allow communication only between neighbors, and this leads to a lower message complexity. This algorithm uses a quite different strategy than those used by others for the same problem. In this algorithm, a node might finish building its part of the routing table much earlier than the others do; if the former node sends messages to any nodes, the messages are guaranteed to be delivered to their destinations through their respective shortest paths. This delivery is assured even if the routing table is not completely built.

In Section 2 we define a model of the system under investigation. This section states some definitions and postulates some assumptions. Dijkstra's sequential algorithm is presented in Section 3. His algorithm is modified in Section 4. We transform the modified algorithm into a distributed algorithm in Section 5. Section 6 concludes the paper.

## 2  Model, Definitions and Assumptions

We represent a computer network, where computers are connected by point-to-point communication links, by an *undirected weighted graph* $G = \langle V, E, W \rangle$, where $V$ is a set of *nodes*, one for each computer; $E$ is a set of *edges*, one for each link; and $W$ is a *weight function* from $E$ to non-zero positive real numbers. There are $n$ nodes and $m$ edges in $G$. An edge is denoted by an unordered pair of nodes. If $u, v \in V$ are connected by an edge, we denote that edge as $uv$ (and $vu$). An edge $uv$ is called an *incident edge* of $u$ (and of $v$), and $u$ and $v$ are *neighbors*, and its *weight* is $W[u, v]$. The weight of $uv$ is equal to the weight of $vu$. A *path* between two nodes $u$ and $v$ is a sequence $p = \langle u = v_0, v_1, v_2, \cdots, v_k = v \rangle$ of nodes such that for each $0 \leq i < k$, $v_i v_{i+1} \in E$, and the weight of the path is $w(p) = \sum_{0 \leq i < k} W[v_i, v_{i+1}]$. The notation $u \overset{p}{\leadsto} v$ denotes that $v$ is reachable from $u$ through path $p$. In this paper we assume that $G$ is connected, that is, there exists a path between each pair of nodes. The *shortest path weight*, also called *distance*, between any pair of nodes $u$ and $v$, denoted $dist(u, v)$, is the minimum weight of all possible paths between them, that is, $dist(u, v) = \min\{w(p) : u \overset{p}{\leadsto} v\}$. A *shortest path* from a node $u$ to another node $v$ is defined as any path $p$ with $w(p) = dist(u, v)$. We also like to determine the neighbor of $u$ in $p$, denoted $via(u, v)$, through which messages for $v$ will be routed.

We assume, as in [9], that each node $u$ knows: (1) the graph size $n$, (2) identities of all nodes, $1, 2, \cdots, n$, (3) identities of its neighbors in *neighbor(u)* and (4) weights of its incident edges. It is assumed that the weights of the edges do not change with time. Messages are delivered to their respective destinations within a finite delay, but they might be delivered out-of-order, that is, the edges are non FIFO. The distributed algorithm presented in this paper allows communication only between neighbors. We assume an asynchronous message passing system, that is, a sender of a message does not (in fact, must not) wait for the receiver to be ready to receive the message. Messages are stored in the edges until they are delivered.

The proposed algorithm uses a table called *routing table* (*RT*, in short) to store the final output of all pairs shortest paths and intermediate results of the computation in some coded form.

The following two properties hold for any graph $G = \langle V, E, W \rangle$.

**Property 1 ([3], Lemma 25.1)** *Subpaths of shortest paths are shortest paths.* □

**Property 2 ([3], Corollary 25.2)** *Let $s$ and $v$ be two nodes in $V$. Suppose a shortest path $p$ from $s$ to $v$ can be decomposed into $s \overset{p'}{\leadsto} u \xrightarrow{uv} v$ for some node $u \in neighbor(v)$ and path $p'$. Then, the weight of a shortest path from $s$ to $v$ is $dist(s, v) = dist(s, u) + W[u, v]$.* □

# 3 Dijkstra's Single Node All Shortest Paths Algorithm

The Dijkstra algorithm for a graph $G = \langle V, E, W \rangle$ is given in Figure 1. It uses greedy method. For a pivotal node $s \in V$, it calculates the distance $dist(s, v)$ for all $v \in V$. For each node $v$, it maintains a distance estimate variable $d[v]$, initialized to $\infty$, which is an upper bound on the weight of a shortest path from $s$ to $v$. The invariant $d[v] \geq dist(s, v)$ is maintained throughout the execution of the algorithm. The technique used by the Dijkstra algorithm is relaxation, a method that repeatedly decreases (relaxes) the upper bound until it becomes equal to the actual shortest weight. It maintains a set $S$ of nodes whose distances from the pivotal node $s$ have already been determined. At the beginning, $d[s] = 0 = dist(s, s)$. It maintains a priority queue $Q$ to contain all nodes in $V - S$. It repeatedly selects the node $u \in Q$ with the minimum distance estimate $d[u]$, inserts $u$ in $S$, and relaxes the distance estimates for all the neighbors of $u$. That is, for each $uv \in E$, the distance estimate $d[v]$ is relaxed with respect to $d[u] + W[u, v]$. In his algorithm each edge is considered twice to relax some distance estimates, that is, a total of $2m$ relaxations. The relaxations cause the distance estimates to decrease monotonically until the estimates become equal to the actual distances. In the Dijkstra algorithm, it is guaranteed that when a node $u$ is visited[1], at that time $d[u] = dist(s, u)$.

For correctness proof, please refer to [3](Theorem 25.10). The Dijkstra algorithm takes $O(n^2)$ time unit if $Q$ is a linear array, $O(m \log n)$ time units if $Q$ is a binary heap, and $O(n \log n + m)$ time units if $Q$ is a Fibonacci heap.

# 4 Modified Algorithm

In the Dijkstra algorithm, when a node $u$ is removed from $Q$ and put in $S$, distance estimates of all the neighbors of $u$ are relaxed. Let $uv$ be an edge. If the node $v$ is already in $S$, there is no need to relax $d[v]$ with respect to $d[u] +$

---

[1] By visiting $u$, we mean that the algorithm is inserting $u$ in $S$ and considering all its incident edges for some distant estimate relaxations.

$W[u, v]$, for the optimal distance $dist(s, v)$ has already been found. To avoid this relaxation, we attach a boolean tag $status[v]$ with each node $v$, initialized to *tentative*, and made *permanent* on reaching the shortest weight for $v$. At the time of visiting $u$, the edge $uv$ will be considered only if $v$ is *tentative*. The modified algorithm is given in Figure 2. In the modified algorithm $Q$ represents a set of frontier nodes during the execution of the algorithm. All nodes in $Q$ are *tentative*. At the beginning all neighbors of the pivotal node $s$ are put in $Q$. For node $v$ in $Q$, $d[v]$ is the tentative distance from $s$ to $v$. The algorithm repeatedly selects the node $u \in Q$ with the minimum distance estimate $d[u]$, makes $u$ permanent (i.e., visits $u$), and relaxes distance estimates of some neighbors of $u$. In the modified algorithm there are $m$ relaxations and $2m$ boolean tests, whereas in the Dijkstra algorithm there are $2m$ relaxations. So, the modified algorithm speeds up the computation by a constant factor if the cost of boolean test is less than half the cost of addition and real comparison. Each node $v$ is inserted in $Q$ exactly once, that is, a total of $n$ insertions in $Q$. In each iteration of the while loop, exactly one node $u$ is removed from $Q$, and $u$ is made *permanent*. (At this moment $d[u]$ is the shortest weight $dist(s, u)$.) So the loop terminates after $n$ iterations. The variable $via[v]$ indicates the neighbor of $s$ through which messages for $v$ should be routed. The correctness proof of this algorithm is very similar to that of Dijkstra's, and omitted from the paper.

# 5  Distributed 'All Pairs Shortest Paths' Algorithm

We first consider a sequential all pairs shortest paths algorithm. The modified algorithm given in Figure 2 could be executed for all nodes of $G$ to obtain the all pairs shortest paths. But, we can do better using the following observation. Let a shortest path $p$, between two nodes $u$ and $v$, run through a node $v' \in neighbor(u)$. Then $dist(u, v) = W[u, v'] + dist(v', v)$ (by looking Property 2 from different angle), and we state this as Property 2'.

**Property 2'** *Suppose a shortest path $p$ from a node $u$ to another node $v$ can be decomposed into $u \xrightarrow{uv'} v' \overset{p'}{\leadsto} v$ for some node $v' \in neighbor(u)$ in $p$ and path $p'$. Then, the weight of a shortest path from $u$ to $v$ is $dist(u, v) = W[u, v'] + dist(v', v)$.* □

Now, instead of calculating both $dist(u, v)$ and $dist(v', v)$ separately and independently, we calculate $dist(v', v)$ first, and then $dist(u, v)$ by one addition operation. But, the question is, how does $u$ know that $p$ runs through $v'$? We answer this question in the following subsection. We present a sequential restricted version of the modified algorithm, for all pairs shortest paths in Section 5.1 first, and then, transform the Restricted algorithm into a distributed algorithm in Section 5.2. (We call the sequential algorithm 'Restricted' because in the algorithm each node accesses the table entries of only its neighbors.)

## 5.1 Restricted Algorithm

Assume $RT$ is the routing table to be built for a given graph $G = \langle V, E, W \rangle$. Each component of the table, $RT[u, v]$, for $u, v \in V$, has four fields: (1) $weight$, (2) $status$, (3) $via$ and (4) $via\text{-}status$. Here, $weight$ is distance estimate between $u$ and $v$; $status$ indicates whether $u$ has visited (directly or indirectly) $v$; $via$ is the neighbor of $u$ in the path used to determine the $weight$; and $via\text{-}status$ indicates $u$'s knowledge of $v$'s $status$ in $via$. The algorithm is given in Figure 3. (In the text, a subcomponent of the routing table, $RT[u, v].field$ is denoted as $field[u, v]$.) The $RT$ table is initialized as follows for all nodes $u$: (1) $weight[u, u] := 0$, $status[u, u] := permanent$, $via[u, u] := u$ and $via\text{-}status[u, u] := permanent$; (2) for all neighbors $v$ of $u$, $weight[u, v] := W[u, v]$, $status[u, v] := tentative$, $via[u, v] := v$ and $via\text{-}status[u, v] := tentative$; and (3) for non adjacent nodes, $weight$s are $\infty$, and $status$ and $via\text{-}status$ values are $tentative$. After the initialization, a repeat-until loop is executed. In each iteration of the loop the following steps are performed for all nodes $u$ (one can imagine that $n$ processors, one for each $u$, are executing the algorithm somewhat synchronously).

Step 1: The node $u$ selects a node $v_u$ with minimum distance estimate $weight[u, v_u]$ from $RT[u, *]$ such that $status[u, v_u] = tentative$. If no tentative $v_u$ is available, $u$ has finished building up its part $RT[u, *]$ of the routing table, and hence, it does not do any effective thing. Otherwise, $u$ executes Step 2.

Step 2: Let the node $v_u$ be reached through a neighbor node $via_u = via[u, v_u]$. If $via\text{-}status[u, v_u]$ is $permanent$ — that is, the node $via_u$ has already visited the node $v_u$ and calculated $dist(via_u, v_u)$, and this visit is also captured in $RT[u, *]$, then $weight[u, v_u] = dist(u, v_u)$ (by Property $2'$) — then $u$ makes $status[u, v_u]$ $permanent$, and loops back to Step 1. Otherwise, it executes Step 3.

Step 3: It checks the table entry $RT[via_u, v_u]$. If $status[via_u, v_u]$ is not $permanent$, — the node $via_u$ has not yet visited the node $v_u$ — then $u$ does nothing and loops back to Step 1. Otherwise, it copies the table entries $RT[via_u, *].\langle weight, status \rangle$ in a local variable $temp_u$ first, and then executes Step 4.

Step 4: The first thing $u$ does is that it makes $status[u, v_u]$ and $via\text{-}status[u, v_u]$ $permanent$. (The node $u$ indirectly visits $v_u$ through the visit of $v_u$ by $via_u$.) Finally, $u$ updates its other table entries whose $status$ are not $permanent$. Consider a node $v'_u$ such that $status[u, v'_u] = tentative$. If $u$ finds $weight[u, v'_u] > W[u, via_u] + temp_u[v'_u].weight$ (note that $temp_u[v'_u].weight$ is the distance estimate between the nodes $via_u$ and $v'_u$), then the $weight[u, v'_u]$ is set to $W[u, via_u] + temp_u[v'_u].weight$, $via[u, v'_u]$ to $via_u$, and $via\text{-}status[u, v'_u]$ to $temp_u[v'_u].status$. Loops back to Step 1.

It is to be noted that we can use the same algorithm without $via\text{-}status$ fields and Step 2. However, they are included to speed up the computation. Note that, in Step 2, when $via\text{-}status[u, v_u]$ is *permanent*, it indicates that the node $via_u$ has visited the node $v_u$, and the effect of this visit is already captured in $RT[u, *]$. So, if $u$ finds $via\text{-}status[u, v_u]$ *permanent*, it need not access the table entries in $RT[via_u, *]$ to update its table entries $RT[u, *]$, and hence executing Step 2 would speed up the computation. The repeat-until loop is executed until no change in $RT$ is observed in an iteration, that is, all nodes have built up their respective parts of $RT$. On termination the *via* fields indicate how messages should be routed. The correctness proof is given in the next subsection.

### 5.1.1 Correctness Proof

We need to show that the Restricted algorithm satisfies two basic properties, namely, (1) *liveness*, that is, the algorithm terminates eventually, and (2) *safety*, that is, on termination of the algorithm, all pairs shortest paths are determined. To show the liveness property, we first show that the algorithm does not lead to any deadlocks, and then, the termination of the algorithm. The safety property is proved at last in Theorem 6.

**Lemma 3** *There is no deadlock.*

*Proof*: Contrarily, assume that there is a deadlock when the algorithm is executed for a graph $G = \langle V, E, W \rangle$. Note that for all $uv \in E$, $W[u, v] > 0$.

Let $u_0, u_1, \cdots, u_k = u_0$ be a sequence of nodes involved in a deadlock. As the algorithm allows each node $u$ to access table components for its neighbors only, then for $0 \leq i < k$, $u_i u_{i \oplus 1} \in E$. (Here $\oplus$ and $\ominus$ are modulo $k$ addition and subtraction operators, respectively. Let $u_i$ be trying to make $status[u_i, v_i]$ *permanent* for some $v_i \in V$, and be waiting for $status[u_{i \oplus 1}, v_i]$ to become *permanent*. That is, the node $u_i$ finds shortest path for $v_i$ through $u_i u_{i \oplus 1}$. (See the following picture.)

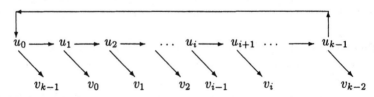

As $u_i$ is trying to make $status[u_i, v_i]$ *permanent* (and $u_i$ has not made $v_{i \ominus 1}$ *permanent* yet in this deadlock situation), we have $weight[u_i, v_{i \ominus 1}] \geq weight[u_i, v_i]$. Note that $u_i$ got the information of $v_i$ through $u_{i \oplus 1}$. Hence, $weight[u_i, v_i] = W[u_i, u_{i \oplus 1}] + weight[u_{i \oplus 1}, v_i]$. Thus we get, $weight[u_0, v_{k-1}] \geq weight[u_0, v_0] = W[u_0, u_1] + weight[u_1, v_0]$. That is, $weight[u_0, v_{k-1}] \geq W[u_0, u_1] + weight[u_1, v_1]$, and hence, $weight[u_0, v_{k-1}] \geq W[u_0, u_1] + W[u_1, u_2] + weight[u_2, v_1]$; and so on. Finally, we have $weight[u_0, v_{k-1}] \geq W[u_0, u_1] + W[u_1, u_2] + \cdots + W[u_{k-1}, u_0] + weight[u_0, v_{k-1}]$. That is, $weight[u_0, v_{k-1}] \geq some\text{-}positive\text{-}number + weight[u_0, v_{k-1}]$. But, a number cannot be greater than itself, and hence, the sequence $u_0, u_1, \cdots, u_k = u_0$ cannot be in a deadlock. This proves the lemma. $\square$

**Corollary 4** *In each iteration of the repeat-until loop, at least one entry in the routing table RT is made permanent. Thus, the algorithm terminates within $n^2 - n$ iterations.* □

Now, before showing the safety property we prove an important lemma. If the algorithm selects $p$ as a shortest path between two nodes $u$ and $v$, then it also selects for all $v'$ in $p$ the subpath of $p$ between $u$ and $v'$ as shortest path between $u$ and $v'$, and the subpath of $p$ between $v'$ and $v$ as shortest path between $v'$ and $v$.

**Lemma 5** *Suppose a node $u$ makes status$[u, v]$ permanent for a node $v$ through a path $p$. Then, for all the nodes $v'$ in $p$,*

*(a) $u$ has made status$[u, v']$ permanent through a subpath of $p$ between $u$ and $v'$, and*

*(b) $v'$ has made status$[v', v]$ permanent through a subpath of $p$ between $v'$ and $v$.*

*Proof:*

($a$) Suppose not. Let $v'$ be a node in $p$ such that $status[u, v']$ is not *permanent*. As $v'$ lies in $p$, we have $weight[u, v'] < weight[u, v]$. Since, both $status[u, v]$ and $status[u, v']$ are tentative, $u$ considers $v'$ before $v$ to make $status[u, v']$ *permanent*. The assertion follows.

($b$) Let $p$ be the sequence $\langle u = u_0, u_1, u_2, \cdots, u_k = v \rangle$ of nodes. The node $u$ makes $status[u, v]$ permanent only if it finds $status[u_1, v] = permanent$; $u_1$ makes $status[u_1, v]$ permanent only if it finds $status[u_2, v] = permanent$; and so on. Ultimately, $u_{k-1}$ makes $status[u_{k-1}, v]$ permanent only if it finds $status[v, v] = permanent$. At the beginning, in the initialization phase, the node $v$ makes its $status[v, v]$ permanent. The assertion follows. □

**Theorem 6** *When the algorithm terminates for a graph $G = \langle V, E, W \rangle$, for all $u, v \in V$, $weight[u, v] = dist(u, v)$.*

*Proof:* Suppose not. Consider any two nodes $u, v \in V$, for which, on termination, $weight[u, v] \neq dist(u, v)$. It is clear from the Restricted algorithm that for any two nodes $u, v \in V$, the $weight[u, v]$ is not changed after the entry $status[u, v]$ is made *permanent*. Then, $weight[u, v] \neq dist(u, v)$ at that time $u$ makes $status[u, v]$ *permanent*.

Without loss of generality, we assume $v$ is the first node whose $status[u, v]$ is made *permanent* by $u$ when $weight[u, v] \neq dist(u, v)$. Let $p$ be the path chosen by $u$ to reach $v$. Let $p$ be the sequence $\langle u = u_0, u_1, u_2, \cdots, u_k = v \rangle$ of nodes.

As $weight[u, v] \neq dist(u, v)$ at that time $u$ makes $status[u, v]$ *permanent*, the path $p$ is not a shortest path. Let $p' = \langle u = u'_0, u'_1, u'_2, \cdots, u'_{k'} = v \rangle$

be a shorter path from $u$ to $v$. We have $w(p') < w(p)$. Then, $W[u, u'_1] = weight[u, u'_1] < weight[u, v]$, and hence, $u$ must make $status[u, u'_1]$ permanent before making $status[u, v]$ permanent. When $u$ makes $status[u, u'_1]$ permanent, it relaxes the distance estimate $weight[u, u'_2]$ with respect to $W[u, u'_1] + W[u'_1, u'_2]$. Thus, $u$ becomes aware of the fact that $weight[u, u'_2] < weight[u, v]$, and hence, $u$ must make $status[u, u'_2]$ permanent before making $status[u, v]$ permanent; and so on. Eventually, $u$ makes $status[u, u'_{k'-1}]$ permanent before making $status[u, v]$ permanent. By virtue of the choice of $v$ to be the first node whose $status[u, v]$ is made permanent when $weight[u, v] \neq dist(u, v)$, we have $weight[u, u'_{k'-1}] = dist(u, u'_{k'-1})$. Let $p'' = \langle u = u''_0, u''_1, u''_2, \cdots, u''_{k''} = u'_{k'-1}\rangle$ be the shortest path chosen by $u$ to reach $u'_{k'-1}$. Then, by Lemma 5, for all $u''_j$ in $p''$, $status[u, u''_j]$ and $status[u''_j, u'_{k'-1}]$ are permanent. The node $u$ can make $status[u, u'_{k'-1}]$ permanent only if $status[u''_1, u'_{k'-1}]$ is permanent. The node $u''_1$ can make $status[u''_1, u'_{k'-1}]$ permanent only if $status[u''_2, u'_{k'-1}]$ is permanent; and so on. At the beginning, in the initialization phase, $status[u'_{k'-1}, u'_{k'-1}]$ is made permanent. When $u''_{k''-1}$ makes $status[u''_{k''-1}, u'_{k'-1}]$ permanent, it relaxes the distance estimate $weight[u''_{k''-1}, v]$ with respect to $W[u''_{k''-1}, u'_{k'-1}] + W[u'_{k'-1}, v]$. When $u''_{k''-2}$ makes $status[u''_{k''-2}, u'_{k'-1}]$ permanent, it relaxes the distance estimate $weight[u''_{k''-2}, v]$ with respect to $W[u''_{k''-2}, u''_{k''-1}] + weight[u''_{k''-1}, v]$, that is, with respect to $W[u''_{k''-2}, u''_{k''-1}] + W[u''_{k''-1}, u'_{k'-1}] + W[u'_{k'-1}, v]$; and so on. When $u$ makes make $status[u, u'_{k'-1}]$ permanent, it relaxes the distance estimate $weight[u, v]$ with respect to $W[u, u''_1] + W[u''_1, u''_2] + \cdots + W[u''_{k''-1}, u'_{k'-1}] + W[u'_{k'-1}, v]$. So, $u$ becomes aware of a shorter path than $p$ to reach $v$. Hence $u$ would not consider the path $p$, contradicting the assumption that it has chosen $p$, and hence $weight[u, v] = dist(u, v)$. The theorem follows. $\square$

It is difficult to determine the exact time complexity of the Restricted algorithm, However, doing that does not provide much insight about how to transform this algorithm into a distributed algorithm.

## 5.2 The Distributed Algorithm

It is to observe that in the Restricted algorithm a node $u$ accesses the $RT$ table entries of its neighbors, and not of any other node in the graph.

In the distributed algorithm, the routing table part $RT[u, *]$ is stored at node $u$. The same Restricted algorithm is executed at each node. The distributed algorithm is given in Figure 4. Here, one point is of particular interest. In the Restricted algorithm of Section 5.1, in Step 3 if a node $u$ finds $status[via_u, v_u]$ is *tentative*, it does nothing; and in the very next iteration it checks the same *status* again. The node $u$ does the checking until the *status* becomes *permanent*, and then, $u$ takes a normal course of actions. In the distributed algorithm, to get the status information for $v_u$ from the neighbor node $via_u$, $u$ sends a message $\langle$*give-me your table-entries* for $v_u\rangle$ to the node $via_u$. Instead of sending a reply containing the $RT[via_u, *].\langle weight, status\rangle$ immediately back to $u$, the node

$via_u$ defers replying until $status[via_u, v_u]$ becomes *permanent*. Since the algorithm does not lead to a deadlock situation, as proved in Section 5.1.1, the node $via_u$ eventually returns a reply message to $u$. On receiving the reply message from $via_u$, $u$ makes $status[u, v_u]$ and $via\text{-}status[u, v_u]$ *permanent*, and updates the other tentative entries in $RT[u, *]$. Thus, to make one table entry status *permanent*, there is an exchange of at most two messages between two neighbors. At the beginning, the entire routing table has $(n^2 - n)$ tentative entries. Hence, the algorithm exchanges at most $2(n^2 - n)$ messages. On termination of the algorithm the *via* fields indicate how messages should be routed. The correctness proof of the distributed algorithm is very much similar to that of the Restricted algorithm, and hence is omitted.

# 6  Conclusion

A fully distributed algorithm to find all pairs shortest paths for an arbitrary connected network has been presented. The algorithm produces the desired output within $2n^2$ messages, where $n$ is the number of nodes in the network. Another achievement in the algorithm is that the message complexity is insensitive to the number of edges in the network. The algorithm allows the communication only between neighbors, and uses two types of messages. The algorithm works with each edge's storage capacity of only two messages for each direction.

In an execution of the algorithm some node $u$ could finish building up its part of the routing table much before the other nodes could do their parts, but they have partially built up their parts. Note that the shortest paths from $u$ to other nodes form a spanning tree rooted at $u$. When $u$ finishes its execution, the spanning tree for $u$ is clearly identified through the *via* fields of the internal nodes in the spanning tree. It is guaranteed that the internal nodes will not change the respective *via* fields through which the messages from $u$ will be transmitted. Hence, it is guaranteed that if $u$ starts sending messages to any nodes, the messages will be delivered to their destinations through their respective shortest paths. This can be done even if some nodes have not finished executing the algorithm completely.

For the sake of simplicity of the presentation, a number of assumptions have been made in Section 2. Some of the assumptions could be relaxed, and the following properties can be obtained by modifying the algorithm a little. If the network size is known to the nodes, a node does not need to exchange messages with its neighbor when it is left with only one node to make *permanent*, leading to a total savings of $2n$ messages. The algorithm works even if the edge weights are different for different directions. The algorithm could easily be modified to correctly build up the routing table in case the network size is unknown to the nodes. The algorithm works even if the network is not connected.

# Acknowledgment

I wish to thank Prof. Jan van Leeuwen who introduced me to the area of routing algorithms. Dr. Gerard Tel has been very cooperative during this work, and provided some useful comments on an earlier draft. His book [8] and the one in [3] have been very useful in preparing this manuscript. I especially thank Prof. K. Vidyasankar who provided some useful comments and many suggestions to simplify the proof of the safety property of the Restricted algorithm.

# References

[1] Y. Afek, M. Ricklin: Sparser: A paradigm for running distributed algorithms. Journal of Algorithms 14, 316–328 (1993)

[2] R. Bellman: Dynamic programming. Princeton University Press (1957)

[3] T.H. Cormen, C.E. Leiserson, R.L. Rivest: Introduction to algorithms. MIT Press, Second printing 1990 (Original 1989)

[4] E.W. Dijkstra: A note on two problems in connection with graphs. Numerische mathematik 1, 269–271 (1959)

[5] J. Edmonds: Matroids and the greedy algorithm. Mathematical Programming 1, 126–136 (1971)

[6] R.W. Floyd: Algorithm 97 (Shortest path). Communications of the ACM 5, 345 (1962)

[7] A. Segall: Distributed network protocols. IEEE Trans. on Information Theory 29, 23–35 (1983)

[8] G. Tel: Introduction to distributed algorithms. INF/DOC-92-05, Department of Computer Science, University of Utrecht, The Netherlands (1992)

[9] S. Toueg: An all-pairs shortest-paths distributed algorithm. Tech Rep RC 8327, IBM T.J. Watson Research Center, Yorktown Heights, NY 10598, USA (1980)

[10] S. Warshall: A theorem on boolean matrices. Journal of the ACM 9, 11–12 (1962)

**Procedure** Dijkstra($G = \langle V, E, W \rangle, s$);
begin
    for all $v \in V$ do $d[v] := \infty$;              {initialize}
    $d[s] := 0$; $S := \{\}$; $Q := V$;

    while $Q \neq \{\}$ do
        remove $u$ from $Q$ with $d[u]$ minimum;   {Extract-min from $Q$}
        $S := S \cup \{u\}$;
        for all $v \in neighbor(u)$ do            {relax distance estimates $d[v]$}
            if $d[v] > d[u] + W[u, v]$ then
                $d[v] := d[u] + W[u, v]$;
    endwhile;
end; {of procedure}

Figure 1: Dijkstra algorithm.

**Procedure** Modified-Dijkstra-Algorithm($G = \langle V, E, W \rangle, s$);
begin
    for all $v \in V$ do $d[v] := \infty$;                {Initilize}
                $status[v] := tentative$;
    $d[s] := 0$; $status[s] := permanent$; $via[s] := s$;   {visit $s$}
    $Q := \{\}$;
    for all $v \in neighbor(s)$ do         {insert all neighbors of $s$ in $Q$}
        $d[v] := W[s, v]$; $via[v] := v$;
        insert $v$ in $Q$;

    {$Q$ contains frontier tentative nodes to be visited}
    while $Q \neq \{\}$ do
        remove $u$ from $Q$ with $d[u]$ minimum; {Extract-min from $Q$}
        $status[u] := permanent$;          {visit $u$; now $d[u] = dist(s, u)$}
        for all $v \in neighbor(u)$, such that $status[v] = tentative$ do
            if $d[v] > d[u] + W[u, v]$ then          {relax the edge $uv$}
                $d[v] := d[u] + W[u, v]$;
                $via[v] := via[u]$;
                if $v \notin Q$ then insert $v$ in $Q$;
            endif;
        endfor;
    endwhile;
end; {of procedure}

Figure 2: Modified algorithm.

**Declarations**

$RT$ : array $[1..n, 1..n]$ of = record

        *weight* : weight-type;

        *status* : (*tentative, permanent*);

        *via* : $1..n$;

        *via-status* : (*tentative, permanent*);

        end;

**Procedure** Restricted-Algorithm($G = \langle V, E, W \rangle$);

begin

  for all $u \in V$ do                               {Initilize}

    for all $v \in V$ do

      if $u = v$ then $RT[u, v] := \langle 0, permanent, u, permanent \rangle$    {for all $u$, $u$ visits itself}

      elseif $uv \in E$ then $RT[u, v] := \langle W[u, v], tentative, v, tentative \rangle$

        else $RT[u, v] := \langle \infty, tentative, ?, tentative \rangle$;       {? - don't care value}

  {there are $n^2 - n$ tentative entries in $RT$}

  repeat

**S1:**  for all $u \in V$ do

      find $v_u$ with $RT[u, v_u].weight$ minimum      {Extract-min from $RT[u, *]$}

        from $RT[u, *]$ such that $RT[u, v_u].status = tentative$;

      {if no $v_u$ is found, such that $status[u, v_u] = tentative$, $u$ has finished building up }

      {its part of $RT$, and it should not execute the remaining part of repeat-until loop}

    endfor;

**S2:**  for all $u \in V$ such that $v_u$ is found at Step S1 do

      let $via_u := RT[u, v_u].via$;      {The node $v_u$ is reached through a neighbor $via_u$ }

      let $via\text{-}status_u := RT[u, v_u].via\text{-}status$;      {status of $via_u$ for $v_u$ found in $RT[u, *]$}

      if $via\text{-}status_u = permanent$ then

        $RT[u, v_u].status := permänent$;      {$via_u$ has already visited $v_u$}

    endfor;

**S3:**  for all $u \in V$ such that $v_u$ is found at Step S1 and

                     $via\text{-}status_u \neq permanent$ at Step S2 do

      if $RT[via_u, v_u].status \neq permanent$ then do-nothing  {$via_u$ has not yet visited node $v_u$}

      else                                   {node $via_u$ has already visited $v_u$}

        let $temp_u := RT[via_u, *].\langle weight, status \rangle$;    {copy all entries in $RT[via_u, *]$}

      endif;

**S4:**  for all $u \in V$ such that $v_u$ is found at Step S1,

                     $via\text{-}status_u \neq permanent$ at Step S2 and

                     $RT[via_u, v_u].status \doteq permanent$ at Step S3 do

    $RT[u, v_u].status := permanent$;                       {indirectly visit $v_u$ through}

    $RT[u, v_u].via\text{-}status := permanent$;             {the visit of $v_u$ by $via_u$.}

    for all $v'_u \in V$ such that $RT[u, v'_u].status = tentative$ do {relax tentative node weights}

      if $RT[u, v'_u].weight > W[u, via_u] + temp_u[v'_u].weight$ then

        $RT[u, v'_u].weight := W[u, via_u] + temp_u[v'_u].weight$;

        $RT[u, v'_u].via := via_u$

        $RT[u, v'_u].via\text{-}status := temp_u[v'_u].status$;

      endif

    endfor;

  until $\langle$no change in $RT$ is performed$\rangle$;

end; {of procedure}

Figure 3: Restricted algorithm.

begin$\{u$ begins here$\}$
  ⟨**atomic action begins**⟩
    for all $v \in V$ do                                    $\{$Initialize $RT[u, *]\}$
      if $u = v$ then $RT[u, v] := \langle 0, permanent, u, permanent \rangle$    $\{$visit itself$\}$
      elseif $v \in neighbor_u$ then $RT[u, v] := \langle W[u, v], tentative, v, tentative \rangle$
        else $RT[u, v] := \langle \infty, tentative, ?, tentative \rangle$;        $\{? $ - don't care value$\}$

    find $v_u$ with $RT[u, v_u].weight$ minimum
               from $RT[u, *]$ such that $RT[u, v_u].status = tentative$;   $\{$Extract-min from $RT[u, *]\}$
    let $via_u := RT[u, v_u].via$;                  $\{$The node $v_u$ is reached through a neighbor $via_u\}$
    send a message $\langle give\text{-}me$ your table-entries for $v_u \rangle$ to the neighbor $via_u$;
  ⟨**atomic action ends**⟩
  wait at $A3$;
end;

$A1$: $\{$arrival of a message $\langle give\text{-}me$ your table-entries for $v \rangle$ from a neighbor $v'$ $\}$
    ⟨**atomic action begins**⟩
    if $RT[u, v].status = permanent$ then
      send a message $\langle table\text{-}entries = RT[u, *].\langle weight, status \rangle \rangle$ to $v'$
    else keep the message pending;
    ⟨**atomic action ends**⟩
    wait at $A3$;

$A2$: $\{$arrival of a reply message $\langle table\text{-}entries$ for $v_u \rangle$ from a neighbor $via_u$ $\}$
    ⟨**atomic action begins**⟩
    $temp_u := receive(RT[via_u, *].\langle weight, status \rangle)$;       $\{$copy all entries of $RT[via_u, *]\}$
    $RT[u, v_u].status := permanent$;    $\{$indirectly visit $v_u$ through the visit of $v_u$ by $via_u$
    $RT[u, v_u].via\text{-}status := permanent$;
    for all $v'_u \in V$ such that $RT[u, v'_u].status = tentative$ do
      if $RT[u, v'_u].weight > W[u, via_u] + temp_u[v'_u].weight$ then $\{$relax$\}$
        $RT[u, v'_u].weight := W[u, via_u] + temp_u[v'_u].weight$;
        $RT[u, v'_u].via := via_u$
        $RT[u, v'_u].via\text{-}status := temp_u[v'_u].status$;
    endfor
    if there are pending messages $\langle give\text{-}me$ your table-entries for $v_u \rangle$ from neighbors $v'$
      for $RT[u, v_u].status$ to become $permanent$, then
      send reply $\langle table\text{-}entries = RT[u, *].\langle weight, status \rangle \rangle$ to $v'$;

    loop: find $v_u$ with $RT[u, v_u].weight$ minimum
               from $RT[u, *]$ such that $RT[u, v_u].status = tentative$; $\{$Extract-min$\}$
         $\{$if no $v_u$ is found, such that $status[u, v_u] = tentative$, $u$ has finished building up$\}$
         $\{$ its part of $RT$, and it should not execute the remaining, and wait at $A3\}$
        let $via_u := RT[u, v_u].via$;     $\{$The node $v_u$ is reached through a neighbor $via_u\}$
        let $via\text{-}status_u := RT[u, v_u].via\text{-}status$;   $\{$status of $via_u$ for $v_u$ found in $RT[u, *]\}$
        if $via\text{-}status_u = permanent$ then $RT[u, v_u].status := permanent$;
                        go back to loop;        $\{$try for other entries$\}$
      else send a message $\langle give\text{-}me$ your table-entries for $v_u \rangle$ to the neighbor $via_u$;
    ⟨**atomic action ends**⟩
    wait at $A3$;

$A3$: $\{$no-work! so, please wait here.$\}$

Figure 4: Distributed algorithm (for a node $u$).

# Linear Layouts of Generalized Hypercubes

Koji Nakano

Advanced Research Laboratory, Hitachi, Ltd., Hatoyama, Saitama 350-03, Japan
e-mail : nakano@harl.hitachi.co.jp

**Abstract.** This paper studies linear layouts of generalized hypercubes, a $d$-dimensional $c$-ary clique and a $d$-dimensional $c$-ary array, and evaluates the bisection width, cut width, and total edge length of them, which are important parameters to measure the complexity of them in terms of a linear layout.

## 1 Introduction

This paper treats two kinds of generalized hypercubes: *a $d$-dimensional $c$-ary clique* (abbreviated as $Cc_d$) and *a $d$-dimensional $c$-ary array* (abbreviated as $Ac_d$). $Cc_d$ has nodes labeled by the $c^d$ integers from 0 to $c^d - 1$. The nodes are connected by the edges if and only if the $c$-ary representations of their labels differ by one and only one digit (Fig. 1). *An $n$-node $c$-ary clique* (abbreviated as $Cc_{(n)}$), which is a more generalized graph, has $n$ nodes labeled by the integers from 0 to $n-1$ and connected in the same way as $Cc_d$. Note that $n$ is not restricted to a power of $c$. $Ac_d$ has the same nodes as $Cc_d$. The nodes are connected if and only if the $c$-ary representations of their labels differ by one and only one digit and the absolute value of the difference in that digit is 1 (Fig. 2).

Several algorithms on parallel computers based on $Cc_d$ and $Ac_d$ topologies have been shown [1, 7]. It is very important to analyze topological properties of them, because they are very attractive as network topologies of future parallel computers. Furthermore, $Cc_d$ and $Ac_d$ include typical topologies which are used for parallel machines: $Cc_1$ corresponds to a $c$-node clique (or a complete graph), $Ac_1$ corresponds to a $c$-node linear array, $Ac_2$ corresponds to a $c \times c$-node 2-dimensional array, $Ac_3$ corresponds to a $c \times c \times c$-node 3-dimensional array, and both $C2_d$ and $A2_d$ correspond to a $d$-dimensional (binary) hypercube. Therefore, the results presented in this paper can be applied to these topologies.

*A linear layout of a graph* $G = (V, E)$ (where $V$ and $E$ are a set of nodes and a set of edges, respectively) is a one-to-one mapping $L : V \to \{0, 1, 2, \ldots, |V| - 1\}$. This means that each $u$ ($\in V$) is assigned to the position $L(u)$ on the baseline. Examples of linear layouts of $C4_2$ and $A4_2$ are illustrated in Figs. 3 and 4, where each node $u$ is assigned to the position $L(u)$, that is, $L(u) = u$ for all $u$. We call such the layout $L$ *the label order layout*. Note that a linear layout can take any permutation (i.e. $|V|!$ permutations), not just the label order layout.

The complexity of $G = (V, E)$ in terms of a linear layout is measured by the following parameters: *the (minimum) bisection width, the cut width,* and *the total edge length.* These parameters are defined as follows. *The cut of a*

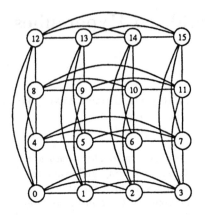

Fig. 1. A 2-dimensional 4-ary array

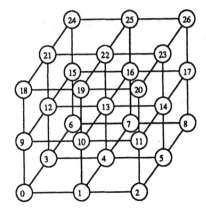

Fig. 2. A 3-dimensional 3-ary array

*graph G under a linear layout L at a gap i* is a set of edges connecting a node at a position less than $i$ and one at a position larger than or equal to $i$, i.e. $C(G, L, i) \stackrel{\text{def}}{=} \{(u, v) \in E | 0 \leq L(u) < i \leq L(v) \leq |V| - 1\}$. *The bisection width of a graph G* is the minimum number of edges in $C(G, L, \lfloor |V|/2 \rfloor)$ over all linear layouts, i.e. $\min_L |C(G, L, \lfloor |V|/2 \rfloor)|$. In other words, the bisection width of a graph is the minimum number of edges which must be removed to separate the graph into two disjoint and equal-sized subgraphs. *The cut width of a graph G under a linear layout L* is the maximum of $|C(G, L, i)|$ over all gaps $i$, i.e. $\max_i |C(G, L, i)|$. *The cut width of a graph G* is the minimum cut width over all linear layouts, i.e. $\min_L \max_i |C(G, L, i)|$. This parameter indicates the number of tracks required by the best linear layout. We will define that *the length of edge (u, v) ∈ E under a linear layout L* is $|L(u) - L(v)|$. Then, *the total edge length of a graph G under a linear layout L* is $\sum_{(u,v) \in E} |L(u) - L(v)|$. Furthermore, *the total edge length of a graph G* is defined as the minimum of this value over all linear layouts, i.e. $\min_L \sum_{(u,v) \in E} |L(u) - L(v)|$. Obviously, the total edge length is equal to the total cut, i.e. $\min_L \sum_{i=1}^{|V|-1} |C(G, L, i)|$

It is very important to compute exact values of them, because they determine the lower bound of the layout area in the VLSI model. For example, the layout area of a processor network is at least $\Omega(B^2)$ if the corresponding graph has bisection width $B$ [6, 13], and the number of tracks of a processor network in a horizontal layouts requires $C$ layers if the corresponding graph has cut width $C$. The total edge length has applications to the coding theory [5] and storage management [12]: Minimizing the total edge length of generalized hypercubes corresponds to minimizing the error of a c-ary channel, and to minimizing the efficiency of managing a d-dimensional data structure in a paging environment. However, the problem to compute the exact values of them are hard problem: For a given graph and an integer $k$, the problem to determine whether the

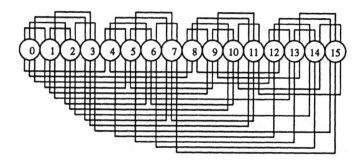

**Fig. 3.** The label order layout of a 2-dimensional 4-ary clique

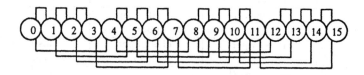

**Fig. 4.** The label order layout of a 2-dimensional 4-ary clique

bisection width of the graph is at most $k$ is NP-complete [4]. Similarly, the problem to determine the cut width is NP-complete even if the degree of the graph is restricted [8].

Several articles have been devoted to the evaluation of them. Brebner [2], Manabe et al. [9], and Nakano et al. [10] have proved that the bisection width of a $d$-dimensional binary hypercube is $2^{d-1}$ using different methods. Leighton [7] showed that the bisection width of $Ac_d$ is $c^{d-1}$ if $c$ is even by embedding a directed complete graph in $Ac_d$. Wada et al. [14] proved that the bisection width of $Cc_d$ is $c^{d+1}/4$ if $c$ is even in a similar way to the Leighton's proof. However, they did not get the exact value of it when $c$ is odd: the bisection width of $Cc_d$ takes a value between $\lceil c^{d+1}/4 - 1/(4c^{d-1}) \rceil$ and $(c+1)(c^d-1)/4$ (inclusive). Nakano et al. [10] also proved that the cut width of $C2_d$ is $\lfloor 2^{d+1}/3 \rfloor$. Wada et al. [15] also proved that the cut width of $Cc_d$ is at most $c^2(c^d - 1)/\{4(c - 1)\}$. Niepel et al. [11] showed that the total edge length of an $n \times 2$-node array is $5n-4$ and conjectured that that of an $n \times m$-node array is $n(m^2+m-1)-m^2$. Harper [5] showed that the total edge length of a $d$-dimensional hypercube is $2^{d-1}(2^d - 1)$. DeMillo et al. [3] showed that the total edge length of 2-dimensional hypercube is at least $n^3/6$.

In this paper, we will evaluate the bisection width, cut width, and total edge length of $Cc_d$ and of $Ac_d$. In Section 2, we consider how many edges a subgraph of $Cc_{(m)}$ with $n$ $(n \leq m)$ nodes may have, and show that $Cc_{(n)}$ has the largest number of edges of all subgraphs with $n$ nodes. In other words, $Cc_{(n)}$ is the maximum subgraph of $Cc_{(m)}$ if $n \leq m$. Section 3 uses this fact to get the exact values of the bisection width, cut width, and total edge length of $Cc_d$. Section 4 presents the method for converting $Cc_d$ into $Ac_d$ and get exact value of the bisection width of $Ac_d$, and nearly exact values of the cut width and the total edge length of $Ac_d$. See Table 1 for comparing our results and previously known results.

**Table 1.** Our results and previously known results

| | Bisection width | Cut width | Total edge length |
|---|---|---|---|
| hypercube | Brebner [2] Manabe [9] Nakano [10] exact | Nakano [10] exact | Harper [5] exact |
| $d$-dimensional $c$-ary clique | Wada [14] exact when $c$ is even | Wada [15] only upper bound | |
| | This paper exact | This paper exact | This paper exact |
| $d$-dimensional $c$-ary array | Leighton [7] exact when $c$ is even | | DeMillo [3] lower bound when $d = 2$§ |
| | This paper exact | This paper nearly exact† | This paper nearly exact‡ |

†The upper bound is about $1 + 2/\{(c + 2)(c - 1)\}$ times as large as the lower bound.
‡The upper bound is about $3c/\{2(c + 1)\}$ times as large as the lower bound.
§DeMillo's lower bound is $c^3/6$, while that of us is about $2c^3/3$.

## 2  Maximum subgraph of $Cc_d$

The main result of this paper is due to the following theorem:

**Theorem 1.** $Cc_{(n)}$ *is a maximum subgraph of* $Cc_{(m)}$ *if* $n \leq m$.

Theorem 1 can be proved by the following lemmas.

**Lemma 2.** *Let* $fc$ *be the function defined as follows:*

$$fc(n) \overset{\text{def}}{=} \begin{cases} n(n-1)/2 & \text{if } n \leq c, \\ \sum_{i=0}^{c-1}\{fc(\lfloor(n+i)/c\rfloor) + (c - i - 1)\lfloor(n+i)/c\rfloor\} & \text{otherwise.} \end{cases}$$

*For all* $n \geq 1$, $Cc_{(n)}$ *has* $fc(n)$ *edges.*

**Lemma 3.** *Let gc be the function defined as follows:*

$$
gc(n) \overset{\text{def}}{=} \begin{cases} n(n-1)/2 & \text{if } n \le c, \\ \max\{\displaystyle\sum_{i=0}^{c-1}\{gc(n_i) + (c-i-1)n_i\}| \\ \quad n_0 \le n_1 \le \cdots \le n_{c-1} < n = \displaystyle\sum_{i=0}^{c-1} n_i\} & \text{otherwise.} \end{cases}
$$

*For any subgraph $G = (V, E)$ of $Cc_{(m)}$, $|E| \le gc(|V|)$ holds.*

**Lemma 4.** $fc = gc$ *holds.*

Note that the division of an integer $n$ into the same $c$ values as equally as possible can be represented as

$$
\lfloor n/c \rfloor, \lfloor (n+1)/c \rfloor, \lfloor (n+2)/c \rfloor, \ldots, \lfloor (n+c-1)/c \rfloor.
$$

In fact, the sequence is $c - r$ $q$'s followed by $r$ $(q + 1)$'s where $n = q \cdot c + r$ $(0 \le r \le c - 1)$. Thus, while $gc(n)$ is evaluated by computing the maximum over all divisions of $n$, $fc(n)$ is evaluated for the equal-sized division of $n$. Therefore, obviously, we have $fc \le gc$. However, Lemma 4 claims $fc = gc$.

Lemma 2 shows the number of edges of $Cc_{(n)}$, and Lemma 3 shows the upper bound of the number of edges of the maximum subgraph. Hence, from Lemma 4, the number of edges of $Cc_{(n)}$ is equal to the number of edges of the maximum subgraph with $n$ nodes. Therefore, these lemmas imply Theorem 1. See the appendix for the proofs of Lemmas 2, 3, and 4.

## 3 Widths and length of $Cc_d$

To get exact evaluations of the widths of $Cc_d$, we first prove the following lemma:

**Lemma 5.** *For any linear layout $L$ and any gap $i$ $(1 \le i \le c^d - 1)$, the cut of $Cc_d$ under $L$ at $i$ is at least as large as that of $Cc_d$ under the label order layout at $i$.*

*Proof.* For a gap $i$ under $L$, divide the edges in $Cc_d$ into $Cc_d^-(L, i)$, $Cc_d^+(L, i)$, and $Cc_d(L, i)$ as follows: $Cc_d^-(L, i)$ (resp. $Cc_d^+(L, i)$) is the set of edges connecting nodes whose positions are less than (resp. larger than or equal to) $i$, and $Cc_d(L, i)$ are the cut under $L$ at a gap $i$. Obviously, we have

1. $Cc_d^-(L, i)$ and $Cc_d^+(L, i)$ are subgraphs of $Cc_d$ with $i$ nodes and with $c^d - i$ nodes, respectively.
2. Let $I$ be the label order layout, i.e. for all $i$, $I(i) = i$ (Fig 3). Since the label order layout of $Cc_d$ is bilateral symmetry, $Cc_d^-(I, i)$ and $Cc_d^+(I, i)$ correspond to the edges of $Cc_{(i)}$ and $Cc_{(c^d - i)}$, respectively.

Hence, from Theorem 1, $|Cc_d^-(L,i)| \le |Cc_d^-(I,i)|$ and $|Cc_d^+(L,i)| \le |Cc_d^+(I,i)|$ hold. Furthermore, obviously,

$$|Cc_d(L,i)| + |Cc_d^-(L,i)| + |Cc_d^+(L,i)| = |Cc_d(I,i)| + |Cc_d^-(I,i)| + |Cc_d^+(I,i)|.$$

Thus, $|Cc_d(L,i)| \ge |Cc_d(I,i)|$ holds. This completes the proof. □

From this lemma, when computing the parameters of $Cc_d$, we do not have to compute the minimum over all linear layouts but only those of the label order layout. In other words, we have

**Lemma 6.** *The bisection width, cut width, and total edge length of $Cc_d$ are equal to those of the label order layout, respectively.*

It is easy to compute the parameters of the label order layout of $Cc_d$. For example, the bisection width of the label order layout (i.e. $|Cc_d(I, \lfloor c^d/2 \rfloor)|$) can be computed as follows: If $c$ is even, since $Cc_d(I, \lfloor c^d/2 \rfloor)$ consists of edges along the $d$th dimension, $|Cc_d(I, \lfloor c^d/2 \rfloor)|$ is equal to $c^d/2 \times c/2 = c^{d+1}/4$. If $c$ is odd, among all edges along each $k$th dimension $(1 \le k \le d)$, $Cc_d(I, \lfloor c^d/2 \rfloor)$ contains $(c^2 - 1)c^{k-1}/4$ edges. By summing up, $Cc_d(I, \lfloor c^d/2 \rfloor)$ has $(c+1)(c^d-1)/4$ edges. As a result, we have the following theorem:

**Theorem 7.** *The bisection width of $Cc_d$ is $c^{d+1}/4$ (if $c$ is even), and $(c+1)(c^d - 1)/4$ (if $c$ is odd).*

Similarly, we can compute the cut width and total edge length of the label order layout and get the following theorems:

**Theorem 8.** *The cut width of $Cc_d$ is $c(c+2)(c^d - 1)/\{4(c+1)\}$ (if $c$ is even and $d$ is even), $c^2\{(c+2)c^{d-1} - 1\}/\{4(c+1)\}$ (if $c$ is even and $d$ is odd), and $(c+1)(c^d-1)/4$ (if $c$ is odd).*

**Theorem 9.** *The total edge length of $Cc_d$ is $(c+1)c^d(c^d-1)/6$.*

## 4  Widths and length of $Ac_d$

Since $Ac_{(n)}$ is not always a maximum subgraph, the method in the previous section cannot be applied to compute the widths of $Ac_d$. Hence, we use a method similar to embedding a directed clique [7]. In other words, $Cc_d$ is embedded in $Ac_d$.

From Theorem 7, the bisection width of a $c$-node clique is $h(c)$, where $h(c)$ is $c^2/4$ (if $c$ is even) and $(c^2 - 1)/4$ (if $c$ is odd). Since each side of $Cc_d$ can be considered as a $c$-node clique, we have

**Lemma 10.** *For any linear layout $L$ and any gap $i$ $(1 \le i \le c^d)$, the cut of $Cc_d$ under the label order layout at $i$ is at most $h(c)$ times as large as the cut of $Ac_d$ under $L$ at $i$.*

*Proof.* Fix a linear layout $L$ and compare $L$ of $Ac_d$ and $L$ of $Cc_d$. It can be considered that each edge in $Ac_d$ corresponds to at most $h(c)$ edges in $Cc_d$ under $L$. Therefore, the cut of $Cc_d$ under $L$ at each gap is at most $h(c)$ times as large as the cut of $Ac_d$ under $L$ at the same gap. Thus, from Lemma 5, the cut of $Cc_d$ under the label order layout at each position is at most $h(c)$ times as large as the cut of $Ac_d$ under $L$ at the same position. □

From this lemma, we have

**Lemma 11.** *The bisection width, cut width, and total edge length of $Ac_d$ are at least as large as those of $Cc_d$ divided by $h(c)$, respectively.*

Therefore, the lower bounds of $Ac_d$ can be obtained from Theorems 7, 8, and 9.

On the other hand, from the definition, we have

**Lemma 12.** *The bisection width, cut width, and total edge length of $Ac_d$ is at most as large as those of $Ac_d$ under the label order layout, respectively.*

From these relation, the upper bounds of $Ac_d$ can be obtained by computing those of the label order layout which can be computed similarly to those of $Cc_d$. Consequently, we have

**Theorem 13.** *The bisection width of $Ac_d$ is $c^{d-1}$ (if $c$ is even), and $(c^d - 1)/(c - 1)$ (if $c$ is odd).*

**Theorem 14.** *The cut width of $Ac_d$ $(c \geq 3)$ is at least $(c + 2)(c^d - 1)/\{c(c+1)\}$ (if $c$ is even and $d$ is even), at least $\{(c + 2)c^{d-1} - 1\}/(c + 1)$ (if $c$ is even and $d$ is odd), at least $(c^d - 1)/(c - 1)$ (if $c$ is odd), and at most $(c^d - 1)/(c - 1)$.*

If $c = 2$, the cut width of $Ac_d$ is equal to that of $Cc_d$.

**Theorem 15.** *The total edge length of $Ac_d$ is at least $2(c + 1)c^{d-2}(c^d - 1)/3$ (if $c$ is even), at least $2c^d(c^d - 1)/\{3(c - 1)\}$ (if $c$ is odd), and at most $c^{d-1}(c^d - 1)$.*

Fortunately, the upper bound of the bisection width is equal to the lower bound. However, the upper bounds of the cut width and total edge length of $Ac_d$ do not match the lower bounds of them. But the difference is not so large; The upper bound of the cut width is at most approximately $1 + 2/\{(c+2)(c-1)\}$ times as large as the lower bound and the upper bound of the total edge length is approximately 1.5 times as large as the lower bound.

## 5 Conclusions

We have presented the exact or nearly exact values of the bisection width, cut width, total edge length of generalized hypercubes. Lemma 5 implies that the label order layout of a $d$-dimensional $c$-ary clique is the optimal layout in the sense that the cut of the label order layout at each gap is smaller than or equal to that of any other layout at the same gap. Similarly to the $Ac_d$ case, this result makes it easy to prove that the upper and lower bounds of the widths and the total length of $Ac_d$ with wraparound edges (referred to as a $d$-dimensional $c$-ary torus [7]) are twice as large as those of $Ac_d$. The exact values of the cut width and total edge length of $Ac_d$ remain to be solved.

# Acknowledgment

The author would like to thank Imrich Vrťo for his valuable comments.

# References

1. L. N. Bhuyan and D. P. Agrawal. Generalized hypercube and hyperbus structures for a computer network. *IEEE Transactions on Computers*, C-33(4), April 1984.
2. G. Brebner. Relating routing and two-dimensinal grids. In P. Bertolazii and F. Luccio, editors, *VLSI: Algorithms and Architectures*, pages 221–231. Elsevier Science Publishers B.V.(North-Holland), 1985.
3. R. A. DeMillo, S. C. Eisenstat, and R. J. Lipton. Preserving average proximity in arrays. *Communications of the ACM*, 21(3):228–231, March 1978.
4. M. R. Garey, D. S. Johnson, and L. Stockmeyer. Some simplified polynomial complete problems. *SIGACT*, pages 47–63, 1974.
5. L. H. Harper. Optimal assignments of numbers to vertices. *J. Soc. Indust. Appl. Math*, 12(1):131–135, March 1964.
6. F. T. Leighton. *Complexity Issues in VLSI: Optimal Layouts for the Shuffle-Exchange Graph and Other Networks*. MIT Press, 1983.
7. F. T. Leighton. *Introduction to Parallel Algorithms and Architectures: Arrays · Trees · Hypercubes*. Morgan Kaufmann, 1992.
8. F. Makedon and I. H. Subdorough. On minimizing width in linear layouts. *Discrete Applied Mathematics*, 23:243–265, 1989.
9. Y. Manabe, K. Hagihara, and N.Tokura. The minimum bisection widths of the cube-connected-cycles graph and cube graph. *Trans. IEICE(D) Japan*, J76-D(6):647–654, June 1984. in Japanese.
10. K. Nakano, W. Chen, T. Masuzawa, K. Hagihara, and N. Tokura. Cut width and bisection width of hypercube graph. *IEICE Transactions*, J73-A(4):856–862, April 1990. in Japanese.
11. L. Niepel and P. Tomasta. Elevation of a graph. *Czechoslovak Mathematical Journal*, 31(106):475–483, 1981.
12. A. L. Rosenberg. Preserving proximity in arrays. *SIAM J. Comput.*, 4(4):443–460, December 1975.
13. C. D. Thompson. Area-time complexity for VLSI. In *Proc. of 11th Symposium on Theory of Computing*, pages 81–88. ACM, 1979.
14. K. Wada and K. Kawaguchi. Optimal bounds of the crossing number and the bisection width for generalized hypercube graphs. In *Proc. of 16th Biennial Symposium on Communications*, pages 323–326, May 1992.
15. K. Wada, H. Suzuki, and K. Kawaguchi. The crossing number of hypercube graphs. In *Proc. of 43rd Convention of IPS Japan*, pages 1–95, 1991. in Japanese.

# Appendix

In the appendix, we will prove Lemmas 2, 3, and 4. First, we will prove Lemma 2.

*Proof.* The proof is by induction on $n$. Obviously for all $n \leq c$, $Cc_{(n)}$ has $fc(n)$ edges. We assume that for all $k \leq n-1$, $Cc_{(k)}$ has $fc(k)$ edges, and will prove that $Cc_{(n)}$ has $fc(n)$ edges. For all $i$ $(0 \leq i \leq c-1)$, let $Vc^i_{(n)}$ be the set of nodes such that the LSD's (Least Significant Digits) of the $c$-ary representations of them are $c - i - 1$. In other words, $Vc^i_{(n)}$ contains the nodes $[\cdots \overset{LSD}{(c - i - 1)}]$. Hence, $Vc^i_{(n)}$ consists of $\lfloor (n+i)/c \rfloor$ nodes. Let $Ec^{ij}_{(n)}$ $(i \leq j)$ be the edges connecting $Vc^i_{(n)}$ and $Vc^j_{(n)}$. Since for all $i$, $Gc^{ii}_{(n)} = (Vc^i_{(n)}, Ec^{ii}_{(n)})$ and $Cc_{(\lfloor (n+i)/c \rfloor)}$ are isomorphic, we have $|Ec^{ii}_{(n)}| = fc(\lfloor (n+i)/c \rfloor)$ from the inductive assumption. For all $i$ and $j$ $(i < j)$, no two edges in $Ec^{ij}_{(n)}$ share a node in $Vc^i_{(n)}$ and every node in $Vc^i_{(n)}$ is connected by an edge in $Ec^{ij}_{(n)}$. Thus $|Ec^{ij}_{(n)}| = |Vc^i_{(n)}| = \lfloor (n+i)/c \rfloor$. Therefore, we have

$$|Ec_{(n)}| = \sum_{i=0}^{c-1} |Ec^{ii}_{(n)}| + \sum_{i<j} |Ec^{ij}_{(n)}| = \sum_{i=0}^{c-1} fc(\lfloor (n+i)/c \rfloor) + \sum_{i<j} \lfloor (n+i)/c \rfloor$$

$$= \sum_{i=0}^{c-1} \{fc(\lfloor (n+i)/c \rfloor) + (c - i - 1)\lfloor (n+i)/c \rfloor\} = fc(n).$$

□

Secondly, we will show the proof of Lemma 3.

*Proof.* The proof is by induction on the number of nodes in $V$. Obviously, for any subgraph $G = (V, E)$, if $|V| \leq c$ then $|E| \leq gc(|V|)$. We assume that $|E| \leq gc(|V|)$ if $|V| \leq n - 1$, and will show that $|E| \leq gc(|V|)$ if $|V| = n$. We select any digit $s$ and divide $V$ into $V^0, V^1, \ldots, V^{c-1}$ as follows: $V^i$ consists of the nodes such that the $s$th digit of the $c$-ary representation of them is $i$. In other words, $V^i$ contains the nodes $[\cdots \overset{s\text{th digit}}{i} \cdots]$. Since we can select $s$ such that there are at least two $V's$ which are not empty, we can assume, for all $i$, $|V^i| < |V|$. Furthermore, by renumbering the indices of $V's$, we can assume that $|V^0| \leq |V^1| \leq \cdots \leq |V^{c-1}| < |V|$ without loss of generality. Let $E^{ij}$ $(i \leq j)$ be the edges in $E$ connecting $V^i$ and $V^j$. Since $|V^0| \leq |V^1| \leq \cdots \leq |V^{c-1}| < n$, we have, for all $i$, $E^{ii} \leq gc(|V^i|)$ from the inductive assumption. Since no two edges in $E^{ij}$ $(i < j)$ share a node in $V^i$, $|E^{ij}|$ is at most as large as $|V^i|$. Therefore, we have

$$|E| = \sum_{i=0}^{c-1} |E^{ii}| + \sum_{i<j} |E^{ij}| \leq \sum_{i=0}^{c-1} gc(|V^i|) + \sum_{i<j} |V^i|$$

$$\leq \sum_{i=0}^{c-1} \{gc(|V^i|) + (n - i - 1)|V^i|\} \leq gc(|V|).$$

□

We have to prove several lemmas as preparation for the proof of Lemma 4. From now on, for given $n_0, n_1, \ldots, n_{c-1}$, let $n = n_0 + n_1 + \cdots + n_{c-1}$ and $n_i = q_i c + r_i$ $(0 \leq r_i \leq c - 1)$. For convenience, let $n_{-1} = -\infty$ and $n_c = +\infty$. Under this notation, the following lemma holds obviously:

**Lemma 16.** *For all* $n_0, n_1, \ldots, n_{c-1}$,

$$n = \sum_{j=0}^{c-1} \lfloor \frac{n+j}{c} \rfloor = \sum_{i=0}^{c-1}\sum_{j=0}^{c-1} \lfloor \frac{n_i+j}{c} \rfloor.$$

Let us consider that for given $n_0, n_1, \ldots, n_{c-1}$ ($n_0 \leq n_1 \leq \cdots \leq n_{c-1}$), the set $\{(i,j)|0 \leq i,j \leq c-1\}$ is sorted by $\lfloor (n_i+j)/c \rfloor$, and let the $(ic+j)$th ($0 \leq i,j \leq c-1$) smallest element[1] and its value be $(a(i,j), b(i,j))$ and $\alpha(i,j)$, respectively. In other words, for all $i$ and $j$, let $\alpha(i,j) = \lfloor (n_{a(i,j)} + b(i,j))/c \rfloor$, and for all $i$, $j$, $i'$, $j'$ such that $ic+j \leq i'c+j'$, we find that $\alpha(i,j) \leq \alpha(i',j')$ holds. Figure 5 illustrates an example of the values of $\lfloor (n_i+j)/c \rfloor$ and $\alpha(i,j)$.

Consider the following procedure that determines two mappings $A, B : \{0, 1, \cdots, c-1\} \times \{0, 1, \cdots, c-1\} \to \{0, 1, \cdots, c-1\}$:

**Step 1** Let $S_j := \lfloor (n+j)/c \rfloor$ for all $j$ ($0 \leq j \leq c-1$), and $i := 0$.

**Step 2** Sort $S_0, S_1, \ldots, S_{c-1}$ by their values and let $p : \{0, \ldots, c-1\} \to \{0, \ldots, c-1\}$ be the one-to-one mapping so that for each $j$, $S_{p(j)}$ is the $j$th smallest element.

**Step 3** For all $j$, determine $A$ and $B$ so that $A(i, p(j)) = a(i,j)$ and $B(i, p(j)) = b(i,j)$.

**Step 4** For all $j$, let $S_{p(j)} := S_{p(j)} - \alpha(i,j)$.

**Step 5** Let $i := i+1$ and if $i < c-1$ then go to Step 2.

| N | $n_i$ | $q_i$ | $r_i$ | $\lfloor (n_i+j)/c \rfloor$ | | | | | | | | $\alpha(i,j)$ | | | | | | | | $\beta(i,j)$ | | | | | | | |
|---|---|---|---|---|---|---|---|---|---|---|---|---|---|---|---|---|---|---|---|---|---|---|---|---|---|---|---|
| | | | | 0 | 1 | 2 | 3 | 4 | 5 | 6 | 7 | 0 | 1 | 2 | 3 | 4 | 5 | 6 | 7 | 0 | 1 | 2 | 3 | 4 | 5 | 6 | 7 |
| 0 | 21 | 2 | 5 | 2 | 2 | 2 | 3 | 3 | 3 | 3 | 3 | 2 | 2 | 2 | 3 | 3 | 3 | 3 | 3 | 2 | 2 | 2 | 3 | 3 | 3 | 3 | 3 |
| 1 | 34 | 4 | 2 | 4 | 4 | 4 | 4 | 4 | 4 | 5 | 5 | 4 | 4 | 4 | 4 | 4 | 4 | 4 | 4 | 4 | 4 | 4 | 4 | 4 | 4 | 4 | 4 |
| 2 | 36 | 4 | 4 | 4 | 4 | 4 | 4 | 5 | 5 | 5 | 5 | 4 | 4 | 5 | 5 | 5 | 5 | 5 | 5 | 5 | 5 | 5 | 4 | 4 | 5 | 5 | 5 |
| 3 | 57 | 7 | 1 | 7 | 7 | 7 | 7 | 7 | 7 | 7 | 8 | 7 | 7 | 7 | 7 | 7 | 7 | 7 | 7 | 7 | 7 | 7 | 7 | 7 | 7 | 7 | 7 |
| 4 | 60 | 7 | 4 | 7 | 7 | 7 | 7 | 8 | 8 | 8 | 8 | 7 | 7 | 7 | 7 | 7 | 7 | 7 | 7 | 7 | 7 | 7 | 7 | 7 | 7 | 7 | 7 |
| 5 | 60 | 7 | 4 | 7 | 7 | 7 | 7 | 8 | 8 | 8 | 8 | 7 | 7 | 8 | 8 | 8 | 8 | 8 | 8 | 7 | 8 | 8 | 8 | 8 | 7 | 8 | 8 |
| 6 | 61 | 7 | 5 | 7 | 7 | 7 | 8 | 8 | 8 | 8 | 8 | 8 | 8 | 8 | 8 | 8 | 8 | 8 | 8 | 8 | 8 | 8 | 8 | 8 | 8 | 8 | 8 |
| 7 | 65 | 8 | 1 | 8 | 8 | 8 | 8 | 8 | 8 | 8 | 9 | 8 | 8 | 8 | 8 | 8 | 8 | 8 | 9 | 9 | 8 | 8 | 8 | 8 | 8 | 8 | 8 |
| | | | | | | | | | | | | | | | | | | | | SUM | 49 | 49 | 49 | 49 | 49 | 49 | 50 | 50 |

**Fig. 5.** An example of the values of $\lfloor (n_i+j)/c \rfloor$, $\alpha(i,j)$, and $\beta(i,j)$

Since $A$ and $B$ are determined one by one from the smallest to the largest element of $\{(i,j)|0 \leq i,j \leq c-1\}$, $A$ and $B$ have the following two properties after completion of the procedure.

**Property 17.** $C(i,j) = \langle A(i,j), B(i,j)\rangle$ *is a one-to-one mapping.*

---

[1] Let the 0th smallest element be the smallest element.

**Property 18.** *Let* $\beta(i,j) = \lfloor (n_{A(i,j)} + B(i,j))/c \rfloor$. *For all* $j$, $\beta(0,j) \le \beta(1,j) \le \cdots \le \beta(c-1,j)$.

These properties can be clearly seen in Fig. 5.

For all $i$, $q_i \le \alpha(i,0) \le \alpha(i,1) \le \cdots \le \alpha(i,c-1) \le q_i + 1$ holds. Hence, after each iteration, for all $j$ and $j'$, $|S_j - S_{j'}| \le 1$ holds. In particular, from Lemma 16 and Property 17, for all $j$, $S_j = 0$ after completion of the procedure. Therefore, $\beta$ has the following property:

**Property 19.** *For all* $j$,

$$\sum_{i=0}^{c-1} \beta(i,j) = \lfloor \frac{n+j}{c} \rfloor.$$

Let $\sum_{i=s}^{t} r_i = q_{s,t}c + r_{s,t}$ $(0 \le r_{s,t} \le c-1)$. For all $s,t$ such that $q_{s-1} < q_s = q_{s+1} = \cdots = q_t < q_{t+1}$, let us imagine the three submatrices which can be obtained by picking up from the $s$-th to the $t$-th row of the matrices in Fig. 5. For example, choose $s = 3$ and $t = 6$ in Fig. 5. The submatrices of $\alpha(i,j)$, $\beta(i,j)$, and $\lfloor (n_i + j)/c \rfloor$ have the following property:

**Property 20.**
- *They have the same number of $q_s$'s and the same number of $q_s + 1$'s.*
- *In the $i$th row of the submatrix of $\lfloor (n_i+j)/c \rfloor$, there are $(c-r_i)$ $q_s$'s followed by $r_i$ $q_s + 1$'s.*
- *In the $(t - q_{s,t})$th row of the submatrix of $\alpha(i,j)$, there are $(c - r_{s,t})$ $q_s$'s followed by $r_{s,t}$ $q_{s,t} + 1$'s, and the rows above and below are filled with $q_s$'s and $q_s + 1$'s, respectively.*
- *In the $(t - q_{s,t})$th row of the submatrix of $\beta(i,j)$, there are $(c - r_{s,t})$ $q_s$'s and $r_{s,t}$ $q_{s,t} + 1$'s, and the rows above and below filled with $q_s$'s and $q_s + 1$'s, respectively.*

From Property 20, we have

**Lemma 21.** *For all $s$ and $t$ such that* $q_{s-1} < q_s = q_{s+1} = \cdots = q_t < q_{t+1}$,

$$\sum_{i=s}^{t} \sum_{j=0}^{c-1} (i+j)\beta(i,j) \le \sum_{i=s}^{t} \sum_{j=0}^{c-1} (i+j)\alpha(i,j) \le \sum_{i=s}^{t} \sum_{j=0}^{c-1} (i+j)\lfloor \frac{n_i + j}{c} \rfloor.$$

From Lemma 21, we have the following corollary:

**Corollary 22.**

$$\sum_{i=0}^{c-1} \sum_{j=0}^{c-1} (i+j)\beta(i,j) \le \sum_{i=0}^{c-1} \sum_{j=0}^{c-1} (i+j)\beta(i,j) \le \sum_{i=0}^{c-1} \sum_{j=0}^{c-1} (i+j)\lfloor \frac{n_i + j}{c} \rfloor.$$

Now, we will prove Lemma 4.

*Proof.* Since $fc \leq gc$ from the definition, it suffices for the lemma to prove $fc \geq gc$. We prove $fc \geq gc$ by induction. Obviously, $fc(n) = gc(n)$ if $n \leq c$. We assume that for all $i (< n)$, $fc(i) \geq gc(i)$ holds, and will prove $fc(n) \geq gc(n)$. For all $n_0, n_1, \ldots, n_{c-1}$ such that $n_0 \leq n_1 \leq \ldots \leq n_{c-1} < n$ and $n > c$, the following relation holds:

$$\sum_{i=0}^{c-1} \{gc(n_i) + (c - i - 1)n_i\}$$

$$\leq \sum_{i=0}^{c-1} fc(n_i) + \sum_{i=0}^{c-1} (c - i - 1)n_i \qquad \text{(from the inductive assumption)}$$

$$= \sum_{i=0}^{c-1} \sum_{j=0}^{c-1} \{fc(\lfloor \frac{n_i + j}{c} \rfloor) + (c - j - 1)\lfloor \frac{n_i + j}{c} \rfloor\} + \sum_{i=0}^{c-1} (c - i - 1)n_i$$

$$= \sum_{i=0}^{c-1} \sum_{j=0}^{c-1} fc(\lfloor \frac{n_i + j}{c} \rfloor) + (2c - 2)n - \sum_{i=0}^{c-1} \sum_{j=0}^{c-1} (i + j)\lfloor \frac{n_i + j}{c} \rfloor$$

(from Lemma 16)

Furthermore, we have

$$fc(n) = \sum_{j=0}^{c-1} \{fc(\lfloor \frac{n + j}{c} \rfloor) + (c - j - 1)\lfloor \frac{n + j}{c} \rfloor\}$$

$$\geq \sum_{j=0}^{c-1} gc(\frac{n + j}{c}) + \sum_{j=0}^{c-1} (c - j - 1)\lfloor \frac{n + j}{c} \rfloor$$

(from the inductive assumption)

$$\geq \sum_{i=0}^{c-1} \sum_{j=0}^{c-1} \{gc(\beta(i,j)) + (c - i - 1)\beta(i,j)\} + \sum_{i=0}^{c-1} \sum_{j=0}^{c-1} (c - j - 1)\beta(i,j)$$

(from Properties 18 and 19)

$$= \sum_{i=0}^{c-1} \sum_{j=0}^{c-1} gc(\lfloor \frac{n_i + j}{c} \rfloor) + (2c - 2)n - \sum_{i=0}^{c-1} \sum_{j=0}^{c-1} (i + j)\beta(i,j).$$

(from Lemma 16 and Property 17)

Thus, from $fc \leq gc$ and Corollary 22, we have:

$$\sum_{i=0}^{c-1} \{gc(n_i) + (c - i - 1)n_i\} \leq fc(n).$$

Therefore, $gc(n) \leq fc(n)$ holds. $\qquad \square$

# Graph Ear Decompositions and
# Graph Embeddings
## (Extended Abstract)

Jianer Chen *    and    Saroja P. Kanchi **

Department of Computer Science, Texas A&M University
College Station  TX 77843-3112, USA

**Abstract.** Ear decomposition of a graph has been extensively studied in relation to graph connectivity. In this paper, a connection of ear decomposition to graph embeddings is exhibited. It is shown that constructing a maximum-paired ear decomposition of a graph and constructing a maximum-genus embedding of the graph are $O(e \log n)$-time equivalent. This gives a polynomial time algorithm for constructing a maximum-paired ear decomposition.

## 1 Introduction

An ear decomposition of a graph is a way of partitioning the edge set of the graph into an ordered collection of edge-disjoint simple paths, called ears. It is well-known that a graph has an ear decomposition if and only if it is 2-edge connected [19].

Ear decomposition of a graph has received much attention recently because of its close relationship to graph connectivity. Lovasz [12] first noted that ear decompositions can be found quickly in parallel. Ear decompositions have been used in designing efficient sequential and parallel algorithms for 2-edge connectivity, 2-vertex connectivity, 3-vertex connectivity [15], and 4-vertex connectivity [11].

Variations of ear decomposition of a graph have also been proposed. Chen and Gross [2] introduced the concept of 2-connected semi-simplicial ear decomposition to study the average genus of a 2-connected simple graph. Chen, Kanevsky, and Tammassia [6] developed a linear time algorithm for constructing a 3-connected ear decomposition, which has applications in graph connectivity and planar graph embeddings. Cheriyan and Maheshwari [7] use nonseparating ear decomposition to develop efficient algorithms finding independent spanning trees in a graph. Ear decompositions for 4-connected graphs have also been studied [5, 18].

The *maximum genus* $\gamma_M(G)$ of a graph $G$ is defined to be the maximum integer $k$ such that there exists a cellular embedding of $G$ into the orientable surface of genus $k$. Since the introductory investigation by Nordhaus, Stewart, and White [13], maximum genus embeddings of a graph have been extensively studied (for a survey, see Ringeisen [16]). Certain graph classes can be precisely characterized by their maximum genus. For example, a graph has maximum genus 0 if and only if it is a cactus [16], and a 2-edge-connected graph has maximum genus 1 if and only if it is a necklace (with five exceptions) [3]. Recent investigations on maximum genus

---

\* Supported in part by the National Science Foundation under Grant CCR-9110824.
\*\* Supported in part by the Engineering Excellence Award from Texas A&M University.

have focused on developing efficient algorithms for maximum genus embeddings of a graph. The first polynomial time algorithm for constructing a maximum genus embedding of a graph was recently developed by Furst, Gross, and McGeoch [8] based on a characterization of the maximum genus of a graph given by Xuong [20] and an efficient matroid parity algorithm by Gabow and Stallmann [9]. A linear-time algorithm for a maximum genus embedding of a graph of bounded maximum genus was developed by Chen [1].

In this paper, we exhibit an interesting connection between ear decompositions and maximum genus embeddings of a graph. We introduce the concept of maximum-paired ear decomposition of a graph (the precise definition will be given in Section 2). We prove that a maximum-paired ear decomposition of a graph $G$ is $k$-paired if and only if the maximum genus of the graph $G$ is $k$. Then we show by developing efficient algorithms that up to an $O(e \log n)$ time computation, constructing a maximum-paired ear decomposition of a graph is equivalent to constructing a maximum genus embedding of the graph. Since a maximum genus embedding of a graph can be constructed in polynomial time [8], our results imply the first polynomial time algorithm for constructing a maximum-paired ear decomposition of a graph.

Our constructions are through a special spanning tree of a graph, the Xuong tree. We first show that constructing a maximum-paired ear decomposition of a graph is linear-time related to constructing a Xuong tree of the graph. Then we exhibit that up to an $O(e \log n)$ time computation, constructing a Xuong tree of a graph is equivalent to constructing a maximum genus embedding of the graph. Our constructions between a Xuong tree and a maximum genus embedding are based on several new techniques developed for representing and constructing a graph embedding.

We point out that our results lose no generality with the restriction that the graph has an ear decomposition, i.e., the graph should be 2-edge-connected. In fact, the maximum genus of a graph is equal to the sum of maximum genera of its 2-edge-connected components [3]. Thus, our results can be applied directly to each 2-edge-connected component of a graph if it is not 2-edge-connected.

## 2 Preliminaries and definitions

It is assumed that the reader is somewhat familiar with the fundamental of graph embeddings. For further description, see Gross and Tucker [10].

A *graph* may have multiple adjacencies or self-adjacencies. An *embedding* must have the "cellularity property" that the interior of every face is simply connected.

A *rotation* at a vertex $v$ is a cyclic permutation of the edge-ends incident on $v$. Thus, a $d$-valent vertex admits $(d - 1)!$ rotations. A list of rotations, one for each vertex of the graph, is called a *rotation system*.

An embedding of a graph $G$ in an orientable surface induces a rotation system, as follows: the rotation at vertex $v$ is the cyclic permutation corresponding to the order in which the edge-ends are traversed in an orientation-preserving tour around $v$. Conversely, by the Heffter-Edmonds principle, every rotation system induces a unique embedding of $G$ into an orientable surface. This bijectivity enables us to study graph embeddings based on graph rotation systems. We will interchangeably use the phrases "an embedding of a graph" and "a rotation system of a graph". In particular, if $\Pi(G)$ is a rotation system of a graph $G$, we will denote by $\gamma(\Pi(G))$ the genus of the corresponding embedding of the graph $G$.

A rotation system of a graph $G$ can be represented by a doubly-connected-edge-list (DCEL) $L$ as follows. Each item in $L$ corresponds to an edge $e = [u, v]$ of $G$ and consists of six components: the two edge-ends $u$ and $v$, the two faces $f_1$ and $f_2$ that have $e$ as a boundary edge ($f_1$ may be identical to $f_2$), the pointers to the items in $L$ that correspond to the edges following $e$ in the rotations at $u$ and $v$, respectively. (For more detailed discussions for DCEL, see [14].) There is a linear time algorithm to trace the boundary walks of all faces in a rotation system given by a DCEL [4, 14]. Therefore, given a rotation system $\Pi(G)$ of a graph $G = (V, E)$, the genus of $\Pi(G)$ can be calculated in time $O(|E|)$ using the Euler polyhedral equation

$$|V| - |E| + |F| = 2 - 2\gamma(\Pi(G))$$

where $|F|$ is the number of faces in the embedding $\Pi(G)$.

Let $T$ be a spanning tree of a graph $G$. The edge complement $G - T$ will be called a *co-tree*. The number of edges in any co-tree is known as the *cycle rank* of $G$, denoted $\beta(G)$. For each edge $e$ in the co-tree $G - T$, the unique cycle in the graph $T \cup \{e\}$ is called *the fundamental cycle* of $e$ with respect to the spanning tree $T$.

The *deficiency* $\xi(G, T)$ of a spanning tree $T$ for a graph $G$ is defined to be the number of components of $G - T$ that have an odd number of edges. The *deficiency* $\xi(G)$ of the graph $G$ is defined to be the minimum of $\xi(G, T)$ over all spanning trees $T$ of $G$. A spanning tree $T$ is a *Xuong tree* if $\xi(G, T) = \xi(G)$. Xuong [20] obtained a characterization of maximum genus of a graph in terms of deficiency of the graph.

**Theorem 1.** [20] *Let $G$ be a connected graph. The maximum genus of $G$ is given by the formula*

$$\gamma_M(G) = \frac{1}{2}(\beta(G) - \xi(G))$$

We now describe the effect of inserting a new edge into an embedded graph. Let $\Pi(G)$ be an embedding of a connected graph $G$. Suppose that we insert a new edge $e = [u, v]$ into the embedding $\Pi(G)$, where $u$ and $v$ are vertices of $G$. There are two possible cases.

If the edge-ends $u$ and $v$ of $e$ are inserted between two corners of the same face $f$, then the new edge $e$ splits the face $f$ into two faces. More precisely, if the boundary walk around the face $f$ in $\Pi(G)$ is of the form $u\alpha v\beta u$, where $\alpha$ and $\beta$ are subwalks, then the new edge $e$ splits the boundary walk of $f$ into two walks: $u\alpha veu$ and $v\beta ueu$, resulting in two new faces. Since both the number of faces and the number of edges are increased by 1, with the number of vertices unchanged, by the Euler polyhedral equation, the embedding genus remains the same.

If the edge-ends $u$ and $v$ of $e$ are inserted between corners of two different faces $f_1$ and $f_2$, then both these faces are merged by $e$ into one larger face. In particular, suppose that the edge $e$ runs from the corner of $u$ in face boundary walk $u\alpha u$ of $f_1$ to the corner of $v$ in face boundary walk $v\beta v$ of $f_2$, then the merged face has boundary walk $uev\beta veu\alpha u$. In this case, since the number of faces is decreased by 1 and the number of edges is increased by 1, with the number of vertices unchanged, by the Euler polyhedral equation, the embedding genus is increased by 1.

We point out that inserting the edge $e$ into the embedding $\Pi(G)$ never decreases the embedding genus.

An *ear decomposition* $D = [P_1, P_2, \cdots, P_r]$ of a graph is a partition of its edge set into an ordered collection of edge-disjoint simple paths $P_1, P_2, \cdots, P_r$ such that $P_1$ is a simple cycle and $P_i$, $i \geq 2$, is a path with only its endpoints in common with $P_1 \cup \cdots, \cup P_{i-1}$. Each $P_i$ is called an *ear*. A *pairing* in an ear decomposition $D = [P_1, P_2, \cdots, P_r]$ is a partition of the ears into pairs of *matched ears* and *single ears* such that each pair of matched ears is of the form $\{P_i, P_{i+1}\}$ where the ear $P_{i+1}$ has an endpoint on the ear $P_i$. A *maximum pairing* of the ear decomposition $D$ is a pairing that maximizes the number of pairs of matched ears. It can be shown that given an ear decomposition $D$, a maximum pairing of $D$ can be constructed in linear time by a greedy algorithm. If a maximum pairing of an ear decomposition $D$ has $k$ pairs of matched ears, we say that the ear decomposition $D$ is *k-paired*. A *maximum-paired ear decomposition* of a graph $G$ is a $k$-paired ear decomposition with $k$ the largest integer over all ear decompositions of $G$.

## 3    Maximum-paired ear decomposition and Xuong tree

In this section, we first develop efficient algorithms to show that constructing a maximum-paired ear decomposition and constructing a Xuong tree are linear-time related. Based on these results, a conclusion is derived that a maximum-paired ear decomposition of a graph is $k$-paired if and only if the maximum genus of the graph is $k$.

Let $H$ be a (not necessarily connected) graph. Two edges in the graph $H$ are *adjacent* if they have an endpoint in common. An *adjacency matching* in $H$ is a partition of edges of $H$ into groups of one or two edges, called *1-groups* and *2-groups*, respectively, such that the two edges in each 2-group are adjacent. We say that the two edges in the same 2-group are *matched*, and the edge in a 1-group is *unmatched*. A *maximum adjacency matching* in $H$ is an adjacency matching that maximizes the number of 2-groups.

**Lemma 2.** *In a maximum adjacency matching of a graph $H$, the number of unmatched edges is equal to the number of components of $H$ that have an odd number of edges.*

We leave to our readers to verify that a maximum adjacency matching of a graph can be constructed in linear time.

Let $G$ be a 2-edge-connected graph and let $T$ be a Xuong tree of $G$. We name each vertex in $G$ by its preorder number in the tree $T$. Given a pair of vertices $u$ and $v$, we denote by $lca(u, v)$ the least common ancestor of $u$ and $v$. We also say that a vertex $w$ is the least common ancestor of a co-tree edge $e = [u, v]$ if $w = lca(u, v)$. Note that the least common ancestor of a co-tree edge $e$ is the smallest vertex in the fundamental cycle of $e$.

**Algorithm**    *Xuong-to-Ears*

Input:   A Xuong tree $T$ of a 2-edge-connected graph $G$
Output:   An ear decomposition of $G$

1. Rename each vertex of $G$ by its preorder number in the tree $T$. Construct the co-tree $H = G - T$.
2. Construct a maximum adjacency matching $\mathcal{M}$ in $H$.
3. Sort all edges in the co-tree $H$ by their least common ancestors. Let the sorted list be $L_1$.
4. We assign each co-tree edge $e$ a number $num(e)$ as follows: for each pair of matched edges $L_1(i)$ and $L_1(j)$ in $\mathcal{M}$, where $i < j$, let $num(L_1(i)) = 2i$ and $num(L_1(j)) = 2i+1$; for each unmatched edge $L_1(k)$ in $\mathcal{M}$, let $num(L_1(k)) = 2k$.
5. Sort all co-tree edges by their new assigned numbers $num(\ )$. Let the sorted list be $L_2$. Now for each co-tree edge $e = L_2(i)$, assign $ear(e) = i$.
6. For each tree edge $e'$, assign $ear(e') = min\{ear(e)\}$, where the minimum is taken over all co-tree edges $e$ whose fundamental cycle contains $e'$.

**Lemma 3.** *If the graph $G$ is 2-edge-connected, then every edge $e$ in $G$ is assigned a unique ear number $ear(e)$ in the algorithm Xuong-to-Ears.*

**Lemma 4.** *The algorithm Xuong-to-Ears constructs a valid ear decomposition for the graph $G$. That is, there is an ear decomposition $D = [P_1, P_2, \cdots, P_r]$ of the graph $G$ such that the ear $P_i$ consists of exactly those edges whose ear number assigned by the algorithm Xuong-to-Ears is $i$.*

*Proof.* (Sketch)    Let the co-tree edges sorted in the list $L_2$ be $e_1, e_2, \cdots, e_r$. First note that by our construction, only edges in the fundamental cycle of the co-tree edge $e_i$ may be assigned an ear number $ear(e_i)$, and that if an edge in the fundamental cycle of $e_i$ has an ear number different from $ear(e_i)$, then the edge must have an ear number less than $ear(e_i)$. In particular, the fundamental cycle of $e_1$ contains the root of the tree $T$ and every edge in it is assigned an ear number $ear(e_1) = 1$. This forms the first ear $P_1$.

Consider the co-tree edge $e_i = [x_i, y_i]$, $i > 1$, with $ear(e_i) = i$. Let the directed tree paths (directed from child to parent) from $x_i$ and $y_i$ to the $lca(x_i, y_i)$ be $Q_x$ and $Q_y$, respectively. If all of the tree edges on $Q_x$ and $Q_y$ are assigned ear number $i$, then the ear $P_i$ is the fundamental cycle of $e_i$. Note that if the vertex $lca(x_i, y_i)$ is not the root of $T$, then it must be contained in the fundamental cycle of another co-tree edge $e_j$, with $j < i$.

Suppose that some edge in the fundamental cycle of $e_i$ has an ear number less than $ear(e_i)$. With a careful analysis, we can show that there are two vertices $v_x$ and $v_y$ on the paths $Q_x$ and $Q_y$, respectively, such that all edges before the vertex $v_x$ (resp. $v_y$) on the path $Q_x$ (resp. $Q_y$) are assigned ear number $ear(e_i) = i$ and all edges after the vertex $v_x$ (resp. $v_y$) on the path $Q_x$ (resp. $Q_y$) are assigned ear number less than $ear(e_i) = i$. Therefore, the path between the vertices $v_x$ and $v_y$ that contains the edge $e_i$ in the fundamental cycle of $e_i$ forms the ear $P_i$.    □

We show that algorithm *Xuong-to-Ears* actually constructs a maximum-paired ear decomposition for the graph $G$.

Let $H = G - T$ be the co-tree and let $\mathcal{M}$ be the maximum adjacency matching in $H$ constructed by the algorithm. By Lemma 2, the number of unmatched edges in $\mathcal{M}$

is equal to the deficiency $\xi(G)$ of the graph $G$. Thus, the number of pairs of matched edges in $\mathcal{M}$ is $(\beta(G) - \xi(G))/2$, which is equal to $\gamma_M(G)$ by Theorem 1. Since each pair of matched edges in $\mathcal{M}$ are adjacent in the list $L_2$, their ear numbers differ by exactly 1. Thus, their corresponding ears are consecutive in the ear decomposition. Moreover, obviously they have at least one vertex in common. This gives us the following lemma immediately.

**Lemma 5.** *The ear decomposition constructed by algorithm Xuong-to-Ears is $k$-paired, where $k \geq \gamma_M(G)$.*

In fact, the integer $k$ in Lemma 5 cannot be larger than $\gamma_M(G)$ because of the following lemma.

**Lemma 6.** *If a graph $G$ has a $k$-paired ear decomposition, the $\gamma_M(G) \geq k$.*

*Proof.* (Sketch) Let $D = [P_1, P_2, \cdots, P_r]$ be a $k$-paired ear decomposition of the graph $G$ and let $\mathcal{P}$ be a maximum pairing of $D$. We construct an embedding of $G$ by first embedding the cycle $P_1$ into the plane then inserting the ears $P_2, \cdots, P_r$ one by one in that order into the embedding. For each pair of matched ears in $\mathcal{P}$, we can always insert them in a way that increases the embedding genus by at least 1. Therefore, this construction will result in an embedding of genus at least $k$. □

Now we are ready for our first theorem in this section.

**Theorem 7.** *The algorithm Xuong-to-Ears constructs a maximum-paired ear decomposition for the graph $G$.*

*Proof.* Lemma 6 shows that the graph $G$ does not have a $k$-paired ear decomposition, with $k > \gamma_M(G)$. While Lemma 5 and Lemma 6 together show that the algorithm Xuong-to-Ears constructs a $\gamma_M(G)$-paired ear decomposition for the graph $G$. □

**Corollary 8.** *A 2-edge-connected graph $G$ has maximum genus $k$ if and only if every maximum-paired ear decomposition of $G$ is $k$-paired.*

We now analyze the algorithm Xuong-to-Ears. Step 1 can be trivially done in linear time. Step 2 can be done in linear time as we discussed. To compute the least common ancestor for each co-tree edge, we use Schieber and Vishkin's algorithm [17], which can compute the least common ancestor of any two vertices in constant time with a linear time preprocessing. All sortings in the algorithm can be implemented using Bucket Sorting that takes linear time. Finally, to assign ear numbers to the tree edges, we pick the co-tree edges sorted in the list $L_2$. For each co-tree edge $e_i = [x_i, y_i]$, we traverse the two tree paths $Q_x$ and $Q_y$ from the vertices $x_i$ and $y_i$, respectively, to the least common ancestor of $e_i$, and stop at a vertex that belongs to an ear of smaller index. By the proof of Lemma 4, exactly those edges traversed in the process should be assigned the ear number $ear(e_i)$. This process can obviously be done in linear time. This gives us the following theorem.

**Theorem 9.** *A maximum-paired ear decomposition of a graph can be constructed from a Xuong tree of the graph in linear time.*

Furst, Gross, and McGeoch [8] have developed a polynomial time algorithm that constructs a Xuong tree for a graph. This result with Theorem 9 gives us a polynomial time algorithm for constructing a maximum-paired ear decomposition given a 2-edge-connected graph.

**Theorem 10.** *There is a polynomial time algorithm that, given a 2-edge-connected graph $G$, constructs a maximum paired ear decomposition for $G$.*

Now we consider the converse of Theorem 9, that is, constructing a Xuong tree from a maximum-paired ear decomposition.

**Lemma 11.** *Let $D = [P_1, P_2, \cdots, P_r]$ be an ear decomposition of a graph $G$. Let $G_D$ be a subgraph of $G$ obtained by deleting one edge from each ear in $D$. Then the graph $G_D$ is a connected spanning subgraph of $G$.*

*Proof.* (Sketch) Since no vertices are deleted, the graph $G_D$ is a spanning subgraph of the graph $G$. To show that the graph $G_D$ is connected, we use induction on the number of ears in the ear decomposition $D$. □

**Corollary 12.** *Let $D = [P_1, P_2, \cdots, P_r]$ be an ear decomposition of a graph $G$. Let $S$ be a subset of ears in $D$. Let $G_S$ be a subgraph of $G$ obtained by deleting one edge from each ear in $S$. Then the graph $G_S$ is a connected spanning subgraph of $G$.*

Now we are ready for the following algorithm.

**Algorithm** *Ears-to-Xuong*

Input: A maximum-paired ear decomposition $D$ of $G$
Output: A Xuong tree $T$ of the graph $G$

1. Construct a maximum pairing $\mathcal{P}$ in the ear decomposition $D$. Let the $\gamma_M(G)$ pairs of matched ears in $\mathcal{P}$ be $\{P_{i_j}, P_{i_j+1}\}$, $1 \le j \le \gamma_M(G)$.
2. For each pair $\{P_{i_j}, P_{i_j+1}\}$, let $v_j$ be a vertex shared by $P_{i_j}$ and $P_{i_j+1}$, delete an edge $e_{i_j}$ in $P_{i_j}$ and an edge $e_{i_j+1}$ in $P_{i_j+1}$ such that both edges are incident on the vertex $v_j$.
3. Let $G_S$ be the subgraph of $G$ after the edge deletions in Step 2, construct a spanning tree $T$ in $G_S$. $T$ is a Xuong tree of $G$.

It is obvious that the algorithm *Ears-to-Xuong* runs in linear time. Moreover, by Corollary 12, the graph $G_S$ is a connected spanning subgraph of $G$. Therefore, the spanning tree $T$ of $G_S$ is also a spanning tree of the graph $G$. Finally, since in the co-tree $G-T$, there are $\gamma_M(G)$ pairs of adjacent edges $e_{i_j}$ and $e_{i_j+1}$, $1 \le j \le \gamma_M(G)$, the number of components in $G - T$ that has an odd number of edges cannot be larger than $\beta(G) - 2\gamma_M(G) = \xi(G)$. We conclude that the spanning tree $T$ is a Xuong tree of the graph $G$.

**Theorem 13.** *A Xuong tree of a graph can be constructed from a maximum-paired ear decomposition of the graph in linear time.*

**Theorem 14.** *Constructing a Xuong tree of a graph and constructing a maximum-paired ear decomposition of the graph are linear-time equivalent.*

In a pairing of an ear decomposition $D$, pairs of matched ears and single ears are in general interlaced. The *canonical pairing* of $D$ is a pairing of $D$ that pairs the first $2k$ ears in the ear decomposition $D$ into $k$ pairs of matched ears and let all other ears be single ears, with $k$ being the largest possible integer. The canonical pairing of $D$ can be written as $D = [P_1, P_1', \cdots, P_k, P_k', S_1, \cdots, S_t]$, where $\{P_i, P_i'\}$ are pairs of matched ears, $1 \leq i \leq k$, and $S_j$ are single ears, $1 \leq j \leq t$.

We show that Corollary 8 is still valid if we restrict to canonical pairings of ear decompositions.

**Theorem 15.** *Every 2-edge-connected graph $G$ has an ear decomposition $D$ whose canonical pairing has $\gamma_M(G)$ pairs of matched ears.*

## 4 Maximum genus embedding and Xuong tree

Furst, Gross, and McGeoch [8] have developed an $O(e^2)$ time algorithm that constructs a maximum genus embedding of a graph from a Xuong tree of the graph. In this section, we first give an improved version of their algorithm, which runs in time $O(e \log n)$. Then we show that a Xuong tree of a graph can also be constructed from a maximum genus embedding of the graph in time $O(e \log n)$. Together with the results in Section 3, we conclude that constructing a maximum genus embedding and constructing a maximum-paired ear decomposition are $O(e \log n)$-time equivalent.

Our algorithm for constructing a maximum genus embedding from a Xuong tree is very similar to the algorithm given in [8], except that we use a more efficient representation for graph embeddings. Moreover, our representation of graph embeddings and the standard DCEL representation of graph embeddings can be transformed from one to the other in linear time.

Each edge $e = [u, v]$ is given two *edge sides* $l(e)$ and $r(e)$, representing the two possible orientations of the edge. In an embedding of the graph $G$, each edge side appears exactly once in the face boundary walks. We use a balanced search tree to represent each face in an embedding so that the operations SPLIT, MERGE, SEARCH, and INSERT can be performed in logarithmic time (see [14]). Moreover, if we traverse the leaves of the balanced search tree from left to right, we are following the order of traversing the boundary walk of the face. We will call this new representation of graph embeddings a *balanced tree representation* of graph embeddings.

Given a standard DCEL representation of a graph embedding, we can construct the balanced tree representation by first tracing all face boundary walks in the embedding then building up a balanced search tree for each face boundary walk. Conversely, given a balanced tree representation of a graph embedding, we can trace the edge sides on each face boundary walk. For each vertex $v$ and an edge side $[u, v]$ in a face boundary walk, the following edge $[v, w]$ in the face boundary walk is the next edge in the corresponding rotation system at vertex $v$.

Now we describe how to construct a maximum genus embedding from a Xuong tree, which is very similar to the one presented in [8].

**Algorithm** *Xuong-to-Embedding*

    Input:   A Xuong tree $T$ of a graph $G$

    Output:   A maximum genus embedding of $G$

1. Construct a maximum adjacency matching $\mathcal{M}$ in the co-tree $G-T$. Let the pairs of matched edges in $\mathcal{M}$ be $\{e_i, e_i'\}$, $1 \le i \le \gamma_M(G)$. Denote by $G_i$ the subgraph $T \cup \{e_1, e_1', \cdots, e_i, e_i'\}$, $0 \le i \le \gamma_M(G)$.
2. Construct an arbitrary embedding $\Pi_0(G_0)$ of the Xuong tree $G_0 = T$. The embedding $\Pi_0(G_0)$ has only one face.
3. For $i = 1, \cdots, \gamma_M(G)$, on the one-face embedding $\Pi_{i-1}(G_{i-1})$ of $G_{i-1}$, arbitrarily insert the edge $e_i$, which splits the face $f$ of $\Pi_{i-1}(G_{i-1})$ into two faces $f_1$ and $f_2$. Now insert the edge $e_i'$ to merge the two faces $f_1$ and $f_2$ into a single face. This gives a one-face embedding $\Pi_i(G_i)$ for the graph $G_i$.
4. Arbitrarily insert unmatched edges in $\mathcal{M}$.

Explanations are needed for the above algorithm. To construct a balanced tree representation for the embedding $\Pi_0(G_0)$ of the Xuong tree $T$, the ordering of a Depth-First search traversing on the tree $T$ is sufficient. To split the single face $f$ of the embedding $\Pi_{i-1}(G_{i-1})$ into two faces $f_1$ and $f_2$ by the edge $e_i = [x_i, y_i]$, suppose that the face boundary walk of $f$ is $x_i \alpha y_i \beta x_i$. We split the list $x_i \alpha y_i \beta x_i$ into two lists and add the two sides of $e_i$ to the two lists: $x_i \alpha y_i l(e_i) x_i$ and $y_i \beta x_i r(e_i) y_i$, which correspond to the two new created faces $f_1$ and $f_2$, respectively. Finally, suppose that the edge $e_i'$ shares the endpoint $x_i$ with the edge $e_i$ and $e_i' = [x_i, z_i]$, then we locate the vertex $z_i$ in one of the faces, then insert the end $x_i$ of $e_i'$ into the other face. More precisely, suppose that we have located $z_i$ in the face $f_1 = x_i \delta_1 z_i \delta_2$, and face boundary walk of $f_2$ is $x_i \mu x_i$, then the resulting merged face is $x_i \delta_1 l(e_i') \mu r(e_i') \delta_2 x_i$. Note that the way we insert the edges $e_i$ and $e_i'$ increases the embedding genus by exactly 1. Since insertion of the pair $\{e_i, e_i'\}$ involves at most constant number of INSERT, SPLIT, MERGE, and SEARCH operations, we conclude that the algorithm *Xuong-to-Embedding* runs in time $O(e \log n)$.

We now consider how to construct a Xuong tree of a graph from a maximum genus embedding of the graph.

Given a maximum genus embedding $\Pi(G)$ of a graph $G$, we construct a Xuong tree of the graph $G$ in two steps. First, we find a spanning subgraph $G'$ of $G$ such that the induced embedding $\Pi'(G')$ of $G'$ from $\Pi(G)$ is of genus $\gamma_M(G)$ and has only one face. Secondly, we construct a Xuong tree of $G$ from the embedding $\Pi'(G')$.

To construct the spanning subgraph $G'$ and the embedding $\Pi'(G')$, we traverse the embedding $\Pi(G)$ and delete those edges that sit on the boundary of two different faces. Note that this process must be done sequentially. For example, suppose that both edges $e_1$ and $e_2$ are on the boundary of two faces $f_1$ and $f_2$, then after deleting the edge $e_1$, the two faces $f_1$ and $f_2$ are merged into a single face $f$ (this is the inverse operation of inserting the edge $e_1$ to split the face $f$ into two faces $f_1$ and $f_2$), the edge $e_2$ is no longer on the boundary of two different faces. Therefore, after deleting an edge on the boundary of two different faces in an embedding, we must retrace the new embedding to find those edges that are on the boundary of two different faces of the new embedding. A straightforward implementation of this algorithm would take time $O(e^2)$. We show below that the spanning subgraph $G'$ can be constructed in linear time.

**Algorithm** *Upper-embeddable-Subgraph*

Input: An maximum genus embedding $\Pi(G)$ of $G$

Output: The induced embedding $\Pi'(G')$ of the spanning subgraph $G'$

1. Trace the face boundary walks of the embedding $\Pi(G)$ and assign each edge a pair $(f_1, f_2)$, where $f_1$ and $f_2$ are the faces to which the two sides of the edge belong, $f_1 \leq f_2$ (suppose the face names are given by the integers $1, 2, \cdots$.)
2. For each pair $(f_1, f_2)$ assigned in Step 1 such that $f_1 < f_2$, pick an edge assigned with that pair. Let $S_0$ be the set of all such edges. Construct a list $L$ such that $L[i]$ is the list of all edges in the set $S_0$ that are assigned a pair of form $(i, f)$.
3. While the list $L$ is not empty, do Step 4 and Step 5.
4. Pick a nonempty list $L[i]$ such that either $i = 1$ or face $i$ has been merged into face 1. For each pair $(i, f)$ in the list $L[i]$, if face $f$ has not yet been merged into face 1, then exclude the corresponding edge from the subgraph $G'$, and mark face $f$ as having been merged into face 1. Delete the pair $(i, f)$ from the list $L[i]$.
5. delete the list $L[i]$ from the list $L$, and go back to Step 3.

The set $S_0$ in Step 2 can be constructed by Bucket sorting all edges of $G$ lexicographically by their assigned pairs $(f_1, f_2)$. To find the list $L[i]$ in Step 4 in constant time, we can link all faces that have been merged into face 1. Therefore, the algorithm *Upper-embeddable-Subgraph* runs in linear time.

Finally, we show how we construct a Xuong tree of the graph $G$ from the maximum genus embedding $\Pi'(G')$ of the maximum upper-embeddable subgraph $G'$.

Let the face boundary walk of the unique face of $\Pi'(G')$ be $l(e_1)l(e_2)\delta_1 r(e_1)\delta_2$, where $\delta_2$ is not empty and no edge has both its edge sides contained in $l(e_2)\delta_1$. Let $G_1 = G' - \{e_1\}$ and $G_2 = G' - \{e_1, e_2\}$. We claim that deleting the edges $e_1$ and $e_2$ from $G'$ results in a connected spanning subgraph of $G$. First of all, all edges that have an edge side in $l(e_2)\delta_1$ are contained in the same connected component of $G_1$, and all edges that have an edge side in $\delta_2$ are contained in the same connected component of $G_1$. Moreover, since no edge has both its edge sides contained in $l(e_2)\delta_1$, the edge $e_2$ has one edge side in $l(e_2)\delta_1$ and one edge side in $\delta_2$. Therefore, all edges in $G_1$ are in the same connected component, i.e., the graph $G_1$ is connected. Deleting the edge $e_1$ from the embedding $\Pi'(G')$ results in a two-face embedding $\Pi_1(G_1)$ of the graph $G_1$ with $l(e_2)\delta_1$ and $\delta_2$ being the two face boundary walks, respectively, where the edge $e_2$ is on the boundary of the two faces. Since no cutedge can be on the boundary of two different faces in any embedding, we conclude that the graph $G_2$ is connected, thus a connected spanning subgraph of the graph $G$. Moreover, removing the edge $e_2$ from the embedding $\Pi_1(G_1)$ results in a one-face embedding $\Pi_2(G_2)$ of the graph $G_2$, and the genus of $\Pi_2(G_2)$ is one less than that of $\Pi'(G')$. Also note that the graph $G'$ is the graph $G_2$ plus two adjacent edges $e_1$ and $e_2$.

If an edge $e$ in $G'$ has a degree-1 endpoint, we can simply ignore it in the later consideration.

We summarize the above discussions into the following algorithm.

**Algorithm**  *Embedding-to-Xuong*

  Input:   An maximum genus embedding $\Pi(G)$ of $G$
  Output:  A Xuong tree $T$ of $G$

1. Construct a one-face embedding $\Pi'(G')$ of the maximum upper-embeddable subgraph $G'$ of the graph $G$.
2. Let $M = \emptyset$.
3. Find an edge $e$ whose edge sides are closest in the face boundary walk of $\Pi'(G')$.
4. If $e$ has a degree-1 endpoint, delete $e$ from $\Pi'(G')$ and go back to Step 3.
5. If the face boundary walk of $\Pi'(G')$ is $l(e)l(e')\delta_1 r(e)\delta_2$, where no edge has both its edge sides in $l(e')\delta_1$, then delete $e$ and $e'$ from $\Pi'(G')$ and add $e$ and $e'$ to the set $M$.
6. If $\Pi'(G')$ is not empty, go back to Step 3.
7. Let $G_0 = G - M$. Construct a spanning tree $T$ for $G_0$. The tree $T$ is a Xuong tree for the graph $G$.

Again, if the embeddings use balanced tree representation, then the algorithm *Embedding-to-Xuong* processes each edge of $G$ with constant number of SEARCH, SPLIT, MERGE, and INSERT operations. Thus, the algorithm *Embedding-to-Xuong* has time complexity $O(e \log n)$.

**Theorem 16.** *Constructing a maximum genus embedding of a graph and constructing a Xuong tree of the graph are $O(e \log n)$-time equivalent.*

By Theorem 14, constructing a Xuong tree of a graph is linear-time related to constructing a maximum-paired ear decomposition. Combining this with Theorem 16, we have

**Theorem 17.** *Constructing a maximum genus embedding of a graph and constructing a maximum-paired ear decomposition of the graph are $O(e \log n)$-time equivalent.*

## 5  Final remarks

We have presented an interesting connection between ear decompositions and graph embeddings, which gives a new characterization of maximum genus of a graph, as well as an efficient algorithm for constructing a maximum-paired ear decomposition.

Maximum genus is one of the most important topological invariants of a graph and is closely related to the isomorphism type of the graph [1, 2]. Moreover, maximum genus is also practically useful because it gives an embedding of the graph with the fewest faces. However, the complexity of computing the maximum genus is not well understood. Our results provide a possible approach that is different from the one by Furst, Gross, and McGeoch [8] to obtain more efficient sequential and parallel algorithms for computing the maximum genus of a graph. Note that it is an open problem whether computing the maximum genus of a graph is in *NC*.

Another open problem is whether counting the number of maximum genus embeddings of a graph is $\#P$-complete [8]. Our result has reduced this problem to counting the number of Xuong trees, which seems a more feasible problem.

# References

1. Chen, J.: A linear time algorithm for isomorphism of graphs of bounded average genus. Lecture Notes in Computer Science **657** (1993) 103-113
2. Chen, J., Gross, J. L.: Limit points for average genus (I): 3-connected and 2-connected simplicial graphs. J. Comb. Theory Ser. B **55** (1992) 83-103
3. Chen, J., Gross, J. L.: Kuratowski-type theorems for average genus. J. Comb. Theory Ser. B **57** (1993) 100-211
4. Chen, J., Gross, J. L., and Rieper, R. G.: Overlap matrices and imbedding distributions. Discrete Mathematics (1993) to appear
5. Chen, J. and Kanevsky, A.: On assembly of 4-connected graphs. Lecture Notes in Computer Science **657** (1993) 158-169
6. Chen, J., Kanevsky, A., Tamassia, R.: Linear time construction of 3-connected ear decomposition of a graph, The 22nd Southeastern International Conference on Combinatorics, Graph Theory, and Computing. Baton Rouge, Feb. 11-15 (1991)
7. Cheriyan, J. and Maheshwari, S. N.: Finding nonseparating induced cycles and independent spanning trees in 3-connected graphs. J. Algorithms **9** (1988) 507-537
8. Furst, M., Gross, J. L., and McGeoch, L. A.: Finding a maximum-genus graph imbedding. J. ACM **35-3** (1988) 523-534
9. Gabow, H. N. and Stallmann, M.: Efficient algorithms for graphic matroid intersection and parity. Lecture Notes in Computer Science **194** (1985) 210-220
10. Gross, J. L., Tucker, T. W.: Topological Graph Theory. Wiley-Interscience, New York (1987)
11. Kanevsky, A. and Ramachandran, V.: Improved algorithms for graph four-connectivity. J. Computer and System Sciences **42** (1991) 288-306
12. Lovasz, L.: Computing ears and branchings in parallel. Proc. 26th Annual IEEE Symposium on Foundations of Computer Science (1985) 464-467
13. Nordhaus, E., Stewart, B., and White, A.: On the maximum genus of a graph. J. Comb. Theory Ser. B **11** (1971) 258-267
14. Preparata, F. P. and Shamos, M. I.: Computational Geometry: An Introduction. Springer-Verlag (1985)
15. Ramachandran, V.: Parallel open ear decomposition with applications to graph biconnectivity and triconnectivity. in Synthesis of Parallel Algorithms, Ed. Reif, Morgan-Kaufmann (1993)
16. Ringeisen, R.: Survey of results on the maximum genus of a graph. J. Graph Theory **3** (1979) 1-13
17. Schieber, B. and Vishkin, U: On finding lowest common ancestors: simplification and parallelization. SIAM J. Computing **17** (1988) 1253-1262
18. Slater, P. J.: A classification of 4-connected graphs. J. Comb. Theory Ser. B **17** (1974) 281-298
19. Whitney, H.: Non-separable and planar graphs. Trans. Amer. Math. Soc. **34** (1932) 339-362
20. Xuong, N. H.: How to determine the maximum genus of a graph. J. Comb. Theory Ser. B **26** (1979) 217-225

# Improved Bounds for the Crossing Numbers on Surfaces of Genus $g$

Farhad Shahrokhi

Department of Computer Science, University of North Texas
P.O.Box 13886, Denton, TX, USA

Laszló A. Székely

Department of Computer Science, Eötvös University
H-1088 Budapest, Hungary

Ondrej Sýkora, Imrich Vrt'o*

Institute for Informatics, Slovak Academy of Sciences
Dubravská 9, 842 35 Bratislava, Slovak Republic

### Abstract

We give drawings of the complete graph on orientable and nonorientable surfaces of genus $g$ and improve the best known upper bounds on the crossing number of a complete graph on these surfaces by a factor $O(\log g)$. Morover, we give a polynomial time algorithm that produces drawings of arbitrary graphs using the drawings of complete graphs. Using our algorithm we establish an upper bound of $O(\frac{m^2 \log^2 g}{g})$ on the crossing number of any graph with $n$ vertices and $m$ edges on an orientable or non-orientable surface of genus $g$. This upper bound is within a factor of $O(\log^2 g)$ from the optimal for many classes of graphs.

## 1  Introduction

Crossing number of a graph is the minimum number of crossings of its edges when the graph is drawn on a surface. Very little is known even for drawings graphs on the sphere [2]. Concerning surfaces of higher genus, there are only a few results for complete and complete bipartite graphs drawn on the torus [4, 5] and octahedral and hypercube graphs drawn on surfaces of special genera [3, 6]. In practice, crossing numbers appear in the fabrication of VLSI circuits. The crossing number of a graph corresponding to the VLSI circuit has strong influence on the area of the layout as well as on the number of wire-contact cuts that should be minimized.

---

*Research of the third and the fourth author was partially supported by Grant No. 88 of Slovak Academy of Sciences and by EC Cooperative action IC1000 "Algorithms for Future Technologies" (Project ALTEC)

This paper reflects our ongoing research on the crossing number problem for drawing graphs on an orientable or non-orientable surface of genus $g \geq 0$ [8, 9, 10, 11, 12]. Our previous results included general lower and upper bounds and their applications to finding nearly optimal bounds for specific graphs like complete graphs, hypercube and their derivatives. In this paper we show that the complete graph $K_n$ can be drawn on the orientable or nonorientable surface of genus $g$ with at most $O(n^4 \log^2 g / g)$ crossings. Thus improving our former result by a factor of $O(\log g)$. The best lower bound is $\Omega(n^4/g)$. We then use this drawing of the complete graph to design a polynomial time algorithm that produces a drawing of any graph on $n$ vertices and $m$ edges with at most $O(\frac{m^2 \log^2 g}{g})$ crossings on an orientable or non-orientable surface of genus $g$. This improves our previous non-constructive upper bound [10] by a factor $O(\log g)$. For many well known classes of graphs our upper bounds is within a multiplicative factor of $O(\log^2 g)$ from the known lower bounds.

In Section 2 we define basic notations. Our main results are proved in Section 3 and we give our conclusions in Section 4.

## 2  Basic Notations

Our basic reference to graph theory is [1]. For any non-negative integer $g$, let $S_g$ and $N_g$ denote the orientable surface and non-orientable surface of genus $g$, respectively. Thus $S_g$ and $N_g$ are topologically equivalent to a sphere with $g$ handles and a sphere with $g$ crosscaps. Let $D$ be a drawing of a graph $G$ on $S_g$ or $N_g$. Let $cr(D)$ denote the number of edge crossings of $G$ in $D$. Let $cr_g(G)$ and $\check{cr}_g(G)$ denote the crossing number of $G$ on $S_g$ and $N_g$, respectively. Note that $cr_0(G)$ is the familiar planar crossing number.
Let $\gamma(G)$ and $\tilde{\gamma}(G)$ denote the orientable and non-orientable genus of $G$, respectively.
Let $G_1 = < V_1, E_1 >$ and $G_2 = < V_2, E_2 >$ be two undirected graphs, $|V_1| \leq |V_2|$. An embedding of $G_1$ in $G_2$ is a pair of injections $\omega = < \phi, \psi >$ satisfying

$\phi : V_1 \rightarrow V_2$ is an injection

$\psi : E_1 \rightarrow \{\text{set of all paths in } G_2\}$,

such that if $uv \in E_1$ then $\psi(uv)$ is a path between $\phi(u)$ and $\phi(v)$. For any $e \in E_2$ and any $u \in V_2$ define

$$\mu_\omega(e) = |\{f \in E_1 : e \in \psi(f)\}| \text{ and}$$
$$m_\omega(u) = |\{a \in E_1 : u \in \psi(a)\}|$$

Moreover, we define,

$$\mu_\omega = \max_{e \in E_2} \{\mu_\omega(e)\}, \text{ and}$$
$$m_\omega = \max_{u \in V_2} \{m_\omega(u)\}.$$

We refer to $\mu_\omega$ and $m_\omega$ as the edge congestion and vertex congestion of $\omega$, respectively.

Let $G_1 =< V_1, E_1 >$ and $G_2 =< V_2, E_2 >$ be two graphs. The cartesian product of $G_1$ and $G_2$, denoted by $G_1 \times G_2$ is a graph with the vertex set $V_1 \times V_2$. Two vertices $(x_1, y_1)$ and $(x_2, y_2)$ are adjacent if and only if $x_1 = x_2$ and $y_1 y_2 \in E_2$ or $y_1 = y_2$ and $x_1 x_2 \in E_1$. Let $2K_n$ denote the complete multigraph obtained from $K_n$ by replacing each edge by two new edges.

Let $PS_n$ denote the perfect shuffle graphs defined as follows [7]: its vertices correspond to $n$-bit binary numbers and two vertices $u$ and $v$ are adjacent if either $u$ and $v$ differ in precisely the last bit, or $u$ is a left or right cyclic shift of $v$.

For $G =< V, E >$, let $U \subseteq V$. Let $E(U)$ denote the set of all edges with both end points in $U$. Let $E^1(U)$ be the set of all edge pairs $e_1, e_2$, so that $e_1 \in E(U), e_2 \in E(U)$ and $e_1$ and $e_2$ are not not incident in $G$. Similarly, define $E^2(U)$ to be the set of all edge pairs $e_1, e_2$, so that $e_1 \in E(U), e_2 \in E(U)$ and $e_1$ and $e_2$ are incident in $G$. Let $\bar{U}$ denote the complement of $U$. For any $x \in V$ let $\Gamma_U(x)$ denote the set of vertices in $U$ which are adjacent to $x$. For $S \subseteq V$ let $\Gamma_U(S)$ denote $\cup_{x \in S} \Gamma_U(x)$.

A lower bound technique based on the concept of graph embedding was introduced by Leighton [7] to estimate plane crossing numbers of bounded degree graphs. Shahrokhi and Székely [9] generalized the Leighton's approach to general graphs. Further generalization was obtained by Shahrokhi, Sýkora, Székely, and Vrt'o [10] who showed the following result.

**Theorem 2.1** Let $G =< V, E >$ and $H =< V', E' >$, be graphs, with $|V'| \leq |V| = n$. Let $\omega$ be an embedding of $H$ into $G$ and $g \geq 0$ be an integer. Then

$$cr_g(G) \geq \frac{cr_g(H)}{\mu_\omega^2} - \frac{n}{2} \left(\frac{m_\omega}{\mu_\omega}\right)^2 \quad \text{and} \quad \check{cr}_g(G) \geq \frac{\check{cr}_g(H)}{\mu_\omega^2} - \frac{n}{2}\left(\frac{m_\omega}{\mu_\omega}\right)^2.$$

The application of Theorem 2.1 requires knowing an embedding of $H$ in $G$ with suitable values for edge and vertex congestion. The theorem can be used in two ways. First, it can be used to derive lower bounds on $cr_g(G)$ and $\check{cr}_g(G)$ in terms of $cr_g(H)$ and $\check{cr}_g(H)$. Second, it can be used in reverse, that is, it can be used to derive upper bounds for $cr_g(H)$ and $\check{cr}_g(H)$ in terms of $cr_g(G)$ and $\check{cr}_g(G)$. This second application of Theorem 2.1 also provides a drawing of $H$, provided that a drawing of $G$ is available.

## 3   Improved Bounds

The following result was obtained in [10]

**Theorem 3.1** (i) For $n \geq 65, g > 0$ and $\frac{g}{(\log g+1)^3} \leq \frac{n}{(\log n)^2}$, we have

$$cr_g(K_n) \quad \text{and} \quad \check{cr}_g(K_n) \leq 23n^4 \frac{(\log g + 1)^3}{g}.$$

(ii) For $G =< V, E >$ we have

$$cr_g(G) \leq \frac{8cr_g(K_n)\binom{m}{2}}{n(n-1)(n-2)(n-3)} \quad \text{and} \quad \check{cr}_g(G) \leq \frac{8\check{cr}_g(K_n)\binom{m}{2}}{n(n-1)(n-2)(n-3)}. \quad \square$$

In this section we present two main results. First, we improve the upper bounds on $cr_g(K_n)$ and $\check{cr}_g(K_n)$ by a factor of $O(\log g)$. Second, we present a deterministic polynomial time algorithm to obtain a drawing of any graph $G$ on $S_g$ or $N_g$ with crossing numbers satisfying the upper bounds (ii) from Theorem 3.1. (The original proof was probabilistic.) We start with one essential lemma which will be used to prove our main results.

**Lemma 3.1** Let $G = Q_l \times PS_t$, that is, $G$ is the cartesian product of $l$ dimensional hypercube and $t$ dimensional perfect shuffle and has $n = 2^{t+l}$ vertices. Then, the following hold.
(i) There is an embedding $\omega$ of $2K_n$ into $G$ so that

$$\mu_\omega = t2^{l+t} = tn.$$

(ii) Let $g = 2^{t+1}$, then

$$cr_g(G) \le 20\frac{n^2}{g} \quad \text{and} \quad \check{cr}_g(G) \le 20\frac{n^2}{g}.$$

**Proof.** To prove (i), we first observe that any embedding of $2K_n$ into $G$ can be viewed as the assignment of 2 paths per any unordered vertex pair or equivalently the assignment of one path per any ordered vertex pair. Let $X = (x_1, x_2)$ and $Y = (y_1, y_2)$ be any two distinct vertices of $G$ and consider the ordered pair $(X, Y)$. Let $M = (y_1, x_2)$. Let path $\bar{p}$ be the unique shortest path in $Q_l$ from $x_1$ to $y_1$ which traverse the dimensions in the ascending order. Then $\bar{p}$ will give rise to a path $p$ in $G$ from $X$ to $M$. Next, consider a unique greedy walk $w$ in $PS_t$ [7] from $x_2$ to $y_2$ and let $\bar{q}$ be a path obtained from $w$ by removing loops. Then $\bar{q}$ will give rise to a path $q$ in $G$ from $M$ to $Y$. Now, define the path from $X$ to $Y$ in $G$ to be the path $p, q$. Standard enumerative arguments can be applied to verify each edge of $G$ is contained in at most $tn$ paths which proves (i). To verify (ii), observe that the number of edges of $PS_t$ is at most $2^{t+1}$, since any vertex of $PS_t$ has a degree at most equal to three. Consequently, $\gamma(PS_t) \le 2^{t+1} = g$, since the number of edges is an upper bound on the genus of any graph. Thus $PS_t$ can be drawn on $S_g$ with no edge crossings; let $D$ denote this drawing of $PS_t$. Next, we construct $D'$ a drawing of $G$ from $D$ as follows. Replace each vertex of $PS_t$ in $D$ by a cube $Q_l$ and add suitable edges between different copies of $Q_l$ to construct $G$. We assume with no loss of generality that the surface is a small disk on a neighborhood of any vertex of $PS_t$, and hence we can place the vertices of $Q_l$ replacing this vertex on a straight line segment inside of this disk and draw the edges of $Q_l$ on one side of this line. There are two types of crossings in this drawing of $G$. First, consider those crossings which are generated inside of each disk and are associated with our drawings of $Q_l$. The total number crossings of this type is at most $2^t.4^l/2$, since we can draw each $Q_l$ with at most $4^l/2$ crossings. Now consider those crossings which are generated due to the existence of edges between different copies of $Q_l$ and are located outside of the small disks. Each vertex of a particular $Q_l$ has at most three edges incident with it outside of a disk containing this particular $Q_l$, since any vertex of $PS_t$ is of degree at most three. Therefore $3.2^l$ many edges leave any particular $Q_l$ and can cause at most $9.4^l$ crossings with each other. Since, we started with a drawing of $PS_t$ with no crossings, only those edges leaving the same copy of $Q_l$ may cross. It follows that the total number of crossings of this type is at most $9.2^t.4^l$ and consequently the total number of crossings is $10\frac{n^2}{2^t} = 20\frac{n^2}{g}$ as claimed. Similarly, one may show the bound for $\check{cr}_g(G)$, by observing that $\tilde{\gamma}(PS_t) \le 2^{t+1}$. $\square$

**Theorem 3.2** *Let $n$ and $g$ be two integers such that $n \geq 4g, g > 0$, then*

$$cr_g(K_n) \leq 200n^4 \frac{\log^2 g}{g} \quad and \quad \bar{cr}_g(K_n) \leq 200n^4 \frac{\log^2 g}{g}.$$

**Proof.** Consider the case $S_g$. Let $k$ be the smallest integer so that $\bar{n} = 2^k \geq n$ and $p$ be largest integer such that $\bar{g} = 2^p \leq g$. First we prove the upper bound on $cr_g(2K_n)$. The rest of the proof follows from the fact that $cr_g(2K_n) = 4cr_g(K_n)$ [6]. Clearly, $cr_g(2K_n) \leq cr_{\bar{g}}(2K_{\bar{n}})$ and $\bar{cr}_g(2K_n) \leq \bar{cr}_{\bar{g}}(2K_{\bar{n}})$, and therefore it suffices to verify our upper bonds for $cr_{\bar{g}}(2K_n)$ and $\bar{cr}_{\bar{g}}(2K_{\bar{n}})$. Let $G = Q_{k+1-p} \times PS_{p-1}$. By Part (i) of Lemma 3.1 there is $\omega$ an embedding of $2K_{\bar{n}}$ into $G$ with congestion $\mu_\omega = \bar{n}(p-1)$ and $m_\omega \leq (k+4-p)\bar{n}(p-1)$. Note that $m_\omega \leq \bar{n}(p-1)(k+4-p)$, since maximum degree times the congestion is an upper bound on the vertex congestion of any embedding. Moreover by Part (ii) of Lemma 3.1 $cr_{\bar{g}}(G) \leq 20\frac{\bar{n}^2}{\bar{g}}$. Now use Theorem 2.1 with $H$ being $2K_{\bar{n}}$, and $g$ being $\bar{g}$ to obtain

$$cr_{\bar{g}}(2K_{\bar{n}}) \leq 20\bar{n}^4 \frac{(p-1)^2}{\bar{g}} + \frac{\bar{n}^3}{2}(k+4-p)^2(p-1)^2 \leq 40\bar{n}^4 \frac{\log^2 g}{g} + \frac{\bar{n}^3}{2}\log^2 g(k+4-p)^2$$

However, $(k+4-p)^2 \leq 10 \times 2^{k-p}$, and hence the error term in the right hand side of the last expression in the above inequality is at most $5\bar{n}^3 2^{k-p} \log^2 g \leq 5\bar{n}^4 \frac{\log^2 g}{\bar{g}} \leq 10\bar{n}^4 \frac{\log^2 g}{g}$. To prove the claim for $cr_g(2K_n)$, observe that $\bar{n} \leq 2n$. The proof for $\bar{cr}_g(K_n)$ is similar. $\square$

Let $G = \langle V, E \rangle$, $|V| = n$, and $D$ be a drawing of $K_n = \langle V', E' \rangle$ on $S_g$, $N_g$, or on a $k$-page book. Assume that $h : V \to V'$ is a bijection. Then we can draw $G$ by drawing any $ij \in E$, using the drawing of $h(i)h(j)$, an edge of $K_n$, in $D$. We denote such a drawing of $G$ by $D_G^h$. Let $U \subseteq V, |U| = t$ and $f : U \to V'$ be a bijection. Let $y$ be any crossing in $D$ produced by a pair of edges $e_1 = ij, e_2 = kl \in E'$. We define a weight $w_U^f(y)$ for $y$ according to the following 6 disjoint cases:

1. $i, j, k, l \in U$, then $w_U^f(y) = 1$, provided that $e_1$ and $e_2$ are the images of two edges of $G$ under $f$; otherwise $w_U^f(y) = 0$.

2. $i, j, k, l \in V' - f(U)$, then, $w_U^f(y) = 8|E^1(\bar{U})|(n-t-4)!/(n-t)!$, provided that $e_1$ and $e_2$ are not incident in $K_n$ ; otherwise $w_U^f(y) = 2|E^2(\bar{U})|(n-t-3)!/n-t)!$.

3. $i, k \in f(U), j, l \in V' - f(U)$, then $w_U^f(y) = (|\Gamma_0(i)| - |\Gamma_0(i,k)|)(|\Gamma_0(k)| - |\Gamma_0(i,k)|)(n-t-2)!/(n-t)!$, provided that $e_1$ and $e_2$ are not incident in $K_n$. Otherwise, $w_U^f(y) = \binom{|\Gamma_0(i,k)|}{2}(n-t-1)!/(n-t)!$, provided that $j = l$, and $w_U^f(y) = \binom{|\Gamma_0(i)|}{2}(n-t-2)!/(n-t)!$, provided that $i = k$.

4. $i, j \in f(U)$ and $k, l \in V' - f(U)$, then $w_U^f(y) = 2|E(\bar{U})|(n-t-2)!/(n-t)!$, provided that $e_1$ is the image of an edge in $E$ under $f$; otherwise $w_U^f(y) = 0$.

5. $i, j \in f(U)$ and $k \in f(U)$ and $l \in V' - f(U)$, then $w_U^f(y) = |\Gamma_0(k)|(n-t-1)!/(n-t)!$, provided that $e_1$ is the image on an edge in $E$ under $f$ ; otherwise $w_U^f(y) = 0$.

6. $i, j \in V' - f(U)$, $k \in f(U)$, and $l \in V' - f(U)$, then $w_U^f(y) = 2|\Gamma_0(k)||E(\bar{U})|(n-t-3)!/(n-t)!$.

Next, we make an observation regarding the properties of the weights assigned to the edge crossings in $D$.

**Lemma 3.2** Let $D$ be a drawing of $K_n$ on $S_g$ or $N_g$. Let $G = < V, E >$, $|V| = n$, $U \subseteq V$, $|U| = t$, and assume that $f : U \to V'$ is bijection. Let

$$\beta_U^f = \sum_{y \in D} w_U^f(y).$$

Consider a random extension of $f$ to a bijection from $V$ to $V'$. That, is randomly and uniformly choosing one of the $(n - t)!$ ways of extending $f$ to a bijection from $V$ to $V'$. Then, $\beta_U^f$ is the expected number of crossings associated with $D_G^f$ in this random extension. Moreover, $\beta_U^f$ can be computed in polynomial time.

**Proof.** Using the six disjoint cases it could be verified that the expected value of number of crossings in a random extension is $\beta_U^f$ and the claim follows. $\square$

Next, we present an algorithms for drawing of graphs on $S_g$ and $N_g$, based on our drawings of complete graphs and as a byproduct we extend our upper bounds for a complete graphs to arbitrary graphs. Our algorithm is a deterministic algorithm. We will use the probabilistic interpretation of the weight assignment to the crossings, as described in Lemma 3.2 to give a short proof of a combinatorial result which is crucial to the performance of the algorithm.

## Algorithm Cross

**Input:** $G = < V, E >$, $|V| = n$, $|E| = m$, and a drawing $D$ of $K_n = < V', E' >$ on $S_g$ and $N_g$.
**Output:** A bijection $f : V \to V'$ and a drawing $D_G^f$ of $G$ on $S_g$ and $N_g$ using $f$.

$U_0 \leftarrow \emptyset$ and $f(U_0) \leftarrow \emptyset$.

**For** $t \leftarrow 1$ **to** $n$ **do**
    **For any** $x \in V' - f(U_{t-1})$ **do**
        $Z_x \leftarrow f(U_{t-1}) \cup \{x\}$
        $f(t) \leftarrow x$
        Compute $\beta_{Z_x}^f$
    **Endfor**
    Let $a \in V' - f(U_{t-1})$ so that $\beta_{Z_a}^f = min_{x \in V' - f(U_{t-1})} \beta_{Z_x}^f$.
    $f(t) \leftarrow a$
    $U_t \leftarrow U_{t-1} \cup \{t\}$.
    $f(U_t) \leftarrow f(U_{t-1}) \cup \{a\}$
    **Endfor**

Output $D_G^f$
**End.**

**Theorem 3.3** During execution of algorithm Cross, we have

$$\beta_{U_n}^f \leq \beta_{U_{n-1}}^f \leq \dots \leq \beta_{U_2}^f \leq \beta_{U_1}^f \leq K = \frac{8\binom{m}{2}}{n(n-1)(n-2)(n-3)}.$$

**Proof.** Using cases 2 and 3 we can verify $\beta^f_{Z_x} \leq K$, for any $x \in V'$ and hence $\beta^f_{U_1} \leq K$. Now observe that by Lemma 3.2 for, $1 \leq t \leq n-1$, $\beta^f_{U_t}$ is the expected value of number of crossings (for $D^f_G$) in a random extension of $f$ when the values of $f$ are fixed for $U_t = \{1, 2, ..., t\}$ and values of $f$ must be selected for $\{t+1, t+2, ..., n\}$ from $V' - f(U_t)$. Moreover, for any $x \in V' - f(U_t)$, $\beta^f_{Z_x}$ is expected value of number of crossings in a random extension of $f$ when values of $f$ are fixed for $U_t \cup \{t+1\}$ with $f(t+1) = x$ and the remaining values of $f$ must be selected from $V' - f(U_t \cup \{x\})$. By the rules of conditional probability we have

$$\frac{\sum_{x \in V' - f(U_t)} \beta^f_{Z_x}}{n-t} = \beta^f_{U_t}.$$

To finish the proof observe that $\beta^f_{U_{t+1}} \leq \beta^f_{Z_x}$ for any $x \in V' - f(U_t)$. $\square$

**Theorem 3.4 (i)** *When algorithm Cross terminates, we have*

$$cr(D^f_G) \leq cr(D)O\left(\frac{m^2}{n^4}\right).$$

(ii) *Algorithm Cross draws $G$ on $S_g$, or $N_g$ with at most $O(\frac{m^2 \log^2 g}{g})$ edge crossings in polynomial time.*

**Proof.** When algorithm terminates $t = n$, $U_n = V$, and $f$ is a bijection from $V$ to $V'$. It follows that $\beta^f_{U_n} = \sum_{y \in D} w^f_V = cr(D^f_G)$. Now, by Theorem 3.3, $\beta^f_{U_n} \leq K$ which implies (i). For (ii) let $D$ be our drawing of $K_n$ in Theorem 3.2. $\square$

## 4 Conclusion

Our constructed upper bounds are within a factor of $O(\log^2 g)$ from the previously announced lower bounds [10] for many classes of graphs including complete graphs, complete bipartite graphs, random graphs, and especially classes of edge-transitive graphs. This is an improvement by a factor of $O(\log g)$ over the previous results.

A main open problem which still remains is to improve our upper bounds for $K_n$, as this will reflect itself in Theorem 3.4 and will result in better upper bounds for any graph. In general an important problem is to determine asymptotically optimal values of $cr_g(K_n)$ and $\tilde{cr}_g(K_n)$.

## References

[1] Chartrand, G. and Lesniak, L., *Graphs and Digraphs*, Wadsworth and Books/Cole Mathematics Series, 1986.

[2] Erdös, P., Guy, R.P., Crossing number problems, *American Mathematical Monthly* **80** (1973), 52-58.

[3] Gross, J.L., On infinite family of octahedral crossing numbers, *J. Graph Theory* **2**(1978), 171-178.

[4] Guy, R. K., Jenkins, T., Schaer, J., The toroidal crossing number of the complete graph, *J. Combinatorial Theory* **4**(1968), 376–390.

[5] Guy, R. K., Jenkins, T., Schaer, J., The toroidal crossing number of $K_{m,n}$, *J. Combinatorial Theory* **6**(1969), 235–250.

[6] Kainen, P. C., A lower bound for crossing number of graphs with applications to $K_n$, $K_{p,q}$ and $Q(d)$, *J. Combinatorial Theory (B)* **12**(1972), 287–298.

[7] Leighton, F. T., Complexity Issues in VLSI, M.I.T. Press, Cambridge, 1983.

[8] Shahrokhi, F., and Székely, L. A., An algebraic approach to the uniform concurrent multicommodity flow problem: theory and applications, Technical Report CRPDC-91-4, Dept. Computer Science, Univ. of North Texas., Denton, 1991.

[9] Shahrokhi, F., and Székely, L. A., Effective lower bounds for crossing number, bisection width and balanced vertex separators in terms of symmetry, in: Proc. *Integer Programming and Combinatorial Optimization, Proceedings of a Conference held at Carnegie Mellon University, May 25-27, 1992, by the Mathematical Programming Society*, eds. E. Balas, G. Cournejols, R. Kannan, 102-113, CMU Press, 1992.

[10] Shahrokhi, F., Sýkora, O., Székely, L.A., Vrt'o, I.: The crossing number of a graph on a compact 2-manifold, *Advances in Mathematics*, to appear.

[11] Sýkora, O., Vrt'o, I., On the crossing number of hypercubes and cube connected cycles, *BIT* **33**(1993), 232-237.

[12] Sýkora, O., Vrt'o, I., Edge separators for graphs of bounded genus with applications, *Theoretical Computer Science* **112**(1993), 419-429.

# Two Algorithms for Finding Rectangular Duals of Planar Graphs*

Goos Kant[1] and Xin He[2]

[1] Department of Computer Science, Utrecht University,
Padualaan 14, 3584 CH Utrecht, the Netherlands. goos@cs.ruu.nl
[2] Department of Computer Science, State University of New York at Buffalo, Buffalo, NY
14260, USA. xinhe@cs.buffalo.edu

**Abstract.** We present two linear-time algorithms for computing a regular edge labeling of 4-connected planar triangular graphs. This labeling is used to compute in linear time a rectangular dual of this class of planar graphs. The two algorithms are based on totally different frameworks, and both are conceptually simpler than the previous known algorithm and are of independent interests. The first algorithm is based on edge contraction. The second algorithm is based on the canonical ordering. This ordering can also be used to compute more compact visibility representations for this class of planar graphs.

## 1 Introduction

The problem of drawing a graph on the plane has received increasing attention due to a large number of applications [3]. Examples include VLSI layout, algorithm animation, visual languages and CASE tools. Vertices are usually represented by points and edges by curves. In the design of floor planning of electronic chips and in architectural design, it is also common to represent a graph $G$ by a *rectangular dual*, defined as follows. A *rectangular subdivision system* of a rectangle $R$ is a partition of $R$ into a set $\Gamma = \{R_1, R_2, \ldots, R_n\}$ of non-overlapping rectangles such that no four rectangles in $\Gamma$ meet at the same point. A *rectangular dual* of a planar graph $G = (V, E)$ is a rectangular subdivision system $\Gamma$ and a one-to-one correspondence $f : V \to \Gamma$ such that two vertices $u$ and $v$ are adjacent in $G$ if and only if their corresponding rectangles $f(u)$ and $f(v)$ share a common boundary. In the application of this representation, the vertices of $G$ represent circuit modules and the edges represent module adjacencies. A rectangular dual provides a placement of the circuit modules that preserves the required adjacencies. Figure 1 shows an example of a planar graph and its rectangular dual.

This problem was studied in [1, 2, 8]. Bhasker and Sahni gave a linear time algorithm to construct rectangular duals [2]. The algorithm is fairly complicated and requires many intriguing procedures. The coordinates of the rectangular dual constructed by it are real numbers and bear no meaningful relationship with the

---

* The work of the first author was supported by the ESPRIT Basic Research Actions program of the EC under contract No. 7141 (project ALCOM II). The work of the second author was supported by National Science Foundation, grant number CCR-9011214.

structure of the graph. This algorithm consists of two major steps: (1) constructing a so-called *regular edge labeling* (REL) of $G$; and (2) constructing the rectangular dual using this labeling. A simplification of step (2) is given in [5]. The coordinates of the rectangular dual constructed by the algorithm in [5] are integers and carry clear combinatorial meaning. However, the step (1) still relies on the complicated algorithm in [2]. (A parallel implementation of this algorithm, working in $O(\log n \log^* n)$ time with $O(n)$ processors, is given in [6].)

In this paper we present two linear time algorithms for finding a regular edge labeling. The two algorithms use totally different approaches and both are of independent interests. The first algorithm is based on the *edge contraction* technique, which was also used for drawing triangular planar graphs on a grid [10]. The second algorithm is based on the *canonical ordering* for 4-connected planar triangular graphs. This technique extends the canonical ordering, which was defined for triangular planar graphs [4] and triconnected planar graphs [7], to this class of graphs. Another interesting representation of planar graphs is the *visibility representation*, which maps vertices into horizontal segments and edges into vertical segments [9, 11]. It turns out that the canonical ordering also gives a reduction of a factor 2 in the width of the visibility representation of 4-connected planar graphs.

The present paper is organized as follows. Section 2 presents the definition of the regular edge labeling and reviews the algorithm in [5] that computes a rectangular dual from a REL. In section 3, we present the edge contraction based algorithm for computing a REL. In section, 4 we present the second REL algorithm based on the canonical ordering. Section 5 discusses the algorithm for the visibility representation and some final remarks.

## 2   The rectangular dual algorithm

Let $G = (V, E)$ be a planar graph with $n$ vertices and $m$ edges. If $(u, v) \in E$, $u$ is a *neighbor* of $v$. $deg(u)$ denotes the number of neighbors of $u$. We assume $G$ is equipped with a fixed plane embedding. The embedding divides the plane into a number of *faces*. The unbounded face is the *exterior face*. Other faces are *interior faces*. The vertices and the edges on the boundary of the exterior face are called *exterior vertices* and *exterior edges*. An interior edge between two exterior vertices is called a *chord*. A path (or a cycle) of $G$ consisting of $k$ edges is called a $k-$path (or a $k-$cycle, respectively). A *triangle* is a 3-cycle. A *quadrangle* is a 4-cycle. A cycle $C$ of $G$ divides the plane into its interior and exterior region. If $C$ contains at least one vertex in its interior, $C$ is called a *separating cycle*.

A *plane triangular graph* is a plane graph all of whose interior faces are triangles. For the rectangular dual problem, as we will see later, we only need to consider plane triangular graphs. Let $G$ be such a graph. Consider an interior vertex $v$ of $G$. We use $N(v)$ to denote the set of neighbors of $v$. If $N(v) = \{u_1, \ldots, u_k\}$ are in counterclockwise order around $v$ in the embedding, then $u_1, \ldots, u_k$ form a cycle, denoted by $Cycle(v)$. The *star* at $v$, denoted by $Star(v)$, is the set of the edges $\{(v, u_i) \mid 1 \leq i \leq k\}$.

We assume the embedding information of $G$ is given by the following data structure. For each $v \in V$, there is a doubly linked circular list $Adj(v)$ containing all vertices of $N(v)$ in counterclockwise order. The two copies of an edge $(u, v)$ (one in $Adj(u)$ and one in $Adj(v)$) are cross-linked to each other. This representation can be constructed as a by-product by using a planarity testing algorithm in linear time.

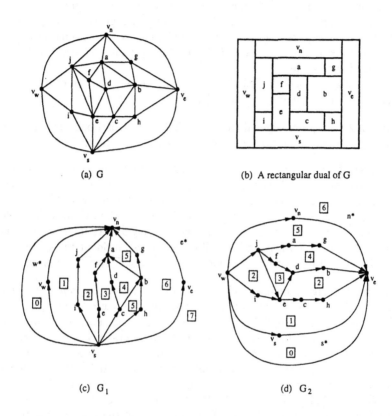

(a) G

(b) A rectangular dual of G

(c) $G_1$

(d) $G_2$

**Fig. 1.** A PTP graph, its rectangular dual, and the $st$-graphs $G_1$ and $G_2$

Consider a plane graph $H = (V, E)$. Let $u_0, u_1, u_2, u_3$ be four vertices on the exterior face in counterclockwise order. Let $P_i$ ($i = 0, 1, 2, 3$) be the path on the exterior face consisting of the vertices between $u_i$ and $u_{i+1}$ (addition is mod 4). We seek a rectangular dual $R_H$ of $H$ such that $u_0, u_1, u_2, u_3$ correspond to the four corner rectangles of $R_H$ and the vertices on $P_0$ ($P_1, P_2, P_3$, respectively) correspond to the rectangles located on the north :(west, south, east, respectively) boundary of $R_H$. In order to simplify the problem, we modify $H$ as follows: Add four new vertices $v_N, v_W, v_S, v_E$. Connect $v_N$ ($v_W, v_S, v_E$, respectively) to every vertex on $P_0$ ($P_1, P_2, P_3$, respectively) and add four new edges $(v_S, v_W), (v_W, v_N), (v_N, v_E), (v_E, v_S)$. Let $G$ be the resulting graph. It's easy to see that $H$ has a rectangular dual $R_H$ if and only if $G$ has a rectangular dual $R_G$ with exactly four rectangles on the boundary of $R_G$ (see Figure 1 (1) and (2)). The following theorem was proved in [1, 8]:

**Theorem 1.** *A planar graph G has a rectangular dual R with four rectangles on the boundary of R if and only if (1) every interior face is a triangle and the exterior face is a quadrangle; (2) G has no separating triangles.*

A graph satisfying the conditions in Theorem 1 is called a *proper triangular planar* (PTP) graph. ¿From now on, we will discuss only such graphs. Note the condition (2) of Theorem 1 implies that $G$ is 4-connected. Since $G$ has no separating triangles, the degree of any interior vertex $v$ of $G$ is at least 4. (If $deg(v) = 3$, $Cycle(v)$ would be a separating triangle.)

The rectangular dual algorithm in [5] heavily depends on the concept of *regular edge labeling* (REL) defined as follows [2, 5]:

**Definition 2.** A regular edge labeling of a PTP graph $G$ is a partition of the interior edges of $G$ into two subsets $T_1, T_2$ of directed edges such that:

1. For each interior vertex $v$, the edges incident to $v$ appear in counterclockwise order around $v$ as follows: a set of edges in $T_1$ leaving $v$; a set of edges in $T_2$ entering $v$; a set of edges in $T_1$ entering $v$; a set of edges in $T_2$ leaving $v$.
2. Let $v_N, v_W, v_S, v_E$ be the four exterior vertices in counterclockwise order. All interior edges incident to $v_N$ are in $T_1$ and entering $v_N$. All interior edges incident to $v_W$ are in $T_2$ and leaving $v_W$. All interior edges incident to $v_S$ are in $T_1$ and leaving $v_S$. All interior edges incident to $v_E$ are in $T_2$ and entering $v_E$.

The regular edge labeling is closely related to *planar st-graphs*. A planar *st*-graph $G$ is a directed planar graph with exactly one source (in-degree 0) vertex $s$ and exactly one sink (out-degree 0) vertex $t$ such that both $s$ and $t$ are on the exterior face and are adjacent. Let $G$ be a planar *st*-graph. For each vertex $v$, the incoming edges of $v$ appear consecutively around $v$, and so do the outgoing edges of $v$. The boundary of every face $F$ of $G$ consists of two directed paths with a common origin, called $low(F)$, and a common destination, called $high(F)$.

Let $G$ be a PTP graph and $\{T_1, T_2\}$ be a REL of $G$. From $\{T_1, T_2\}$, we can construct two planar *st*-graphs as follows. Let $G_1$ be the graph consisting of the edges of $T_1$ plus the four exterior edges (directed as $v_S \rightarrow v_W$, $v_W \rightarrow v_N$, $v_S \rightarrow v_E$, $v_E \rightarrow v_N$), and a new edge $(v_S, v_N)$. Then $G_1$ is a planar *st*-graph with source $v_S$ and sink $v_N$. For each vertex $v$, the face of $G_1$ that separates the incoming edges of $v$ from the outgoing edges of $v$ in the clockwise direction is denoted by $left(v)$. The other face of $G_1$ that separates the incoming and the outgoing edges of $v$ is denoted by $right(v)$.

Let $G_2$ be the graph consisting of the edges of $T_2$ plus the four exterior edges (directed as $v_W \rightarrow v_S$, $v_S \rightarrow v_E$, $v_W \rightarrow v_N$, $v_N \rightarrow v_E$), and a new edge $(v_W, v_E)$. Then $G_2$ is a planar *st*-graph with source $v_W$ and sink $v_E$. For each vertex $v$, the face of $G_2$ that separates the incoming edges of $v$ from the outgoing edges of $v$ in the clockwise direction is denoted by $above(v)$. The other face of $G_2$ that separates the incoming and the outgoing edges of $v$ is denoted by $below(v)$.

The dual graph $G_1^*$ of $G_1$ is defined as follows. Every face $F_k$ of $G_1$ is a node $v_{F_k}$ in $G_1^*$, and there exists an edge $(v_{F_i}, v_{F_k})$ in $G_1^*$ if and only if $F_i$ and $F_k$ share a common edge in $G_1$. We direct the edges of $G_1^*$ as follows: if $F_l$ and $F_r$ are the left and the right face of an edge $(v, w)$ of $G_1$, direct the dual edge from $F_l$ to $F_r$

if $(v, w) \neq (v_S, v_N)$ and from $F_r$ to $F_l$ if $(v, w) = (v_S, v_N)$. $G_1^*$ is a planar $st$-graph whose source and sink are the right face (denoted by $w^*$) and the left face (denoted by $e^*$) of $(v_S, v_N)$, respectively. For each node $F$ of $G_1^*$, let $d_1(F)$ denote the length of the longest path from $w^*$ to $F$. Let $D_1 = d_1(e^*)$. For each interior vertex $v$ of $G$, define: $x_{\text{left}}(v) = d_1(left(v))$, and $x_{\text{right}}(v) = d_1(right(v))$. For the four exterior vertices, define: $x_{\text{left}}(v_W) = 0$; $x_{\text{right}}(v_W) = 1$; $x_{\text{left}}(v_E) = D_1 - 1$; $x_{\text{right}}(v_E) = D_1$; $x_{\text{left}}(v_S) = x_{\text{left}}(v_N) = 1$; $x_{\text{right}}(v_S) = x_{\text{right}}(v_N) = D_1 - 1$.

The dual graph $G_2^*$ of $G_2$ is defined similarly. For each node $F$ of $G_2^*$, let $d_2(F)$ denote the length of the longest path from the source node of $G_2^*$ to $F$. Let $D_2$ be the length of the longest path from the source node to the sink node of $G_2^*$. For each interior vertex $v$ of $G$, define: $y_{\text{low}}(v) = d_2(below(v))$, and $y_{\text{high}}(v) = d_2(above(v))$. For the four exterior vertices, define: $y_{\text{low}}(v_W) = y_{\text{low}}(v_E) = 0$; $y_{\text{high}}(v_W) = y_{\text{high}}(v_E) = D_2$; $y_{\text{low}}(v_S) = 0$; $y_{\text{high}}(v_S) = 1$; $y_{\text{low}}(v_N) = D_2 - 1$; $y_{\text{high}}(v_N) = D_2$.

The rectangular dual algorithm relies on the following theorem [5].

**Theorem 3.** *Let $G$ be a PTP graph and $\{T_1, T_2\}$ be a REL of $G$. For each vertex $v$ of $G$, assign $v$ the rectangle $f(v)$ bounded by the four lines $x = x_{\text{left}}(v)$, $x = x_{\text{right}}(v)$, $y = y_{\text{low}}(v)$, $y = y_{\text{high}}(v)$. Then the set $\{f(v)|v \in V\}$ form a rectangular dual of $G$.*

Figure 1 shows an example of the theorem. Figure 1 (3) shows the $st$-graph $G_1$. The small squares in the figure represent the nodes of $G_1^*$ and the integers in the squares represent their $d_1$ values. Figure 1 (4) shows the graph $G_2$. Figure 1 (2) shows the rectangular dual constructed as in Theorem 3. The algorithm for computing a rectangular dual is as follows [5]:

**Algorithm 1**: Rectangular Dual (Input: a PTP graph $G = (V, E)$).

1. Construct a regular edge labeling $\{T_1, T_2\}$ of $G$.
2. Construct from $\{T_1, T_2\}$ the planar $st$-graphs $G_1$ and $G_2$.
3. Construct the dual graph $G_1^*$ from $G_1$ and $G_2^*$ from $G_2$.
4. Compute $d_1(F)$ for nodes in $G_1^*$ and $d_2(F)$ for nodes in $G_2^*$.
5. Assign each vertex $v$ of $G$ a rectangle $f(v)$ as in Theorem 3.

The steps 2 through 5 of Algorithm 1 can be easily implemented in linear time [5]. In next two sections we present two algorithms for constructing a REL of PTP graphs.

## 3 Algorithm based on edge contraction

In this section, we present our first algorithm for computing a REL of a PTP graph $G$. The basic technique is *edge contraction* and *edge expansion*. We begin with the definition of edge contraction. Let $e = (v, u)$ be an interior edge of $G$. Let $C_1$ and $C_2$ be the two faces with $e$ as the common boundary. Let $e_1$ and $e_2$ be the other two edges and $y$ the third vertex of $C_1$. Let $e_3$ and $e_4$ be the two other edges and $z$ the third vertex of $C_2$. The operation of *contracting* $e$ deletes $e$ and merges $u$ and $v$ into a new vertex $o_e$. The edges incident to $u$ and $v$ (except $e_1, e_2, e_3, e_4$) are incident to the new vertex $o_e$ in the resulting graph. $e_1$ and $e_2$ are replaced by a new edge $(y, o_e)$.

$e_3$ and $e_4$ are replaced by a new edge $(z, o_e)$. (See Figure 2.) The resulting *contracted graph* is denoted by $G/e$. The edges $e_1, e_2, e_3, e_4$ are called the *surrounding edges e.* The edges $(y, o_e)$ and $(z, o_e)$ are called the *residue edges* of $e$.

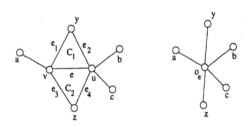

**Fig. 2.** Edge contraction

The graph $G' = G/e$ has a plane embedding inherited from the embedding of $G$. Since $G$ is a PTP graph, $e$ is not on any separating triangle. Thus $G'$ has no multiple edges. It's easy to see that $G'$ with the inherited embedding is a plane triangular graph. If $e$ is on a separating quadrangle of $G$, then $G'$ has a separating triangle. If $e$ is **not** on any separating quadrangle of $G$, it is called a *contractible edge*. For any contractible edge $e$, $G/e$ is a PTP graph.

The following equivalent definition of contractible edges is useful in our discussion. Consider a vertex $v$ and a neighbor $u$ of $v$. Let $y$ and $z$ be the two neighbors of $v$ that are consecutive with $u$ in $N(v)$. The edge $(u, v)$ is contractible if and only if for any neighbor $x$ $(x \neq y, z)$ of $v$, the only common neighbors of $u$ and $x$ are $v$ and possibly $y$ or $z$. In this case, $u$ is called a *contractible neighbor* of $v$.

**Lemma 4.** *Let $G$ be a PTP graph and $v$ be an interior vertex of $G$. If $deg(v) = 4$, then $v$ has at least two contractible neighbors. If $deg(v) = 5$, then $v$ has at least one contractible neighbor.*

Let $e$ be a contractible edge of a PTP graph $G$. Suppose a REL $\{T_1', T_2'\}$ of $G' = G/e$ has been found. Then we can *expand $e$* and obtain a REL $\{T_1, T_2\}$ of $G$ from $\{T_1', T_2'\}$ as follows. Let $e_1, e_2, e_3, e_4$ be the surrounding edges of $e$. For any edge $e'$ of $G$ that is not $e$ and not a surrounding edge of $e$, the label of $e'$ with respect to $\{T_1, T_2\}$ is the same as its label with respect to $\{T_1', T_2'\}$. We need to specify proper labels of $e, e_1, e_2, e_3, e_4$ with respect to $\{T_1, T_2\}$. Depending on the labels of the edges in $Star(o_e)$ with respect to $\{T_1', T_2'\}$, there are six cases (up to the rotation of the edges around $o_e$) as shown in Figure 3. These figures shows the labels of relevant edges before and after the expansion.

We assume $(o_e, y)$ is in $T_1'$ and directed as $o_e \to y$. Other cases are similar by rotating the edges in $Star(o_e)$. Consider the label of $(o_e, z)$ with respect to $\{T_1', T_2'\}$. If $z \to o_e \in T_1'$, the situation is shown in Fig 4.1. The case $o_e \to z \in T_1'$ is shown in Fig 4.2. Suppose $o_e \to z \in T_2'$. Let $(o_e, x)$ be the first edge in $Star(o_e)$ following $(o_e, y)$ in clockwise order. Depending on the label of $(o_e, x)$ with respect to $\{T_1', T_2'\}$,

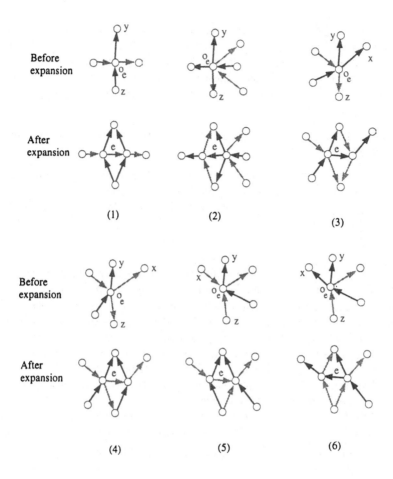

**Fig. 3.** Edge expansion

there are two cases as shown in Fig 4.3 and 4.4. Suppose $z \to o_e \in T_2'$. Let $(o_e, x)$ be the first edge in $Star(o_e)$ following $(o_e, y)$ in counterclockwise order. Depending on the label of $(o_e, x)$ with respect to $\{T_1', T_2'\}$, there are two cases as shown in Fig 4.5 and 4.6. Note that the conditions of the six cases are completely determined by the labels of at most six edges in $Star(o_e)$: the two residue edges $(o_e, y), (o_e, z)$ and the four edges that are consecutive with $(o_e, y), (o_e, z)$ in $Star(o_e)$.

The basic idea of our algorithm is as follows. Since the minimum degree of $G$ is at most 5, we pick a degree-4 or a degree-5 vertex $v$ and select a contractible neighbor $u$ of $v$. Then contract $e = (v, u)$ and recursively find a REL for the graph $G' = G/e$. Finally expand $e$ to obtain a REL for $G$. In order to find the contractible neighbors of $v$, however, we need to check, for each pair $u$ and $w$ of $v$'s neighbors, if $u$ and $w$ share a common neighbor or not. Since the degree of $u$ and $w$ can be large, this checking can be too expensive. In order to achieve linear time, we will only consider special *good* degree-4 and degree-5 vertices defined as follows. Let $V_i = \{v \in V | deg(v) = i\}$ and $V_{[i,j]} = \{v \in V | i \leq deg(v) \leq j\}$. Define $n_i = |V_i|$ and $n_{[i,j]} = |V_{[i,j]}|$. The vertices in $V_{[4,19]}$ are called *light* vertices. The vertices in $V_{[20,\infty)}$ are called *heavy* vertices. A degree-5 vertex $v$ is *good* if $v$ has at most one heavy neighbor. A degree-4 vertex $v$ is *good* if either $v$ has at most one heavy neighbor, or $v$ has two heavy neighbors which are not consecutive in $N(v)$.

**Lemma 5.** *Any PTP graph $G = (V, E)$ with at least one heavy vertex has at least 7 good vertices.*

*Proof.* Since the exterior face of $G$ is a quadrangle and all interior faces of $G$ are triangles, we have $|E| = 3n - 7$ by Euler's formula. Hence $4n_4 + 5n_5 + 6n_{[6,19]} + 20n_{[20,\infty)} \leq \sum_{4 \leq i} i n_i = \sum_{v \in V} deg(v) = 2|E| = 6n - 14 = 6(n_4 + n_5 + n_{[6,19]} + n_{[20,\infty)}) - 14$. This gives: $14n_{[20,\infty)} + 2n_6 \leq 2n_4 + n_5 + 2n_6 - 14 \leq 2n - 14$. Hence:

$$7n_{[20,\infty)} + n_6 \leq n - 7 \tag{1}$$

Let $p_4$ ($p_5$, respectively) be the number of good degree-4 (degree-5, respectively) vertices. So there are $n_4 - p_4$ bad degree-4 vertices and $n_5 - p_5$ bad degree-5 vertices. Define $S = \sum_{v \in V_{[20,\infty)}} deg(v)$. Since each bad degree-5 vertex $v$ has at least two heavy neighbors, it contributes at least 2 to $S$. Consider a bad degree-4 vertex $v$. If $v$ has at least three heavy neighbors, then $v$ contributes at least 3 to $S$. Suppose $v$ has two heavy neighbors $u$ and $w$ which are consecutive in $N(v)$. The edges $(v, u)$ and $(v, w)$ contribute 2 to $S$. The edge $(u, w)$ also contributes 2 to $S$. But since $(u, w)$ is shared with one other face, just half of the contribution can be apportioned to $v$. So the contribution of $v$ to $S$ is at least 3. Thus $3(n_4 - p_4) + 2(n_5 - p_5) \leq S$, which gives $3n_4 + 2n_5 - (3p_4 + 2p_5) \leq \sum_{v \in V_{[20,\infty)}} deg(v)$. This in turn implies: $3n_4 + 2n_5 + \sum_{v \in V_{[4,5]}} deg(v) + \sum_{v \in V_{[6,19]}} deg(v) - (3p_4 + 2p_5) \leq \sum_{v \in V} deg(v) = 2|E| = 6n - 14 = 6(n_4 + n_5 + n_6 + n_{[7,19]} + n_{[20,\infty)}) - 14$. Simplifying this inequality, we get: $n_4 + n_5 + n_{[7,19]} - (3p_4 + 2p_5) \leq 6n_{[20,\infty)} - 14$. Hence:

$$3p_4 + 2p_5 \geq n - (n_6 + 7n_{[20,\infty)}) + 14 \tag{2}$$

From (1) and (2) we have: $3(p_4 + p_5) \geq 3p_4 + 2p_5 \geq n - (n - 7) + 14 = 21$. This proves the lemma.

We are now ready to present our first REL construction algorithm.

**Algorithm 2:** REL (Input: A PTP graph $G = (V, E)$).

1. Compute the degrees of the vertices of $G$.
2. Collect all good degree-4 and degree-5 interior vertices into a list $L$.
3. $i \leftarrow n$.
4. While $G$ has more than one interior vertex do:
   **4.1** Remove a vertex $v$ from $L$. Mark $v$ as $w_i$. Decrease $i$ by 1. Record the neighborhood structure of $v$.
   **4.2** Find a contractible neighbor $u$ of $v$. Contract the edge $(v, u)$. (The new vertex is still denoted by $u$.) Modify the adjacency lists and the degrees of the vertices affected by the contraction. If any of the affected vertices becomes a good vertex, put it into $L$.

End While (the last marked vertex is $w_6$).

**5.** $G$ has only one interior vertex now. Construct the trivial REL for $G$.

**6.** For $i = 6$ to $n$ do:

Put $w_i$ back into $G$. Expand the corresponding contracted edge.

**Theorem 6.** *Algorithm 2 computes a REL of a PTP graph in $O(n)$ time.*

*Proof.* The correctness of the algorithm follows from the above discussion. We only need to analyze its complexity. Step 1 clearly takes $O(n + m) = O(n)$ time. Since good vertices have degree at most 5, each of them can be determined and put into $L$ in $O(1)$ time. By Lemma 5, $L$ will never be empty during the execution of the while loop.

Since the degree of a good vertex $v$ is at most 5, the neighborhood structure of $v$ can be recorded in $O(1)$ time. Other operations of Step 4.1 can be easily done in $O(1)$ time also. The only non-trivial part is Step 4.2. We need to find a contractible neighbor of $v$ in $O(1)$ time. Suppose $deg(v) = 5$ and $u_i$ $(0 \leq i \leq 4)$ are $v$'s neighbors. If $v$ has no heavy neighbor or has one heavy neighbor (say $u_0$), we can check, for each pair $u_i$ and $u_j$ $(1 \leq i, j \leq 4)$, if they share a common neighbor. Since the degrees of $u_i$ and $u_j$ are bounded by 19, this takes $O(1)$ time. If none of $u_i$ $(1 \leq i \leq 4)$ is contractible, then $u_0$ is contractible by Lemma 4. Now suppose $deg(v) = 4$ with neighbors $u_0, u_1, u_2, u_3$. If $v$ has at most one heavy neighbor, the situation is the same as the degree-5 case. If $v$ has two heavy neighbors, then they are not consecutive in $N(v)$. Suppose they are $u_0$ and $u_2$. We can check if $u_1$ and $u_3$ share a common neighbor in $O(1)$ time. If $u_1$ and $u_3$ have no common neighbors, then both of them are contractible. Otherwise $u_0$ and $u_2$ are contractible.

After selecting a contractible neighbor $u$ for $v$, the operation of contracting $(v, u)$ affects the vertices in $N(v)$. The adjacency lists and the degrees of these vertices are modified. Since $deg(v) \leq 5$, this can be done in $O(1)$ time by using the cross-linked adjacency lists data structure. New good vertices can be detected and inserted into $L$ in $O(1)$ time.

Finally, the edge expansion operation only involves 5 edges adjacent to the corresponding contracted edge. This can be done in $O(1)$ time by using the neighborhood structure recorded at Step 4.1.

# 4 Algorithm based on canonical ordering

In this section we consider 4-connected planar triangular graphs (all of whose face, including the exterior face, are triangles). We introduce the *canonical ordering* for such graphs, which is the basis for our second algorithm for finding a REL of a PTP graph $G$. Note that adding an edge connecting two non-adjacent exterior vertices of a PTP-graph $G$ leads to a 4-connected planar triangular graph. The applications of the canonical ordering to other classes of planar graphs have been studied in [4, 7].

## 4.1 The canonical ordering of 4-connected planar triangular graphs

Let $G$ be a 4-connected planar triangular graph with three exterior vertices $u, v, w$.

**Theorem 7.** *There exists a labeling of the vertices $v_1 = u, v_2 = v, v_3, \ldots, v_n = w$ of $G$ meeting the following requirements for every $4 \leq k \leq n$:*

1. *The subgraph $G_{k-1}$ of $G$ induced by $v_1, v_2, \ldots, v_{k-1}$ is biconnected and the boundary of its exterior face is a cycle $C_{k-1}$ containing the edge $(u, v)$.*
2. *$v_k$ is in the exterior face of $G_{k-1}$, and its neighbors in $G_{k-1}$ form a (at least 2-element) subinterval of the path $C_{k-1} - \{(u, v)\}$. If $k \leq n - 2$, $v_k$ has at least 2 neighbors in $G - G_{k-1}$.*

*Proof.* The vertices $v_n, v_{n-1}, \ldots, v_3$ are defined by reverse induction. Number the three exterior vertices $u, v, w$ by $v_1, v_2$ and $v_n$. Let $G_{n-1}$ be the subgraph of $G$ after deleting $v_n$. By 4-connectivity of $G$, $G_{n-1}$ is triconnected, and its exterior face $C_{n-1}$ is a cycle and, hence, admits the constraints of the theorem. Let $v_{n-1} \neq v_1$ be the vertex of $C_{n-1}$ adjacent to both $v_2$ and $v_n$ in $G$. By the 4-connectivity, $G - \{v_n, v_{n-1}\}$ is biconnected and its exterior face $C_{n-1}$ is a cycle and, hence, admits the constraints.

Let $k < n - 1$ be fixed and assume that $v_i$ has been determined for every $i > k$ such that the subgraph $G_i$ induced by $V - \{v_{i+1}, \ldots, v_n\}$ satisfies the constraints of the theorem. Let $C_k$ denote the boundary of the exterior face of $G_k$. Assume first that $C_k$ has no interior chords. Suppose $v_1, c_{k_1}, \ldots, c_{k_p}, v_2$ are the vertices of $C_k$ in this order between $v_1$ and $v_2$. Then it follows by the 4-connectivity of $G$ that $p \geq 2$. If all vertices $c_{k_1}, \ldots, c_{k_p}$ have only one edge to the vertices in $G - G_k$, then since $G$ is a planar triangular graph, they are adjacent to the same vertex $v_j$ for some $k < j < n$. In this case we also have $(v_1, v_j), (v_2, v_j) \in G$. But then $\{(v_1, v_j), (v_j, v_2), (v_2, v_1)\}$ would be a separating triangle. Hence at least one vertex, say $c_{k_\alpha}$, has at least 2 neighbors in $G - G_k$. $c_{k_\alpha}$ is the next vertex $v_k$ in our ordering.

Next assume $C_k$ has interior chords. Let $(c_a, c_b)$ $(b > a + 1)$ be a chord such that $b - a$ is minimal. Let also $(c_d, c_e)$ be a chord with $e > d \geq b$ such that $e - d$ is minimal. (If there is no such a chord, let $(c_a, c_b) = (c_d, c_e)$ and number the vertices in clockwise order around $C_k$ such that $a = 1 < b = d$ and $e = 1$.) Assume, without loss of generality, that $v_1, v_2 \notin \{c_{a+1}, \ldots, c_{b-1}\}$. If all vertices $c_{a+1}, \ldots, c_{b-1}$ have only one edge to the vertices in $G - G_k$, then since $G$ is a triangular graph, they are adjacent to the same vertex $v_j$, and we also have $(v_a, v_j), (v_b, v_j) \in G$. But then $\{(v_a, v_j), (v_j, v_b), (v_b, v_a)\}$ would be a separating triangle. Hence there is at least one vertex $c_\alpha, a < \alpha < b$, having at least two neighbors in $G - G_k$ and having no incident chords. $c_\alpha$ is the next vertex $v_k$ in our ordering.

**Theorem 8.** *The canonical ordering can be computed in linear time.*

*Proof.* We label each vertex $v$ by $Interval(v)$, which can have the following values: (a): not yet visited, (b): visited once, or (p): visited more than once and the visited edges form $p$ intervals in $Adj(v)$. We also maintain a variable $Chords(v)$ for each vertex $v$ on the exterior face, denoting the number of incident chords of $v$.

We start with $v_n$ and $v_{n-1}$ and initialize the labels of their neighbors. We compute the ordering in reverse order and update the labels after choosing a vertex $v_k$ as follows: we visit each neighbor $v$ of $v_k$ along the edge connecting them. Let $c_i, \ldots, c_j$ $(j > i)$ be the neighbors (in this order) of $v_k$ in $G_{k-1}$. If $j = i + 1$, then there was a chord $(c_i, c_j)$ in $G_{k-1}$, hence we decrease $Chords(c_i)$ and $Chords(c_j)$ by one, since $(c_i, c_j)$ becomes part of $C_{k-1}$. If $j > i + 1$, then for each $c_l$ $(i < l < j)$, we compute $Chords(c_l)$. If $c_l$ has a chord to $v$, then we also increase $Chords(v)$ with one. This is done by marking the vertices that are part of the exterior face. For each $c_l$ ($i \leq l \leq j$), we update $Interval(c_l)$: if $c_l$ has label $Interval(c_l) = (a)$, label $(b)$ replaces label $(a)$. If $Interval(c_l)$ has label $(b)$ and $v_k$ in $Adj(c_l)$ is adjacent to a previous visited vertex in $Adj(c_l)$, then $Interval(c_l)$ becomes $(1)$, otherwise it becomes $(2)$. Otherwise assume $Interval(c_l) = (p)$, with $p \geq 1$. If the two incident vertices $v'$ and $v''$ of vertex $v_k$ in $Adj(c_l)$ are already visited, then $Interval(c_l)$ becomes $(p-1)$. If none of $v'$ and $v''$ is visited, then $Interval(c_l)$ becomes $(p+1)$, else $Interval(c_l)$ is not changed. It is clear that $Interval(c_l) = (p)$ means that the vertices already visited and incident to $c_l$ are composed of $p$ intervals in $Adj(c_l)$.

By Theorem 7 it follows that, if $k \geq 3$, then there is a vertex $v$ with $Interval(v) = 1$ and $Chords(v) = 0$, and this can be chosen as the next vertex $v_k$ in our ordering. We mark $v$ as being visited. Since there are only a linear number of edges, we can find the canonical ordering in linear time.

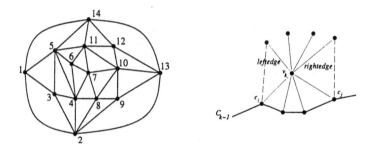

**Fig. 4.** The canonical ordering from the graph of Figure 1.

## 4.2 From a canonical ordering to a REL

To compute a REL of a PTP graph $G$, we first add an edge connecting two non-adjacent exterior vertices of $G$. This gives a 4-connected planar triangular graph $G'$. We compute a canonical numbering of $G'$ and then delete the added edge. The four exterior vertices of $G$ are now numbered as $v_1, v_2, v_{n-1}, v_n$, respectively. Next we show that a REL of $G$ can be easily derived from the canonical ordering.

First, for each edge $(v_i, v_j)$ of $G$, direct it from $v_i$ to $v_j$, if $i < j$. Define the *base-edge* of a vertex $v_k$ to be the edge $(v_l, v_k)$ for which $l < k$ is minimal. The vertex $v_k$ has incoming edges from $c_i, \ldots, c_j$ belonging to $C_{k-1}$ (the exterior face of $G_{k-1}$), assuming in this order from left to right. We call $c_i$ the *leftpoint* of $v_k$ and $c_j$ the *rightpoint* of $v_k$. Let $v_{k_1}, \ldots, v_{k_l}$ be the higher-numbered neighbors of $v_k$, in this order from left to right. We call $(v_k, v_{k_1})$ the *leftedge* and $(v_k, v_{k_l})$ the *rightedge* of $v_k$.

407

**Lemma 9.** *A base-edge cannot be a leftedge or a rightedge.*

**Lemma 10.** *An edge is either a leftedge, a rightedge or a base-edge.*

We construct a REL for $G$ as follows: all leftedges belong to $T_1$, all rightedges belong to $T_2$. The base-edge $(c_\alpha, v_k)$ of $v_k$ is added to $T_1$, if $\alpha = j$, to $T_2$, if $\alpha = i$, and otherwise arbitrary to either $T_1$ or $T_2$. (The four exterior edges belong to neither $T_1$ nor $T_2$.)

**Lemma 11.** $\{T_1, T_2\}$ *forms a regular edge labeling for* $G$.

*Proof.* Let $v_{k_1}, \ldots, v_{k_d}$ be the outgoing edges of the vertex $v_k$ ($3 \leq k \leq n - 2$). It follows from Theorem 7 that $d \geq 2$. Then $(v_k, v_{k_1})$ is the leftedge of $v_k$ and is in $T_1$. $(v_k, v_{k_d})$ is the rightedge of $v_k$ and is in $T_2$. The edges $(v_k, v_{k_2}), \ldots, (v_k, v_{k_{d-1}})$ are the base-edges of $v_{k_2}, \ldots, v_{k_{d-1}}$, respectively. Let the vertex $v_{k_\beta}$ ($1 \leq \beta \leq d$) be the highest-numbered neighbor of $v_k$. Then all vertices from $v_{k_1}$ to $v_{k_\beta}$ have a monotone increasing number, as well as the vertices from $v_{k_d}$ to $v_{k_\beta}$. Otherwise there was a vertex $v_{k_l}$ such that $v_{k_{l-1}}$ and $v_{k_{l+1}}$ are numbered higher than $v_{k_l}$. But this implies that $v_k$ is the only lower-numbered neighbor of $v_{k_l}$, which is a contraction with the canonical ordering of $G$. Hence for every $v_{k_l}$ ($1 < l < d, l \neq \beta$), either $k_{l-1} < k_l < k_{l+1}$ or $k_{l-1} > k_l > k_{l+1}$. Thus, by the construction of $T_1$ and $T_2$, the edges $(v_k, v_{k_l})$ are added to $T_1$, if $1 \leq l < \beta$, and to $T_2$, if $\beta < l \leq d$. The edge $(v_k, v_{k_\beta})$ is arbitrarily added to either $T_1$ or $T_2$. This completes the proof that the edges appear in counterclockwise order around $v_k$ as follows: a set of edges in $T_2$ entering $v_k$; a set of edges in $T_1$ entering $v_k$; a set of edges in $T_2$ leaving $v_k$; a set of edges in $T_1$ leaving $v_k$.

Let $v_{1_1}, \ldots, v_{1_d}$ be the higher numbered neighbors of $v_1$ from left to right. Then $v_{1_1} = v_n$ and $v_{1_d} = v_2$, and by the argument described above, $(v_1, v_{1_2}), \ldots, (v_1, v_{1_{d-1}})$ belong to $T_2$. Similarly, all outgoing edges of $v_2$ belong to $T_1$. All incoming edges of $v_{n-1}$ belong to $T_2$, and all incoming edges of $v_n$ belong to $T_1$. This completes the proof.

Since the construction of $\{T_1, T_2\}$ from the canonical numbering can be easily done in $O(n)$ time, Theorem 8 and Lemma 11 constitute our linear time REL algorithm. See Figure 4 for the construction of a REL from a canonical ordering.

## 5  Algorithm for visibility representation

The *visibility representation* of a planar graph $G$ maps the vertices of $G$ to horizontal line segments and edges of $G$ to vertical line segments [9, 11]. In this section, we show that the canonical ordering can be used to construct a more compact visibility representation for a 4-connected planar triangular graph $G$.

First let the edges of $G$ be directed as $v_i \to v_j$, if $i < j$. $G$ is a planar $st$-graph and every vertex (except $v_1, v_2, v_{n-1}$ and $v_n$) has at least 2 incoming and 2 outgoing edges. Let $d(v)$ denote the length of the longest path from the source $v_1$ of $G$ to $v$. We construct the dual graph $G^*$ of $G$ and direct the edges of $G^*$ as follows: if $F_l$ and $F_r$ are the left and the right face of some edge $(v, w)$ of $G$, direct the dual edge from $F_l$ to $F_r$ if $(v,w) \neq (v_1,v_n)$ and from $F_r$ to $F_l$ if $(v,w) = (v_1,v_n)$. $G^*$ is a planar $st$-graph. For each node $F$ of $G^*$, let $d^*(F)$ denote the length of the longest path from the source node of $G^*$ to $F$. The algorithm for constructing the visibility representation of Rosenstiehl & Tarjan [9] and Tamassia & Tollis [11] is almost identical to the rectangular dual algorithm.

**Algorithm 3:** Visibility Representation
Input: A 4-connected planar triangular graph $G$.

1. Compute a canonical ordering of $G$.
2. Construct the planar $st$-graphs $G$ and its dual $G^*$.
3. Compute $d(v)$ for the vertices of $G$ and $d^*(F)$ for the nodes of $G^*$.
4. For each vertex $v$ of $G$ do:
   Draw horizontal line between $(d^*(left(v)), d(v))$ and $(d^*(right(v)) - 1, d(v))$.
5. For each edge $(u,v)$ of $G$ do:
   Draw vertical line between $(d^*(left(u,v)), d(u))$ and $(d^*(left(u,v)), d(v))$.

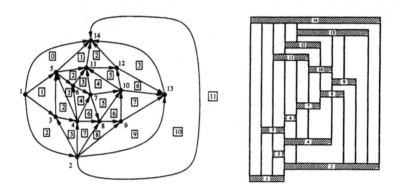

**Fig. 5.** The canonical ordering leads to a compact visibility representation.

**Theorem 12.** VISIBILITY($G$) *constructs a visibility representation of $G$ on a grid of size at most* $(n-1) \times (n-1)$.

*Proof.* The correctness of VISIBILITY($G$) is shown in [9, 11]. We show that the grid size is at most $(n-1) \times (n-1)$. This follows directly for the height, since the length of the longest path from $v_1$ to $v_n$ is at most $n-1$.

Let $s^*$ be the source node of $G^*$ and $t^*$ be the sink node of $G^*$. Every vertex $v$ of $G$ corresponds to a face $F_v$ of $G^*$. If $v \neq v_1, v_2, v_{n-1}, v_n$, then $v$ has $\geq 2$ incoming and $\geq 2$ outgoing edges, hence the two directed paths from $low(F_v)$ to $high(F_v)$ both have length $\geq 2$. Let $G^{*'}$ be the graph obtained from $G^*$ by removing the sink node $t^*$ and its incident edges. (In Figure 5, $t^*$ is the node represented by the square labeled by 11.) This merges the faces $F_{v_1}, F_{v_2}$ and $F_{v_n}$ of $G^*$ into one face $F'$. Note that for every face $F \neq F_{v_{n-1}}$ of $G^{*'}$, the two directed paths of $F$ between $low(F)$ and $high(F)$ in $G^{*'}$ have length $\geq 2$.

Let $s^{*'}$ be the source of $G^{*'}$ and let $t^{*'}$ be the sink of $G^{*'}$. Notice that $s^{*'} = s^* = low(F')$ and $t^{*'} = left((v_2, v_n)) = high(F')$. (In Figure 5, $t^{*'}$ is the node represented by the square labeled by 10.) Clearly, there are at least two edges $e$ with $F_{v_{n-1}} = left(e)$, and the only edge $e$ with $right(e) = F_{v_{n-1}}$ has endpoint $t^{*'}$. Let $P_{long}$ be any longest path from $s^{*'}$ to $t^{*'}$. Then the length of any longest path from $s^*$ to $t^*$ in $G^*$ is 1 plus the length of $P_{long}$.

We claim that $P_{long}$ has at most one consecutive sequence of edges in common with any face $F$ of $G^{*'}$. Toward a contradiction assume the claim is not true. Suppose that $P_{long}$ visits some nodes of $F$, assume that $w_1$ is the last one, then $l \geq 1$ nodes $u_1, \ldots, u_l \notin F$, then some nodes of $F$ again, let $w_d$ be the first one. Let $w_2, \ldots, w_{d-1}$ be the nodes, in this order, of $F$, which are not visited by $P_{long}$ (see Figure 6.) Suppose $F = right((w_1, w_2))$. (If $F = left((w_1, w_2))$, the proof is similar.) Let $F_1 = left((w_1, w_2))$. Notice that $w_1 = low(F_1)$. The directed path of $F_1$, starting with edge $(w_1, w_2)$, has length $\geq 2$. Hence $w_2$ has an outgoing edge to a node of $F_1$, and an outgoing edge to $w_3$. Thus $w_2 = low(F_2)$, with $F_2 = left((w_2, w_3))$. Repeating this argument it follows that $w_{d-1} = low(F_{d-1})$, with $F_{d-1} = left((w_{d-1}, w_d))$. However it is easy to see that $w_d = high(F_{d-1})$. This means that one of the two directed paths of $F_{d-1}$ has length 1. This contradiction proves the claim.

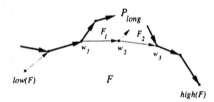

**Fig. 6.** Example of the proof of Theorem 5.1.

When traversing an edge $e$ of $P_{long}$, we visit either $left(e)$ or $right(e)$ (or both) for the first time. We *assign* each edge $e$ to the face $F$, with $e \in F$, which we visit for the first time now. $G^{*'}$ has $n-2$ faces. To every face $F$ of $G^{*'}$, by the claim, at most one edge $e \in P_{long}$ is assigned. Hence the longest path from $s^*$ to $t^*$ in $G^*$ has length $\leq n-1$.

VISIBILITY$(G)$ can be applied to a general 4-connected planar graph by first triangulating it. (The triangulation of a 4-connected planar graph is clearly still 4-connected.) Since the worst-case bounds for visibility representation by applying an arbitrary $st$-numbering is $(2n - 5) \times (n - 1)$ [9, 11], our algorithm reduces the width of the visibility representation by a factor 2 in the case of 4-connected planar graphs. Maybe this approach can be used to obtain better grid bounds in general, by splitting the graph into 4-connected components. Consider for this problem a planar triangular graph $G$. Let $C$ be a separating triangle, such that there are no separating triangles inside $C$. The subgraph inside $C$ yields a 4-connected component, say $B_1$. $B_1$ can be drawn within the required bounds. Drawing $B_1$ inside $G - B_1$ may increase the drawing of $G - B_1$ by at most $|B_1| - 1$ in height and width. If the face $F$ on the vertices $u, v, w$ is not a rectangle in the visibility representation of $G - B_1$, then this is no problem. The difficult case is when $F$ is a rectangle. Solving this remaining problem gives an important improvement in the visibility representations, which plays a major role in a lot of practical commercial environments.

The canonical ordering, presented in this paper, implies an acyclic orientation of the graph, in which every vertex (except $v_1, v_2, v_{n-1}, v_n$) has $\geq 2$ incoming and $\geq 2$ outgoing edges. This extends the results for the $st$-ordering for biconnected planar graphs [9] (in which every vertex $v, v \neq v_1, v_n$, has $\geq 1$ incoming and $\geq 1$ outgoing edge in the acyclic orientation), and the canonical ordering for planar triangular graphs [7] (in which every vertex $v, v \neq v_1, v_2, v_n$, has $\geq 2$ incoming and $\geq 1$ outgoing edge in the acyclic orientation). Another observation is that the canonical ordering, presented in section 4, gives a simple algorithm to test whether a planar triangular graph is 4-connected.

An interesting research field is to problem of computing a canonical ordering of a 4-connected planar graph such that $v_{i+1}$ is a neighbor of $v_i$. This would yield a simple algorithm for constructing hamiltonian circuits in 4-connected triangular planar graphs. We leave this question open for the interested reader.

# References

1. Bhasker, J., and S. Sahni, A linear algorithm to check for the existence of a rectangular dual of a planar triangulated graph, Networks 7 (1987), pp. 307-317.
2. Bhasker, J., and S. Sahni, A linear algorithm to find a rectangular dual of a planar triangulated graph, *Algorithmica* 3 (1988), pp. 247-178.
3. Eades, P., and R. Tamassia, *Algorithms for Automatic Graph Drawing: An Annotated Bibliography*, Dept. of Comp. Science, Brown Univ., Technical Report CS-89-09, 1989.
4. Fraysseix, H. de, J. Pach and R. Pollack, How to draw a planar graph on a grid, *Combinatorica* 10 (1990), pp. 41-51.
5. He, X., On finding the rectangular duals of planar triangulated graphs, *SIAM J. Comput.*, to appear.
6. He, X., *Efficient Parallel Algorithms for two Graph Layout Problems*, Technical Report 91-05, Dept. of Comp. Science, State Univ. of New York at Buffalo, 1991.
7. Kant, G., Drawing planar graphs using the *lmc*-ordering, *Proc. 33th Ann. IEEE Symp. on Found. of Comp. Science*, Pittsburgh, 1992, pp. 101-110.
8. Koźmiński, K., and E. Kinnen, Rectangular dual of planar graphs, *Network* 5 (1985), pp. 145-157.
9. Rosenstiehl, P., and R. E. Tarjan, Rectilinear planar layouts and bipolar orientations of planar graphs, *Discr. and Comp. Geometry* 1 (1986), pp. 343-353.
10. Schnyder, W., Embedding planar graphs on the grid, in: *Proc. 1st Annual ACM-SIAM Symp. on Discr. Alg.*, San Francisco, 1990, pp. 138-147.
11. Tamassia, R., and I. G. Tollis, A unified approach to visibility representations of planar graphs, *Discr. and Comp. Geometry* 1 (1986), pp. 321-341.

# A More Compact Visibility Representation*

Goos Kant

Dept. of Computer Science
Utrecht University
Padualaan 14
3584 CH Utrecht
the Netherlands
goos@cs.ruu.nl

**Abstract.** In this paper we present a linear time and space algorithm for constructing a visibility representation of a planar graph on an $(\lfloor \frac{3}{2}n \rfloor - 3) \times (n - 1)$ grid, thereby improving the previous bound of $(2n - 5) \times (n - 1)$. To this end we build in linear time the 4-block tree of a triangulated planar graph.

## 1 Introduction

The problem of "nicely" drawing a graph in the plane has received increasing attention due to the large number of applications [3]. Examples include VLSI layout, algorithm animation, visual languages and CASE tools. Several criteria to obtain a high aesthetic quality have been established. Typically, vertices are represented by distinct points in a line or plane, and are sometimes restricted to be grid points. (Alternatively, vertices are sometimes represented by line segments.) Edges are often constrained to be drawn as straight lines or as a contiguous set of line segments (e.g., when bends are allowed). The objective is to find a layout for a graph that optimizes some cost function, such as area, minimum angle, number of bends, or satisfies some other constraint (see [3] for an up to date overview).

One of the most beautiful ways for drawing $G$ is by using a *visibility representation*. In a visibility representation every vertex is mapped to a horizontal segment, and every edge is mapped to a vertical line, only touching the two vertex segments of its endpoints. It is clear that this leads to a nice and readable picture, and it therefore gains a lot of interest. It has been applied in several industrial applications, for representing electrical diagrams and schemas (Rosenstiehl, personal communication). See Figure 1 for an example. Otten & Van Wijk [15] showed that every planar graph admits such a representation, and a linear time algorithm for constructing it is given by Rosenstiehl & Tarjan [18] (independently, Tamassia & Tollis [19] came up with the same algorithm). The size of the required grid is $(2n - 5) \times (n - 1)$, with $n$ the number of vertices. The algorithm is based on a so called *st-numbering*: a numbering $v_1, \ldots, v_n$ of the vertices such that $(v_1, v_n) \in G$ and every vertex $v_i$

---

* This work was supported by ESPRIT Basic Research Action No. 7141 (project ALCOM II: *Algorithms and Complexity*). Part of this work was done while visiting the Graph Theory workshop at the Bellairs Research Institute of McGill University (Montreal), Feb. 12-19, 1993.

$(1 < i < n)$ has neighbors $v_j$ and $v_k$ with $j < i < k$. The *height* of the drawing is the longest path from $v_1$ to $v_n$, which has length at most $n - 1$. The *width* of the drawing is the longest path in the dual graph, which is $f - 1$, where $f$ is the number of faces in $G$ (by Euler's formula: $m \leq 3n - 6$ and $f = m - n + 2$).

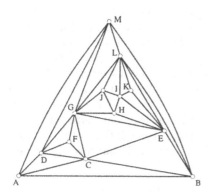 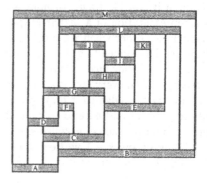

**Fig. 1.** Example of a visibility representation.

The algorithm is used in several drawing algorithms. We mention here the algorithm of Tamassia & Tollis [20] for constructing an orthogonal drawing, and the work of Di Battista, Tamassia & Tollis [5] for computing *constrained* visibility representations. Rosenstiehl & Tarjan also discuss the open problems concerning the grid size of visibility representations [18]. The requirement of using a small area seems to become a core area in the research field of graph drawing, due to the important applications in VLSI-design and chip layout (e.g., see the titles and contents of [1, 4, 10]).

In this paper we show that every planar graph can be represented by a visibility representation on a grid of size at most $(\lfloor \frac{3}{2}n \rfloor - 3) \times (n - 1)$. This improves all previous bounds considerably. An outline of the algorithm to achieve this is as follows. Assume the input graph $G$ is triangulated (otherwise a simple linear time algorithm can be applied to make it so [12, 16]). Then we split $G$ into its 4-connected components, and construct the 4-block tree of $G$. We show that we can do this in linear time for triangulated planar graphs, thereby improving the $O(n \cdot \alpha(m,n) + m)$ time algorithm of Kanevsky et al. [9] for this special case. To each 4-connected component the algorithm of Kant & He is applied, who showed that if the planar graph is 4-connected, then a visibility representation of it can be constructed with grid size at most $(n - 1) \times (n - 1)$ [13]. The representations of the 4-connected components are combined into one entire drawing, leading to the desired width.

The paper is organized as follows. In Section 2 we deliver the necessary definitions, theorems and the algorithm for constructing a visibility representation. In Section 3 we give our framework for computing a more compact visibility representation. In Section 4 we show that constructing a 4-block tree of a triangulated planar graph can be constructed in linear time. Section 5 contains some final remarks and open questions.

## 2 Definition and Backgrounds

Let $G = (V, E)$ be a planar graph with $n$ vertices and $m$ edges. A graph is called *planar* if it can be drawn without any pair of crossing edges. A *planar embedding* is a representation of a planar graph in which at every vertex all edges are sorted in clockwise order when visiting them around the vertex with respect to the planar drawing. The embedding divides the plane into a number of *faces*. The unbounded face is the *exterior face* or *outerface*. A cycle of length 3 is a *triangle*. A planar graph is *triangulated* if every face is a triangle. A triangulated planar graph has $3n - 6$ edges and adding any edge to it destroys the planarity. A cycle $C$ of $G$ divides the plane into its interior and exterior region. If $C$ contains at least one vertex in its interior and its exterior, then $C$ is called a *separating cycle*. A graph $G$ is called $k$-*connected*, if deleting of any $k - 1$ vertices does not disconnect $G$. In our algorithms we need the following theorem of Kant & He.

**Theorem 1 [13].** *There exists a labeling of the vertices $v_1 = u, v_2 = v, v_3, \ldots, v_n = w$ of a triangulated 4-connected planar graph $G$ with outerface $u, v, w$, meeting the following requirements for every $4 \leq k \leq n$:*

1. *The subgraph $G_{k-1}$ of $G$ induced by $v_1, v_2, \ldots, v_{k-1}$ is 2-connected and the boundary of its exterior face is a cycle $C_{k-1}$ containing the edge $(u, v)$.*
2. *$v_k$ is in the exterior face of $G_{k-1}$, and its neighbors in $G_{k-1}$ form a (at least 2-element) subinterval of the path $C_{k-1} - \{(u, v)\}$. If $k \leq n - 2$, $v_k$ has at least 2 neighbors in $G - G_{k-1}$.*

We will call this ordering the *canonical 4-ordering*. It is shown in [13] that a canonical 4-ordering can be constructed in linear time for a 4-connected triangular planar graph $G$. Moreover, the ordering can be made such that $v_{n-1}$ is a neighbor of both $v_1$ and $v_n$. Every canonical 4-ordering of $G$ is also an $st$-numbering of $G$. Let all edges $(v_i, v_j)$ be directed $v_i \rightarrow v_j$ if $j > i$. Then all incoming edges of any vertex $v$ in $G$ appear consecutively around $v$, as do all outgoing edges, in any planar embedding of $G$. $in(v)$ and $out(v)$ denote the number of incoming and outgoing edges of $v$, respectively. Let $v$ be a vertex with incoming edges from vertices $u_1, \ldots, u_{in(v)}$ (from left to right) and outgoing edges to vertices $w_1, \ldots, w_{out(v)}$. We call $u_1$ the *leftvertex* of $v$ and $u_{in(v)}$ the *rightvertex* of $v$. $w_1$ is called the *leftup* of $v$ and $w_{out(v)}$ is called the *rightup* of $v$ (see Figure 2). The following lemma is useful:

**Lemma 2.** *Let $v_1, v_2, \ldots, v_n$ be a canonical 4-ordering of a 4-connected triangular planar graph $G$, such that $v_{n-1}$ is a neighbor of both $v_1$ and $v_n$. Then the numbering $u_i = v_{n-i+1}$ (with $1 \leq i \leq n$) is also a correct canonical 4-ordering $u_1, \ldots, u_n$ of $G$.*

**Proof.** By 4-connectivity of $G$, $v_1, v_{n-1}$ and $v_n$ form one face in $G$, hence the vertices $u_1, u_2$ and $u_n$ are forming one face. Every vertex $u_i$ ($2 < i < n - 1$) has at least 2 incoming and 2 outgoing edges, since vertex $v_{n-i+1}$ has at least 2 outgoing and 2 incoming edges. From this observation also the 2-connectivity of $G_i$, the induced subgraph on $u_1, \ldots, u_i$, follows, which completes the proof.

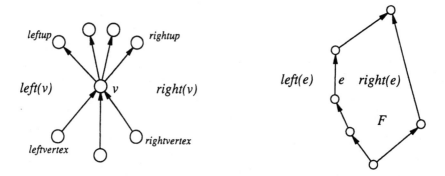

**Fig. 2.** Properties of planar $st$-graphs.

The boundary of every face consists of exactly two directed paths in $G$ [18, 19]. We define $left(e)$ ($right(e)$) to be the face to the left (right) of $e$. The face separating the incoming from the outgoing edges in the clockwise direction is called $left(v)$ and the other separating face is called $right(v)$.

Vertex $v_1$ has no incoming edges (source) and $v_n$ has no outgoing edges (sink). Let $d(v)$ denote the length of the longest path from $v_1$ to $v$. Let $D = d(v_n)$. $G^*$ denotes the dual graph of $G$: every face of $G$ is directed by a vertex in $G^*$, and there is an edge between two vertices in $G^*$, if the corresponding faces in $G$ share an edge. We direct the edges of $G^*$ as follows: if $F_l$ and $F_r$ are the left and the right face of some edge $(v, w)$ of $G$, direct the dual edge from $F_l$ to $F_r$ if $(v, w) \neq (v_1, v_n)$ and from $F_r$ to $F_l$ if $(v_r w) = (v_1, v_n)$. $G^*$ is also a planar $st$-graph. For each node $F$ of $G^*$, let $d^*(F)$ denote the number of *nodes* on the longest path from the source node of $G^*$ to $F$. Let $D^*$ denote the maximum of $d^*(F)$. The algorithm for constructing the visibility representation of Rosenstiehl & Tarjan [18] and Tamassia & Tollis [19] can now be described as follows:

$\text{VISIBILITY}(G)$;
    compute an $st$-numbering for $G$;
    construct the planar $st$-graph $G$ and its dual $G^*$;
    compute $d(v)$ for all vertices in $G$ and $d^*(F)$ for all vertices in $G^*$;
    **for** each vertex $v \neq s, t$ **do**
        draw a horizontal line between $(d^*(left(v)), d(v))$ and $(d^*(right(v)) - 1, d(v))$;
    **rof**;
    **for** vertex $s$, draw a horizontal line between $(0, 0)$ and $(D^* - 1, 0)$;
    **for** vertex $t$, draw a horizontal line between $(0, D)$ and $(D^* - 1, D)$;
    **for** each edge $(u, v) \neq (s, t)$ **do**
        draw a line between $(d^*(left(u, v)), d(u))$ and $(d^*(left(u, v)), d(v))$;
    **rof**;
    **for** edge $(s, t)$, draw a line between $(0, 0)$ and $(0, D)$;
$\text{END VISIBILITY}(G)$;

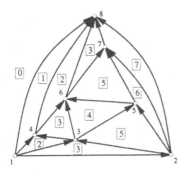 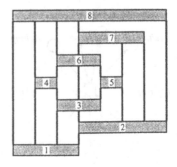

**Fig. 3.** The canonical 4-ordering and corresponding visibility representation.

In the remaining part of this paper, $\Gamma(G)$ denotes the drawing, obtained by applying VISIBILITY$(G)$. $y(v)$ denotes the $y$-coordinate of the segment of vertex $v$, and $x(u,v)$ denotes the $x$-coordinate of edge $(u,v)$ in $\Gamma(G)$. Notice that $x(v_1,v_n) < x(v_1,v_2) < x(v_2,v_n)$ in $\Gamma(G)$. The size of the drawing is the size of the smallest rectangle with sides parallel to the $x$- and $y$-axis that covers the drawing. See Figure 2 for a 4-connected triangular planar graph $G$ with a canonical 4-ordering, and the corresponding visibility representation $\Gamma(G)$. (The small squares in the figure represent the vertices of $G^*$ and the integers in the squares represent their $d^*$ values.)

**Theorem 3 [13].** *If $G$ is a 4-connected triangular planar graph and we take the canonical 4-ordering as st-ordering, then the algorithm* VISIBILITY *constructs a visibility representation of $G$ on a grid of size at most $(n-1) \times (n-1)$ in linear time and space.*

## 3 A General Compact Visibility Representation

In this section we show how we can construct a visibility representation of a planar graph $G$ on a grid, yielding a width smaller than $2n-5$. We assume that $G$ is triangulated (otherwise apply an arbitrary linear time and space triangulation algorithm [12]). To apply the result of Theorem 2.3 we split $G$ into its 4-connected components. Since $G$ is triangulated, a 4-connected component of $G$ is a 4-connected triangulated planar subgraph of $G$. From this we construct the *4-block tree* $T$ of $G$: every 4-connected component $G_b$ of $G$ is represented by a node $b$ in $T$. There is an edge between two nodes $b$ and $b'$ in $T$, if the separating triplet belongs to both $G_b$ and $G_{b'}$. By planarity every triplet of three vertices is a separating triangle and belongs to precisely two 4-connected components. The separating triplet is an interior face in $G_b$ and the exterior face of $G_{b'}$. We show in Section 4 that $T$ can be computed in linear time and space for triangulated planar graphs. See Figure 3 for the 4-block tree of the graph in Figure 1.

We root $T$ at an arbitrary node $b$. We compute a canonical ordering for $G_b$, as defined in Theorem 2.1, and direct the edges accordingly. In the algorithm, we

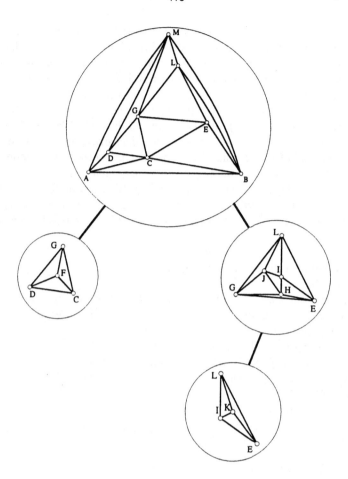

**Fig. 4.** The 4-block tree of the graph in Figure 1.

traverse $T$ top-down and visit the corresponding 4-connected components. Let $b'$ be a child of $b$ in $T$, and let $V(G_{b'})$ denote the set of vertices of $G_{b'}$. Let $u, v, w$ be the separating triplet (triangle) of $G_{b'}$. Assume the edges are directed $u \to v$ and $v \to w$ in $G_b$. We-define $c(G_{b'}) = v$. Using Lemma 2, we have two possibilities for computing a canonical ordering in $G_{b'}$ (let $n' = |V(G_{b'})|$):

NORMAL($G_{b'}$) We set $u = v_1, v = v_2$ and $w = v_{n'}$ and compute a canonical ordering $v_1, v_2, \ldots, v_n$ for $G_{b'}$,

REVERSE($G_{b'}$) We set $u = u_{n'}, v = u_2$ and $w = u_1$, and compute the canonical ordering $u_1, u_2, \ldots, u_{n'}$ to $G_{b'}$. Then the ordering is reversed: we set $v_i := u_{n'-i+1}$, for all $i$ with $1 \leq i \leq n'$.

In NORMAL($G_{b'}$), $v$ has number $v_2$, in REVERSE($G_{b'}$), $v$ has number $v_{n'-1}$. In both orderings, $u = v_1$ and $w = v_{n'}$. See also Figure 3.

Both numberings will be applied in the algorithm to achieve a more compact visibility representation. We introduce a label $l(v)$ for each vertex $v$, which can have the value *up*, *down* or *unmarked*. If $l(v) = $ *unmarked*, then $v$ is called unmarked, otherwise $v$ is called marked. Assume we visit node $b'$ in $T$, and we have to compute a canonical ordering for $G_{b'}$. Let $v = c(G_{b'})$. The value of $l(v)$ implies whether we use NORMAL$(G_{b'})$ or REVERSE$(G_{b'})$: if $l(v) = up$, we apply NORMAL$(G_{b'})$, if $l(v) = down$, we apply REVERSE$(G_{b'})$, otherwise we can do both. We will show later that using these marks, the increase of the width when drawing $G_{b'}$ "inside" $G_b$ is at most $n' - 3$ instead of $n' - 2$, when $l(v) = up$ or $down$ ($b$ the parent-node of $b'$ in $T$).

This method is applied to all 4-connected components of $G$. After directing the edges, this yields a directed acyclic graph, and applying a topological ordering yields an $st$-numbering of the vertices. Applying the algorithm VISIBILITY$(G)$ now gives the entire drawing. The complete algorithm can now be described more precisely as follows:

COMPACTVISIBILITY (input: a planar graph $G$)
    triangulate $G$;
    construct the 4-block tree $T$ of $G$, and root $T$ at arbitrary node $b$;
    compute the canonical 4-ordering for $G_b$ and direct the edges of $G_b$;
    let $n' = |V(G_b)|$; $l(v_2) = down$, $l(v_{n'-1}) := up$;
    $l(v_i) := unmarked$ for all $v_i \in G_b, i \neq 2, i \neq n' - 1$;
    **for** every child $b'$ of $b$ **do** DRAWCOMPONENT$(G_{b'})$ **rof**;
    compute an $st$-numbering in the directed graph $G$;
    apply VISIBILITY to $G$;
END COMPACTVISIBILITY

**procedure** DRAWCOMPONENT$(G')$;
    **begin**
      **Case** $l(c(G'))$ **of**
        *unmarked* : NORMAL$(G')$; $l(c(G')) := down$; $l(v_{|V(G')|-1}) := up$;
        *up*      : NORMAL$(G')$; $l(c(G')) := unmarked$; $l(v_{|V(G')|-1}) := up$;
        *down*    : REVERSE$(G')$; $l(c(G')) := unmarked$; $l(v_2) := down$;
      **for** every $v_i \in G'$ with $2 < i < |V(G')| - 1$ **do** set $l(v_i) := unmarked$ **rof**;
      direct the edges of $G'$ $v_i \to v_j$ iff $j > i$;
      **for** every child $b''$ of current node $b'$ in $T$ **do** DRAWCOMPONENT$(G_{b''})$ **rof**;
    **end**;

Since $G$ is triangulated, the following lemma can easily be verified:

**Lemma 4.** *Let $G$ be a triangular planar graph, and let $u = rightvertex(v)$ and $w = rightup(v)$. Setting $x(u, v) := x(v, w) := \max\{x(u, v), x(v, w)\}$ does not increase the width of $\Gamma(G)$.*

Let now $b'$ be a (non-root) node in $T$ with parent $b$ in $T$. Let $G'$ be the subgraph of $G$, consisting of all visited 4-connected components in COMPACTVISIBILITY. Let $u, v, w$ be the outerface of $G_{b'}$, with $u \to v$ and $v \to w$ in $G_b$. The following lemma follows ($n' = |V(G_{b'})|$) :

**Lemma 5.** *If $x(u, w) < x(u, v) < x(v, w)$ in $\Gamma(G')$, then applying* NORMAL$(G_{b'})$ *has the result that the width of $\Gamma(G' \cup G_{b'})$ is at most the width of $\Gamma(G') + n' - 3$.*

*Proof.* In $\Gamma(G_{b'})$, $x(u, v) - x(u, w) \leq n' - 2$ and $x(v, w) - x(u, w) \leq n' - 1$ by Theorem 3. In $\Gamma(G')$, $x(u, v) - x(u, w) \geq 1$ and $x(v, w) - x(u, w) \geq 2$. Hence the width of $\Gamma(G' \cup G_{b'})$ is at most the width of $\Gamma(G') + (n' - 1) - 2$.

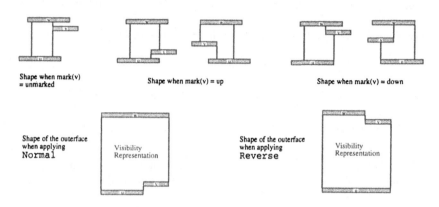

**Fig. 5.** The shape of the faces with respect to $l(v)$, NORMAL$(G_{b'})$ and REVERSE$(G_{b'})$.

The same can be proved for $x(v, w) < x(u, v) < x(u, w)$ and applying NORMAL, or when $x(u, w) < x(v, w) < x(u, v)$ or $x(u, v) < x(v, w) < x(u, w)$ and applying REVERSE. See Figure 3 for an illustration of this. Assume now that $G'$, $b$ and $b'$ are defined as in the previous lemma. The following lemma can now be proved. Let $v = c(G_{b'})$.

**Lemma 6.** *If $v$ is marked, then after applying* NORMAL$(G_{b'})$ *if $l(v) = up$ and* REVERSE$(G_{b'})$ *if $l(v) = down$, the width of $\Gamma(G' \cup G_{b'})$ is at most the width of $\Gamma(G') + n' - 3$,*

*Proof.* Let $u, v, w$ be the separating triplet of $G_{b'}$, with $u \to v$ and $v \to w$ in $G_b$, thus $v = c(G_{b'})$. If $out(v) = 1$ in $G_b$, then $l(v) = up$. Hence either $x(v, w) < x(u, v) < x(u, w)$ in $\Gamma(G')$ or we can change $\Gamma(G')$ without increasing the width (by Lemma 4) such that $x(u, w) < x(u, v) < x(v, w)$ in $\Gamma(G')$. Applying Lemma 5 yields the desired result. The same follows when $in(v) = 1$ in $G_b$, thus when $l(v) = down$.

The remaining case is when $v$ was $c(G_{b''})$ for some $b'' \neq b'$, and at the moment of visiting $b''$, $l(v) = unmarked$. Let the separating triplet of $G_{b''}$ be $u', v, w'$, with $u' \to v$ and $v \to w'$. Let $G''$ be the subgraph of $G$, consisting of the visited 4-connected components at the moment of visiting $G_{b''}$. By Lemma 4 we may assume that $x(u', v) = x(v, w')$ in $\Gamma(G'')$.

Consider the case $x(u', w') < x(u', v)$ (the case $x(u', v) < x(u', w')$ goes similar). In COMPACTVISBILITY NORMAL$(G_{b''})$ is applied. Since $in(v) = 1$ in $G_{b''}$, we can set $x(u',v)$ to $x(v,w')$ in $\Gamma(G_{b''})$ without increasing the width (see Lemma 4). This has the result that $x(v,leftup(v)) < x(u',v)$ in $\Gamma(G')$. If $w = leftup(v)$ then the proof is completed by observing that $x(u,w) < x(v,w) < x(u,v)$ in $\Gamma(G')$ and applying Lemma 5.

If $w \neq leftup(v)$, then $w = rightup(v)$ and $u = rightvertex(v)$. By Lemma 4 we may assume that both $x(u',v) = x(v,w')$ and $x(rightvertex(v),v) = x(v,rightup(v))$ holds in $\Gamma(G'')$. But since $x(u',v) < x(v,w')$ holds in $\Gamma(G_{b''})$ it directly follows that $x(rightvertex(v),v) < x(v,rightup(v))$ in $\Gamma(G'' \cup G_{b''})$. Again the proof is completed by observing that $x(u,v) < x(v,w) < x(u,w)$ in $\Gamma(G')$ and applying Lemma 5.

See Figure 6 for an illustration of the proof of Lemma 6.

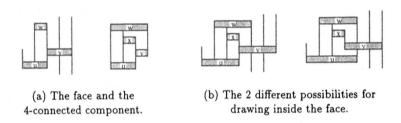

(a) The face and the 4-connected component.

(b) The 2 different possibilities for drawing inside the face.

**Fig. 6.** Illustration of Lemma 6.

**Lemma 7.** *The width of the visibility representation of $G$ is at most $\lfloor \frac{3}{2}n \rfloor - 3$.*

*Proof.* Let $b_1, \ldots, b_p$ be the nodes of $T$ in visiting order. Let $K_i$ be the number of marked vertices after visiting $b_i$ $(1 \leq i \leq p)$. Let $K_0$ be the initial number of marked vertices. $K_0 = 2$, since initially only $v_2$ and $v_{n'-1}$ are marked. When we visit $G_{b_i}$, then vertex $v_2$ or $v_{n'-1}$ is added to the current graph and is unmarked. If $l(c(G_{b_i})) \neq unmarked$, then the increase in width is at most $|V(G_{b_i})|-3$ and $l(c(G_{b_i}))$ becomes *unmarked*, i.e., $K_i = K_{i-1}$. If $l(c(G_{b_i})) = unmarked$, then the increase in width is at most $|V(G_{b_i})|-2$ and $l(c(G_{b'}))$ becomes *down*, i.e, $K_i = K_{i-1}+2$. Hence in both cases when visiting $G_{b_i}$, the width of the drawing increases by at most $|V(G_{b_i})| - 3 + \frac{K_i - K_{i-1}}{2}$. $|V(G_{b'})| - 3$ is also precisely the number of added vertices of $G_{b'}$.

Since the width of $\Gamma(G_{b_1})$ is at most $|V(G_{b_1})|-1$ and $K_p$ is even and $K_p \leq n-2$ (the source and the sink of $G$ never get marked), it follows that the total width of $\Gamma(G)$ is at most $n - 1 + \lfloor \frac{K_p - K_0}{2} \rfloor = n - 1 + \lfloor \frac{n-2-2}{2} \rfloor \leq \lfloor \frac{3}{2}n \rfloor - 3$.

Regarding the time complexity we show in Section 4 that the 4-block tree can be computed in linear time. Computing a canonical 4-ordering also requires linear time [13]. We maintain the direction of the edges of the visited 4-connected components, and from this $c(G_{b'})$ can be determined directly in $O(1)$ time. Finally VISIBILITY($G$) is applied, which requires linear time [18, 19]. This completes the following theorem.

**Theorem 8.** *There is a linear time and space algorithm for computing a visibility representation of a planar graph $G$ on a grid of size at most $(\lfloor \frac{3}{2}n \rfloor - 3) \times (n - 1)$.*

Consider the graph of Figure 1. In Figure 3 the 4-block tree is given. The visibility representation of the root-block is given in Figure 2. Drawing the other 4-connected components inside and applying VISIBILITY leads to the drawing as given in Figure 1. Notice that $l(D), l(G)$ and $l(I)$ are *down*, $l(F), l(J)$ and $l(K)$ are *up*, all the other vertices of the graph are unmarked. Hence 6 vertices are marked, and the total width is at most $n - 1 + \frac{6}{2} = 15$. (The width in the drawing in Figure 1 is 12.)

## 4   Constructing the 4-block tree

In this section we show a method for constructing the 4-block tree of a triangulated planar graph. Since this class of graphs has some special properties, there is no need to use the complicated algorithm of Kanevsky et al., which builds a 4-block tree of a general graph in $O(n \cdot \alpha(m,n) + m)$ time [9]. In our case a separating triplet is a separating triangle, which forms the basis for the algorithm.

For determining the separating triangles, we use the algorithm of Chiba & Nishizeki [2] for determining triangles in a graph. (In [17], Richards describes another linear time algorithm.) Chiba & Nishizeki first sort the vertices in $v_1, \ldots, v_n$ in such a way that $deg(v_1) \geq deg(v_2) \geq \ldots \geq deg(v_n)$. Observing that each triangle containing vertex $v_i$ corresponds to an edge joining two neighbors of $v_i$, they first mark all vertices $u$ adjacent to $v_i$ (for the current index $i$). For each marked $u$ and each vertex $w$ adjacent to $u$ they test whether $w$ is marked. If so, a triangle $v_i, u, w$ is listed. After this test is completed for each marked vertex $u$, they delete $v_i$ from $G$ and repeat the procedure with $v_{i+1}$. Starting with $v_1$, this algorithm lists all triangles without duplication in $n - 2$ steps.

In our case, we are looking for *separating* triangles. If a triangle is not separating, then it is a face in the triangulated planar graph. To test this, we store the embedding of the original graph also in adjacency lists, say in adjacency lists $Adj(v)$ for all $v \in G$. If a triangle $u, v, w$ is not separating, (i.e., a face), then $u$ and $w$ appear consecutively in $Adj(v)$, which can be tested in $O(1)$ time by maintaining crosspointers. To compute the time complexity of this algorithm, Chiba & Nishizeki use the *arboricity* of $G$, defined as the minimum number of edge-disjoint forests into which $G$ can be decomposed, and denoted by $a(G)$ [7].

**Lemma 9 [2].** $\sum_{(u,v) \in E} \min\{deg(u), deg(v)\} \leq 2 \cdot a(G) \cdot m.$

Using this lemma Ghiba & Nishizeki show that the time complexity of the algorithm is $O(a(G) \cdot m)$. If $G$ is planar then $a(G) \leq 3$ ([7]), so the algorithm runs in $O(n)$ time in case $G$ is planar.

To obtain the 4-block tree we introduce now the following data structure: Let $L$ be the list of separating triangles. $L(u, v, w)$ denotes the record in $L$, containing separating triangle $u, v, w$. $L(u, v, w)$ contains the edges $(u, v), (v, w)$ and $(w, u)$, and there are crosspointers between $L$ and the edges and vertices in $G$.

We now want to "sort" the separating triangles, containing edge $(u, v)$. Hereto we do the following: Let $Adj(v) = w_0, w_1, \ldots, w_{d-1}$ (in clockwise order around $v$ with respect to a planar embedding). We sort the separating triangles $v, w_i, w_j$ stored at $v$ in order with respect to $w_0, \ldots, w_{d-1}$. If there are separating triangles $v, w_i, w_j$ and $v, w_i, w_k$ then we set a pointer from edge $(v, w_i)$ in $L(v, w_i, w_j)$ to edge $(v, w_i)$ in $L(v, w_i, w_k)$, if $k > j$ (addition modulo $d$). We do this for every vertex $v \in G$. Of course, when visiting vertex $w_i$ and considering separating triangles $v, w_i, w_j$ and $v, w_i, w_k$, there is no need to place another directed edge between $L(v, w_i, w_j)$ and $L(v, w_i, w_k)$.

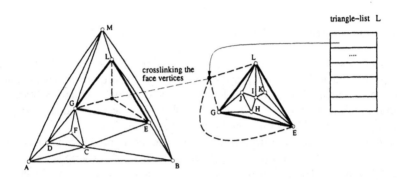

**Fig. 7.** The data structure for constructing the 4-block tree.

Observe now that when edge $(u, v)$ in $L(u, v, w)$ has no outgoing edge, then this means that when we split $G$ at $(u, v), (v, w), (w, u)$ into two subgraphs, say $G_1$ and $G_2$, then all other separating triangles, containing $(u, v)$, belong to either $G_1$ or $G_2$. We start at an arbitrary separating triangle in $L$, say $u, v, w$, where each edge in $L(u, v, w)$ has either an incoming or an outgoing edge. We split the graph at $(u, v), (v, w)$ and $(w, u)$ into two subgraphs, say $G_1$ and $G_2$. Let $Adj(v) = w_0, \ldots, w_{d-1}$ and let $u = w_0$ and $w = w_i$, $0 < i < d$. To obtain $G_1$ and $G_2$, we split $Adj(v)$ into two adjacency lists, say $Adj_1(v)$ and $Adj_2(v)$ with $Adj_1(v) = w_0, \ldots, w_i$ and $Adj_2(v) = w_i, \ldots, w_{d-1}, w_0$ (similar for $Adj(u)$ and $Adj(w)$). Let $Adj_1(v)$ correspond to $G_1$. If all other separating triangles, containing $(u, v)$, belong to $G_1$, then we introduce $w_0$ in $Adj_2(v)$. This yields that all other separating triangles in $L$, containing $(u, v)$, still point to the right edge in the data structure, viz. in the adjacency list of $G_1$. Testing whether the other separating triangles, containing $(u, v)$, belong to $G_1$ or $G_2$ can be tested by checking whether $(u, v)$ in $L(u, v, w)$ has an incoming or an outgoing edge. we introduce $w_0$ in $Adj_1(v)$. Similar is done for the edges $(v, w)$ and $(w, u)$ with respect to $Adj(w)$ and $Adj(u)$. We mark $L(u, v, w)$ as visited and delete the incoming and outgoing edges of $L(u, v, w)$, and we continue until all separating triangles in $L$ are visited.

To construct the 4-block tree, we apply a simple traversal through the datastructure for determining the connected components. The connections via the face vertices give the connections in the 4-block tree. For every 4-connected component we add pointers to its three vertices on the outerface. The complete algorithm can now be described as follows:

CONSTRUCT 4-BLOCK TREE
    enumerate all separating triangles and store them in $L$;
    sort the separating triangles and add directed edges in $L$;
    **while** not every triangle in $L$ is visited **do**
        Let $L(u, v, w)$ be a record in $L$ such that each edge $(u, v), (v, w)$ and $(w, u)$
            in $L(u, v, w)$ has either an incoming or outgoing edge;
        split the graph at edges $(u, v), (u, w), (v, w)$;
        set a pointer between the two corresponding faces;
    **od**;
    determine the connected components and construct the 4-block tree;
END CONSTRUCT 4-BLOCK TREE

**Theorem 10.** *The 4-block tree of a triangulated planar graph can be constructed in linear time and space.*

*Proof.* Determining and storing the separating cycles requires $O(n)$ time, because every planar graph has at most $n - 4$ separating triangles. Sorting the separating triangles at vertex $v$ can be done in $O(deg(v))$ time by using (double) bucket-sort, since vertex $v$ belongs to at most $deg(v) - 2$ separating triangles. Hence the total sorting time is $O(n)$. By maintaining a sublist of $L$ where we store all separating triangles, which can be visited next, we can find the next separating triangle in $O(1)$ time. Since there are crosspointers between the edges and vertices in $G$ and the corresponding entry in $L$, we can split the graph at the separating triangle in $O(1)$ time. Determining the connected components and building up the 4-block tree is achieved by a simple traversal through the graph, which completes the proof.

## 5   Final Remarks

In this paper a new and rather simple method for constructing a visibility representation is given, based on the canonical ordering for 4-connected triangular planar graphs, the 4-block tree, and the general algorithm for constructing a visibility representation, introduced by Rosenstiehl & Tarjan [18] and Tamassia & Tollis [19]. We decreased the width from $2n - 5$ to $\lfloor \frac{3}{2}n \rfloor - 3$. This bound is tight, since there exist planar graphs, for which the algorithm COMPACTVISIBILITY indeed requires an $(\lfloor \frac{3}{2}n \rfloor - 3) \times (n - 1)$ grid (see Figure 5). For 4-connected planar graphs, it was already known that a visibility representation can be constructed on an $(n - 1) \times (n - 1)$ grid [13].

    The important open question is whether the bound of $\lfloor \frac{3}{2}n \rfloor - 3$ for the width is also a lower bound, or in other words: do there exist planar graphs with $n$ vertices, for which any visibility representation requires a grid with width at least $\frac{3}{2}n + O(1)$? This is a hard problem, and it is even difficult to construct a class of planar graphs with $n$ vertices, for which any drawing requires a width of size $c \cdot n$ with $c > 1$.

    Another (difficult) technique for optimization is the following: From the algorithm it directly follows that for every face $u, v, w$ in the visibility representation of the visited components so far, not drawn as a rectangle, we can set $l(v) = up$ or $down$ (according to the shape of the face), when $u \rightarrow v$ and $v \rightarrow w$. By the algorithm VISIBILITY it follows that $u$ is the rightvertex of $v$ and $w$ is the rightup of $v$. The problem which arises now is: how can we dynamically maintain the longest paths in the dual graph of $G$, while inserting the 4-connected components?

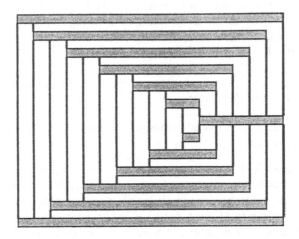

**Fig. 8.** A graph requiring an $(\lfloor \frac{3}{2}n \rfloor - 3) \times (n - 1)$ grid by COMPACTVISIBILITY.

We also presented an algorithm for the construction of the 4-block tree. A linear time implementation in the case of triangulated planar graphs is presented. When the planar graph is not triangulated, then the complexity of constructing a 4-block tree is still $O(n \cdot \alpha(m,n) + m)$, by applying the general algorithm [9]. The open question is whether determining the 4-connected components of a planar graph can be computed in $O(n)$ time. From this it would not be too difficult to compute the 4-block tree in linear time. The problem is already linear time solvable in case of 2- and 3-connected components (see e.g., Hopcroft & Tarjan [8]), hence solving this open problem yields a nice generalization.

As a last subject we consider the method of triangulating planar graphs. It would be interesting to triangulate $G$ such that it is 4-connected. Indeed, $G$ may have separating triangles, in which case it is not possible. Suppose now $G$ has no separating triangles. (This can be tested in linear time.) Can we triangulate $G$ such that the triangulation is 4-connected? Or in general, can we triangulate $G$ such that no new separating triangles are introduced? Can we construct a very compact visibility representation of $G$ without triangulating $G$? These problems have practical applications for our algorithm for constructing visibility representations, but also for constructing the rectangular duals in [13]. We leave these questions for the interested reader.

# References

1. Bertolazzi, P., R.F. Cohen, G. Di Battista, R. Tamassia and I.G. Tollis, How to draw a series-parallel graph, in: O. Nurmi and E. Ukkonen (Eds.), *Proc. Scand. Workshop*

*on Algorithm Theory (SWAT'92)*, Lecture Notes in Computer Science 621, Springer-Verlag, 1992, pp. 272–283.

2. Chiba, N., and T. Nishizeki, Arboricity and subgraph listing algorithms, *SIAM J. Comput.* 14 (1985), pp. 210–223.

3. Di Battista, G., Eades, P., and R. Tamassia, I.G. Tollis, *Algorithms for Automatic Graph Drawing: An Annotated Bibliography*, Dept. of Comp. Science, Brown Univ., Technical Report, 1993.

4. Di Battista, G., R. Tamassia and I.G. Tollis, Area requirement and symmetry display in drawing graphs, *Discrete and Comp. Geometry* 7 (1992), pp. 381–401.

5. Di Battista, G., R. Tamassia and I.G. Tollis, Constrained visibility representations of graphs, *Inform. Process. Letters* 41 (1992), pp. 1–7.

6. Fraysseix, H. de, J. Pach and R. Pollack, How to draw a planar graph on a grid, *Combinatorica* 10 (1990), pp. 41–51.

7. Harary, F., *Graph Theory*, Addison-Wesley Publishing Company, Inc., Reading, Mass., 1972.

8. Hopcroft, J., and R.E. Tarjan, Dividing a graph into triconnected components, *SIAM J. Comput.* 2 (1973), pp. 135–158.

9. Kanevsky, A., R. Tamassia, G. Di Battista and J. Chen, On-line maintenance of the four-connected components of a graph, in: *Proc. 32th Annual IEEE Symp. on Found. of Comp. Science*, Puerto Rico, 1991, pp. 793–801.

10. Kant, G., Drawing planar graphs using the *lmc*-ordering, *Proc. 33th Ann. IEEE Symp. on Found. of Comp. Science*, Pittsburgh, 1992, pp. 101–110.
    Revised and extended version in:

11. Kant, G., *Algorithms for Drawing Planar Graphs*, PhD thesis, Dept. of Computer Science, Utrecht University, 1993.

12. Kant, G., On triangulating planar graphs, submitted to *Information and Computation*, 1993.

13. Kant, G., and X. He, Two Algorithms for Finding Rectangular Duals of Planar Graphs, Tech. Report RUU-CS-92-41, Dept. of Computer Science, Utrecht University, 1992.

14. Nummenmaa, J.; Constructing compact rectilinear planar layouts using canonical representation of planar graphs, *Theoret. Comp. Science* 99 (1992), pp. 213–230.

15. Otten, R.H.J.M., and J.G. van Wijk, Graph representation in interactive layout design, in: *Proc. IEEE Int. Symp. on Circuits and Systems*, 1978, pp. 914–918.

16. Read, R.C., A new method for drawing a graph given the cyclic order of the edges at each vertex, *Congr. Numer.* 56 (1987), pp. 31–44.

17. Richards, D., Finding short cycles in planar graphs using separators, *J. Alg.* 7 (1986), pp. 382–394.

18. Rosenstiehl, P., and R.E. Tarjan, Rectilinear planar layouts and bipolar orientations of planar graphs, *Discr. and Comp. Geometry* 1 (1986), pp. 343–353.

19. Tamassia, R., and I.G. Tollis, A unified approach to visibility representations of planar graphs, *Discr. and Comp. Geometry* 1 (1986), pp. 321–341.

20. Tamassia, R., and I.G. Tollis, Planar grid embedding in linear time, *IEEE Trans. Circuits and Systems* 36 (1989), pp. 1230–1234.

# List of Participants

H. L. Bodlaender
Dept. of Computer Science
Utrecht University
P.O. Box 80.089
3508 TB Utrecht
the Netherlands
hansb@cs.ruu.nl

J.E. Boillat
Ingenieurschule Bern
P.O. Box 5557
CH-3001 Berne 1
Switzerland
boillat@iam.unib.ch

A. Brandstädt
FB 11 Mathematik
FG Informatik
UNI-GH-Duisburg
Postfach 101503
D-47048 Duisburg 1
Germany
ab@marvin.uni-duisburg.de

J. Chen
Dept. of Computer Science
Texas A&M University
College Station, TX 77843-3112
USA
chen@cs.tamu.edu

Z. Czech
Institute of Computer Science
Silesia University of Technology
Pstrowskiego 16
44-100 Gliwice
Poland
zjc@gleto2.gliwice.edu.pl

E. Dahlhaus
Basser Dept. of Computer Science
University of Sydney
New-South-Wales 2006
Australia
dahlhaus@cs.su.oz.au

J. Desel
Institut für Informatik
TU München
Postfach 202420
D-80290 München
Germany
desel@informatik.tu-muenchen.de

A. Dessmark
Dept. of Computer Science
Lund University
Box 118
S-221 00 Lund
Sweden
Anders.Dessmark@dna.lth.se

H. Dörr
Institut für Informatik
Freie Universität Berlin
Takustrasse 9
D-14195 Berlin 33
Germany
doerr@inf.fu-berlin.de

S. Fischer
Department of Mathematics
and Computer Science
University of Amsterdam
Plantage Muidergracht 24
1018 TV Amsterdam
the Netherlands
sophie@fwi.uva.nl

B. de Fluiter
Dept. of Computer Science
Utrecht University
Padualaan 14
3584 CH Utrecht
the Netherlands
babette@cs.ruu.nl

A. Franzke
Institut für Informatik
Universität Koblenz-Landau
Rheinau 1
D-56075 Koblenz
Germany
franzke@informatik.uni-koblenz.de

O. Garrido
Dept. of Computer Science
Lund University
Box 118
S-221 00 Lund
Sweden
Oscar.Garrido@dna.lth.se

A. Gibbons
Dept. of Computer Science
University of Warwick
Gibbett Hill Road
Coventry CV4 7AL
England
amg@dcs.warwick.ac.uk

S. Haldar
(visiting at: Dept of Computer Science
Utrecht University)
Theoretical Computer Science Group
Tata Institute of Fundamental Research
Homi Bhabha Road
Bombay 400 005
India
haldar@tcs.tifr.res.in

G. Havas
Dept. of Computer Science
University of Queensland
Queensland 4072
Australia
havas@cs.uq.oz.au

N.W. Holloway
Dept. of Computer Science
University of Warwick
Gibbett Hill Road
Coventry CV4 7AL
England

G. Ivanyos
Computer and Automation Institute
Hungarian Academy of Sciences
Lágymányosi u.11
1111 Budapest
Hungary
ig@ilab.sztaki.hu

Z. Ivković
Department of Computer
and Information Sciences
University of Delaware
Newark, DE 19716
USA
ivkovich@dewey.udel.edu

K. Jansen
FB 11 Mathematik
FG Informatik
Universität Duisburg
Postfach 110503
D-47048 Duisburg 1
Germany
klaus@marvin.uni-duisburg.de

G. Kant
Dept. of Computer Science
Utrecht University
P.O. Box 80.089
Padualaan 14
3584 CH Utrecht
the Netherlands
goos@cs.ruu.nl

H. Kaplan
Dept. of Computer Science
Tel-Aviv University
Tel-Aviv 69978
Israel
haimk@math.tau.ac.il

M. Kaufmann
W. Schickard Institut f. Informatik
Universität Tübingen
Sand 13
D-72076 Tübingen
Germany
mk@informatik.uni-tuebingen.de

K. Kawaguchi
Nagoya Institute of Technology
Gokiso-cho Syowa-ku
Nagoya, Aichi 466
Japan
kawaguchi@laser.elcom.nitech.ac.jp

R. Klasing
FB 17 Mathematik-Informatik
Universität-GH Paderborn
Warburgerstrasse 100
D-33095 Paderborn
Germany
klasing@uni-paderborn.de

W. Knödel
Institut für Informatik
Universität Stuttgart
Breitwiesenstrasse 20-22
D-70565 Stuttgart 80
Germany

D. Krizanc
School of Computer Science
Carleton University
Ottawa, Ontario K1S 5B6
Canada
krizanc@scs.carleton.ca

L. Kučera
Dept. of Applied Mathematics
Charles University
Malostranské nam. 25
118 00 Prague
Czech Republik
ludek@altec.ms.mff.cuni.cz

J. van Leeuwen
Dept. of Computer Science
Utrecht University
P.O. Box 80.089
Padualaan 14
3584 CH Utrecht
the Netherlands
jan@cs.ruu.nl

S. Leonardi
Dipart. di Inform. e Sistemistica
Universita di Roma 'La Sapienza'
Via Salaria 113
I-00198 Roma
Italy
leonardi@disparcs.ing.uniroma1.it

N-W. Lin
Dept. of Computer Science
University of Arizona
Tucson, AZ 85721
USA
naiwei@cs.arizona.edu

Z. Lonc
Instytut Mathematyki
Warsaw University of Technology
Pl. Politechniki 1
Warsaw, 00-661
Poland
zblonc@plwatu21.bitnet

M. Marathe
Dept. of Computer Science
University at Albany
SUNY
Albany, NY 12222
USA
madhav@cs.albany.edu

E.W. Mayr
Institut für Informatik
TU München
Postfach 202420
D-80290 München 2
Germany
mayr@informatik.tu-muenchen.de

M. Nagl
Lehrstuhl Informatik III
RWTH Aachen
Ahornstrasse 55
D-52056 Aachen
Germany
nagl@rwthi3.informatik.rwth-aachen.de

K. Nakano
Advanced Research Laboratory
Hitachi Ltd.
Hatoyama
Saitama, 350-03
Japan
nakano@harl.hitachi.co.jp

S. Nikoletseas
Department of Computer Science
and Engineering
Patras University
Patras 26110
Greece
nikole@grpatvx1.bitnet

S. Öhring
Lehrstuhl für Informatik I
Universität Würzburg
Am Hubland
D-97074 Würzburg
Germany
oehring@informatik.uni-wuerzburg.de

S. Olariu
Dept. of Computer Science
Old Dominion University
Norfolk, VA 23529-0162
USA
olariu@cs.odu.edu

M. Punt
WG'93 secretary
Dept. of Computer Science
Utrecht University
Padualaan 14
3584 CH Utrecht
the Netherlands
margje@cs.ruu.nl

S. Ravindran
Dept. of Computer Science
University of Warwick
Gibbet Hill Road
Coventry CV4 7AL
England

H. Rohnert
Siemens AG
ZFE BT SE 32
Otto-Hahn-Ring 6
D-8000 Munich
Germany
rohnert@km21.zfe.siemens.de

I. Schiermeyer
Lehrstuhl C für Mathematik
RWTH Aachen
Templergraben 55
D-52056 Aachen
Germany
LN010SC@dacth11.bitnet

G. Schmidt
Fakultät für Informatik
Universität der BW München
Werner-Heisenberg-Weg 39
D-85577 Neubiberg
Germany
schmidt@informatik.unibw-muenchen.de

J.F. Sibeijn
Max-Planck-Institut für Informatik
Universität des Saarlandes
Im Stadtwald
D-66123 Saarbrücken
Germany
jopsi@mpi-sb.mpg.de

P. Spirakis
Computer Technology Institute
P.O. Box 1122
26 110 Patras
Greece
SPIRAKIS@GRPATVX1.bitnet

B. von Stengel
Informatik 5
Universität der BW München
Werner-Heisenberg-Weg 39
D-85577 Neubiberg
Germany
i51bbvs@rz.unibw-muenchen.de

O. Sýkora
Institute for Informatics
Slovak Academy of Sciences
Dúbravská 9
842 35 Bratislava
Slovak Republic
sykorao@savba.cs

R.B. Tan
Dept. of Mathematics and Comp. Science
Univ. of Sciences and Arts of Oklahoma
P.O. Box 82345
Chickasha, OK 73018-0001
USA
rtan@mercur.usao.edu
also: rbtan@cs.ruu.nl

J. Wiedermann
Institute of Computer Science
Academy of Sciences of the Czech Republic
Pod vodarenskou vezi 2
182 07 Prague 8
Czech Republic
wieder@uivt.cas.cz

G. Tinhofer
Institut für Mathematik
T.U. München
Postfach 202420
D-80290 München
Germany
Gottfried.Tinhofer@
statistik.tu-muenchen.dbp.de

H.-J. Voss
TU Dresden
Abteilung Mathematik
Mommsenstrasse 13
D-01062 Dresden
Germany

K. Wada
Dept.of Electrical and Computer Engineer-
ing
Nagoya Institute of Technology
Gokiso-cho Syowa-ku
Nagoya, Aichi 466
Japan
wada@laser.elcom.nitech.ac.jp

E. Wanke
Gesellschaft für Math. und Datenverarb.
Schloss Birlinghoven
D-53757 Sankt Augustin
Germany
wanke@gmd.de

I. Wegener
FB Informatik (Lehrstuhl II)
Universität Dortmund
Postfach 500500
D-44221 Dortmund
Germany
wegener@cantor.informatik.uni-dortmund.de

# Authors Index

# Lecture Notes in Computer Science

For information about Vols. 1–719
please contact your bookseller or Springer-Verlag

# Springer-Verlag
## and the Environment

We at Springer-Verlag firmly believe that an international science publisher has a special obligation to the environment, and our corporate policies consistently reflect this conviction.

We also expect our business partners – paper mills, printers, packaging manufacturers, etc. – to commit themselves to using environmentally friendly materials and production processes.

The paper in this book is made from low- or no-chlorine pulp and is acid free, in conformance with international standards for paper permanency.